Tracking Control of Linear Systems

Tracking Control of Linear Systems

Lyubomir T. Gruyitch

CRC Press
Taylor & Francis Group
Boca Raton London New York

CRC Press is an imprint of the
Taylor & Francis Group, an **informa** business

CRC Press
Taylor & Francis Group
6000 Broken Sound Parkway NW, Suite 300
Boca Raton, FL 33487-2742

First issued in paperback 2019

© 2013 by Taylor & Francis Group, LLC
CRC Press is an imprint of Taylor & Francis Group, an Informa business

No claim to original U.S. Government works

ISBN-13: 978-1-4665-8751-9 (hbk)
ISBN-13: 978-0-367-37999-5 (pbk)

This book contains information obtained from authentic and highly regarded sources. Reasonable efforts have been made to publish reliable data and information, but the author and publisher cannot assume responsibility for the validity of all materials or the consequences of their use. The authors and publishers have attempted to trace the copyright holders of all material reproduced in this publication and apologize to copyright holders if permission to publish in this form has not been obtained. If any copyright material has not been acknowledged please write and let us know so we may rectify in any future reprint.

Except as permitted under U.S. Copyright Law, no part of this book may be reprinted, reproduced, transmitted, or utilized in any form by any electronic, mechanical, or other means, now known or hereafter invented, including photocopying, microfilming, and recording, or in any information storage or retrieval system, without written permission from the publishers.

For permission to photocopy or use material electronically from this work, please access www.copyright.com (http://www.copyright.com/) or contact the Copyright Clearance Center, Inc. (CCC), 222 Rosewood Drive, Danvers, MA 01923, 978-750-8400. CCC is a not-for-profit organization that provides licenses and registration for a variety of users. For organizations that have been granted a photocopy license by the CCC, a separate system of payment has been arranged.

Trademark Notice: Product or corporate names may be trademarks or registered trademarks, and are used only for identification and explanation without intent to infringe.

Library of Congress Cataloging-in-Publication Data

Gruiich, L. T. (Liubomir Tikhomira)
Tracking control of linear systems / author, Lyubomir T. Gruyitch.
pages cm
Includes bibliographical references and index.
ISBN 978-1-4665-8751-9 (hardback)
1. Production control. 2. Linear control systems. 3. Automatic tracking. I. Title.

TS157.G78 2013
658.5--dc23 2013008825

Visit the Taylor & Francis Web site at
http://www.taylorandfrancis.com

and the CRC Press Web site at
http://www.crcpress.com

Contents

List of Figures

Acknowledgments

After the third invited public lecture on Lyapunov-like tracking theory at the Department of Electrical Engineering, Louisiana State University, Baton Rouge, September 1989, a young gentleman told me how he was delighted to see somebody speaking about the real control issue. He continued saying that the presented Lyapunov-type control algorithms appeared too complicated for the engineering applications. He asked whether I had simpler ones. My reply was affirmative. "I would like to see them", he said and introduced himself as

WILLIAM PRATT MOUNFIELD, Junior.

Our tracking control algorithms need the same information for the control generation and implementation as that used by the nature to create and to realize control. We call this tracking control:

NATURAL TRACKING CONTROL (*NTC*).

Several researchers gave up, after unsuccessful trials, to realize the digital simulation of the *NTC*. Dr. Mounfield faced also, at the beginning, the difficulty to simulate effectively the *NTC*. The existence of the positive unit feedback without delay in the controller seemed the unsolvable difficulty for the digital simulation.

Dr. Mounfield continued with strong enthusiasm and deep interest. He succeeded to simulate effectively thousands of examples of the *NTC* of various technical plants and processes such as *NTC* of chemical processes, electrical motors, planes, robots and ships. He proposed the *High-Gain NTC*, the published results on which are in [249] - [254].

I have been grateful to my dear colleague, the cocreator of the NTC, and true friend

WILLIAM PRATT MOUNFIELD, Junior, Ph. D.

His wonderful cooperation, the great time I enjoyed in his company and his readiness to help my family and me in difficult situations have been memorable.

Belgrade, September 29, 2012 — Lyubomir T. Gruyitch

Preface

0.1 On the book

We can argue that all real dynamic systems in general, and control systems in particular, are (essentially) nonlinear. The principal argument is the boundedness of energy and matter sources available for the systems work, for their functioning. Besides, there are often geometric and kinematic limitations. Furthermore, some physical processes are inherently nonlinear (e.g., friction). Sophisticated control algorithms are mainly nonlinear. The study of nonlinear systems has been attracting more and more research efforts, which have been decreasing in the investigations of the linear systems.

The study of their linearized mathematical models considered as the linear systems has been important for providing the first insight in responses, sensitivity, stability, controllability, and optimality of the systems in case the variations of values of variables are (sufficiently) small to justify the linearization. The linear systems theory introduces various systems and control concepts and explains dynamic properties of the systems, which the nonlinear systems theory adopts and usually generalizes. Therefore, the study and teaching of the linear systems theory has been the indispensable background of the dynamic and control systems research and education. The common attitude is that the fundamentals of the linear systems theory have been completed. We will show that they lack the crucial linear systems dynamic characteristic that is their *full transfer function matrix* $F(s)$.

The linearized continuous-*time* mathematical descriptions of real (biological, economical, technical) systems can be in the *time* domain or in the *complex* domain. All system variables are expressed in terms of *time* t in the former case. All system variables are represented by their Laplace transforms in terms of the *complex variable* s in the latter case. The linearized continuous-*time* mathematical models of real (biological, economical, technical) systems will be called the *systems* in the sequel.

The fundamental characteristic of the Single-Input-Single-Output (*SISO*) linear systems is their (scalar) transfer function $G(s)$ and that of the Multiple-Input-Multiple-Output (*MIMO*) linear systems is their transfer function matrix denoted herein also by $G(s)$. It has been very useful tool to study or to design the linear systems. This is effective if, and only if, *all initial conditions are equal to zero.* This is due to the definition of $G(s)$. The usage of $G(s)$ is effective in studying system observability, Lyapunov stability of completely controllable and observable system, and Bounded-Input-Bounded-Output (*BIBO*) stability. Although controllability concerns the system behavior under nonzero initial conditions, the controllability criteria can be expressed in terms of $G(s)$.

The initial conditions express the influence of the system past on its current and future behavior. They are most often arbitrary and unpredictable. Their ignorance in studying and/or in designing systems makes the system design and study crucially incomplete. The discovery of the existence of *the system full transfer function matrix* $F(s)$ overcomes and resolves this drawback and problem. It has the same characteristics as the system transfer function matrix $G(s)$:

- $F(s)$ *describes in the complex domain how the system transfers in the course of time the influences of both the input vector action and of all initial conditions on the system dynamic behavior, and*

- *it is completely determined by the system itself meaning its full independence of both the input vector and the vector of all initial conditions.*

We will define and determine precisely **the system full (complete) transfer function matrix** $F(s)$ for the following two different classes of the systems:

- Input-Output (*IO*) control systems,

and

- Input-State-Output (*ISO*) control systems

by referring to [148] for the general definitions.

The usage of *the system full transfer function matrix* $F(s)$ permits us to refine and advance studies of system dynamic properties (e.g., of the complete system response; of system equivalence, realization and minimal realization; of Lyapunov stability; of *BI* stability properties under arbitrary bounded initial conditions, the characterization of which is given in [148], as well as of control system tracking, of trackability and of tracking control synthesis, which represent the core topics of this book). It shows exactly when poles and zeros may (not) be cancelled.

The use of the system full transfer function matrix $F(s)$ requires the same knowledge of mathematics as for the application of the system transfer function matrix $G(s)$. Nothing more.

The author introduced *the system full transfer function matrix* $F(s)$ to the senior students through the undergraduate courses on linear dynamic systems and on linear control systems first at the Department of Electrical Engineering, University of Natal in Durban (*UND*), R. South Africa, 1993, and at the National Engineering School (Ecole Nationale d'Ingénieurs de Belfort, *ENIB*) in Belfort, France, 1994 through 1999. Then the author started to lecture it to the freshmen or juniors of the new University of Technology Belfort-Montbeliard (*UTBM*), which was created in 1999 as the union of the *ENIB* and the Polytechnic Institute of Sevenans. The author made the lecture notes [81] available to the *UND* students during the course. The author's lecture notes [70], [71], [78], [150], [154], [168], containing the topic on *the system full transfer function matrix* $F(s)$, were immediately after the classes, available to the students in the copy center of *ENIB/UTBM*.

The book [148] presents the development of the system full transfer function matrix $F(s)$, which we will show herein. It will serve as the basis for the definitions and the determination of the full transfer function matrix $F(s)$ of the *IO* plant, of the *IO* controller and of the *IO* feedback control system, as well as of the *ISO* plant, of the *ISO* controller and of the *ISO* feedback control system. These results will be used in this book.

The primary control purpose is for control to force ***an object/a plant*** *to behave exactly as demanded, or at least sufficiently closely to the demanded behavior.* The object desired (internal and/or output) behavior expresses its demanded dynamic (internal and/or output) behavior, respectively. Its desired dynamic output behavior is mathematically described by the desired *time* evolution $\mathbf{Y}_d(t)$ of the plant real output vector $\mathbf{Y}$. This means that the object real output response $\mathbf{Y}(t)$ should **follow/track** its desired output response $\mathbf{Y}_d(t)$ sufficiently closely, i.e., control is to force the object to realize/exhibit an appropriate kind of ***tracking***. Such a control is ***tracking control***. It is clear that **tracking** and **tracking control synthesis** are the fundamental control issues.

Attacking the tracking and the tracking control synthesis problems we meet another fundamental control problem:

Problem 1 ***The fundamental tracking problem***

Do the properties of the object enable the existence of tracking control for all initial conditions from a neighborhood of $\mathbf{Y}_{d0}$, *[i.e., of* $\mathbf{Y}_d(t)$ *at the initial moment* $t = 0$*], for all permitted disturbances* $\mathbf{D}(.) \in \mathfrak{D}$ *and for every object desired output behavior* $\mathbf{Y}_d(.) \in \mathfrak{Y}_d$*? If, and only if, they do, then the object is* ***trackable*** *over* $\mathfrak{D}x\mathfrak{Y}_d$.

The goal of the book is to contribute to the advancement of the linear control systems theory and the corresponding university courses, to open new directions for research in this theory and its applications. It represents a further development of the existing linear control systems theory that will not be repeated herein.

The author's hope is that the monograph will achieve this goal effectively in the framework of *time*-invariant continuous *time* linear control systems.

The author consulted in particular the books by the following authors in the course of writing this book: B. D. O. Anderson and J. B. Moore [6], P. J. Antsaklis and A. N. Michel [8], P. Borne et al. [19], W. L. Brogan [21], G. S. Brown and D. P. Campbell [22], F. M. Callier and C. A. Desoer [23], [24], C.-T. Chen [29], H. Chestnut and R. W. Mayer [33], J. J. D'Azzo and C. H. Houpis [46], C. A. Desoer [49], C. A. Desoer and M. Vidyasagar [52], T. Kailath

[198], B. C. Kuo [214], [215], H. Kwakernaak and R. Sivan [216], H. Lauer, R. Lesnick and L. E. Matson [219], A. M. Lyapunov [232], L. A. MacColl [233], J. M. Maciejowski [234], J. L. Melsa and D. G. Schultz [240], K. Ogata [268], [269], D. H. Owens [270], H. M. Power and R. J. Simpson [282], H. H. Rosenbrock [291], R. E. Skelton [300], J. C. West [317], D. M. Wiberg [318], W. A. Wolovich [319] and W. M. Wonham [320]. This book is complementary to them and/or extends their parts that are related to the system transfer functions and/or to the stability issues and/or to the plant-control relationship and/or to the control synthesis.

Scientific Work Place (*SWP*) of MacKichan Company, USA, is a fully adequate program for typing scientific books and papers. The text of the book is typed in Scientific Work Place and figures are drawn in Power Point.

0.2 In gratitude

The author is grateful to Mr. George Pearson with MacKichan Company for his very kind and effective assistance to resolve various problems related to the SWP application among which are the problem of figure conversion from Power Point through Scientific Work Place to PDF and the problem of effective application of SWP to generate simultaneously the Author Index and Subject Index.

The author is also thankful to
Ms. Nora Konopka, Publisher of Engineering & Environmental Sciences, for leading the publication process elegantly and effectively,
Ms. Michele Dimont, Project Editor, for leading the editing process with great care and patience.
Ms. Michele Smith, Editorial Assistant – Engineering, for very useful assistance,
Ms. Jessica Vakili, Senior Project Coordinator, Editorial Project Development, for very useful assistance,
The copyeditor, for very careful book editing and useful comments,
Mr. John Gandour, Cover Designer, for the illustrative cover design,
All of CRC Press/Taylor & Francis.

Belgrade, September 29 and November 23, 2012, February 5 and April 4, 2013.

Lyubomir T. Gruyitch
http://www.truth-science.info

Part I

ON CONTROL SYSTEMS CLASSIC FUNDAMENTALS

Chapter 1

Introduction

1.1 *Time*, physical variables and systems

Time is not only the basic constituent of the existence, but it is also the crucial physical variable for every process, every motion, for the work of every dynamic system, hence for every control system. The behavior of every plant, of its controller and of its control system occurs in *time*. The physical reality, the human experience with it, the human understanding, the accumulated human knowledge lead to the following definition of *time* [170], [171]:

Definition 2 *[148]* ***Time***

Time *(i.e.,* ***the temporal variable****)* t *(or* τ*):*

- *is an independent scalar physical variable,*
- *its value called* ***instant*** *or* ***moment*** *determines uniquely* ***when*** *somebody or something started/ceased to exist,*
- *its values determine uniquely* ***since when and until when*** *somebody or something existed/exists or will exist,*
- *its values specify uniquely* ***how long*** *somebody or something existed, exists or will exist,*
- *its values reflect uniquely whether an event* E_1 *occurs* ***then when*** *another event* E_2 *has not yet happened, or the event* E_1 *took/takes/will take place just* ***then when*** *the event* E_2 *was/is/will be happening, or the event* E_1 *occurs* ***then when*** *the event* E_2 *has already happened,*
- *its value* ***occupies (covers, encloses, is over and in) equally*** *everybody and everything* ***everywhere and always****, and*
- *its value has been, is, and will be* ***permanently changing smoothly, strictly monotonously continuously, equally*** *in all spatial directions and their senses, in and around everybody and everything,* ***independently*** *of everybody and everything.*

The physical nature, the phenomenon, the meaning and the sense of *time* cannot be explained, cannot be expressed in terms of other well-defined notions

[170], [171], e.g., in terms of energy, of matter, of space, or of another physical phenomena or variable. *Time* possesses its own, original, self-contained nature. It is simply *the nature of time, i.e., the temporal nature* or *the time nature.*

The letter t (or τ) denotes **the value of *time*** t (τ), i.e., of **instant** or of **moment**. It is an **instantaneous (momentous)** and **elementary *time* value.** It can take place *exactly once* and then it is *the same everywhere for, and in, everybody and everything.* It is *not repeatable.* It is *untouchable. Nobody and nothing can influence either the value of time or the flow of time values.*

Let T stand for *time.* Then $[T]$ denotes the physical dimension of *time.* It *is impossible to express the physical dimension of time in terms of the physical dimension of another variable.* The physical dimension of *time* is one of the basic physical dimensions. It enables us to define and to determine the physical dimensions of most of the physical variables. The selection of the unit 1_t of *time* is free. Once it is chosen, then it is fixed. It can be *second* s so that $1_t = s$, which is denoted as $t\ \langle 1_t \rangle = t\ \langle s \rangle$.

There exists *exactly one* (denoted by $\exists!$) real number that corresponds to the chosen moment (instant), and vice versa. The numerical value of the moment t, denoted by **num** t, is a real number, *num* $t \in \mathfrak{R}$. The set $\mathfrak{R}$ is the set of all real numbers.

Theorem 3 *[170, Theorem 45, p. 98], [171, Theorem 45, p. 98]* ***Universal time speed law***

Time is the unique physical variable such that the speed v_t (v_τ) of the evolution (of the flow) of its value and of its numerical value:

a) is invariant with respect to a choice of a relative zero moment t_{zero}, of an initial moment t_0, of a time scale and of a time unit 1_t, i.e., invariant relative to a choice of a time axis, invariant relative to a selection of spatial coordinates, invariant relative to everybody and everything,

and

b) its value (its numerical value) is invariant relative to everybody and everything, and equals one arbitrary time unit per the same time unit (equals one), respectively,

$$v_t = \frac{dt}{dt} = 1[TT^{-1}]\left\langle 1_t 1_t^{-1} \right\rangle = 1[TT^{-1}]\left\langle 1_\tau 1_\tau^{-1} \right\rangle = \frac{d\tau}{d\tau} = v_\tau,$$
$$num v_t = num v_\tau = 1, \tag{1.1}$$

relative to arbitrary time axes T and T_τ, i.e., its numerical value equals 1 (one) with respect to all time axes (with respect to any accepted relative zero instant t_{zero}, any chosen initial instant t_0, any time scale and any selected time unit 1_t), with respect to all spatial coordinate systems, with respect to all beings and all objects.

Time set $\mathcal{T}$ is the set of all moments. It is an open, unbounded and connected set. It is in the biunivoque correspondence with the set $\mathfrak{R}$ of all real numbers. Formally mathematically:

$$\mathfrak{T} = \{t : num\ t \in \mathfrak{R},\ dt > 0,\ v_t = \frac{dt}{dt} = t^{(1)} \equiv 1\},$$
$$\forall t \in \mathfrak{T},\ \exists! x \in \mathfrak{R} \Longrightarrow x = num\ t$$
$$and$$
$$\forall x \in \mathfrak{R},\ \exists! t \in \mathfrak{T} \Longrightarrow num\ t = x,$$
$$num\ inf\mathfrak{T} = num\ t_{\inf} = -\infty\ and\ num\ sup\mathfrak{T} = num\ t_{\sup} = \infty. \tag{1.2}$$

The rule of the correspondence between $\mathfrak{T}$ and $\mathfrak{R}$ determines the **accepted relative zero numerical *time* value** t_{zero}, the adopted ***time* scale** and the used ***time* unit** 1_t (or 1_τ). Isaac Newton [264, p. 8: "Scholium"] correctly noted that there exist various *time* units such as: nanosecond, microsecond, milisecond, second, minute, hour, day, week, month, century, millennium, etc. He clearly explained the sense and the meaning of *relative time* [170], [171].

The *time* set $\mathfrak{T}$ and *the set* $\mathfrak{R}$ *of the real numbers* are crucially different. Their essential difference is the dynamic nature of the former and the static nature of the latter. The instants (which are the elements of $\mathfrak{T}$) flow permanently, monotonously, continuously. This expresses the dynamic nature of the *time* set $\mathfrak{T}$. The real numbers do not move.

Let, as in [148], [170], [171], the relative zero moment t_{zero} have the zero numerical value, $num\ t_{zero} = 0$. Also, let it be adopted that it is the initial moment t_0, $t_0 = t_{zero}$, $num\ t_0 = 0$. This is permissible because the systems to be studied are *time*-invariant. This choice of t_0, $t_0 = t_{zero}$, determines the subset $\mathfrak{T}_0$ of $\mathfrak{T}$:

$$\mathfrak{T}_0 = \{t : t \in \mathfrak{T}, numt \in [0, \infty[\}.$$

Note 4 *The numerical value num t of the instant t is a real number without a physical dimension. The instant t is a temporal value that has the physical dimension-the temporal dimension T of time. This opens the problem of the mutual dimensional incompatibility of time and other physical variables. Let the normalized, dimensionless, mathematical temporal variable be denoted by* $\bar{t}$ *and defined by*

$$\bar{t} = \frac{t}{1_t}[-].$$

The set $\overline{\mathfrak{T}}$ *should replace the time set* $\mathfrak{T}$*:*

$$\overline{\mathfrak{T}} = \{\bar{t}[-] : \bar{t} = num\bar{t} = num\ t \in \mathfrak{R},\ d\bar{t} > 0,\ \bar{t}^{(1)} \equiv 1\}.$$

Knowing this, we continue to use the letter t also for $\bar{t}$*, and* $\mathfrak{T}$ *also for* $\overline{\mathfrak{T}}$*. Hence,*

$$t[-] = numt[-].$$

The *time* set $\mathfrak{T}$ is **continuum** meaning that between any two different instants $t_1 \in \mathfrak{T}$ and $t_2 \in \mathfrak{T}$, there is a third instant $t_3 \in \mathfrak{T}$, either $t_1 < t_3 < t_2$ or $t_2 < t_3 < t_1$. The *time* set $\mathfrak{T}$ is **the continuous-*time* set.** *Time* possesses some general properties valid for all physical variables. The following principles, [170, pp. 131-136], [171, pp. 141-146] express them.

Principle 5 ***Physical Continuity and Uniqueness Principle (*****PCUP*****): scalar form***

A physical variable can change its value from one value to another one only by passing through every intermediate value, and it possesses a unique local instantaneous real value in any place (in any being or in any object) at any moment.

Principle 6 ***Physical Continuity and Uniqueness Principle (*****PCUP*****): matrix and vector form***

A vector physical variable or a matrix (vector) of physical variables can change, respectively, its vector or matrix (vector) value from one vector or matrix (vector) value to another one only by passing elementwise through every intermediate vector or matrix (vector) value, and it possesses a unique local instantaneous real vector or matrix (vector) value in any place [i.e., in any being or in any object] at any moment, respectively.

Principle 7 ***Physical Continuity and Uniqueness Principle (*****PCUP*****): system form***

The system physical variables (including those of their derivatives or integrals, which are also physical variables) can change, respectively, their (scalar or vector or matrix) values from one (scalar or vector or matrix) value to another one only by passing elementwise through every intermediate (scalar or vector or matrix) value, and they possess unique local instantaneous real (scalar or vector or matrix) values in any place at any moment.

The *PCUP* is inherent for an accurate modeling physical systems.

Corollary 8 *[170], [171]* ***Mathematical model of a physical variable, mathematical model of a physical system and PCUP***

a) For a mathematical (scalar or vector) variable to be, respectively, an adequate description of a physical (scalar or vector) variable it is necessary that it obeys the Physical Continuity and Uniqueness Principle.

b) For a mathematical model of a physical system to be an adequate description of the physical system it is necessary that all its system variables obey the Physical Continuity and Uniqueness Principle, i.e., that the mathematical model obeys the Physical Continuity and Uniqueness Principle.

The synthesis of the properties of *time* and of the common properties of the physical variables expressed by *PCUP* (Principle 5 through Principle 7) results in

Principle 9 ***Time Continuity and Uniqueness Principle (TCUP)***

Any (scalar or vector) physical variable and any vector/matrix of physical variables can change, respectively, its scalar/vector/matrix value from one

scalar/vector/matrix value to another one only continuously in time by passing (elementwise) through every intermediate scalar/vector/matrix value, and it possesses a unique local instantaneous real scalar/vector/matrix value in any place (in any being or in any object) at any moment.

Definition 10 *[170], [171] The system form of the $TCUP$ means that all system variables satisfy the $TCUP$.*

The effective application of the $TCUP$ to the stability study of dynamic systems and to their control synthesis is in [151], [158], [164], [165], [169], [172], [174].

Corollary 11 *[170], [171]* ***Mathematical representation of a physical variable, mathematical model of a physical system and TCUP***

a) For a mathematical (scalar or vector) variable to be, respectively, an adequate description of a physical (scalar or vector) variable, it is necessary that it obeys the Time Continuity and Uniqueness Principle.

b) For a mathematical model of a physical system to be an adequate description of the physical system, it is necessary that its system variables obey the Time Continuity and Uniqueness Principle or, equivalently, that the mathematical model obeys the Time Continuity and Uniqueness Principle.

c) For a mathematical model of a physical system to be an adequate description of the physical system, it is necessary that its solutions are unique and continuous in time.

The complete study of *time* and the fundamentals of the novel, consistent, physical and mathematical relativity theory, its relationship to systems, and its importance for control is in the books [155], [170], [171].

1.2 Systems and complex domain

Let $\mathfrak{C}$ be *the set of all complex numbers* s, σ and ω be real numbers, or real valued scalar complex variable s, and $j = \sqrt{-1}$ be the imaginary unit,

$$s = (\sigma + j\omega) \in \mathfrak{C},\ \sigma \in \mathfrak{R},\ \omega \in \mathfrak{R}. \tag{1.3}$$

The domain of the complex number s, or of the complex variable s, has appeared more appropriate than the temporal domain for the mathematical treatment of the *time*-invariant continuos-*time* linear dynamic, hence also control, systems. Laplace transform $\mathcal{L}\{.\}$/inverse Laplace transform $\mathcal{L}^{-1}\{.\}$ permits the mathematical passage from one of these domains to another one. It is well known that these systems possess an important input-output dynamic and structural characteristic in the complex domain. It is *the system transfer function* $G(.)$ *for the SISO system*, and its generalization-*the matrix transfer function* denoted also by $G(.)$ *for the MIMO system*. Their complex values are *the system transfer function* $G(s)$ *for the SISO system*, and its generalization-*the transfer function matrix* denoted also by $G(s)$ *for the MIMO system*. The

SISO system transfer function and the *MIMO* system transfer function matrix describe in the complex domain $\mathfrak{C}$ how the system temporally transfers actions of the input variables $I_{(.)}$ (of the input vector $\mathbf{I}$) on the output variables $Y_{(.)}$ (on the output vector $\mathbf{Y}$) *exclusively under all zero initial conditions.*

The following notions and notations will be used in the sequel:

For the $SISO$ systems we will write I instead of I_1, $\mathbf{I} = (I_1) = (I) \in \mathfrak{R}^1$, and Y instead of Y_1, $\mathbf{Y} = (Y_1) = (Y) \in \mathfrak{R}^1$, or simply $I \in \mathfrak{R}$ and $Y \in \mathfrak{R}$, respectively.

$\mathfrak{R}^1$ is the *one dimensional real vector space*, the elements of which are one dimensional real valued vectors, while the elements of $\mathfrak{R}$ are scalars (real numbers). The division of elements of $\mathfrak{R}^1$ is not defined, while it is defined for the elements of $\mathfrak{R}$ except for the division by zero.

The system linearity, expressed by the superposition principle, enables the separate mathematical treatment of the action of the input vector and the influences of all initial conditions on the system dynamic behavior. The problems of the pole-zero cancellation, of the system realization, of $BIBO$ (Bounded-Input-Bounded-Output) stability, and, most often, of tracking have been studied by accepting a priori all zero initial conditions. This has permitted the use of the related system *transfer function* (*matrix*) $G(s)$ and *the block diagram* technique induced by the properties of Laplace transform $\mathcal{L}\{.\}$ and of $G(s)$. It has been utilized to investigate Lyapunov stability properties of the systems despite their being defined only for zero input vector and for nonzero initial conditions. The system response to the input vector action has been analyzed also, mainly under all zero initial conditions.

The influence of the past, (i.e., of the history) of the dynamic system on its present and future behavior is unavoidable. The initial conditions express and transfer, in the very clear and condensed form, the permanent influence of the system past on the system future behavior if the system is without a memory and without a *time* delay. The past is untouchable; the initial conditions cannot be selected or predicted.

In order to carry out a complete study of the real system dynamic behavior and properties, it is necessary to treat the influence of the initial conditions. Understanding this, O. I. Elgerd [56], H. M. Power and R. J. Simpson [282], and R. E. Skelton [300] introduced various system transfer function matrices. Skelton defined the transfer function matrix relative to the initial state of the ISO system. He introduced also the block diagram of the state-space system description with the initial state vector. By following the main stream of the system and control theories, he continued to use only the system transfer function matrix $G(s)$ and the classical block diagram technique, which are valid only under all zero initial conditions.

The following examples, as elaborated in [148], of trivial $SISO$ systems explain simply the crucial difference between the system transfer function $G(s)$ and the system **full (complete) transfer function (matrix)** $F(s)$ that is valid *under arbitrary initial conditions.* They show that the use of the latter is indispensable.

Example 12 *Let two simplest systems be analyzed,*

$$A)\ \frac{dY}{dt} = I,$$

and

$$B)\ \frac{dY}{dt} = \frac{dI}{dt}.$$

Y is the output variable, I is the input variable.

The application of left (−), or right (+) Laplace transform $\mathcal{L}^{\mp}\{.\}$*, or just Laplace transform* $\mathcal{L}\{.\}$*, to the preceding equations yields, respectively,*

$$A)\ Y^{\mp}(s) = \frac{1}{s}I^{\mp}(s) + \frac{1}{s}Y(0^{\mp}),$$

$$B)\ Y^{\mp}(s) = \frac{s}{s}I^{\mp}(s) - \frac{1}{s}I(0^{\mp}) + \frac{1}{s}Y(0^{\mp}).$$

The more compact vector-matrix form of these equations reads

$$a)\ Y^{\mp}(s) = \underbrace{\begin{bmatrix} \overbrace{\frac{1}{s}}^{G_A(s)} & \vdots & \overbrace{\frac{1}{s}}^{G_{A0}(s)} \end{bmatrix}}_{F_A(s)} \underbrace{\begin{bmatrix} I^{\mp}(s) \\ Y(0^{\mp}) \end{bmatrix}}_{\mathbf{V}_A(s)} = F_A(s)\mathbf{V}_A(s),$$

$$F_A(s) = \begin{bmatrix} \frac{1}{s} & \vdots & \frac{1}{s} \end{bmatrix} = \begin{bmatrix} G_A(s) \vdots G_{A0}(s) \end{bmatrix},\ \mathbf{V}_A(s) = \begin{bmatrix} I^{\mp}(s) \\ C_{0A}^{\mp} \end{bmatrix},\ C_{0A}^{\mp} = Y(0^{\mp}),$$

$$B)\ Y^{\mp}(s) = \underbrace{\begin{bmatrix} \overbrace{\frac{s}{s}}^{G_B(s)} & \vdots & \underbrace{\overbrace{-\frac{1}{s}}^{G_{Bi0}(s)} \ \vdots \ \overbrace{\frac{1}{s}}^{G_{By0}(s)}}_{G_{B0}(s)} \end{bmatrix}}_{F_B(s)} \underbrace{\begin{bmatrix} I^{\mp}(s) \\ I(0^{\mp}) \\ Y(0^{\mp}) \end{bmatrix}}_{\mathbf{V}_B(s)} = F_B(s)\mathbf{V}_B(s),$$

$$F_B(s) = \begin{bmatrix} \frac{s}{s} & \vdots & -\frac{1}{s} & \vdots & \frac{1}{s} \end{bmatrix} = \begin{bmatrix} G_B(s) \vdots G_{Bi0}(s) \vdots G_{By0}(s) \end{bmatrix} =$$

$$= \begin{bmatrix} G_B(s) \vdots G_{B0}(s) \end{bmatrix},\ \mathbf{V}_B(s) = \begin{bmatrix} I^{\mp}(s) \\ \mathbf{C}_{0B}^{\mp} \end{bmatrix},\ \mathbf{C}_{0B}^{\mp} = \begin{bmatrix} I(0^{\mp}) \\ Y(0^{\mp}) \end{bmatrix}.$$

The complex function $F_{(..)}(.) : \mathfrak{C} \longrightarrow \mathfrak{C}^{1xq}$ *describes fully (completely) the manner in which the system temporally transfers influences of all actions: of the history through the initial conditions and of the input variable, on the system output behavior, where* $q = 2$ *in the case A), and* $q = 3$ *in the case B). The function* $F_{(..)}(.)$ *is a matrix function despite the simplicity of the systems. They are scalar and of the first order. The function* $F_{(..)}(.)$ *is* ***the full (complete) matrix transfer function*** *of the system. It permits the extension of the notion of the system block, Fig. 1.1 and Fig. 1.2, and, by following R. E. Skelton*

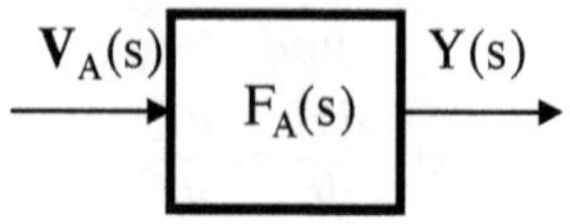

Figure 1.1: The full block of the system under A) in general for the nonzero initial output, $Y_0 \neq 0$.

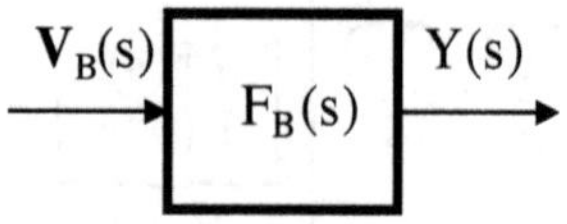

Figure 1.2: The full block of the system under B) in general for the nonzero initial input, $I_0 \neq 0$ and for nonzero initial output, $Y_0 \neq 0$.

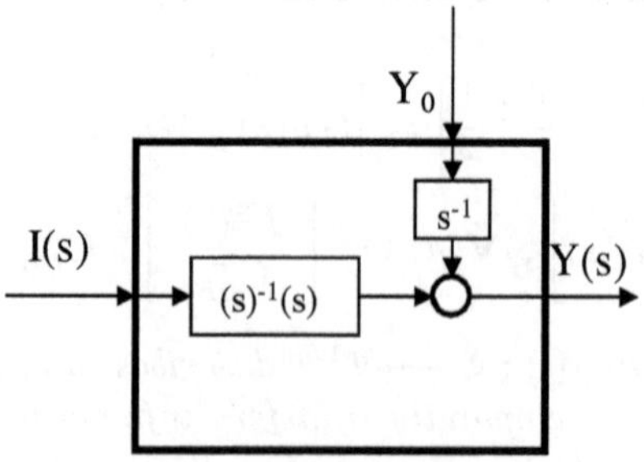

Figure 1.3: The full block diagram of the system under A).

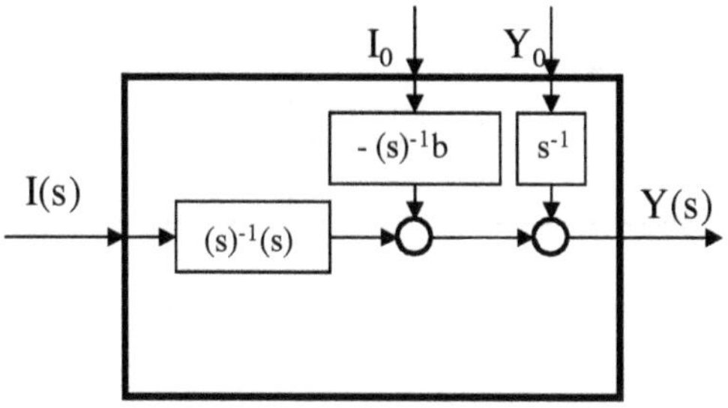

Figure 1.4: The full block diagram of the system under B).

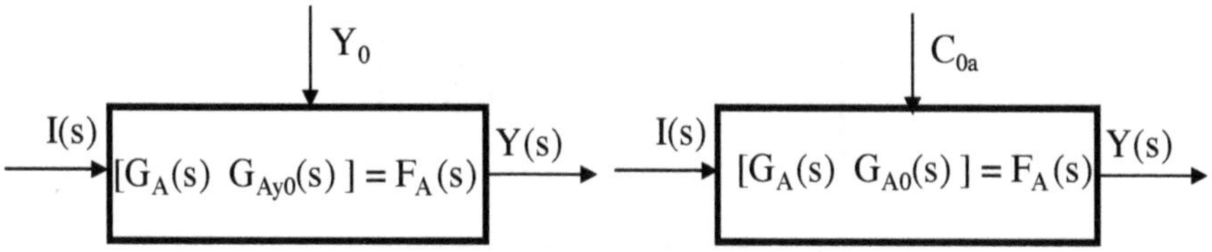

Figure 1.5: The system full block with the submatrices of the full system transfer function matrix $F(s)$ of the system under A).

[300], we can use it to extend the notion of its block diagram, Fig. 1.3 and Fig. 1.4, respectively.

There exist several transmissions and transformations of different influences through the system on its output. The corresponding transfer functions $G_{(..)}(.) : \mathfrak{C} \longrightarrow \mathfrak{C}$ *reflect them and describe them in the complex domain. They are the scalar entries of* $F_{(..)}(s)$*, Fig. 1.3 through Fig. 1.8.*

The irreducible transfer function of the system under B) is constant, $G_A(s) = 1 = const.$ *It yields the minimal system realization* $Y(t) = I(t)$ *under the zero initial condition, i.e., the transfer function realization. The same result follows from* $Y(s) = G_B(s)I(s)$ *regardless of the form of* $G_B(s)$ *(either reducible,*

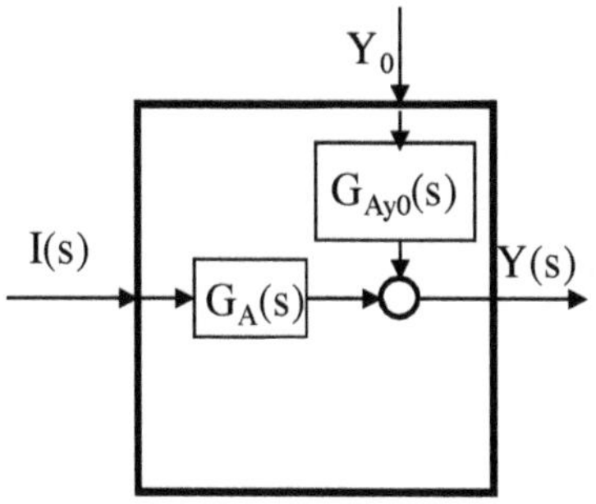

Figure 1.6: The system full block diagram with the submatrices of the full system transfer function matrix $F(s)$ of the system under A).

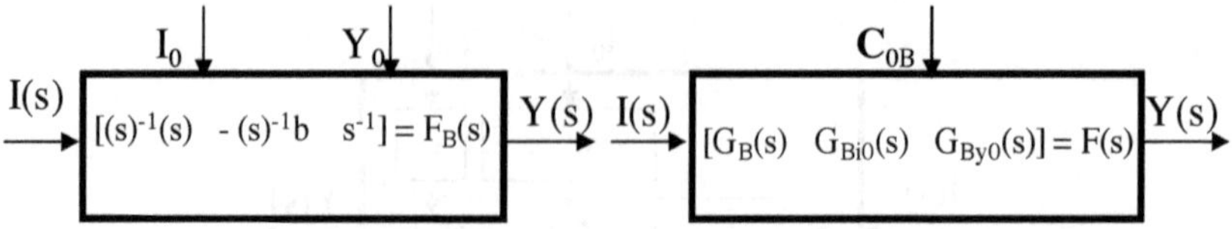

Figure 1.7: The system full block with the submatrices of the full system transfer function matrix $F(s)$ of the system under B)

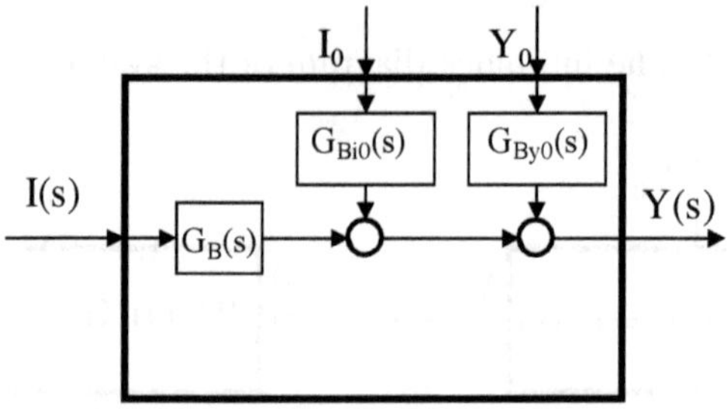

Figure 1.8: The system full block diagram with the submatrices of the full system transfer function matrix $F(s)$ of the system under B).

$G_B(s) = (s)^{-1}s$, or irreducible, $G_B(s) = 1$) under zero initial conditions. This is incorrect in general. The equation $Y(t) = I(t)$ corresponds to a static system, the behavior of which does not depend on initial conditions. Dynamic problems (e.g., controllability, observability, stability) do not exist for such a system. However, the correct relationship between output and input in general is $Y(t) = [I(t) - I(0^-)] + Y(0^-)$. It results from the system IO differential equation after its integration. It shows the output response of a dynamic system. Its equilibrium state $X = Y = 0$ is stable, but not attractive (hence, not asymptotically stable), i.e., the system is not stable. The same result follows if we use the full transfer function matrix $F_B(s)$ and the vector $\mathbf{C}_{0B}^-$ of all initial conditions, or simply if we use $\mathbf{V}_B(s)$, in the expression for $Y^-(s)$,

$$Y^-(s) = F_B(s)\mathbf{V}_B(s),\ \mathbf{V}_B(s) = \begin{bmatrix} I^{\mp}(s) \\ \mathbf{C}_{0B}^- \end{bmatrix},\ \mathbf{C}_{0B}^- = \begin{bmatrix} I(0^{\mp}) \\ Y(0^{\mp}) \end{bmatrix}$$

and when we apply the inverse of left Laplace transform to this equation. The denominator polynomial of $F_B(s)$ is its characteristic polynomial $\Delta(s)$ and, in this case, its minimal polynomial $m(s)$, $\Delta(s) = m(s) = s$. The cancellation of the zero s_1^0 and the pole s_1^* at the origin, $s_1^0 = s_1^* = 0$, of $G_B(s)$ is not possible in $F_B(s)$ even though it is possible in the transfer function $G_B(s)$,

$$G_b(s) = \frac{s}{s},\ F_b(s) = \left[\frac{s}{s} \,\vdots\, -\frac{1}{s} \,\vdots\, \frac{1}{s}\right].$$

The use of $G_B(s)$ for the pole-zero cancellation or for Lyapunov stability test is wrong. We should use instead the full transfer function matrix $F_B(s)$.

Example 13 *The consequence of the ignorance of the initial conditions in the complex domain can be severe when an unstable pole of the reducible form of the system transfer function (matrix) is cancelled with the equal zero. Let*

$$Y^{(2)} - 4Y = -2I + I^{(1)},$$

so that

$$Y^{\mp}(s) = \underbrace{\left[\overbrace{\frac{s-2}{s^2-4}}^{G(s)} \vdots \overbrace{\frac{-1}{s^2-4}}^{G_{I_0}(s)} \vdots \overbrace{\frac{s}{s^2-4}}^{G_{Y_0}(s)} \vdots \overbrace{\frac{1}{s^2-4}}^{G_{Y_0^{(1)}}(s)} \right]}_{F(s)} \underbrace{\left[\begin{array}{c} I^{\mp}(s) \\ I(0^{\mp}) \\ Y(0^{\mp}) \\ Y^{(1)}(0^{\mp}) \end{array} \right]}_{\mathbf{V}(s)},$$

$$Y^{\mp}(s) = F(s)\mathbf{V}(s),\ \mathbf{V}(s) = \left[\begin{array}{c} I^{\mp}(s) \\ \mathbf{C}_0 \end{array} \right],\ \mathbf{C}_0 = \left[\begin{array}{c} I(0^{\mp}) \\ Y(0^{\mp}) \\ Y^{(1)}(0^{\mp}) \end{array} \right].$$

The system transfer function $G(s) = \left(s^2-4\right)^{-1}(s-2)$ is reducible. From its irreducible form $G(s) = (s+2)^{-1}$ follows its minimal realization $Y^{(1)} + 2Y = I$. It is not the system minimal realization. The irreducible form $(s+2)^{-1}$ of $G(s)$ should not be used either to test Lyapunov stability properties of the system, or to test system BIBO stability under bounded nonzero initial conditions, or to determine the system output response under nonzero initial conditions. The cancellation of the zero $s^0 = 2$ and the equal unstable pole $s^* = 2$ in the reducible transfer function $G(s) = \left(s^2-4\right)^{-1}(s-2)$ is impossible in $F(s)$.

The application of $F(s)$ yields all correct results on the pole-zero cancellation, on the system minimal polynomial, on the (minimal) system realization, on the system complete output response, on Lyapunov stability properties, on *BIBO* stability under bounded arbitrary initial conditions, as shown and proved in [148], and on tracking under arbitrary initial conditions, as explained in the sequel. Moreover, the properties of $F(s)$ lead to the generalization of the block diagram technique, i.e., they imply *the full (complete) block diagram technique* (established in [148]).

The author introduced, defined and determined the full transfer function matrix $F(s)$ for *time*-invariant continuous-*time* linear systems in [81], and for *time*-invariant discrete-*time* linear systems in [154]. He used it in these references, as well as in [71], [78], [150], [174], for the analysis of the complete system output response. The purpose of what follows is to show how the use of $F(s)$ enables the resolution (in the complex domain) of the problems of trackability, of tracking and of tracking control synthesis. For its advantages in studying system minimal realization, the zero-pole cancellation, Lyapunov stability and *BIBO* stability under nonzero initial conditions see [148]. It can be effectively exploited also for stabilizing and/or optimal control synthesis. However, these issues exceed the scope of this work.

1.3 Notational preliminaries

Lower case ordinary letters denote scalars, bold (lower case and capital, Greek and Roman) letters signify vectors, capital italic letters stand for matrices, and capital $\mathfrak{Fraktur}$ letters are used for sets and spaces.

Note 14 ***On the new notation*** *[81]*

In order to define and use effectively the system full transfer function matrix $F(s)$, we need new, simple and elegant notation. For example, instead of using

$$\mathbf{Y}^{\mp}(s) = F(s)\begin{bmatrix} I^{\mp}(s) \\ I(0^{\mp}) \\ Y(0^{\mp}) \\ Y^{(1)}(0^{\mp}) \end{bmatrix},$$

we can use

$$\mathbf{Y}^{\mp}(s) = F(s)\begin{bmatrix} I^{\mp}(s) \\ I(0^{\mp}) \\ \mathbf{Y}^{1}(0^{\mp}) \end{bmatrix}, \ \mathbf{Y}^{1}(0^{\mp}) = \begin{bmatrix} Y(0^{\mp}) \\ Y^{(1)}(0^{\mp}) \end{bmatrix},$$

by introducing the general compact vector notation

$$\mathbf{Y}^{k} = \begin{bmatrix} \mathbf{Y} \\ \mathbf{Y}^{(1)} \\ \dots \\ \mathbf{Y}^{(k)} \end{bmatrix} = \begin{bmatrix} \mathbf{Y}^{(0)} \\ \mathbf{Y}^{(1)} \\ \dots \\ \mathbf{Y}^{(k)} \end{bmatrix} \in \mathfrak{R}^{(k+1)N}, \ k \in \{0, 1, ...\}, \ \mathbf{Y}^{0} = \mathbf{Y},$$

which is different from

$$\mathbf{Y}^{(k)} = \frac{d^{k}\mathbf{Y}}{dt^{k}} \in \mathfrak{R}^{N}, \ k \in \{0, 1, ...\}.$$

The extended system matrix $A^{(\nu)}$ is induced by the system matrices $A_i \in \mathfrak{R}^N$, $i \in \{0, 1, ...\}$,

$$A^{(\nu)} = \left[A_0 \vdots A_1 \vdots ... \vdots A_\nu\right] \in \mathfrak{R}^{(\nu+1)N}.$$

We use the matrix function $S_i^{(k)}(.) : \mathfrak{C} \longrightarrow \mathfrak{C}^{\,i(k+1)\text{x}i}$ of s,

$$S_i^{(k)}(s) = \left[s^0 I_i \vdots s^1 I_i \vdots s^2 I_i \vdots ... \vdots s^k I_i\right]^T \in \mathfrak{C}^{\,i(k+1)\text{x}i},$$
$$(k, i) \in \{(\mu, M), \ (\nu, N)\}, \tag{1.4}$$

in order to set

$$\sum_{i=0}^{i=\nu} A_i s^i$$

into the compact form $A^{(\nu)}S_N^{(\nu)}(s)$,

$$\sum_{i=0}^{i=\nu} A_i s^i = A^{(\nu)} S_N^{(\nu)}(s).$$

Note 15 *Higher system order and/or higher system dimension, more advantageous the new notation.*

We will use the symbolic vector notation and operations in the elementwise sense as follows:

- *the zero and unit vectors,*

$$\mathbf{0}_N = [0 \;\; 0 \;...0]^T \in \mathfrak{R}^N, \;\; \mathbf{1}_N = [1 \;\; 1 \;...1]^T \in \mathfrak{R}^N,$$

- *the matrix* E *is associated elementwise with a vector* $\boldsymbol{\varepsilon}$,

$$\boldsymbol{\varepsilon} = [\varepsilon_1 \;\; \varepsilon_2 \;\; . \; . \; . \;\; \varepsilon_N]^T = \boldsymbol{\varepsilon}^{(0)} \Longrightarrow E = diag\,\{\varepsilon_1 \;\; \varepsilon_2 \;\; . \; . \; . \;\; \varepsilon_N\} = E^{(0)},$$

- *the matrix* E^k *is associated elementwise with a vector* $\boldsymbol{\varepsilon}^k$,

$$\boldsymbol{\varepsilon}^{(i)} = \left[\varepsilon_1^{(i)} \;\; \varepsilon_2^{(i)} \;\; . \; . \; . \;\; \varepsilon_N^{(i)}\right]^T \Longrightarrow \boldsymbol{\varepsilon}^k = \left[\boldsymbol{\varepsilon}^{(0)T} \;\; \boldsymbol{\varepsilon}^{(1)T} \;\; . \; . \; . \;\; \boldsymbol{\varepsilon}^{(k)T}\right]^T \Longrightarrow$$

$$E^{(i)} = diag\left\{\varepsilon_1^{(i)} \;\; \varepsilon_2^{(i)} \;\; . \; . \; . \;\; \varepsilon_N^{(i)}\right\} \Longrightarrow E^k = blocdiag\left\{E^{(0)} \;\; E^{(1)} \;\; . \; . \; . \;\; E^{(k)}\right\},$$

- *the vector and matrix absolute values hold elementwise,*

$$|\boldsymbol{\varepsilon}| = [|\varepsilon_1| \;\; |\varepsilon_2| \;\; . \; . \; . \;\; |\varepsilon_N|]^T, \;\; |E| = diag\,\{|\varepsilon_1| \;\; |\varepsilon_2| \;\; . \; . \; . \;\; |\varepsilon_N|\},$$

- *the elementwise vector inequality,*

$$\mathbf{w} = [w_1 \;\; w_2 \;\; . \; . \; . \;\; w_N]^T,$$
$$\mathbf{w} \neq \boldsymbol{\varepsilon} \Longleftrightarrow w_i \neq v_i, \;\; \forall i = 1, 2, ..., N.$$

We define the following sign function:

- $sign(.) : \mathfrak{R} \rightarrow \{-1, 0, 1\}$ *the scalar signum function,*

$$sign(v) = |v|^{-1}\, v \;\; if \;\; v \neq 0, \;\; and \;\; sign(0) = 0.$$

Other new notation is defined at its first use and in Appendix A.

Chapter 2

Control Systems

2.1 *IO* control systems

2.1.1 General *IO* system description

What follows in general is the basis for the next subsections.

The mathematical modeling of many physical systems results (after a possible linearization) in the *time*-invariant linear vector *Input-Output* (*IO*) differential equation (2.1) to be called *the IO system*,

$$\sum_{k=0}^{k=\nu} A_k \mathbf{Y}^{(k)}(t) = \sum_{k=0}^{k=\mu} B_k \mathbf{I}^{(k)}(t),\ detA_\nu \neq 0,\ \forall t \in \mathfrak{T},\ \nu \geq 1,\ 0 \leq \mu \leq \nu,$$
$$\mathbf{Y}^{(k)}(t) = \frac{d^k \mathbf{Y}(t)}{dt^k},\ A_k \in \mathfrak{R}^{N\text{x}N},\ B_k \in \mathfrak{R}^{N\text{x}M},\ k = 0, 1, .., \nu,$$
$$\mu < \nu \Longrightarrow B_i = O,\ i = \mu + 1,\ \mu + 2, ..., \nu. \tag{2.1}$$

This mathematical description can be the general *IO* mathematical description of an object/plant, of a controller and of a whole control system.

Let $\mathfrak{C}^k$ be the k-dimensional complex vector space, $\mathfrak{R}^k$ be the k-dimensional real vector space, $O_{M\text{x}N}$ be the zero matrix in the MxN-dimensional real matrix space $\mathfrak{R}^{M\text{x}N}$, and O_N be the zero matrix in $\mathfrak{R}^{N\text{x}N}$, $O_N = O_{N\text{x}N}$. Analogously, let $\mathbf{0}_k \in \mathfrak{R}^k$ be the zero vector in $\mathfrak{R}^k$. Let $\mathbf{I} = [I_1\ I_2\ ...\ I_M]^T \in \mathfrak{R}^M$ be the input vector, and $\mathbf{Y} = [Y_1\ Y_2\ ...\ Y_N]^T \in \mathfrak{R}^N$ be the output vector. The values I_k and Y_m are measured with respect to the total zeros of these variables if they have total zeros. If a variable does not have a total zero, then some its value is accepted to play the role of its total zero. Temperature has the total zero that is Kelvin zero. Position does not have the total zero.

Note 16 *The condition $detA_\nu \neq 0$ is a sufficient condition, but not a necessary condition, for all the output variables of the system (2.1) to have the same order ν of their highest derivatives. In case the order k of the highest derivative of* *an output variable Y_i is lower than the highest derivative order ν of some other output variable Y_j, then all entries of the i-th column of $A_{k+1}, \ldots, A_\nu$ are equal to zero implying their singularity: $detA_m = 0$, $m = k+1, ..., \nu$.*

Example 17 *Let*

$$M = 1,\ N = 2,\ \nu = 1,\ \mu = 0,$$

$$\underbrace{\begin{bmatrix} -3 & 2 \\ 3 & -2 \end{bmatrix}}_{A_1} \mathbf{Y}^{(1)}(t) + \underbrace{\begin{bmatrix} 4 & -6 \\ -4 & 6 \end{bmatrix}}_{A_0} \mathbf{Y}(t) = \begin{bmatrix} 4 \\ 2 \end{bmatrix} I(t) \Longrightarrow$$

$$detA_1 = \begin{vmatrix} -3 & 2 \\ 3 & -2 \end{vmatrix} = 0,\ det\left(A_1 s + A_0\right) = \begin{vmatrix} -3s+4 & 2s-6 \\ 3s-4 & -2s+6 \end{vmatrix} \equiv 0.$$

Y_1 and Y_2 have the first order highest derivatives. A_1 is singular. Evidently, the condition $detA_\nu \neq 0$ is not necessary for all the output variables of the system (2.1) to have the same order of their highest derivatives. There does not exist a solution to the given vector differential equation because $det\left(A_1 s + A_0\right) \equiv 0$. We show this by considering the scalar form of the mathematical model of the system,

$$-3Y_1^{(1)}(t) + 2Y_2^{(1)}(t) + 4Y_1(t) - 6Y_2(t) = 4I(t),$$
$$3Y_1^{(1)}(t) - 2Y_2^{(1)}(t) - 4Y_1(t) + 6Y_2(t) = 2I(t).$$

We multiply the second equation by -1. The result is

$$-3Y_1^{(1)}(t) + 2Y_2^{(1)}(t) + 4Y_1(t) - 6Y_2(t) = 4I(t),$$
$$-3Y_1^{(1)}(t) + 2Y_2^{(1)}(t) + 4Y_1(t) - 6Y_2(t) = -2I(t).$$

The left-hand sides of these equations are the same. Their right-hand sides are different. They have a solution only for $I(t) = 0$. The solution is then trivial, $Y_1(t) = Y_2(t) = 0$.

Laplace transform $\mathcal{L}\{.\}$ of the system mathematical model for all zero initial conditions reads

$$\left(A_1 s + A_0\right)\mathbf{Y}(s) = \begin{bmatrix} -3s+4 & 2s-6 \\ 3s-4 & -2s+6 \end{bmatrix} \mathbf{Y}(s) = \begin{bmatrix} 4 \\ 2 \end{bmatrix} I(s).$$

$\mathbf{Y}(s)$ and $I(s)$ are Laplace transforms of $\mathbf{Y}(t)$ and of $I(t)$, $\mathbf{Y}(s) = \mathcal{L}\{\mathbf{Y}(t)\}$ and $I(s) = \mathcal{L}\{I(t)\}$. The preceding vector equation is not solvable in $\mathbf{Y}(s)$ because $det\left(A_1 s + A_0\right) \equiv 0$.

Condition 18 *The matrix A_ν of the IO system (2.1) is nonsingular,*

$$detA_\nu \neq 0. \tag{2.2}$$

This condition, i.e., (2.2), ensures that the characteristic polynomial $f(s)$ is not identically equal to zero,

$$\exists s \in \mathfrak{C} \Longrightarrow f(s) = \det\left(\sum_{k=0}^{k=\nu} A_k s^k\right) \neq 0. \tag{2.3}$$

It enables solvability of Laplace transform of (2.1) in $\mathbf{Y}(s)$, see (I.4) in Appendix I.

Note 19 ***The validity of Condition 18 holds in the sequel.***

The complexity of the requirements for the high quality tracking need the following compact notation for the extended matrices [81]; (Note 14 in *Notational preliminaries* 1.3 herein):

$$A^{(\nu)} = \left[A_0 \vdots\ A_1 \vdots ... \vdots A_\nu\right] \in \mathfrak{R}^{N\mathrm{x}(\nu+1)N},$$
$$B^{(\mu)} = \left[B_0 \vdots B_1 \vdots ... \vdots B_\mu\right] \in \mathfrak{R}^{N\mathrm{x}(\mu+1)M}, \tag{2.4}$$

$$\mathbf{I}^{\mu}(t) = \left[\mathbf{I}^{T}(t) \vdots \mathbf{I}^{(1)T}(t) \vdots ... \vdots \mathbf{I}^{(\mu)T}(t)\right]^{T} \in \mathfrak{R}^{(\mu+1)M}, \tag{2.5}$$

$$\mathbf{Y}^{\nu}(t) = \left[\mathbf{Y}^{T}(t) \vdots \mathbf{Y}^{(1)T}(t) \vdots ... \vdots \mathbf{Y}^{(\nu)T}(t)\right]^{T} \in \mathfrak{R}^{(\nu+1)N}. \tag{2.6}$$

The corresponding initial vectors are

$$\mathbf{I}_{0\mp}^{\mu-1} = \mathbf{I}^{\mu-1}(0^{\mp}) = \left[\mathbf{I}_{0(\mp)}^{T} \vdots \mathbf{I}_{0(\mp)}^{(1)T} \vdots ... \vdots \mathbf{I}_{0(\mp)}^{(\mu-1)T}\right]^{T} \in \mathfrak{R}^{\mu M}, \tag{2.7}$$

$$\mathbf{Y}_{0\mp}^{\nu-1} = \mathbf{Y}^{\nu-1}(0^{\mp}) = \left[\mathbf{Y}_{0(\mp)}^{T} \vdots \mathbf{Y}_{0(\mp)}^{(1)T} \vdots ... \vdots \mathbf{Y}_{0(\mp)}^{(\nu-1)T}\right]^{T} \in \mathfrak{R}^{\nu N}. \tag{2.8}$$

The superscript ν is in the parentheses in $A^{(\nu)}$ in order to distinguish $A^{(\nu)}$ from the ν-*th* power A^{ν} of A,

$$A^{(\nu)} = \left[A_0 \vdots\ A_1 \vdots ... \vdots A_\nu\right] \neq A^{\nu} = \underbrace{AA....A}_{\nu-times}.$$

The superscript μ is not in the parentheses in $\mathbf{I}^{\mu}(t)$ because $\mathbf{I}^{(\mu)}(t)$ denotes the μ-*th* derivative $d^{\mu}\mathbf{I}(t)/dt^{\mu}$ of $\mathbf{I}(t)$,

$$\mathbf{I}^{\mu}(t) = \left[\mathbf{I}^{T}(t) \vdots \mathbf{I}^{(1)T}(t) \vdots ... \vdots \mathbf{I}^{(\mu)T}(t)\right]^{T} \neq \mathbf{I}^{(\mu)}(t) = \frac{d^{\mu}\mathbf{I}(t)}{dt^{\mu}}.$$

This compact notation permits us to set the IO vector differential equation (2.1) into the following simple compact form:

$$A^{(\nu)}\mathbf{Y}^{\nu}(t) = B^{(\mu)}\mathbf{I}^{\mu}(t),\ t \in \mathfrak{T}. \tag{2.9}$$

The use of (2.9) instead of (2.1) is simpler and more elegant.

In order to complete the notation in this framework let us be reminded of the complex matrix function $S_i^{(k)}(.) : \mathfrak{C} \longrightarrow \mathfrak{C}^{\,i(k+1)\mathrm{x}i}$, (1.4) (Section 1.3),

$$S_i^{(k)}(s) = \left[s^0 I_i \,\vdots\, s^1 I_i \,\vdots\, s^2 I_i \,\vdots\, ... \,\vdots\, s^k I_i\right]^T \in \mathfrak{C}^{\,i(k+1)\mathrm{x}i},$$
$$(k,i) \in \{(\mu, M),\ (\nu, N)\}, \tag{2.10}$$

The dimensions of the matrices $A_k \in \mathfrak{R}^{N\mathrm{x}N}$ and $B_k \in \mathfrak{R}^{N\mathrm{x}M}$, $k = 0, 1, .., \nu$, and Condition 18 furnish

$$deg\left[det\left(\sum_{k=0}^{k=\nu} A_k s^k\right)\right] = deg\left[det\left(A^{(\nu)}S_N^{(\nu)}(s)\right)\right] = \eta,\ \eta = \nu N,$$
$$deg\left[adj\left(\sum_{k=0}^{k=\nu} A_k s^k\right)\right] = deg\left[adj\left(A^{(\nu)}S_N^{(\nu)}(s)\right)\right] = \sigma,\ \sigma = (\nu - 1)\,N,$$
$$deg\left(\sum_{k=0}^{k=\mu} B_k s^k\right) = deg\left(B^{(\mu)}S_M^{(\mu)}(s)\right) = \mu. \tag{2.11}$$

In these equations $deg\left[adj\left(A^{(\nu)}S_N^{(\nu)}(s)\right)\right]$ and $deg\left(B^{(\mu)}S_M^{(\mu)}(s)\right)$ denote the greatest power of s over all elements of

$$adj\left(A^{(\nu)}S_N^{(\nu)}(s)\right)\ and\ B^{(\mu)}S_M^{(\mu)}(s),$$

respectively.

In general

$$deg\left[det\left(A^{(\nu)}S_N^{(\nu)}(s)\right)\right] = \eta,\ 0 \le \eta \le \nu N,$$
$$deg\left[adj\left(A^{(\nu)}S_N^{(\nu)}(s)\right)\right] = \sigma,\ 0 \le \sigma \le (\nu - 1)\,N.$$

By referring to [148] we state the following:

Definition 20 ***A realization of the IO system (2.1), i.e., (2.9), for an arbitrary input vector function and for arbitrary input and output initial conditions is*** *the quadruple* $(\nu,\ \mu,\ A^{(\nu)},\ B^{(\mu)})$.

Comment 21 *[148] The realization* $(\nu,\ \mu,\ A^{(\nu)},\ B^{(\mu)})$ *of the IO system (2.1), i.e., (2.9), is* ***its IO realization.***

The essential differences between the system transfer function matrix $G(s)$ and the system full transfer function matrix $F(s)$ need the following two definitions.

Definition 22 *The characteristic polynomial of the system full transfer function matrix $F(s)$ is **the system characteristic polynomial** $f(s)$.*

The characteristic polynomial $f(s)$ of the IO system (2.1), i.e., (2.9), is in Equation (2.3).

Definition 23 *The minimal polynomial of the system full transfer function matrix $F(s)$ is **the system minimal polynomial** $m(s)$.*

Note 24 *The system minimal polynomial and the minimal polynomial of the system transfer function matrix $G(s)$ can be different.*

The order of a system is different from its dimension and from its dynamic dimension in general.

Definition 25 *The number ν of the highest derivative of the output vector function $\mathbf{Y}(.)$ in (2.1), i.e., (2.9), is **the order of the IO system (2.1), i.e., (2.9).***

Definition 26 ***The dimension of the IO system (2.1), i.e., (2.9),** denoted by dim_{IO}, is the dimension N of its output vector $\mathbf{Y}$ in (2.1), i.e., in (2.9), $dim_{IO} = dim\mathbf{Y} = N$.*

Example 27 *The IO system*

$$\underbrace{\begin{bmatrix} 0 \vdots 0 \vdots 4 \\ 0 \vdots 0 \vdots 0 \\ 0 \vdots 0 \vdots 0 \end{bmatrix}}_{A_2} \mathbf{Y}^{(2)}(t) + \underbrace{\begin{bmatrix} 0 \vdots 0 \vdots 0 \\ 0 \vdots 0 \vdots 0 \\ 3 \vdots 2 \vdots 0 \end{bmatrix}}_{A_1} \mathbf{Y}^{(1)}(t) + \underbrace{\begin{bmatrix} 0 \vdots 0 \vdots 0 \\ 0 \vdots 6 \vdots 0 \\ 0 \vdots 0 \vdots 0 \end{bmatrix}}_{A_0} \mathbf{Y}(t) =$$

$$= \begin{bmatrix} 1 \vdots 0 \\ 0 \vdots 3 \\ 4 \vdots 0 \end{bmatrix} \mathbf{I}(t) + \begin{bmatrix} 0 \vdots 2 \\ 7 \vdots 0 \\ 2 \vdots 3 \end{bmatrix} \mathbf{I}^{(2)}(t)$$

yields

$$\nu = 2,\ \mu = 2,\ N = 3,\ \ M = 2,\ detA_{\nu} = detA_2 = det \begin{bmatrix} 0 \vdots 0 \vdots 4 \\ 0 \vdots 0 \vdots 0 \\ 0 \vdots 2 \vdots 0 \end{bmatrix} = 0,$$

$$deg\left[det\left(\sum_{k=0}^{k=\nu=2} A_k s^k\right)\right] = deg\begin{vmatrix} 0 & 0 & 4s^2 \\ 0 & 6 & 0 \\ 3s & 2s & 0 \end{vmatrix} = deg\left(-72s^3\right) = 3 = \eta > \nu = 2,$$

$$deg\left[adj\left(\sum_{k=0}^{k=\nu=2} A_k s^k\right)\right] = deg\begin{bmatrix} 0 & 8s^3 & -24s^2 \\ 0 & -12s^3 & 0 \\ -18s & 0 & 0 \end{bmatrix} = deg\left(s^3\right) = 3 = \sigma,$$

$$deg\left(\sum_{k=0}^{k=\mu=2} B_k s^k\right) = deg\begin{bmatrix} 1 & 2s^2 \\ 7s^2 & 3 \\ 4+2s^2 & 3s^2 \end{bmatrix} = deg\left(s^2\right) = 2 = \mu,$$

$$\sigma = \eta = 3 > \nu = 2,\ \eta < \nu N = 2\text{x}3 = 6.$$

For this second order system, $\nu = 2$, it is found that its dimension equals $N = 3$, $dim\mathbf{Y} = N = 3$. Their product, $\nu N = 6 \geq 3 = \eta$, is bigger than the degree η of the system characteristic polynomial. Only the third output variable Y_3 has the second derivative, the first derivative exists of both the first and the second output variable, Y_1 and Y_2 in the system mathematical model.

Example 28 *The IO system*

$$\begin{bmatrix} 1 \vdots 0 \vdots 3 \\ 2 \vdots 0 \vdots 0 \\ 0 \vdots 2 \vdots 2 \end{bmatrix} \mathbf{Y}^{(2)}(t) + \begin{bmatrix} 0 \vdots 0 \vdots 0 \\ 0 \vdots 3 \vdots 0 \\ 0 \vdots 0 \vdots 0 \end{bmatrix} \mathbf{Y}^{(1)}(t) + \begin{bmatrix} 0 \vdots 0 \vdots 0 \\ 0 \vdots 0 \vdots 0 \\ 0 \vdots 0 \vdots 2 \end{bmatrix} \mathbf{Y}(t) =$$

$$= \begin{bmatrix} 1 \vdots 0 \\ 0 \vdots 2 \\ 4 \vdots 0 \end{bmatrix} \mathbf{I}(t) + \begin{bmatrix} 2 \vdots 0 \\ 0 \vdots 5 \\ 6 \vdots 4 \end{bmatrix} \mathbf{I}^{(2)}(t)$$

induces

$$\nu = 2,\ \mu = 2,\ N = 3,\ M = 2,\ detA_\nu = detA_2 =$$

$$= det\begin{bmatrix} 1 \vdots 0 \vdots 3 \\ 2 \vdots 0 \vdots 0 \\ 0 \vdots 2 \vdots 2 \end{bmatrix} = 12 \neq 0,$$

$$deg\left[det\left(\sum_{k=0}^{k=\nu=2} A_k s^k\right)\right] = deg\left[\begin{vmatrix} s^2 & 0 & 3s^2 \\ 2s^2 & 3s & 0 \\ 0 & 2s^2 & 2s^2+2 \end{vmatrix}\right] =$$

$$= deg\left(12s^6 + 6s^5 + 6s^3\right) = 6 = \eta = 2\text{x}3 = \nu N,$$

$$deg\left[adj\left(\sum_{k=0}^{k=\nu=2} A_k s^k\right)\right] = deg\begin{bmatrix} 6s^3+6s & 6s^4 & -9s^3 \\ -4s^4-4s^2 & 2s^4+2s^2 & 6s^4 \\ 4s^4 & -2s^4 & 3s^3 \end{bmatrix} =$$
$$= deg\left(s^4\right) = 4 = \sigma,$$

$$deg\left(\sum_{k=0}^{k=\mu=2} B_k s^k\right) = deg\begin{bmatrix} 1+s^2 & 0 \\ 0 & 2+5s^2 \\ 4+6s^2 & 4s^2 \end{bmatrix} = deg\left(s^2\right) = 2 = \mu,$$
$$\sigma = 4 = 2\text{x}2 = 2\text{x}\,(3-1) = \nu\,(N-1) < \eta = \nu N = 6.$$

For this second order system of the dimension three, the product of the system order ($\nu = 2$) and of the system dimension ($N = 3$) is equal to the degree ($\eta = 6$) of the system characteristic polynomial, $\nu N = 6 = \eta$.

We will accept the definitions of *the dynamic dimension* and of *the least dimension* by following H. H. Rosenbrock, [291, pp. 30, 47, 48] and [148], and by noting that he used the term *order* in the sense of *dimension.* However, we accepted (Definition 25, Definition 26) to distinguish *the dimension of the system* from *the order of the system* that we use in the classical mathematical sense of the order of a differential equation that describes a physical dynamic system (Definition 25 and Definition 26). Also, we will define the (minimal) dynamic dimension of a system (realization) in the same sense as the (minimal) system dimension, respectively:

Definition 29 *(a) The number of the initial conditions that determine uniquely the output response of the system (realization) to an arbitrary input vector function* $\mathbf{I}(.)$ *and to arbitrary initial conditions, or equivalently, the degree of the characteristic polynomial of the system (realization), is* ***the dynamic dimension of the system (realization)*** *denoted by ddim, respectively. For the IO system its dynamic dimension is denoted by* $ddim_{IO}$*.*

(b) The number of the independent initial conditions that must be known (i.e., the minimal number of the initial conditions that should be known) in order to determine uniquely the output response of the system (realization) to an arbitrary input vector function $\mathbf{I}(.)$ *and to arbitrary initial conditions, or equivalently, the degree of the minimal polynomial of the system (realization), is* ***the least (the minimal) dynamic dimension of the IO system (realization)*** *denoted by mddim, respectively. For the IO system its minimal dynamic dimension is denoted by* $mddim_{IO}$*.*

Note 30 *The dimension N of the output vector* $\mathbf{Y}$ *of the IO system (2.1), i.e., (2.9), and its order ν determine jointly the dynamic dimension* $ddim_{IO}$ *of the system (realization),*

$$\nu \leq ddim_{IO} \leq \nu N.$$

Example 31 *The degree of the system characteristic polynomial,*

$$deg\left[det\left(\sum_{k=0}^{k=\nu=2} A_k s^k\right)\right] = deg\begin{vmatrix} 0 & 0 & 4s^2 \\ 0 & 6 & 0 \\ 3s & 2s & 0 \end{vmatrix} = deg\left(-72s^3\right) = 3 = \eta > \nu = 2,$$

determines the dynamic dimension $ddim_{IO}$ of the three-dimensional second order IO system (Example 27)

$$\underbrace{\begin{bmatrix} 0 \vdots 0 \vdots 4 \\ 0 \vdots 0 \vdots 0 \\ 0 \vdots 0 \vdots 0 \end{bmatrix}}_{A_2} \mathbf{Y}^{(2)}(t) + \underbrace{\begin{bmatrix} 0 \vdots 0 \vdots 0 \\ 0 \vdots 0 \vdots 0 \\ 3 \vdots 2 \vdots 0 \end{bmatrix}}_{A_1} \mathbf{Y}^{(1)}(t) + \underbrace{\begin{bmatrix} 0 \vdots 0 \vdots 0 \\ 0 \vdots 6 \vdots 0 \\ 0 \vdots 0 \vdots 0 \end{bmatrix}}_{A_0} \mathbf{Y}(t) =$$

$$= \begin{bmatrix} 1 \vdots 0 \\ 0 \vdots 3 \\ 4 \vdots 0 \end{bmatrix} \mathbf{I}(t) + \begin{bmatrix} 0 \vdots 2 \\ 7 \vdots 0 \\ 2 \vdots 3 \end{bmatrix} \mathbf{I}^{(2)}(t).$$

The dynamic dimension $ddim_{IO}$ is equal to 3, $ddim_{IO} = 3$. It is bigger than the system order (2), and it is equal to the system dimension ($dim_{IO} = 3$). Therefore, it is less than their product $2x3 = 6$.

Example 32 *For the second-order three-dimensional IO system (Example 28)*

$$\begin{bmatrix} 1 \vdots 0 \vdots 3 \\ 2 \vdots 0 \vdots 0 \\ 0 \vdots 2 \vdots 2 \end{bmatrix} \mathbf{Y}^{(2)}(t) + \begin{bmatrix} 0 \vdots 0 \vdots 0 \\ 0 \vdots 3 \vdots 0 \\ 0 \vdots 0 \vdots 0 \end{bmatrix} \mathbf{Y}^{(1)}(t) + \begin{bmatrix} 0 \vdots 0 \vdots 0 \\ 0 \vdots 0 \vdots 0 \\ 0 \vdots 0 \vdots 2 \end{bmatrix} \mathbf{Y}(t) =$$

$$= \begin{bmatrix} 1 \vdots 0 \\ 0 \vdots 2 \\ 4 \vdots 0 \end{bmatrix} \mathbf{I}(t) + \begin{bmatrix} 2 \vdots 0 \\ 0 \vdots 5 \\ 6 \vdots 4 \end{bmatrix} \mathbf{I}^{(2)}(t)$$

it was found that

$$deg\left[det\left(\sum_{k=0}^{k=\nu=2} A_k s^k\right)\right] = deg\left[\begin{vmatrix} s^2 & 0 & 3s^2 \\ 2s^2 & 3s & 0 \\ 0 & 2s^2 & 2s^2+2 \end{vmatrix}\right] =$$

$$= deg\left(12s^6 + 6s^5 + 6s^3\right) = 6 = \eta = 2x3 = \nu N.$$

The dynamic dimension is equal to 6, $ddim_{IO} = 6$. The product of the system order (2) and of the system dimension ($dim_{IO} = 3$) is equal to the degree of the system characteristic polynomial, i.e., it is equal to the system dynamic dimension, $\nu N = 2x3 = 6 = \eta = ddim_{IO}$.

Definition 29 leads to the following statement, as observed in [148].

Proposition 33 *A system realization is* ***the minimal system realization*** *if, and only if, its characteristic polynomial is its minimal polynomial. It is also called* ***the irreducible system realization.***

In order to prove the tracking conditions for acceptable input vector functions $\mathbf{I}(.) : \mathfrak{T}_0 \to \mathfrak{R}^M$, let $\mathbf{I}(.)$ belong to the class $\mathfrak{L}$ of *time*-dependent bounded functions having Laplace transforms in the form of strictly proper real rational vector functions,

$$\mathfrak{L} = \left\{ \mathbf{I}(.) : \left(\begin{array}{c} \exists \gamma(\mathbf{I}) \in \mathfrak{R}^+ \Longrightarrow \|\mathbf{I}(t)\| < \gamma(\mathbf{I}),\ \forall t \in \mathfrak{T}_0, \\ \mathcal{L}^{\mp}\{\mathbf{I}(t)\} = \mathbf{I}^{\mp}(s) = \left[I_1^{\mp}(s)\ \ I_2^{\mp}(s)\ \ \ldots\ \ I_M^{\mp}(s) \right]^T, \\ I_k^{\mp}(s) = \dfrac{\sum_{j=0}^{j=\zeta_k} a_{kj} s^j}{\sum_{j=0}^{j=\psi_k} b_{kj} s^j}, 0 \le \zeta_k < \psi_k,\ \forall k = 1, 2, ..., M, \end{array} \right) \right\}. \tag{2.12}$$

In order to ensure that the original $\mathbf{I}(t)$ does not contain an impulse component because such component is unbounded, left Laplace transform $\mathbf{I}^-(s)$, or right Laplace transform $\mathbf{I}^+(s)$, or just Laplace transform $\mathbf{I}(s)$ of the input vector function $\mathbf{I}(.) \in \mathfrak{L}$, should be strictly proper.

It is clear that the zero input vector function $\mathbf{I}(.)$, $\mathbf{I}(t) \equiv \mathbf{0}_M$, belongs to $\mathfrak{L}$.

$\mathfrak{C}^{ki} = \mathfrak{C}^k(\mathfrak{R}^i)$ is *the family of all functions defined and k-times continuously differentiable on* $\mathfrak{R}^i$, and

$\mathfrak{C}^k = \mathfrak{C}^k(\mathfrak{T}_0)$ is *the family of all functions defined, continuous and k-times continuously differentiable on* $\mathfrak{T}_0$, $\mathfrak{C} = \mathfrak{C}^0(\mathfrak{T}_0)$;

$\mathfrak{C}^{k-}\left(\mathfrak{R}^N\right)$ is *the family of all functions defined everywhere and k-times continuously differentiable on* $\mathfrak{R}^N$-$\{\mathbf{0}_N\}$, *which have defined and continuous derivatives at the origin* $\mathbf{0}_N$ of $\mathfrak{R}^N$ up to the order $(k-1)$ and which are defined and continuous at the origin $\mathbf{0}_N$ and have defined the left and the right *k-th* order derivative at the origin $\mathbf{0}_N$*;*

$\mathfrak{I}^k$ is a given, or to be determined, family of all bounded and *k-times* continuously differentiable permitted input vector functions $\mathbf{I}(.) \in \mathfrak{C}^k \cap \mathfrak{L}$,

$$\mathfrak{I}^k \subset \mathfrak{C}^k \cap \mathfrak{L}. \tag{2.13}$$

$\mathfrak{I}^0 = \mathfrak{I}$ is the family of all bounded continuous permitted input vector functions $\mathbf{I}(.) \in \mathfrak{I}$,

$$\mathfrak{I} = \mathfrak{C} \cap \mathfrak{L}. \tag{2.14}$$

$\mathfrak{I}^k_-$ is a subfamily of $\mathfrak{I}^k$, $\mathfrak{I}^k_- \subset \mathfrak{I}^k$, such that the real part of every pole of Laplace transform $\mathbf{I}(s)$ of every $\mathbf{I}(.) \in \mathfrak{I}^k_-$ is negative, $\mathfrak{I}_- = \mathfrak{I}^0_-$.

2.1.2 Input-output (*IO*) description of a plant

The compact form of the *IO* differential equation (2.1) (Subsection 2.1.1) of a plant $\mathcal{P}$ to be controlled by control $\mathbf{U}(.) : \mathfrak{T}\text{x}... \longrightarrow \mathfrak{R}^r$, or which is controlled, and which is subjected to the action of the disturbance vector $\mathbf{D}(t) \in \mathfrak{R}^d$, reads

$$
\begin{gathered}
A_P^{(\nu)}\mathbf{Y}^{\nu}(t) = C_{Pu}^{(\mu_{Pu})}\mathbf{U}^{\mu_{Pu}}(t) + D_{Pd}^{(\mu_{Pd})}\mathbf{D}^{\mu_{Pd}}(t),\ \det A_{P\nu} \neq 0,\ \forall t \in \mathfrak{T}, \\
A_P^{(\nu)} = \left[A_{P0} \vdots\ A_{P1} \vdots ... \vdots A_{P\nu}\right] \in \mathfrak{R}^{N\text{x}(\nu+1)N}, \\
C_{Pu}^{(\mu_{Pu})} = \left[C_{P_u0} \vdots\ C_{P_u1} \vdots ... \vdots C_{P_u\mu_{Pu}}\right] \in \mathfrak{R}^{N\text{x}(\mu_{Pu}+1)r}, \\
D_{Pd}^{(\mu_{Pd})} = \left[D_{P_d0} \vdots\ D_{P_d1} \vdots ... \vdots D_{P_d\mu_{Pd}}\right] \in \mathfrak{R}^{N\text{x}(\mu_{Pd}+1)d}, \\
\nu \geq \max\{\mu_{Pd},\ \mu_{Pu}\}.
\end{gathered}
\tag{2.15}
$$

We will call the plant $\mathcal{P}$ (2.1), (2.15), *the IO plant.*

$\mathfrak{D}^k$ is a given, or to be determined, *family of all bounded and k-times continuously differentiable permitted disturbance vector functions* $\mathbf{D}(.) \in \mathfrak{L}$,

$$\mathfrak{D}^k \subseteq \mathfrak{C}^k \cap \mathfrak{L}, \tag{2.16}$$

and

$\mathfrak{D}^0 = \mathfrak{D}$ is *the family of all bounded continuous permitted disturbance vector functions* $\mathbf{D}(.)$. Their Laplace transforms are strictly proper real rational complex functions,

$$\mathfrak{D} \subseteq \mathfrak{C} \cap \mathfrak{L}. \tag{2.17}$$

$\mathfrak{D}_-^k$ is a subfamily of $\mathfrak{D}^k$, $\mathfrak{D}_-^k \subset \mathfrak{D}^k$, such that *the real part of every pole of Laplace transform* $\mathbf{D}(s)$ *of every* $\mathbf{D}(.) \in \mathfrak{D}_-^k$ *is negative,* $\mathfrak{D}_- = \mathfrak{D}_-^0$.

$\mathfrak{U}^k$ is a given, or to be determined, *family of all bounded and k-times continuously differentiable realizable control vector functions* $\mathbf{U}(.)$, $\mathfrak{U}^k \subseteq \mathfrak{C}^k$.

$\mathfrak{U}^0 = \mathfrak{U}$ is the *family of all bounded continuous realizable control vector functions* $\mathbf{U}(.)$, $\mathfrak{U}^0 \subseteq \mathfrak{C}^0$.

$\mathfrak{Y}_d^k$ is a given, or to be determined, *family of all bounded and k-times continuously differentiable realizable desired output vector functions* $\mathbf{Y}_d(.) \in \mathfrak{L}$,

$$\mathfrak{Y}_d^k \subseteq \mathfrak{C}^k \cap \mathfrak{L}. \tag{2.18}$$

$\mathfrak{Y}_d^0 = \mathfrak{Y}_d$ is a *family of all bounded continuous realizable desired output vector functions* $\mathbf{Y}_d(.)$. Their Laplace transforms are strictly proper real rational complex functions,

$$\mathfrak{Y}_d \subseteq \mathfrak{C} \cap \mathfrak{L}. \tag{2.19}$$

$\mathfrak{Y}_{d-}^k$ is a subfamily of $\mathfrak{Y}_d^k$, $\mathfrak{Y}_{d-}^k \subset \mathfrak{Y}_d^k$, such that *the real part of every pole of Laplace transform* $\mathbf{Y}_d(s)$ *of every* $\mathbf{Y}_d(.) \in \mathfrak{Y}_{d-}^k$ *is negative,* $\mathfrak{Y}_{d-} = \mathfrak{Y}_{d-}^0$.

The preceding families include the corresponding zero vectors,

$$\mathbf{0}_{iM} \in \mathfrak{D}^i,\ \mathbf{0}_{kN} \in \mathfrak{Y}_d^k.$$

Let us introduce

$$\mathbf{I}^{(k)}(t) = \left[\begin{array}{c} \mathbf{D}^{(k)}(t) \\ \mathbf{U}^{(k)}(t) \end{array}\right] \in \mathfrak{R}^{d+r},\ B_{Pk} = \left[D_{Pdk} \vdots C_{Puk}\right] \in \mathfrak{R}^{N\times(d+r)}$$

$$k = 0,1,..,\mu; \qquad B_{P}^{(\mu)} = \left[B_{P0} \vdots B_{P1} \vdots ... \vdots B_{P\mu}\right] \in \mathfrak{R}^{N\times(\mu+1)(d+r)},$$

$$\mu_{Pu} < \mu \Longrightarrow C_{Puk} = O_{Nd},\ k = \mu_{Pu} + 1,..,\mu,$$
$$\mu_{Pd} < \mu \Longrightarrow D_{Pdk} = O_{Nr},\ k = \mu_{Pd} + 1,..,\mu. \qquad (2.20)$$

These notations transform the IO differential equation (2.15) of the plant $\mathcal{P}$ into the general compact form (2.9),

$$A_{P}^{(\nu)}\mathbf{Y}^{\nu}(t) = B_{P}^{(\mu)}\mathbf{I}^{\mu}(t),\ \forall t \in \mathfrak{T}. \qquad (2.21)$$

2.1.3 Input-output (*IO*) description of a feedback controller

The compact form of the IO differential equation of a controller ($\mathcal{CR}$) either disconnected from the plant in which case $\mathbf{Y}^{\mu_{Cy}}(t) \equiv \mathbf{0}_{\mu_{Cy}+1}$, or interconnected with the plant in a feedback (i.e., closed-loop) control system, reads

$$A_{CR}^{(\nu)}\mathbf{U}^{\nu}(t) = P_{CR}^{(\mu_{Cyd})}\mathbf{Y}_{d}^{\mu_{Cyd}}(t) - Q_{CR}^{(\mu_{Cy})}\mathbf{Y}^{\mu_{Cy}}(t),\ \det A_{CR\nu} \neq 0,\ \forall t \in \mathfrak{T}$$

$$P_{CR}^{(\mu_{Cyd})} = \left[P_{CR0} \vdots\ P_{CR1} \vdots ... \vdots P_{CR\mu_{Cyd}}\right] \in \mathfrak{R}^{r\times(\mu_{Cyd}+1)N},$$

$$Q_{CR}^{(\mu_{Cy})} = \left[Q_{CR0} \vdots\ Q_{CR1} \vdots ... \vdots Q_{CR\mu_{Cy}}\right] \in \mathfrak{R}^{r\times(\mu_{Cy}+1)N},$$

$$\mathbf{Y}_{d}(t) \in \mathfrak{R}^{N} \textit{ is the plant desired output vector at } t \in \mathfrak{T}. \qquad (2.22)$$

The controller $\mathcal{CR}$ (2.22) is the IO controller. We can put it into the general form (2.1) (Subsection 2.1.1) by applying the following notation:

$$\mathbf{I}^{(k)}(t) = \left[\begin{array}{c} \mathbf{Y}_{d}^{(k)}(t) \\ \mathbf{Y}^{(k)}(t) \end{array}\right] \in \mathfrak{R}^{2N},\ B_{CRk} = \left[P_{CRk} \vdots - Q_{CRk}\right] \in \mathfrak{R}^{r\times 2N},$$

$$k = 0,1,..,\mu; \qquad B_{CR}^{(\mu)} = \left[B_{CR0} \vdots\ B_{CR1} \vdots ... \vdots B_{CR\mu}\right] \in \mathfrak{R}^{r\times 2N(\mu+1)},$$

$$\mu_{Cyd} < \mu \Longrightarrow P_{CRk} = O_{rN},\ k = \mu_{Cyd} + 1,..,\mu,$$
$$\mu_{Cy} < \mu \Longrightarrow Q_{CRk} = O_{rN},\ k = \mu_{Cy} + 1,..,\mu,$$
$$A_{CR}^{(\nu)}\mathbf{U}^{v}(t) = B_{CR}^{(\mu)}\mathbf{I}^{\mu}(t),\ \forall t \in \mathfrak{T}. \qquad (2.23)$$

(2.23) is the general form of the IO controller.

In a special case, $P_{CRk} \equiv Q_{CRk} = E_{CRk}$ and $\mu_{Cyd} = \mu_{Cy} = \mu$; hence, $P_{CR}^{(\mu)} = Q_{CR}^{(\mu)} = E_{CR}^{(\mu)}$. We can use the output error vector $\boldsymbol{\varepsilon}$,

$$\boldsymbol{\varepsilon} = \mathbf{Y}_{d} - \mathbf{Y}, \qquad (2.24)$$

of the plant $\mathcal{P}$ for the input vector of the controller $\mathfrak{C}$ so that (2.22) and (2.23) then become

$$A_{CR}^{(\nu)}\mathbf{U}^{\upsilon}(t) = E_{CR}^{(\mu)}\boldsymbol{\varepsilon}^{\mu}(t),\ \forall t \in \mathfrak{T}. \tag{2.25}$$

2.1.4 Input-output (*IO*) description of a feedback control system

The output vector $\mathbf{Y}_P = \mathbf{Y}$ of the plant is simultaneously the output vector of the IO control system, $\mathbf{Y}_{CS} = \mathbf{Y}_P = \mathbf{Y}$. The overall IO differential equation of the IO control system $\mathcal{CS}$ is the following:

$$A_{CS}^{(\nu)}\mathbf{Y}^{\nu}(t) = D_{CS}^{(\mu_{Pd})}\mathbf{D}_{Pd}^{\mu_{Pd}}(t) + W_{CS}^{(\mu_{Cyd})}\mathbf{Y}_{d}^{\mu_{Cyd}}(t),\ \det A_{cs\nu} \neq O_r, \forall t \in \mathfrak{T},$$
$$D_{CS}^{(\mu_{Pd})} = D_{Pd}^{(\mu_{Pd})},\ (2.15)\ \text{Subsection 2.1.2},$$
$$W_{CS}^{(\mu_{Cyd})} = \left[W_{CS0} \vdots\ W_{CS1} \vdots \dots \vdots W_{CS\mu_{Cyd}}\right] \in \mathfrak{R}^{Nx(\mu_{Cyd}+1)N}. \tag{2.26}$$

We can put it into the general compact form (2.9) (Subsection 2.1.1) by using the following notation:

$$\mathbf{I}^{(k)}(t) = \begin{bmatrix} \mathbf{D}^{(k)}(t) \\ \mathbf{Y}_d^{(k)}(t) \end{bmatrix} \in \mathfrak{R}^{d+N},\ B_{CSk} = \left[D_{CSk} \vdots W_{CSk}\right] \in \mathfrak{R}^{Nx(d+N)},$$
$$k = 0, 1, .., \mu,\ B_{CS}^{(\mu)} = \left[B_{CS0} \vdots\ B_{CS1} \vdots \dots \vdots B_{CS\mu}\right] \in \mathfrak{R}^{Nx(\mu+1)(d+N)},$$
$$\mu_{Pd} < \mu \Longrightarrow D_{CSk} = O_N,\ k = \mu_{Pd} + 1, .., \mu,$$
$$\mu_{CSyd} < \mu \Longrightarrow W_{CSk} = O_N,\ k = \mu_{CSyd} + 1, .., \mu, \tag{2.27}$$

so that

$$A_{CS}^{(\nu)}\mathbf{Y}^{\nu}(t) = B_{CS}^{(\mu)}\mathbf{I}^{\mu}(t),\ \forall t \in \mathfrak{T}. \tag{2.28}$$

This is the compact IO mathematical model of the IO control system.

2.2 *ISO* control systems

2.2.1 General *ISO* system

This subsection is the basis for the next subsections.

The *Input-State-Output* (ISO) *systems* described by *the state equation* (2.29) and by *the output equation* (2.30),

$$\frac{d\mathbf{X}(t)}{dt} = A\mathbf{X}(t) + B\mathbf{I}(t),\ \forall t \in \mathfrak{T}, \tag{2.29}$$
$$\mathbf{Y}(t) = C\mathbf{X}(t) + D\mathbf{I}(t),\ \forall t \in \mathfrak{T}, \tag{2.30}$$
$$A \in \mathfrak{R}^{nxn},\ \mathbf{X} \in \mathfrak{R}^{n},\ B \in \mathfrak{R}^{nxM},$$
$$C \in \mathfrak{R}^{Nxn},\ D \in \mathfrak{R}^{NxM},\ n \geq N,$$
$$(\textit{either } C \neq O_{N,n} \textit{ or } D \neq O_{N,M} \textit{ or both}),$$

enabled the development of the fundamental control concepts (e.g., of controllability, observability and the system equivalence), and the establishment of the related complete results in the theory of linear control systems. These equations can be a general mathematical model of an object/a plant, of a controller, or of a whole control system. We commonly refer to these equations, for short, as *the ISO system* (2.29), (2.30) or as *the state-space system.*

The presentation in this subsection is equivalent to the corresponding presentation in [148].

Note 34 ***The IO system and the ISO system equivalence***
Appendix B contains the transformation of the IO system (2.1) (Section 2.1) into the equivalent ISO system (2.29), (2.30). Appendix C presents the inverse transformation of the ISO system (2.29), (2.30) into the IO system (2.1).

Definition 35 *The order of the highest derivative of the state vector function* $\mathbf{X}(.)$ *in (2.29), i.e., one* (1)*, is* ***the order of the ISO system (2.29), (2.30).***

This definition holds for all *ISO* systems. They are the first-order systems. The form of the state equation (2.29) is *Cauchy form.* It is also called *the normal form.* There is only the first derivative $\mathbf{X}^{(1)}(t)$ of the state vector $\mathbf{X}(t)$ in (2.29). It does not contain any derivative or integral of the input vector $\mathbf{I}(t)$.

Definition 36 *(a) The dimension* n *of the state vector* $\mathbf{X}$, $dim_{ISO} = dim\mathbf{X} = n$*, of the ISO system (2.29), (2.30) is its* ***dimension,*** *denoted by* dim_{ISO}.

(b) The dimension n *of the ISO system (2.29), (2.30) is also its* ***dynamic dimension*** $ddim_{ISO}$, $ddim_{ISO}=n$.

(c) The dimension n *of the ISO system (2.29), (2.30) is also its* ***minimal dynamic dimension*** $mddim_{ISO}$, $mddim_{ISO}=n$.

The well known definition of the realization of the transfer function matrix of the *ISO* system holds only under zero initial conditions. Its generalization to the systems with an arbitrary input vector function $\mathbf{I}(.)$ and with an arbitrary initial state vector $\mathbf{X}_{0\mp}$ follows.

Definition 37 ***A realization of the ISO system (2.29), (2.30)*** *for an arbitrary input vector function* $\mathbf{I}(.)$ *and for an arbitrary initial state vector* $\mathbf{X}_{0\mp}$ *is the quadruple* $(A,\ B,\ C,\ D)$.

Comment 38 *The realization* $(A,\ B,\ C,\ D)$ *of the ISO system (2.29), (2.30) is* ***its ISO realization.*** *Its IO realization is elaborated in Appendix C.*

2.2.2 *ISO* plant

If the system is a plant ($\mathcal{P}$), then its input vector $\mathbf{I}$ (2.29), (2.30) comprises its disturbance vector $\mathbf{D}$ and its control vector $\mathbf{U}$,

$$\mathbf{I}=\begin{bmatrix}\mathbf{D}\\ \mathbf{U}\end{bmatrix},\ \mathbf{D}\in R^{d},\ \mathbf{U}\in R^{r}. \tag{2.31}$$

The matrices B and D (2.29), (2.30) are then partitioned accordingly,

$$B = \begin{bmatrix} B_P \vdots L_P \end{bmatrix} \in R^{n_P x(r+d)},\ B_P \in R^{n_P xr},\ L_P \in R^{n_P xd},$$

$$\mathbf{X} = \mathbf{X}_P \in R^{n_P},\ n = n_P,$$

$$D = \begin{bmatrix} H_P \vdots D_P \end{bmatrix} \in R^{Nx(d+r)},\ H_P \in R^{Nxr},\ D_P \in R^{Nxd},\ \mathbf{Y}_P \in R^N, \quad (2.32)$$

so that the *ISO* description (2.29), (2.30) of the plant, which we will call *the ISO plant*, takes the following specific form:

$$\frac{d\mathbf{X}_P(t)}{dt} = A_P\mathbf{X}_P(t) + B_P\mathbf{U}(t) + L_P\mathbf{D}(t),\ \forall t \in \mathfrak{T},\ \mathbf{X}_P \in R^{n_P}, \quad (2.33)$$

$$\mathbf{Y}_P(t) = C_P\mathbf{X}_P(t) + H_P\mathbf{U}(t) + D_P\mathbf{D}(t),\ \forall t \in \mathfrak{T},\ C_P \in R^{Nxn_P}. \quad (2.34)$$

2.2.3 *ISO* feedback controller

If the system is an *ISO* controller ($\mathcal{CR}$), then the plant desired output $\mathbf{Y}_d$ and the plant real output $\mathbf{Y}$ compose the controller input $\mathbf{I}$,

$$\mathbf{I} = \begin{bmatrix} \mathbf{Y}_d \\ \mathbf{Y} \end{bmatrix},\ \mathbf{Y}_d \in R^N,\ \mathbf{Y} \in R^N. \quad (2.35)$$

The matrices B and D (2.29), (2.30) are then partitioned as follows:

$$B = \begin{bmatrix} B_C \vdots L_C \end{bmatrix} \in R^{n_C x2N},\ B_C \in R^{n_C xN},\ L_c \in R^{n_C xN},$$

$$\mathbf{X} = \mathbf{X}_C \in R^{n_C},\ n = n_C,\ C_C \in R^{rxn_C}$$

$$D = \begin{bmatrix} H_C \vdots D_C \end{bmatrix} \in R^{rx2N},\ H_C \in R^{rxN},\ D_C \in R^{rxN},\ \mathbf{Y}_C \in R^r, \quad (2.36)$$

so that the *ISO* description (2.29), (2.30) of the controller takes the following specific form, in which the *ISO* controller output vector $\mathbf{Y}_C$ is the control vector $\mathbf{U}$ of the plant:

$$\frac{d\mathbf{X}_C(t)}{dt} = A_C\mathbf{X}_C(t) + B_C\mathbf{Y}_d(t) - L_C\mathbf{Y}(t),\ \forall t \in \mathfrak{T},\ \mathbf{X}_C \in R^{n_C}, \quad (2.37)$$

$$\mathbf{Y}_C(t) = \mathbf{U}(t) = C_C\mathbf{X}_C(t) + H_C\mathbf{Y}_d(t) - D_C\mathbf{Y}(t),\ \forall t \in \mathfrak{T}. \quad (2.38)$$

In the case when $B_C = L_C$ and $H_C = D_C$, then the controller input is the output error vector (2.24) (Subsection 2.1.3) of the plant,

$$\mathbf{I} = \boldsymbol{\varepsilon},\ \boldsymbol{\varepsilon} \in R^N. \quad (2.39)$$

The matrices B and D become then

$$B = B_C = L_C \in R^{n_C xN},$$

$$D = H_C = D_C \in R^{rxN}, \quad (2.40)$$

so that the *ISO* description (2.29), (2.30) of the controller takes the following specific form:

$$\frac{d\mathbf{X}_C(t)}{dt} = A_C\mathbf{X}_C(t) + B_C\varepsilon(t), \ \forall t \in \mathfrak{T}, \tag{2.41}$$

$$\mathbf{Y}_C(t) = \mathbf{U}(t) = C_C\mathbf{X}_C(t) + H_C\varepsilon(t), \ \forall t \in \mathfrak{T},. \tag{2.42}$$

2.2.4 *ISO* feedback control system

If the system is an overall feedback *ISO* control system ($\mathcal{CS}$), then its input vector $\mathbf{I}$ comprises the plant disturbance vector $\mathbf{D}$ and the plant desired output vector $\mathbf{Y}_d$,

$$\mathbf{I} = \begin{bmatrix} \mathbf{D} \\ \mathbf{Y}_d \end{bmatrix}, \mathbf{D} \in R^d, \ \mathbf{Y}_d \in R^N. \tag{2.43}$$

The matrices B and D (2.29), (2.30), are then partitioned accordingly,

$$B = \begin{bmatrix} L_{CS} \vdots P_{CS} \end{bmatrix} \in R^{n_{CS}\text{x}(d+N)}, \ L_{CS} \in R^{n_{CS}\text{x}d}, \ P_{CS} \in R^{n_{CS}\text{x}N},$$

$$\mathbf{X} = \mathbf{X}_{CS} \in R^{n_{CS}}, \ n = n_{CS}, \ C_{CS} \in R^{N\text{x}n_{CS}}$$

$$D = \begin{bmatrix} D_{CS} \vdots Q_{CS} \end{bmatrix} \in R^{N\text{x}(d+N)}, \ D_{CS} \in R^{N\text{x}d}, \ Q_{CS} \in R^{N\text{x}N}, \ \mathbf{Y}_{CS} \in R^N, \tag{2.44}$$

so that the *ISO* description (2.29), (2.30) of the *ISO* control system of the plant takes the following specific form, in which the real output vector $\mathbf{Y}$ of the plant is simultaneously the output vector $\mathbf{Y}_{CS}$ of the control system:

$$\frac{d\mathbf{X}_{CS}(t)}{dt} = A_{CS}\mathbf{X}_{CS}(t) + L_{CS}\mathbf{D}(t) + P_{CS}\mathbf{Y}_d(t), \ \forall t \in \mathfrak{T}, \tag{2.45}$$

$$\mathbf{Y}_{CS}(t) = \mathbf{Y} = C_{CS}\mathbf{X}_{CS}(t) + D_{CS}\mathbf{D}(t) + Q_{CS}\mathbf{Y}_d(t), \ \forall t \in \mathfrak{T}. \tag{2.46}$$

so that the LSQ description (2.29), (2.30) of the controller takes the following form:

$$\frac{dX_2}{dt} = A_2 X_2(t) + B_2 s(t), \quad \forall t \ge 0, \quad [illegible]$$

$$Y_2(t) = C_2 X_2(t) + D_2 s(t), \quad \forall t \ge 0, \quad [illegible]$$

2.2.3 (SS) feedback control system

In the [illegible] feedback [illegible] control system [illegible] vector [illegible] the total disturbance vector D and the plant [illegible] vector Y_1,

$$[illegible]$$

[illegible]

$$[illegible]$$

$$[illegible]$$

$$[illegible]$$

[illegible] (2.29) [illegible] and [illegible] the [illegible] the vector Y of the [illegible] control system:

$$\frac{dX}{dt} = A_0 X(t) + [illegible]$$

$$Y(t) = C_0 X(t) + [illegible]$$

Chapter 3

System Regimes

3.1 System regime meaning

We refer to [148] for what follows.

The manner of the temporal evolution of a process, of a work, of a movement of the system, or the manner of the temporal evolution of the system response, determines the system behavior. It depends on

a) *the system properties,*

and

b) *the actions upon the system.*

There are two categories of the system properties: *quantitative* and *qualitative. A quantitative system property* is, for example, the overshoot of the system response. The quantitative system properties reflect the system behavior under particular external and internal conditions. Controllability, observability, stability, and trackability (Chapter 9.3)([85]-[87], [132]-[140], [141], [157], [161], [164], [166], [167], [176]-[180], [249]-[258]) are examples of *qualitative system properties.* They relate the system behavior to a set of external and/or internal conditions, the set of which can be finite or infinite, bounded or unbounded, but not a singleton that is characteristic of the quantitative system properties.

Two different principal *actions on the system* are the following:

∘ *Actions on the system, which were created during the system history by past (external and/or internal) influences on the system.* ***Initial conditions*** express these actions. Initial conditions can be those of input variables, or of internal dynamics variables in general or of state variables in particular, and of output variables. This book treats the system behavior under *arbitrary initial conditions.*

∘ *Actions that influence the system behavior on* $\mathfrak{T}_0$. These actions are ***the external actions***. They are ***the input variables*** if, and only if, they influence essentially the system behavior.

Definition 39 *The set of all (initial and exterior) conditions under which the system operates and the type of its behavior (i.e., the type of the temporal evo-*

lution: of a process, of a work, of a movement of the system and/or of its response) represent ***a system regime.***

The following criteria determine various system regimes:

- *The existence (the nonexistence) of the initial conditions.* Their values different from (equal to) zero express their existence (nonexistence), respectively.
- *The existence (the nonexistence) of the external actions* (Section 3.2).
- *The realization of the system (plant) demanded behavior.* If the system is an object/a plant, its demanded behavior is called its *desired behavior* and it is defined by its *desired response* (or, more precisely, by its *desired output response*) denoted by $\mathbf{Y}_d(.\ ;\ .\ ;\ .) = \mathbf{Y}_d(.)$ (Section 3.3).
- *The type of the system (plant) behavior* is important for stability studies [148], which are beyond the scope of this book.

3.2 Forced and free regimes

3.2.1 Introduction

The notation $(\sigma, \infty[$ means either $]\sigma, \infty[$ or $[\sigma, \infty[$,

$$(\sigma, \infty[\in \{]\sigma, \infty[,\ [\sigma, \infty[\}.$$

In this case the criterion for the classification of the system regimes is *the existence of the exterior actions.* They are largely referred to in the literature, but their definitions are rare. What follows is equivalent to the corresponding definitions in [148].

Definition 40 *(a) A system is in* ***a forced regime*** *on $(\sigma, \infty[$ if, and only if, there is a moment τ in $(\sigma, \infty[$ when the input vector different from the zero vector acts on the system:*

$$\exists \tau \in (\sigma, \infty[\Longrightarrow \mathbf{I}(\tau) \neq \mathbf{0}_M.$$

(b) A system is in ***a free regime*** *on $(\sigma, \infty[$ if, and only if, its input vector is equal to the zero vector for every $t \in (\sigma, \infty[$:*

$$\mathbf{I}(t) = \mathbf{0}_M,\ \forall t \in (\sigma, \infty[.$$

The omission of the expression "on $(\sigma, \infty[$" is acceptable if, and only if, $\sigma = 0$, i.e., $(\sigma, \infty[= \mathfrak{T}_0$.

Definition 41 *A system behavior is* ***trivial*** *if, and only if, the system movement is always equal to the zero vector. Otherwise, it is* ***nontrivial.***

In order for the system behavior to be trivial in a free regime it is necessary and sufficient that all initial conditions are equal to zero. This means that for the physical behavior of the system in a free regime to be nontrivial it is necessary and sufficient that there is an accumulated energy in the system at the initial moment.

Lyapunov stability properties, and linear system observability, reflect the system behavior in a free regime. $BIBO$ stability, practical stability, controllability, trackability, and tracking reflect the system behavior in a forced regime either under zero initial conditions ($BIBO$ stability) or under arbitrary initial conditions.

3.2.2 Basic problem

Only the system order, dimension, parameters and structure determine the link of Laplace transform of the system output vector and Laplace transform of the system input vector, under all zero initial conditions, whatever is the form of the input vector function. What if the initial conditions are different from zero?

Problem 42 ***The basic problem***

What is the complex domain description of the system, which determines the relationship between Laplace transform of the output vector and Laplace transform of the input vector only in terms of the system parameters, whatever the form of the input vector function, and for arbitrary initial conditions?

The subsequent presentations will show the solutions to this problem for the IO and ISO systems. They will discover the existence of the dynamic system characteristic that generalizes the system transfer function matrix $G(s)$. The new system characteristic is ***the system full (complete) transfer function matrix*** $F(s)$ ([78], [81], [148]), [150], [168], [174, Theorem 2.3.1, Theorem 2.3.2, pp. 26 - 29]). Its application permits us to treat fully and correctly in the complex domain many qualitative dynamical properties as well as quantitative dynamical characteristics of the systems.

3.3 Desired regime

3.3.1 Introduction

The analysis of a desired regime is fully meaningful only if the system is a plant or its control system. The demanded plant (output) response $\mathbf{Y}_d(.)$ determines the plant desired regime, and vice versa. The study of the nominal behavior of the linear dynamic system in general in [148] will be applied to desired behavior of the plant in what follows.

Definition 43 ***Desired regime***

A plant is in ***a desired regime on*** $\mathfrak{T}_0$ *(for short: in* ***a desired regime****) if, and only if, it realizes its desired (output) response* $\mathbf{Y}_d(t)$ *all the time,*

$$\mathbf{Y}(t) = \mathbf{Y}_d(t), \ \forall t \in \mathfrak{T}_0. \tag{3.1}$$

A necessary condition for a plant to be in a desired (nominal, nonperturbed) regime follows directly from this definition.

Proposition 44 *The initial real output vector should be equal to the initial desired output vector,*

$$\mathbf{Y}_0 = \mathbf{Y}_{d0},$$

in order for the plant to be in a desired regime,

$$\mathbf{Y}(t) = \mathbf{Y}_d(t),\ \forall t \in \mathfrak{T}_0 \Longrightarrow \mathbf{Y}_0 = \mathbf{Y}_{d0}.$$

If the initial real output vector of the plant is different from the initial desired output vector, then it is impossible for the plant to be in a nominal regime (on $\mathfrak{T}_0$),

$$\mathbf{Y}_0 \neq \mathbf{Y}_{d0} \Longrightarrow \exists \sigma \in \mathfrak{T}_0 \Longrightarrow \mathbf{Y}(\sigma) \neq \mathbf{Y}_d(\sigma).$$

The real initial output vector $\mathbf{Y}(0) = \mathbf{Y}_0$ is mainly different from the desired initial output vector $\mathbf{Y}_d(0) = \mathbf{Y}_{d0}$. The plant is mainly in a *nondesired* regime.

Definition 45 ***Nominal input***

An input vector function $\mathbf{I}^*(.)$ *of a system/plant is* ***nominal with respect to its desired response*** $\mathbf{Y}_d(.)$, *which is denoted by* $\mathbf{I}_N(.)$, *if, and only if,* $\mathbf{I}(.) = \mathbf{I}^*(.)$ *guarantees that the induced real response* $\mathbf{Y}(.) = \mathbf{Y}^*(.)$ *of the system/plant obeys* $\mathbf{Y}^*(t) = \mathbf{Y}_d(t)$ *always as soon as all the internal and the output system/plant initial conditions are demanded/desired,*

$$\mathbf{I}^*(.) = \mathbf{I}_N(.) \Longleftrightarrow \langle \mathbf{Y}^*(t) = \mathbf{Y}_d(t),\ \forall t \in \mathfrak{T}_0 \rangle . \tag{3.2}$$

This is general definition. It permits us to define the nominal input vector for different classes of systems.

Note 46 *If an input vector function* $\mathbf{I}^*(.)$ *is nominal with respect to the desired response* $\mathbf{Y}_{d1}(.)$ *of a plant, it does not guarantee that it is nominal relative to another desired response* $\mathbf{Y}_{d2}(.)$ *of the plant. The notion "nominal relative to the desired response* $\mathbf{Y}_d(.)$*" has the relative sense and validity.*

Comment 47 *The nominal input vector function* $\mathbf{I}_N(.)$ *of the plant incorporates both* ***the nominal perturbation vector function*** *denoted by* $\mathbf{D}_N(.)$ *and* ***the nominal control vector function*** $\mathbf{U}_N(.)$*:*

$$\mathbf{I}_N(.) = \begin{bmatrix} \mathbf{D}_N(.) \\ \mathbf{U}_N(.) \end{bmatrix}.$$

Definition 48 ***Nominal control***

A control vector function $\mathbf{U}^*(.)$ *of the plant is* ***nominal relative to*** $[\mathbf{D}(.), \mathbf{Y}_d(.)]$, *which is denoted by* $\mathbf{U}_N(.)$,

$$\mathbf{U}_N(t) \equiv [\mathbf{U}_N(t; \mathbf{D}; \mathbf{Y}_d)], \tag{3.3}$$

if, and only if, $\mathbf{U}(.) = \mathbf{U}^*(.)$ *ensures that the corresponding real response* $\mathbf{Y}(.) = \mathbf{Y}^*(.)$ *of the system obeys* $\mathbf{Y}^*(t) = \mathbf{Y}_d(t)$ *all the time as soon as all the internal and the output system initial conditions are desired (nominal).*

Comment 49 *This definition specifies the nominal control vector function* $\mathbf{U}_N(.)$ *relative to chosen both* $\mathbf{D}(.)$ *and* $\mathbf{Y}_d(.)$*, (3.3).*

Definition 50 ***Nominal input pair***

An input vector functional pair $[\mathbf{D}^*(.), \mathbf{U}^*(.)]$ *of the system/plant is* ***nominal relative to its demanded/desired response*** $\mathbf{Y}_d(.)$*, which is denoted by* $[\mathbf{D}_N(.), \mathbf{U}_N(.)]$*, if, and only if,* $[\mathbf{D}(.), \mathbf{U}(.)] = [\mathbf{D}^*(.), \mathbf{U}^*(.)]$ *ensures that the corresponding real response* $\mathbf{Y}(.) = \mathbf{Y}^*(.)$ *of the plant obeys* $\mathbf{Y}^*(t) = \mathbf{Y}_d(t)$ *all the time as soon as all the internal and the output system initial conditions are desired (nominal),*

$$[\mathbf{D}^*(.), \mathbf{U}^*(.)] = [\mathbf{D}_N(.), \mathbf{U}_N(.)] \Longleftrightarrow \langle \mathbf{Y}^*(t) = \mathbf{Y}_d(t),\ \forall t \in \mathfrak{T}_0 \rangle . \tag{3.4}$$

Comment 51 *This definition specifies the nominal vector functions* $\mathbf{D}_N(.)$ *and* $\mathbf{U}_N(.)$ *of both* $\mathbf{D}(.)$ *and* $\mathbf{U}(.)$ *relative to* $\mathbf{Y}_d(.)$*:*

$$[\mathbf{D}_N(t), \mathbf{U}_N(t)] \equiv [\mathbf{D}_N(t; \mathbf{Y}_d), \mathbf{U}_N(t; \mathbf{Y}_d)] . \tag{3.5}$$

Definition 52 $\mathbf{Y}_d(.)$ ***realizable under the action of the given*** $\mathbf{D}(.)$

The desired output behavior $\mathbf{Y}_d(.)$ *of the plant is* ***realizable under the action of the given*** $\mathbf{D}(.)$ *if, and only if, there exists the plant nominal control vector function* $\mathbf{U}_N(.)$ *relative to* $[\mathbf{D}(.), \mathbf{Y}_d(.)]$ *(3.3).*

This definition determines realizability of a specific $\mathbf{Y}_d(.)$ for a single given $\mathbf{D}(.)$. We broaden it to realizability of a specific $\mathbf{Y}_d(.)$ in $\mathfrak{D}^i$.

Definition 53 $\mathbf{Y}_d(.)$ ***realizable in*** $\mathfrak{D}^i$

The desired response $\mathbf{Y}_d(.)$ *of the plant is* ***realizable in*** $\mathfrak{D}^i$ *if, and only if, there exist both* $\mathbf{D}^*(.) \in \mathfrak{D}^i$ *and* $\mathbf{U}^*(.)$ *such that the pair* $[\mathbf{D}^*(.), \mathbf{U}^*(.)]$ *is the nominal pair relative to* $\mathbf{Y}_d(.)$ *,*

$$[\mathbf{D}^*(.), \mathbf{U}^*(.)] = [\mathbf{D}_N(.), \mathbf{U}_N(.)] \ \ \textit{relative to } \mathbf{Y}_d(.).$$

Equivalently,

Definition 54 $\mathbf{Y}_d(.)$ ***realizable in*** $\mathfrak{D}^i$

The desired output behavior $\mathbf{Y}_d(.)$ *of the plant is* ***realizable in*** $\mathfrak{D}^i$ *if, and only if, there exists the plant nominal vector functional pair* $[\mathbf{D}_N(.), \mathbf{U}_N(.)]$*,* $\mathbf{D}_N(.) \in \mathfrak{D}^i$*, relative to* $\mathbf{Y}_d(.)$*, (3.5).*

Comment 55 *This definition does not require that for every* $\mathbf{D}(.) \in \mathfrak{D}^i$ *there exists the nominal control vector function* $\mathbf{U}_N(.)$ *relative to* $\mathbf{Y}_d(.)$*.*

Definition 56 $\mathbf{Y}_d(.)$ ***realizable on*** $\mathfrak{D}^i$

The desired response $\mathbf{Y}_d(.)$ *of the plant is* ***realizable on*** $\mathfrak{D}^i$ *if, and only if, for every* $\mathbf{D}(.) \in \mathfrak{D}^i$ *there exists the nominal control vector function* $\mathbf{U}_N(.)$ *relative to* $[\mathbf{D}(.), \mathbf{Y}_d(.)]$*, (3.3).*

Comment 57 *This definition demands that for every* $\mathbf{D}(.) \in \mathfrak{D}^i$ *there exists the nominal control vector function* $\mathbf{U}_N(.)$ *relative to a specific, given,* $\mathbf{Y}_d(.)$.

Comment 58 *[148] The realizability of* $\mathbf{Y}_d(.)$ *in* $\mathfrak{D}^i$ *is necessary, but not sufficient, for the realizability of* $\mathbf{Y}_d(.)$ *on* $\mathfrak{D}^i$. *The realizability of* $\mathbf{Y}_d(.)$ *on* $\mathfrak{D}^i$ *is sufficient, but not necessary, for the realizability of* $\mathbf{Y}_d(.)$ *in* $\mathfrak{D}^i$.

Comment 59 *The nominal control vector function* $\mathbf{U}_N(.)$ *for the plant relative to its desired output* $\mathbf{Y}_d(.)$ *is simultaneously the desired (nominal) output vector function of the controller. It is to be determined from the condition that it should be the nominal control vector function of the plant relative to its desired output vector function* $\mathbf{Y}_d(.)$.

Comment 60 *If the controller is a feedback controller in a closed loop control system of the plant, then it has two input vector functions: the desired* $\mathbf{Y}_d(.)$ *and the real* $\mathbf{Y}(.)$ *output vector functions of the plant, or equivalently, the error vector function* $\boldsymbol{\varepsilon}(.)$, *or the deviation vector function* $\mathbf{y}(.)$,

$$\boldsymbol{\varepsilon}(.) = -\mathbf{y}(.) = \mathbf{Y}_d(.) - \mathbf{Y}(.). \tag{3.6}$$

The nominal input vector function of the controller is therefore

$$\mathbf{I}_N(.) = \begin{bmatrix} \mathbf{Y}_d(.) \\ \mathbf{Y}_d(.) \end{bmatrix},$$

which implies the nominal (desired) output error vector $\boldsymbol{\varepsilon}_N$ *as the nominal input vector to the controller described in terms of the error vector* $\boldsymbol{\varepsilon}$,

$$\boldsymbol{\varepsilon}_N(t) \equiv \mathbf{Y}_d(t) - \mathbf{Y}_d(t) \equiv \mathbf{0}_N.$$

Comment 61 *If a system represents an overall closed-loop, feedback, control system, then the nominal input vector function* $\mathbf{I}_N(.)$ *of the whole control system incorporates both the nominal perturbation vector function* $\mathbf{D}_N(.)$ *of the controlled plant and the plant desired output vector function* $\mathbf{Y}_d(.)$:

$$\mathbf{I}_N(.) = \begin{bmatrix} \mathbf{D}_N(.) \\ \mathbf{Y}_d(.) \end{bmatrix}.$$

Since the control system desired output vector function is the desired (output) response $\mathbf{Y}_d(.)$ *of the plant, then only the nominal perturbation function* $\mathbf{D}_N(.)$ *relative to* $\mathbf{Y}_d(.)$ *is to be determined.*

Control system designer's crucial interest is in a solution of the following.

Problem 62 *Under what conditions does there exist a nominal control vector function* $\mathbf{U}(.)$ *relative to the plant desired (nominal) output response* $\mathbf{Y}_d(.)$, *or equivalently, under what conditions is the plant desired output response* $\mathbf{Y}_d(.)$ *realizable in* $\mathfrak{D}^i$ *and/or realizable on* $\mathfrak{D}^i$*?*

Some qualitative system properties (e.g., trackability properties and tracking properties) have a sense if, and only if, there exists an affirmative solution to the preceding problem.

3.3.2 *IO* control systems

Definition 43 and Definition 45 in general, and Definition 48 and Definition 50 in particular (Subsection 3.3.1) will be applied to the *IO* plant (2.1) (Section 2.1).

In order to present the complex domain condition for an input vector function to be nominal for the system relative to its desired output vector response, we use the complex matrix function $S_i^{(k)}(.) : \mathfrak{C} \longrightarrow \mathfrak{C}^{\ i(k+1)xi}$ (1.4) (Section 1.3) and (2.10) (Section 2.1),

$$S_i^{(k)}(s) = \left[s^0 I_i \vdots s^1 I_i \vdots s^2 I_i \vdots ... \vdots s^k I_i \right]^T \in \mathfrak{C}^{\ i(k+1)xi},$$
$$(k,i) \in \{(\mu, M), \ (\nu, N)\}, \tag{3.7}$$

in which I_i is the *i-th* order identity matrix, and the complex matrix function $Z_k^{(\varsigma-1)}(.) : \mathfrak{C} \to \mathfrak{C}^{(\varsigma+1)kx\varsigma k}$,

$$Z_k^{(\varsigma-1)}(s) = \begin{bmatrix} O_k & O_k & O_k & ... & O_k \\ s^0 I_k & O_k & O_k & ... & O_k \\ ... & ... & ... & ... & ... \\ s^{\varsigma-1} I_k & s^{\varsigma-2} I_k & s^{\varsigma-3} I_k & ... & s^0 I_k \end{bmatrix}, \ \varsigma \geq 1,$$
$$Z_k^{(\varsigma-1)}(s) \in \mathfrak{C}^{(\varsigma+1)kx\varsigma k}, \ (\varsigma, k) \in \{(\mu, M), \ (\nu, N)\} \tag{3.8}$$

They will also enable us to resolve effectively the basic problem 42 (Subsection 3.2.2). [148]

Note 63 *[148]If $\varsigma = 0$, then the matrix $Z_k^{(\varsigma-1)}(s) = Z_k^{(-1)}(s)$ should be completely omitted rather than replaced by the zero matrix. The matrix $Z_k^{(\zeta-1)}(s)$ is not defined for $\zeta \leq 0$ and should be treated as the nonexisting one. Derivatives exist only for natural numbers, i.e., $\mathbf{Y}^{(\varsigma)}(t)$ can exist only for $\varsigma \geq 1$. Matrix function $Z_k^{(\zeta-1)}(.)$ is related to Laplace transform of derivatives only.*

Theorem 64 *[148] In order for a vector function $\mathbf{I}^*(.)$ to be nominal for the IO plant (2.1), i.e., for (2.9), relative to its desired response $\mathbf{Y}_d^{\nu}(.)$: $\mathbf{I}^*(.) = \mathbf{I}_N(.)$, it is necessary and sufficient that 1) and 2) hold:*

1) rank $B^{(\mu)} = N \leq M$, equivalently

$$rank \sum_{k=0}^{k=\mu} B_k s^k = rank B^{(\mu)} S_M^{(\mu)}(s) = N \leq M,$$

and

2) any one of the following equations is valid:

$$\sum_{k=0}^{k=\mu} B_k \mathbf{I}^{*(k)}(t) = \sum_{k=0}^{k=\nu} A_k \mathbf{Y}_d^{(k)}(t), \ \forall t \in \mathfrak{T}_0, \tag{3.9}$$

$$B^{(\mu)} \mathbf{I}^{*\mu}(t) = A^{(\nu)} \mathbf{Y}^{\nu}(t), \ \forall t \in \mathfrak{T}_0, \tag{3.10}$$

or equivalently in the complex domain:

$$\mathbf{I}^*(s) = \left(\sum_{k=0}^{k=\mu} B_k s^k\right)^{-1} \bullet$$
$$\bullet \left\langle \sum_{k=0}^{k=\mu} B_k \left[\sum_{i=1}^{i=k} s_d^{k-i} \mathbf{I}^{*(i-1)}(0)\right] + \sum_{k=0}^{k=\nu} A_k \left[s^k \mathbf{Y}^\nu(s) - \sum_{i=1}^{i=k} s^{k-i} \mathbf{Y}_d^{(i-1)}(0)\right]\right\rangle . \tag{3.11}$$

i.e.,

$$\mathbf{I}^*(s) = \left[B^{(\mu)} S_M^{(\mu)}(s)\right]^{-1} \bullet$$
$$\bullet \left\langle B^{(\mu)} Z_M^{(\mu-1)}(s) \mathbf{I}^{*\mu-1}(0) + A^{(\nu)} \left[S_N^{(\nu)}(s) \mathbf{Y}^\nu(s) - Z_N^{(\nu-1)}(s) \mathbf{Y}_d^{\nu-1}(0)\right]\right\rangle . \tag{3.12}$$

The proof of this theorem, which is general, is in Appendix D. It holds for all *IO* systems described by (2.1), i.e., (2.9).

Theorem 65 *For the desired output response* $\mathbf{Y}^\nu(.)$ *of the IO plant (2.15) to be realizable in* $\mathfrak{D}^{\mu_{Pd}}$ *it is necessary and sufficient that there is an input vector functional pair* $[\mathbf{D}^*(.), \mathbf{U}^*(.)]$, $\mathbf{D}^*(.) \in \mathfrak{D}^{\mu_{Pd}}$, *which obeys the following differential equation:*

$$D_{Pd}^{(\mu_{Pd})} \mathbf{D}^{*\mu_{Pd}}(t) + C_{Pu}^{(\mu_{Pu})} \mathbf{U}^{*\mu_{Pu}}(t) = A_P^{(\nu)} \mathbf{Y}_d^\nu(t), \forall t \in \mathfrak{T}_0,$$
$$\textit{under the condition } \mathbf{Y}_0^{\nu-1} = \mathbf{Y}_{d0}^{\nu-1}. \tag{3.13}$$

Such functional pair $[\mathbf{D}^*(.), \mathbf{U}^*(.)]$ *is nominal for the plant (2.15) relative to its desired output response* $\mathbf{Y}_d^\nu(.)$.

Proof. *Necessity.* Let the desired output response $\mathbf{Y}_d^\nu(.)$ of the *IO* plant (2.15) be realizable in $\mathfrak{D}^{\mu_{\mathfrak{P}\mathfrak{d}}}$. Definition 54 holds. It implies the existence of the nominal input vector functional pair $[\mathbf{D}_N(.), \mathbf{U}_N(.)]$ for the *IO* plant (2.15) relative to its desired response $\mathbf{Y}_d^\nu(.)$. Definition 50 is valid. It and (2.15) imply

$$D_{Pd}^{(\mu_{Pd})} \mathbf{D}_N^{\mu_{Pd}}(t) + C_{Pu}^{(\mu_{Pu})} \mathbf{U}_N^{\mu_{Pu}}(t) = A_P^{(\nu)} \mathbf{Y}_d^\nu(t), \forall t \in \mathfrak{T}_0,$$
$$\textit{under the condition } \mathbf{Y}_0^{\nu-1} = \mathbf{Y}_{d0}^{\nu-1}.$$

This equation becomes (3.13) for $[\mathbf{D}^*(.), \mathbf{U}^*(.)] = [\mathbf{D}_N(.), \mathbf{U}_N(.)]$.

Sufficiency. Let the conditions of the theorem be valid. Let $[\mathbf{D}(.), \mathbf{U}(.)] = [\mathbf{D}^*(.), \mathbf{U}^*(.)]$. The equation (2.15) takes the following form:

$$A_P^{(\nu)} \mathbf{Y}^\nu(t) = C_{Pu}^{(\mu_{Pu})} \mathbf{U}^{*\mu_{Pu}}(t) + D_{Pd}^{(\mu_{Pd})} \mathbf{D}^{*\mu_{Pd}}(t), \; \forall t \in \mathfrak{T}_0.$$

After subtracting (3.13) from the preceding equation, and after using $\mathbf{y}(.) = \mathbf{Y}(.) - \mathbf{Y}_d(.)$ (3.6) (Subsection 3.3.1), the result is the following:

$$A_P^{(\nu)} \mathbf{y}^\nu(t) = \mathbf{0}_N, \; \forall t \in \mathfrak{T}_0.$$

This homogenous differential equation has the unique trivial solution: $\mathbf{y}(t) = \mathbf{Y}(t) - \mathbf{Y}_d(t) = \mathbf{0}_N$, $\forall t \in \mathfrak{T}_0$, because it is linear with the constant coefficients and with the zero initial conditions due to $\mathbf{y}_0^{\nu-1} = \mathbf{0}_N$ in view of $\mathbf{Y}_0^{\nu-1} = \mathbf{Y}_{d0}^{\nu-1}$ (3.13). Hence,

$$\mathbf{y}^{\nu-1}(t) = \mathbf{0}, \ \forall t \in \mathfrak{T}_0,$$

or equivalently,

$$\mathbf{Y}^{\nu-1}(t) = \mathbf{Y}_d^{\nu-1}(t), \ \forall t \in \mathfrak{T}_0.$$

This and Definition 54 (Subsection 3.3.1) show that such pair $[\mathbf{D}^*(.), \mathbf{U}^*(.)]$ is nominal for the plant (2.1) relative to its desired output $\mathbf{Y}_d^{\nu}(.)$, which completes the proof ■

This theorem explains how we can determine an input vector functional pair $[\mathbf{D}^*(.), \mathbf{U}^*(.)]$ of the plant (2.15) to be nominal relative to the plant single desired response $\mathbf{Y}_d^{\nu}(.)$. We should solve only the differential equation (3.13) for $[\mathbf{D}^*(t), \mathbf{U}^*(t)]$. Initial vector values of the functional pair $[\mathbf{D}^*(.), \mathbf{U}^*(.)]$ and of its derivatives should satisfy (3.13) at $t = 0$.

Theorem 66 *For the desired output response $\mathbf{Y}^{\nu}(.)$ of the IO plant (2.15) to be realizable on $\mathfrak{D}^{\mu_{Pd}}$ it is necessary and sufficient that for every $\mathbf{D}(.) \in \mathfrak{D}^{\mu_{Pd}}$ there is a control vector function $\mathbf{U}^*(.)$ that obeys the following differential equation:*

$$C_{Pu}^{(\mu_{Pu})}\mathbf{U}^{*\mu_{Pu}}(t) = -D_{Pd}^{(\mu_{Pd})}\mathbf{D}^{\mu_{Pd}}(t) + A_P^{(\nu)}\mathbf{Y}^{\nu}(t), \forall t \in \mathfrak{T}_0,$$
$$\textit{under the condition } \mathbf{Y}_0^{\nu-1} = \mathbf{Y}_{d0}^{\nu-1}. \tag{3.14}$$

Such control vector function $\mathbf{U}^*(.)$ is nominal for the IO plant (2.15) on $\mathfrak{D}^{\mu_{Pd}}$ relative to its desired output response $\mathbf{Y}_d^{\nu}(.)$.

Proof. *Necessity.* Let the desired output response $\mathbf{Y}_d^{\nu}(.)$ of the *IO* plant (2.15) be realizable on $\mathfrak{D}^{\mu_{\mathfrak{P}\mathfrak{d}}}$. Definition 56 holds. It implies the existence of the nominal control vector function $\mathbf{U}_N(.)$ for the *IO* plant (2.15) relative to its desired response $\mathbf{Y}_d^{\nu}(.)$ for every $\mathbf{D}(.) \in \mathfrak{D}^{\mu_{Pd}}$. Definition 48 is valid. It and (2.15) imply

$$C_{Pu}^{(\mu_{Pu})}\mathbf{U}_N^{\mu_{Pu}}(t) = -D_{Pd}^{(\mu_{Pd})}\mathbf{D}^{\mu_{Pd}}(t) + A_P^{(\nu)}\mathbf{Y}_d^{\nu}(t), \forall t \in \mathfrak{T}_0,$$
$$\forall \mathbf{D}(.) \in \mathfrak{D}^{\mu_{Pd}}, \textit{ under the condition } \mathbf{Y}_0^{\nu-1} = \mathbf{Y}_{d0}^{\nu-1}.$$

This equation becomes (3.14) for $\mathbf{U}^*(.) = \mathbf{U}_N(.)$.

Sufficiency. We accept that the conditions of the theorem are valid. We choose arbitrary $\mathbf{D}(.) \in \mathfrak{D}^{\mu_{Pd}}$ and $\mathbf{U}(.) = \mathbf{U}^*(.)$. The equation (2.15) takes the following form:

$$A_P^{(\nu)}\mathbf{Y}^{\nu}(t) = C_{Pu}^{(\mu_{Pu})}\mathbf{U}^{*\mu_{Pu}}(t) + D_{Pd}^{(\mu_{Pd})}\mathbf{D}^{\mu_{Pd}}(t), \ \forall t \in \mathfrak{T}_0.$$

We subtract (3.14) from this equation and we apply $\mathbf{y}(.) = \mathbf{Y}(.) - \mathbf{Y}_d(.)$. The result is again

$$A_P^{(\nu)}\mathbf{y}^{\nu}(t) = \mathbf{0}, \ \forall t \in \mathfrak{T}_0.$$

This homogenous differential equation has the unique trivial solution: $\mathbf{y}(t) = \mathbf{Y}(t) - \mathbf{Y}_d(t) = \mathbf{0}$, $\forall t \in \mathfrak{T}_0$, because it is linear with the constant coefficients and the initial conditions are all equal to zero due to (3.14). Hence,

$$\mathbf{y}^{\nu-1}(t) = \mathbf{0}, \ \forall t \in \mathfrak{T}_0,$$

or equivalently,

$$\mathbf{Y}^{\nu-1}(t) = \mathbf{Y}_d^{\nu-1}(t), \ \forall t \in \mathfrak{T}_0.$$

This and Definition 48 (Subsection 3.3.1) complete the proof ■

The preceding theorem permits us to resolve in the complex domain the problem of the realizability of $\mathbf{Y}_d^{\nu}(.)$ on $\mathfrak{D}^{\mu_{Pd}}$ for the plant (2.15).

The following theorem by W. A. Wolovich [319, p. 162, Theorem 5.5.3] is effective for the easy verification of the complex domain realizability criterion presented in the next theorem (Theorem 68):

Theorem 67 *[319, p. 162, Theorem 5.5.3] A pxm rational transfer matrix, $F(s)$, has a left (right) inverse if, and only if,*

$$rankF(s) = m(= p).$$

Theorem 68 *For the desired output response $\mathbf{Y}_d^{\nu}(.)$ of the IO plant (2.15) to be realizable on $\mathfrak{D}^{\mu_{Pd}}$ it is necessary and sufficient that both*

1) $N \leq r$,

and

2) $rankC_{Pu}^{(\mu_{Pu})} = N$.

The solution for $\mathbf{U}(s)$ is determined by

$$\mathbf{U}(s) = \left[C_{Pu}^{(\mu_{Pu})} S_r^{(\mu_{Pu})}(s)\right]^T \left\langle \left[C_{Pu}^{(\mu_{Pu})} S_r^{(\mu_{Pu})}(s)\right] \left[C_{Pu}^{(\mu_{Pu})} S_r^{(\mu_{Pu})}(s)\right]^T \right\rangle^{-1} \bullet$$

$$\bullet \left\{ \begin{array}{c} \left[A_P^{(\nu)} S_N^{(\nu)}(s)\right] \mathbf{Y}_d(s) - A_P^{(\nu)} Z_N^{(\nu-1)}(s)\mathbf{Y}_0^{(\nu-1)} + \\ + \left[D_{Pd}^{(\mu_{Pd})} S_d^{(\mu_{Pd})}(s)\right] \mathbf{D}(s) - D_{Pd}^{(\mu_{Pd})} Z_d^{(\mu_{Pd}-1)}(s)\mathbf{D}_0^{(\mu_{Pd}-1)} + \\ C_{Pu}^{(\mu_{Pu})} Z_r^{(\mu_{Pu}-1)}(s)\mathbf{U}_0^{*(\mu_{Pu}-1)} \end{array} \right\}. \quad (3.15)$$

Appendix E contains the proof of this theorem different from the proof of Theorem 67 in [319].

This theorem presents the solution to the problem of the existence of a nominal input vector function $\mathbf{U}_N(.)$ for the *IO* plant (2.15) relative to $\mathbf{Y}_d^{\nu}(.)$. It is simultaneously the solution to the problem of the realizability of the system desired output $\mathbf{Y}_d^{\nu}(.)$. The condition is expressed exclusively in terms of the system properties and not in terms of properties of the disturbance vector function $\mathbf{D}(.)$, or in terms of features of the control vector function $\mathbf{U}(.)$ or of the desired output vector function $\mathbf{Y}_d^{\nu}(.)$.

Comment 69 *The realizability of the desired output response $\mathbf{Y}_d^{\nu}(.)$ of the IO plant (2.15) on $\mathfrak{D}^{\mu_{Pd}}$ takes into account all disturbances $\mathbf{D}(.) \in \mathfrak{D}^{\mu_{Pd}}$, while the output function controllability is defined only for $\mathbf{D}(t) = \mathbf{0}_d$, $\forall t \in \mathfrak{T}_0$.*

Note 70 *We will treat in the sequel only IO objects that satisfy the condition $N \leq r$ for the realizability of their desired output vector functions $\mathbf{Y}_d^{\nu}(.)$.*

3.3.3 *ISO* control systems

We specify now Definition 45 (Subsection 3.3.1) in the framework of the *ISO* objects (2.29), (2.30) (Section 2.2).

Definition 71 *[148] A functional vector pair* $[\mathbf{I}^*(.), \mathbf{X}^*(.)]$ *is* ***nominal for the ISO plant (2.29), (2.30) relative to its desired response*** $\mathbf{Y}_d(.)$, *which is denoted by* $[\mathbf{I}_N(.), \mathbf{X}_N(.)]$, *if, and only if,* $[\mathbf{I}(.), \mathbf{X}(.)] = [\mathbf{I}^*(.), \mathbf{X}^*(.)]$ *ensures that the corresponding real response* $\mathbf{Y}(.) = \mathbf{Y}^*(.)$ *of the system obeys* $\mathbf{Y}^*(t) = \mathbf{Y}_d(t)$ *all the time,*

$$[\mathbf{I}^*(.), \mathbf{X}^*(.)] = [\mathbf{I}_N(.), \mathbf{X}_N(.)] \Longleftrightarrow \langle \mathbf{Y}^*(t) = \mathbf{Y}_d(t),\ \forall t \in \mathfrak{T}_0 \rangle .$$

The time evolution $\mathbf{X}_N(t; \mathbf{X}_{N0}; \mathbf{I}_N)$, $\mathbf{X}_N(0; \mathbf{X}_{N0}; \mathbf{I}_N) \equiv \mathbf{X}_{N0}$, *of the nominal state vector* $\mathbf{X}_N$ *is* ***the desired motion*** $\mathbf{X}_d(.; \mathbf{X}_{d0}; \mathbf{I}_N)$ ***of the ISO plant (2.29), (2.30) relative to its desired response*** $\mathbf{Y}_d(.)$, *for short:* ***the desired motion,***

$$\mathbf{X}_d(t; \mathbf{X}_{d0}; \mathbf{I}_N) \equiv \mathbf{X}_N(t; \mathbf{X}_{N0}; \mathbf{I}_N),\ \mathbf{X}_d(0; \mathbf{X}_{d0}; \mathbf{I}_N) \equiv \mathbf{X}_{d0} \equiv \mathbf{X}_{N0}. \tag{3.16}$$

I is the identity matrix of the dimension n: $I_n = I$. The matrix

$$\begin{bmatrix} -B & sI - A \\ D & C \end{bmatrix}$$

is $(N+n) \times (M+n)$ matrix.

Theorem 72 *[148] In order for a functional vector pair* $[\mathbf{I}^*(.), \mathbf{X}^*(.)]$ *to be nominal for the ISO plant (2.29),(2.30) relative to its desired response* $\mathbf{Y}_d(.)$, $[\mathbf{I}^*(.), \mathbf{X}^*(.)] = [\mathbf{I}_N(.), \mathbf{X}_d(.)]$, *it is necessary and sufficient that it obeys the following equations:*

$$-B\mathbf{I}^*(t) + \frac{d\mathbf{X}^*(t)}{dt} - A\mathbf{X}^*(t) = \mathbf{0}_n,\ \forall t \in \mathfrak{T}_0, \tag{3.17}$$

$$D\mathbf{I}^*(t) + C\mathbf{X}^*(t) = \mathbf{Y}_d(t),\ \forall t \in \mathfrak{T}_0, \tag{3.18}$$

or equivalently,

$$\begin{bmatrix} -B & sI - A \\ D & C \end{bmatrix} \begin{bmatrix} \mathbf{I}^*(s) \\ \mathbf{X}^*(s) \end{bmatrix} = \begin{bmatrix} \mathbf{X}_0{}^* \\ \mathbf{Y}_d(s) \end{bmatrix}. \tag{3.19}$$

For the proof see Appendix F.

The initial state vector $\mathbf{X}^*(0)$ rests free for the choice. There are $(M+n)$ unknown scalar variables in $(N+n)$ equations (3.17), (3.18), or equivalently, of (3.19). The unknown variables are the entries of $\mathbf{I}^*(s) \in \mathfrak{C}^M$ and of $\mathbf{X}^*(s) \in \mathfrak{C}^n$.

Case 73 *[148]* $N > M$

If $N > M$, *then the equations (3.17), (3.18), or equivalently (3.19), do not have a solution. The number of the unknown scalar variables is less than the number of available equations.*

Case 74 *[148] $N \leq M$*
If $N \leq M$ and

$$rank \begin{bmatrix} -B & sI-A \\ D & C \end{bmatrix} = N+n \leq M+n$$

for all complex numbers s for which $\det(sI-A) \neq 0$,

then

$$\det\left\{ \begin{bmatrix} -B & sI-A \\ D & C \end{bmatrix} \begin{bmatrix} -B & sI-A \\ D & C \end{bmatrix}^T \right\} \neq 0.$$

The equations (3.17), (3.18), or equivalently (3.19), have the solution determined by

$$\begin{bmatrix} \mathbf{I}^*(s) \\ \mathbf{X}^*(s) \end{bmatrix} = \begin{bmatrix} -B & sI-A \\ D & C \end{bmatrix}^T \bullet$$

$$\bullet \left\{ \begin{bmatrix} -B & sI-A \\ D & C \end{bmatrix} \begin{bmatrix} -B & sI-A \\ D & C \end{bmatrix}^T \right\}^{-1} \begin{bmatrix} \mathbf{X}_0{}^* \\ \mathbf{Y}_d(s) \end{bmatrix}.$$

Case 75 *[148] $N = M$*
If $N = M$ and

$$rank \begin{bmatrix} -B & sI-A \\ D & C \end{bmatrix} = N+n = M+n$$

for all complex numbers s for which $\det(sI-A) \neq 0$,

then the equations (3.17), (3.18), or equivalently (3.19), have the unique solution determined by

$$\begin{bmatrix} \mathbf{I}^*(s) \\ \mathbf{X}^*(s) \end{bmatrix} = \begin{bmatrix} -B & sI-A \\ D & C \end{bmatrix}^{-1} \begin{bmatrix} \mathbf{X}_0{}^* \\ \mathbf{Y}_d(s) \end{bmatrix}.$$

Conclusion 76 *In order for a nominal functional vector pair $[\mathbf{I}_N(.), \mathbf{X}_d(.)]$ for the plant (2.29), (2.30) relative to its desired response $\mathbf{Y}_d(.)$ to exist, it is necessary and sufficient that the conditions of Case 74 hold. Then, the functional vector pair $[\mathbf{I}_N(.), \mathbf{X}_d(.)]$ is nominal relative to the desired response $\mathbf{Y}_d(.)$ of the system (2.29), (2.30) [148].*

The preceding results present the solution for the problem of the existence of a nominal functional vector pair $[\mathbf{I}_N(.), \mathbf{X}_d(.)]$ for the *ISO* plant (2.29), (2.30) relative to its desired response $\mathbf{Y}_d(.)$.

Note 77 *Only the ISO plants that satisfy the conditions of Case 74 for the realizability of their desired output vector functions $\mathbf{Y}_d(.)$ will be treated herein.*

We will study realizability of the desired output response $\mathbf{Y}_d(.)$ of the *ISO* plant (2.33), (2.34) (Subsection 2.2.2) in details.

The general definitions 52 through 54 and 56 (Subsection 3.3.1) take the following forms in the framework of the *ISO* plants:

Definition 78 *A functional vector pair* $[\mathbf{U}^*(.), \mathbf{X}_P^*(.)]$ *of the ISO plant (2.33), (2.34) is* ***nominal relative to*** $[\mathbf{D}(.), \mathbf{Y}_d(.)]$,*which is denoted by* $[\mathbf{U}_N(.), \mathbf{X}_{Pd}(.)]$,

$$[\mathbf{U}_N(t), \mathbf{X}_{Pd}(t)] \equiv [\mathbf{U}_N(t; \mathbf{D}; \mathbf{Y}_d), \mathbf{X}_{Pd}(t; \mathbf{D}; \mathbf{Y}_d)], \tag{3.20}$$

if, and only if, $[\mathbf{U}(.), \mathbf{X}_P(.)] = [\mathbf{U}^*(.), \mathbf{X}_P^*(.)]$ *ensures that the corresponding real response* $\mathbf{Y}(.) = \mathbf{Y}^*(.)$ *of the plant obeys* $\mathbf{Y}^*(t) = \mathbf{Y}_d(t)$ *all the time as soon as all the state and the output system initial conditions are desired (nominal),*

$$[\mathbf{U}^*(.), \mathbf{X}_P^*(.), \mathbf{X}_{P0}^*, \mathbf{Y}_{P0}^*] = [\mathbf{U}_N(.), \mathbf{X}_{Pd}(.), \mathbf{X}_{Pd0}, \mathbf{Y}_{d0}] \Longleftrightarrow \\ \Longleftrightarrow \langle \mathbf{Y}^*(t) = \mathbf{Y}_d(t), \ \forall t \in \mathfrak{T}_0 \rangle . \tag{3.21}$$

Comment 79 *The preceding definition specifies the nominal functional vector pair* $[\mathbf{U}_N(.),\ \mathbf{X}_{Pd}(.)]$ *relative to chosen both* $\mathbf{D}(.)$ *and* $\mathbf{Y}_d(.)$ *(3.20).*

Definition 80 *The desired output behavior* $\mathbf{Y}_d(.)$ *of the ISO plant (2.33), (2.34) is* ***realizable for the given*** $\mathbf{D}(.)$ *if, and only if, there exists the plant nominal vector functional pair* $[\mathbf{U}_N(.), \mathbf{X}_{Pd}(.)]$ *relative to* $[\mathbf{D}(.), \mathbf{Y}_d(.)]$, *(3.20).*

This definition determines realizability of a specific $\mathbf{Y}_d(.)$ for a single, given $\mathbf{D}(.)$. We broaden it to realizability of a specific $\mathbf{Y}_d(.)$ **in** $\mathfrak{D}$.

Definition 81 *A vector functional triplet* $[\mathbf{D}^*(.), \mathbf{U}^*(.), \mathbf{X}_P^*(.)]$ *of the ISO plant (2.33), (2.34) is* ***nominal relative to its desired response*** $\mathbf{Y}_d(.)$, *which is denoted by* $[\mathbf{D}_N(.),\ \mathbf{U}_N(.),\ \mathbf{X}_{Pd}(.)]$, *if, and only if,* $[\mathbf{D}(.), \mathbf{U}(.), \mathbf{X}_P(.)] = [\mathbf{D}^*(.), \mathbf{U}^*(.), \mathbf{X}_P^*(.)]$ *ensures that the corresponding real response* $\mathbf{Y}(.) = \mathbf{Y}^*(.)$ *of the system obeys* $\mathbf{Y}^*(t) = \mathbf{Y}_d(t)$ *all the time as soon as all the state and the output system initial conditions are desired (nominal),*

$$[\mathbf{D}^*(.), \mathbf{U}^*(.), \mathbf{X}_P^*(.), \mathbf{X}_{P0}^*, \mathbf{Y}_{P0}^*] = [\mathbf{D}_N(.), \mathbf{U}_N(.), \mathbf{X}_{Pd}(.), \mathbf{X}_{Pd0}, \mathbf{Y}_{d0}] \Longleftrightarrow \\ \Longleftrightarrow \langle \mathbf{Y}^*(t) = \mathbf{Y}_d(t), \ \forall t \in \mathfrak{T}_0 \rangle .$$

Comment 82 *This definition determines the nominal vector functions* $\mathbf{D}_N(.)$, $\mathbf{U}_N(.)$ *and* $\mathbf{X}_{Pd}(.)$ *of* $\mathbf{D}(.)$, $\mathbf{U}(.)$ *and* $\mathbf{X}_P(.)$ *relative to* $\mathbf{Y}_d(.)$*:*

$$[\mathbf{D}_N(t), \mathbf{U}_N(t), \mathbf{X}_{Pd}(t)] \equiv [\mathbf{D}_N(t; \mathbf{Y}_d), \mathbf{U}_N(t; \mathbf{Y}_d), \mathbf{X}_{Pd}(t; \mathbf{Y}_d)] . \tag{3.22}$$

Definition 83 *The desired response* $\mathbf{Y}_d(.)$ *of the plant (2.33), (2.34) is* ***realizable in*** $\mathfrak{D}$ *if, and only if, there exist both* $\mathbf{D}^*(.) \in \mathfrak{D}$ *and* $[\mathbf{U}^*(.), \mathbf{X}_P^*(.)]$ *such that the triplet* $[\mathbf{D}^*(.), \mathbf{U}^*(.), \mathbf{X}_P^*(.)]$ *is nominal relative to* $\mathbf{Y}_d(.)$,

$$[\mathbf{D}^*(.), \mathbf{U}^*(.), \mathbf{X}_P^*(.)] = [\mathbf{D}_N(.), \mathbf{U}_N(.), \mathbf{X}_{Pd}(.)] \ \textit{relative to } \mathbf{Y}_d(.).$$

Equivalently,

Definition 84 *The desired output behavior* $\mathbf{Y}_d(.)$ *of the plant (2.33), (2.34) is* ***realizable in*** $\mathfrak{D}$ *if, and only if, there exists the plant nominal vector functional triplet* $[\mathbf{D}_N(.), \mathbf{U}_N(.), \mathbf{X}_{Pd}(.)]$, $\mathbf{D}_N(.) \in \mathfrak{D}$, *relative to* $\mathbf{Y}_d(.)$ *(3.22).*

Comment 85 *This definition does not require that for every* $\mathbf{D}(.) \in \mathfrak{D}$ *there exists the nominal control vector function* $\mathbf{U}_N(.)$ *relative to* $\mathbf{Y}_d(.)$.

Theorem 86 *For the desired output response* $\mathbf{Y}_d(.)$ *of the* ISO *plant (2.33), (2.34) to be realizable in* $\mathfrak{D}$ *it is necessary and sufficient that there is a vector functional triplet* $[\mathbf{D}^*(.), \mathbf{U}^*(.), \mathbf{X}_P^*(.)]$, $\mathbf{D}^*(.) \in \mathfrak{D}$, *which obeys the following equations in the time domain:*

$$-L_P\mathbf{D}^*(t) - B_P\mathbf{U}^*(t) + \frac{d\mathbf{X}_P^*(t)}{dt} - A_P\mathbf{X}_P^*(t) = \mathbf{0}_n, \ \forall t \in \mathfrak{T}_0,$$
$$D_P\mathbf{D}^*(t) + H_P\mathbf{U}^*(t) + C_P\mathbf{X}_P^*(t) = \mathbf{Y}_d(t), \ \forall t \in \mathfrak{T}_0, \tag{3.23}$$

or equivalently in the complex domain,

$$\begin{bmatrix} -L_P & -B_P & (sI_n - A_P) \\ D_P & H_P & C_P \end{bmatrix} \begin{bmatrix} \mathbf{D}^*(s) \\ \mathbf{U}^*(s) \\ \mathbf{X}_P^*(s) \end{bmatrix} = \begin{bmatrix} \mathbf{X}_{P0}^* \\ \mathbf{Y}_d(s) \end{bmatrix}, \ \mathbf{X}_{P0} \in \mathfrak{R}^n. \tag{3.24}$$

Such vector functional triplet $[\mathbf{D}^*(.), \mathbf{U}^*(.), \mathbf{X}_P^*(.)]$, $\mathbf{D}^*(.) \in \mathfrak{D}$, *is nominal for the* ISO *plant (2.33), (2.34) relative to its desired output response* $\mathbf{Y}_d(.)$, $[\mathbf{D}^*(.), \mathbf{U}^*(.), \mathbf{X}_P^*(.)] = [\mathbf{D}_N(.), \mathbf{U}_N(.), \mathbf{X}_{Pd}(.)]$.

Proof. *Necessity.* Let the desired output response $\mathbf{Y}_d(.)$ of the ISO plant (2.33), (2.34) be realizable in $\mathfrak{D}$. Definition 83 holds. It implies the existence of the nominal vector functional triplet $[\mathbf{D}_N(.), \mathbf{U}_N(.), \mathbf{X}_{Pd}(.)]$, $\mathbf{D}_N(.) \in \mathfrak{D}$, for the ISO plant (2.33), (2.34) relative to its desired response $\mathbf{Y}_d(.)$. Definition 81 is valid. It and (2.33), (2.34) imply

$$\frac{d\mathbf{X}_{Pd}(t)}{dt} = A_P\mathbf{X}_{Pd}(t) + B_P\mathbf{U}_N(t) + L_P\mathbf{D}_N(t), \ \forall t \in \mathfrak{T}_0, \tag{3.25}$$
$$\mathbf{Y}_d(t) = C_P\mathbf{X}_{Pd}(t) + H_P\mathbf{U}_N(t) + D_P\mathbf{D}_N(t), \ \forall t \in \mathfrak{T}_0. \tag{3.26}$$

These equations are Equations (3.23) in another form when we set $[\mathbf{D}^*(.), \mathbf{U}^*(.), \mathbf{X}_P^*(.)] = [\mathbf{D}_N(.), \mathbf{U}_N(.), \mathbf{X}_{Pd}(.)]$. Equation (3.24) represents in the vector form their Laplace transform.

Sufficiency. We accept that the conditions of the theorem are valid. We choose $[\mathbf{D}(.), \mathbf{U}(.)] = [\mathbf{D}^*(.), \mathbf{U}^*(.)]$, $\mathbf{D}^*(.) \in \mathfrak{D}$, and $\mathbf{X}_{P0}^* \in \mathfrak{R}^n$ so that

$$C_P\mathbf{X}_{P0}^* = \mathbf{Y}_{d0} - D_P\mathbf{D}_0^* - H_P\mathbf{U}_0^*.$$

Equations (2.33), (2.34) take the following forms:

$$\frac{d\mathbf{X}_P(t)}{dt} = A_P\mathbf{X}_P(t) + B_P\mathbf{U}^*(t) + L_P\mathbf{D}^*(t), \ \forall t \in \mathfrak{T}_0,$$
$$\mathbf{Y}(t) = C_P\mathbf{X}_P(t) + H_P\mathbf{U}^*(t) + D_P\mathbf{D}^*(t), \ \forall t \in \mathfrak{T}_0.$$

We subtract (3.23) from the preceding equations, we use $\mathbf{x}_P^*(.) = \mathbf{X}_P(.) - \mathbf{X}_P^*(.)$ and $\mathbf{y}(.) = \mathbf{Y}(.) - \mathbf{Y}_d(.)$. Let $\mathbf{X}_{P0} = \mathbf{X}_{P0}^*$ so that $\mathbf{x}_{P0}^* = \mathbf{0}_n$. The results are the following:

$$\frac{d\mathbf{x}_P^*(t)}{dt} = A_P\mathbf{x}_P^*(t),\ \forall t \in \mathfrak{T}_0,\ \mathbf{x}_{P0}^* = \mathbf{0}_n,$$
$$\mathbf{y}(t) = C_P\mathbf{x}_P^*(t),\ \forall t \in \mathfrak{T}_0.$$

The preceding homogenous differential equation has the unique trivial solution: $\mathbf{x}_P^*(t) = \mathbf{X}_P(t) - \mathbf{X}_P^*(t) = \mathbf{0}_n$, i.e., $\mathbf{X}_P(t) = \mathbf{X}_P^*(t)$, $\forall t \in \mathfrak{T}_0$, because it is linear with the constant coefficients and with the zero initial condition $\mathbf{x}_{P0}^* = \mathbf{0}_n$. Hence,

$$\mathbf{y}(t) = \mathbf{0}_N,\ \forall t \in \mathfrak{T}_0,$$

or equivalently,

$$\mathbf{Y}(t) = \mathbf{Y}_d(t),\ \forall t \in \mathfrak{T}_0,$$

hence, $\mathbf{X}_P(t) \equiv \mathbf{X}_P^*(t) \equiv \mathbf{X}_{Pd}(t)$, which completes the proof in view of Definitions 81 and 83 ■

Let

$$F(s) = \left[G_d(s) \vdots G_u(s) \vdots G_{xo}(s)\right],$$
$$G_d(s) = C_P\left(sI_n - A_P\right)^{-1}L_P + D_P,$$
$$G_u(s) = C_P\left(sI_n - A_P\right)^{-1}B_P + H_P,$$
$$G_{xo}(s) = C_P\left(sI_n - A_P\right)^{-1}. \tag{3.27}$$

$F(s)$ in (3.27) is *the full transfer function matrix* of the plant (2.33), (2.34) (for details see Subsection 7.2.2 in the sequel).

Theorem 87 *For the desired output response $\mathbf{Y}_d(.)$ of the ISO plant (2.33), (2.34) to be realizable in $\mathfrak{D}$ it is necessary and sufficient that there is a vector functional triplet $[\mathbf{D}^*(.), \mathbf{U}^*(.), \mathbf{X}_P^*(.)]$, $\mathbf{D}^*(.) \in \mathfrak{D}$, which obeys the following:*

1) $N \leq r$

and

2)

$$\exists s \in \mathfrak{C} \Longrightarrow rank\begin{bmatrix} -B_P & (sI_n - A_P) \\ H_P & C_P \end{bmatrix} = N, \tag{3.28}$$

equivalently

$$\exists s \in \mathfrak{C} \Longrightarrow rank G_u(s) = N. \tag{3.29}$$

The solution is determined by

$$\begin{bmatrix} -B_P & (sI_n - A_P) \\ H_P & C_P \end{bmatrix}\begin{bmatrix} \mathbf{U}^*(s) \\ \mathbf{X}_P^*(s) \end{bmatrix} = \begin{bmatrix} \mathbf{X}_{P0}^* + L_P\mathbf{D}^*(s) \\ \mathbf{Y}_d(s) - D_P\mathbf{D}^*(s) \end{bmatrix},\ \mathbf{X}_{P0} \in \mathfrak{R}^n, \tag{3.30}$$

equivalently by

$$\mathbf{U}^*(s) = G_u^T(s)\left[G_u(s)G_u^T(s)\right]^{-1}\left[\mathbf{Y}_d(s) - G_d(s)\mathbf{D}^*(s) - G_{xo}(s)\mathbf{X}_{P0}^*\right],$$
$$\mathbf{X}_P^*(s) = \left(sI_n - A_P\right)^{-1}\left[B_P\mathbf{U}^*(s) + L_P\mathbf{D}^*(s) + \mathbf{X}_{P0}^*\right], \tag{3.31}$$

for any $(\mathbf{D}^*(.),\ \mathbf{X}_{PN0}) \in \mathfrak{D}x\mathfrak{R}^n$.

Proof. We accept any $\mathbf{D}^*(.) \in \mathfrak{D}$ because there are three unknown vector variables: $\mathbf{D}^*(.)$, $\mathbf{U}^*(.)$, and $\mathbf{X}_P^*(.)$, i.e., there are $(d+r+n)$ unknown scalar variables and only $(r+n)$ scalar equations. We apply Theorem 86 and use (3.24). Since $\mathbf{D}^*(.) \in \mathfrak{D}$ is chosen, we put (3.24) into the following form:

$$\begin{bmatrix} -B_P \vdots (sI_n - A_P) \\ H_P \vdots \quad C_P \end{bmatrix} \begin{bmatrix} \mathbf{U}^*(s) \\ \mathbf{X}_P^*(s) \end{bmatrix} = \begin{bmatrix} L_P\mathbf{D}^*(s) + \mathbf{X}_{P0}^* \\ -D_P\mathbf{D}^*(s) + \mathbf{Y}_d(s) \end{bmatrix}.$$

For this equation to have a solution it is necessary and sufficient that $N \leq r$ and that (3.28) holds. From

$$\begin{bmatrix} O_{n,N} & (sI_n - A_P) \\ I_N & C_P \end{bmatrix} \bullet$$

$$\bullet \begin{bmatrix} \underbrace{C_P\left(sI_n - A_P\right)^{-1} L_P + D_P}_{G_d(s)} & \underbrace{C_P\left(sI_n - A_P\right)^{-1} B_P + H_P}_{G_u(s)} & O_{N,n} \\ -\left(sI_n - A_P\right)^{-1} L_P & -\left(sI_n - A_P\right)^{-1} B_P & I_n \end{bmatrix} =$$

$$= \begin{bmatrix} -L_P & -B_P & (sI_n - A_P) \\ D_P & H_P & C_P \end{bmatrix}$$

we get

$$\begin{bmatrix} O_{n,N} & (sI_n - A_P) \\ I_N & C_P \end{bmatrix} \begin{bmatrix} G_u(s) & O_{N,n} \\ -\left(sI_n - A_P\right)^{-1} B_P & I_n \end{bmatrix} =$$

$$= \begin{bmatrix} -B_P & (sI_n - A_P) \\ H_P & C_P \end{bmatrix}.$$

For every $s \in \mathfrak{C}$ different from the eigenvalues of A,

$$rank \begin{bmatrix} -B_P & (sI_n - A_P) \\ H_P & C_P \end{bmatrix} = rank\ G_u(s).$$

This proves that (3.28) and (3.29) are equivalent.

From (3.24) follows (3.30) that can be rewritten as

$$-L_P\mathbf{D}^*(s) - B_P\mathbf{U}^*(s) + (sI_n - A_P)\mathbf{X}_P^*(s) = \mathbf{X}_{P0}^*,$$
$$D_P\mathbf{D}^*(s) + H_P\mathbf{U}^*(s) + C_P\mathbf{X}_P^*(s) = \mathbf{Y}_d(s),$$

We determine straightforward the solutions of these equations as given in (3.31) ■

This theorem does not guarantee the existence of the nominal regime for every $\mathbf{D}(.) \in \mathfrak{D}$.

Definition 88 *A vector functional pair* $[\mathbf{U}^*(.), \mathbf{X}_P^*(.)]$ *of the ISO plant (2.33), (2.34) is* ***nominal relative to its desired response*** $\mathbf{Y}_d(.)$ ***on*** $\mathfrak{D}$*, which is denoted by* $[\mathbf{U}_N(.), \mathbf{X}_{Pd}(.)]$*, if, and only if,* $[\mathbf{U}(.), \mathbf{X}_P(.)] = [\mathbf{U}^*(.), \mathbf{X}_P^*(.)]$ *ensures that the corresponding real response* $\mathbf{Y}(.) = \mathbf{Y}^*(.)$ *of the system obeys* $\mathbf{Y}^*(t) = \mathbf{Y}_d(t)$ *for every* $\mathbf{D}(.) \in \mathfrak{D}$ *and all the time as soon as all the state and the output system initial conditions are desired (nominal).*

Definition 89 *The desired response* $\mathbf{Y}_d(.)$ *of the plant is* ***realizable on*** $\mathfrak{D}$ *if, and only if, there exists the nominal vector functional pair* $[\mathbf{U}_N(.), \mathbf{X}_{Pd}(.)]$ *on* $\mathfrak{D}$ *relative to* $\mathbf{Y}_d(.)$ *(3.20).*

Theorem 90 *For the desired output response* $\mathbf{Y}_d(.)$ *of the ISO plant (2.33), (2.34) to be realizable on* $\mathfrak{D}$*, it is necessary and sufficient that for every* $\mathbf{D}(.) \in \mathfrak{D}$ *there is a vector functional pair* $[\mathbf{U}^*(.), \mathbf{X}_P^*(.)]$*, which, for every* $\mathbf{D}(.) \in \mathfrak{D}$*, obeys the following equations in the time domain:*

$$-B_P\mathbf{U}^*(t) + \frac{d\mathbf{X}_P^*(t)}{dt} - A_P\mathbf{X}_P^*(t) = L_P\mathbf{D}(t), \ \forall t \in \mathfrak{T}_0, \ \forall \mathbf{D}(.) \in \mathfrak{D},$$
$$H_P\mathbf{U}^*(t) + C_P\mathbf{X}_P^*(t) = \mathbf{Y}_d(t) - D_P\mathbf{D}(t), \ \forall t \in \mathfrak{T}_0, \ \forall \mathbf{D}(.) \in \mathfrak{D}, \tag{3.32}$$

or equivalently in the complex domain,

$$\begin{bmatrix} -B_P & (sI_n - A_P) \\ H_P & C_P \end{bmatrix} \begin{bmatrix} \mathbf{U}^*(s) \\ \mathbf{X}_P^*(s) \end{bmatrix} = \begin{bmatrix} L_P & I_n & O_{n,N} \\ -D_P & O_{N,n} & I_N \end{bmatrix} \begin{bmatrix} \mathbf{D}(s) \\ \mathbf{X}_{P0}^* \\ \mathbf{Y}_d(s) \end{bmatrix},$$
$$\forall \mathbf{D}(.) \in \mathfrak{D}, \ \mathbf{X}_{P0}^* \in \mathfrak{R}^n, \tag{3.33}$$

Such vector functional pair $[\mathbf{U}^*(.), \mathbf{X}_P^*(.)]$ *is nominal for every* $\mathbf{D}^*(.) \in \mathfrak{D}$ *for the ISO plant (2.33), (2.34) relative to its desired output response* $\mathbf{Y}_d(.)$*,* $[\mathbf{U}^*(.), \mathbf{X}_P^*(.)] = [\mathbf{U}_N(.), \mathbf{X}_{Pd}(.)]$.

Proof. *Necessity.* Let the desired output response $\mathbf{Y}_d(.)$ of the *ISO* plant (2.33), (2.34) be realizable on $\mathfrak{D}$. Definition 89 holds. It implies the existence of the nominal vector functional pair $[\mathbf{U}_N(.), \mathbf{X}_{Pd}(.)]$ on $\mathfrak{D}$ for the *ISO* plant (2.33), (2.34) relative to its desired response $\mathbf{Y}_d(.)$. Definition 88 is valid. It and (2.33), (2.34) imply

$$\frac{d\mathbf{X}_{Pd}(t)}{dt} = A_P\mathbf{X}_{Pd}(t) + B_P\mathbf{U}_N(t) + L_P\mathbf{D}_N(t), \ \forall t \in \mathfrak{T}_0,$$
$$\mathbf{Y}_d(t) = C_P\mathbf{X}_{Pd}(t) + H_P\mathbf{U}_N(t) + D_P\mathbf{D}_N(t), \ \forall t \in \mathfrak{T}_0. \tag{3.34}$$

When we accept $[\mathbf{U}^*(.),\mathbf{X}_P^*(.)] = [\mathbf{U}_N(.),\mathbf{X}_{Pd}(.)]$, then Equations (3.34) represent Equations (3.32) in another form. Their Laplace transform is Equation (3.33).

Sufficiency. We accept that the conditions of the theorem are valid. We choose arbitrarily $\mathbf{D}(.) \in \mathfrak{D}$, and we accept $\mathbf{U}(.) = \mathbf{U}^*(.)$ and $\mathbf{X}_{P0}^* \in \mathfrak{R}^n$ so that

$$C_P\mathbf{X}_{P0}^* = \mathbf{Y}_{d0} - D_P\mathbf{D}_0 - H_P\mathbf{U}_0^*.$$

Equations (2.33), (2.34) take the following forms:

$$\frac{d\mathbf{X}_P(t)}{dt} = A_P\mathbf{X}_P(t) + B_P\mathbf{U}^*(t) + L_P\mathbf{D}(t), \ \forall t \in \mathfrak{T}_0,$$
$$\mathbf{Y}(t) = C_P\mathbf{X}_P(t) + H_P\mathbf{U}^*(t) + D_P\mathbf{D}(t), \ \forall t \in \mathfrak{T}_0.$$

We subtract (3.32) from the preceding equations; we use $\mathbf{x}_P^*(.) = \mathbf{X}_P(.) - \mathbf{X}_P^*(.)$ and $\mathbf{y}(.) = \mathbf{Y}(.) - \mathbf{Y}_d(.)$. We select $\mathbf{X}_{P0} = \mathbf{X}_{P0}^*$ so that $\mathbf{x}_{P0}^* = \mathbf{0}_n$. The results are the following:

$$\frac{d\mathbf{x}_P^*(t)}{dt} = A_P \mathbf{x}_P^*(t),\ \forall t \in \mathfrak{T}_0,\ \mathbf{x}_{P0}^* = \mathbf{0}_n,$$
$$\mathbf{y}(t) = C_P \mathbf{x}_P^*(t),\ \forall t \in \mathfrak{T}_0.$$

The preceding homogenous differential equation has the unique trivial solution: $\mathbf{x}_P^*(t) = \mathbf{X}_P(t) - \mathbf{X}_P^*(t) = \mathbf{0}_N$, i.e., $\mathbf{X}_P(t) = \mathbf{X}_P^*(t)$, $\forall t \in \mathfrak{T}_0$, because it is linear with the constant coefficients and with the zero initial condition $\mathbf{x}_{P0}^* = \mathbf{0}_n$. Hence,

$$\mathbf{y}(t) = \mathbf{0},\ \forall t \in \mathfrak{T}_0,$$

or equivalently,

$$\mathbf{Y}(t) = \mathbf{Y}_d(t),\ \forall t \in \mathfrak{T}_0;$$

hence $\mathbf{X}_P(t) \equiv \mathbf{X}_P^*(t) \equiv \mathbf{X}_{Pd}(t)$, which completes the time-domain proof in view of Definitions 88 and 89 ■

This theorem enables us to solve more effectively in the complex domain the problem of the realizability of $\mathbf{Y}_d(.)$ on $\mathfrak{D}$ for the *ISO* plant (2.33), (2.34).

Theorem 91 *For the desired output response $\mathbf{Y}_d(.)$ of the ISO plant (2.33), (2.34) to be realizable on $\mathfrak{D}$, it is necessary and sufficient that both*

1) $N \leq r,$

and

2)

$$\exists s \in \mathfrak{C} \Longrightarrow rank \begin{bmatrix} -B_P & (sI_n - A_P) \\ H_P & C_P \end{bmatrix} = N, \tag{3.35}$$

or equivalently

$$\exists s \in \mathfrak{C} \Longrightarrow rank G_u(s) = N \tag{3.36}$$

The solution is determined by

$$\begin{bmatrix} \mathbf{U}^*(s) \\ \mathbf{X}_P^*(s) \end{bmatrix} = \begin{bmatrix} -B_P & (sI_n - A_P) \\ H_P & C_P \end{bmatrix}^T \bullet$$
$$\bullet \left\langle \begin{bmatrix} -B_P & (sI_n - A_P) \\ H_P & C_P \end{bmatrix} \begin{bmatrix} -B_P & (sI_n - A_P) \\ H_P & C_P \end{bmatrix}^T \right\rangle^{-1} \bullet$$
$$\bullet \begin{bmatrix} L_P & I_n & O_{n,N} \\ -D_P & O_n & I_N \end{bmatrix} \begin{bmatrix} \mathbf{D}(s) \\ \mathbf{X}_{P0}^* \\ \mathbf{Y}_d(s) \end{bmatrix},$$
$$\mathbf{X}_{P0}^* \in \mathfrak{R}^n, \tag{3.37}$$

or equivalently by

$$\mathbf{U}^*(s) = G_u^T(s) \left[G_u(s) G_u^T(s)\right]^{-1} \bullet$$
$$\bullet \left\{\mathbf{Y}_d(s) - G_d(s)\mathbf{D}(s) - G_{xo}(s)\mathbf{X}_{P0}^*\right\},$$
$$\mathbf{X}_P^*(s) = (sI_n - A_P)^{-1}\left[B_P \mathbf{U}^*(s) + L_P \mathbf{D}(s) + \mathbf{X}_{P0}^*\right]. \tag{3.38}$$

Such vector functional pair $[\mathbf{U}^*(.), \mathbf{X}_P^*(.)]$ *is nominal on* $\mathfrak{D}$ *for the ISO plant (2.33), (2.34) relative to its desired output response* $\mathbf{Y}_d(.)$, $[\mathbf{U}^*(.), \mathbf{X}_P^*(.)] = [\mathbf{U}_N(.), \mathbf{X}_{Pd}(.)]$.

Appendix G contains the proof of this theorem.

Theorem 91 establishes the conditions for the realizability of the desired output response $\mathbf{Y}_d(.)$ of the *ISO* plant (2.33), (2.34) on $\mathfrak{D}$, which are determined only by the dimensions N of the plant output vector $\mathbf{Y}$ and r of the control vector $\mathbf{U}$, and by the parameter matrices A_P, B_P, C_P and H_P of the plant, i.e., by the rank of its system matrix $P_{ISO}(s)$ (see [148] for details),

$$P_{ISO}(s) = \begin{bmatrix} -B_P & (sI_n - A_P) \\ H_P & C_P \end{bmatrix} \in \mathfrak{C}^{(n+N)\mathrm{x}(n+r)}.$$

Simultaneously, the theorem determines the nominal vector functional pair $[\mathbf{U}^*(.), \mathbf{X}_P^*(.)]$ for every $\mathbf{D}^*(.) \in \mathfrak{D}$ for the *ISO* plant (2.33), (2.34) relative to its desired output response $\mathbf{Y}_d(.)$.

Chapter 4

Transfer function matrix $G(\mathrm{s})$

4.1 On definitions of $G(\mathrm{s})$

There exist two approaches to define the system transfer function $G(s)$ of the $SISO$ system and its generalization, the transfer function matrix $G(s)$ of the $MIMO$ system. Laplace transform with its properties is the basis for both. Its application to derivatives of a (scalar or vector, either input or output) variable introduces the initial conditions (of the variable and of its derivatives) in the complex domain. Since system (scalar or matrix) parameters multiply the variable and its derivatives, therefore, the same parameters multiply the corresponding Laplace transforms of the variable and of its derivatives. The result is a sum of Laplace transform of the variable and of the (second) sum of the products of the parameters and initial values. The second sum introduces (altogether) the double sum of the products of the parameters and initial values as soon as the system order is higher than one. This holds for both $SISO$ and $MIMO$ systems.

The double sum in the initial conditions of the input variable and its derivatives together with the double sum of the initial conditions of the output variable and its derivatives appeared as a mathematical obstacle to determine the linear homogeneous relationship between Laplace transform of the output variable and Laplace transform of the input variable together with all initial conditions. The accepted exit from this mathematical complication was to accept the unjustifiable assumption that all initial conditions are equal to (scalar or vector) zero. This is common to both approaches to define the system transfer function (matrix) $G(s)$.

The older approach defines ([21], [46], [62], [214]- [216], [234]-[269], [270], [282], [300]) the transfer function $G(s)$ as the ratio of left Laplace transform $Y^{-}(s)$ of the output variable $Y(t)$ and of left Laplace transform $I^{-}(s)$ of the input variable $I(t)$ under all (input and output) initial conditions (at $t = 0^{-}$)

equal to zero:

$$G(s) = \frac{Y^{-}(s)}{I^{-}(s)}, \ \mathbf{I}^{\mu-1}(0^{-}) = \mathbf{0}_{\nu}, \ \mathbf{Y}^{\nu-1}(0^{-}) = \mathbf{0}_{\mu}, \tag{4.1}$$

for the ν-*th* order *SISO* system. It enables the linear homogenous relationship between $Y^{-}(s)$ and $I^{-}(s)$,

$$Y^{-}(s) = G(s)I^{-}(s), \ \mathbf{I}^{\mu-1}(0^{-}) = \mathbf{0}_{\nu}, \ \mathbf{Y}^{\nu-1}(0^{-}) = \mathbf{0}_{\mu}. \tag{4.2}$$

This takes the vector-matrix form for the N dimensional ν-*th* order *MIMO* system,

$$\mathbf{Y}^{-}(s) = G(s)\mathbf{I}^{-}(s), \ \mathbf{I}^{\mu-1}(0^{-}) = \mathbf{0}_{\nu}, \ \mathbf{Y}^{\nu-1}(0^{-}) = \mathbf{0}_{\mu}, \tag{4.3}$$

where $G(s)$ is the matrix composed of all transfer functions of the system, which is the system transfer function matrix. It is the matrix value of the complex matrix function $G(.)$ that relates in the linear homogeneous form $\mathbf{Y}^{-}(s)$ to $\mathbf{I}^{-}(s)$ under *all zero initial conditions.* If all input variables are Dirac impulse, of which Laplace transform is one, then (4.2) becomes

$$Y^{-}(s) = G(s), \ I^{-}(s) = 1, \mathbf{I}^{\mu-1}(0^{-}) = \mathbf{0}_{\nu}, \ \mathbf{Y}^{\nu-1}(0^{-}) = \mathbf{0}_{\mu}, \tag{4.4}$$

and (4.3) takes the following form:

$$\mathbf{Y}^{-}(s) = G(s)\mathbf{1}_{M}, \ \mathbf{I}^{-}(s) = \mathbf{1}_{M}, \ \mathbf{I}^{\mu-1}(0^{-}) = \mathbf{0}_{\nu}, \ \mathbf{Y}^{\nu-1}(0^{-}) = \mathbf{0}_{\mu}. \tag{4.5}$$

Equations (4.4) and (4.5) provide the explanation of the physical meaning of $G(s)$ that it is left Laplace transform of the output unit impulse response of the system under *all zero initial conditions.*

The latter approach ([8], [23], [29], [198], [318]-[320]) defines the transfer function $G(s)$ as left Laplace transform of the *SISO* system output response to the unit impulse input under *all zero initial conditions.* The transfer function matrix $G(s)$ of the *MIMO* system is then left Laplace transform of the system output vector response to the unit impulse action of all input variables under *all zero initial conditions.*

4.2 On importance of G(s)

The transfer function (matrix) $G(s)$ has become the crucial dynamic characteristic of the *time*-invariant continuous-*time* linear dynamic (hence, control) systems. It is independent of both the system (scalar or vector) input variable and, naturally, of all initial conditions. Its independence of all initial conditions is the consequence of its definition to hold only under *all zero initial conditions.* Only the system dimension, order, structure and parameters determine (uniquely) $G(s)$.

The transfer function (matrix) $G(s)$ is the basis of the block diagram technique that has become very effective for both analysis and synthesis of the systems in the complex domain under the condition that *all initial conditions are null.* It is then a very effective tool for the control engineers to design the overall control system.

We have used the transfer function (matrix) $G(s)$ to test Lyapunov stability properties even though they are defined for nonzero initial conditions and for zero input (vector). It has then become crucial for system stabilization.

Controllability and observability tests are expressed in terms of $G(s)$. It permits the system optimization together with its relative stabilization.

There is the link between $G(s)$ and the system matrix that is important for the system equivalence.

It is the fact that Laplace transform introduces initial conditions in the complex domain. Therefore, the argument that the existence of the initial conditions in the complex domain is not natural does not hold. This raises the following question and imposes the following problem:

Can we deal with arbitrary initial conditions in the complex domain equally effectively as we have been doing for zero initial conditions?

The reply is positive. The next part of the book shows how we reply affirmatively to the question and effectively resolve the problem.

Part II

NOVEL SYSTEM FUNDAMENTAL: FULL TRANSFER FUNCTION MATRIX $F(\mathrm{s})$

Chapter 5

Problem statement

P. J. Antsaklis and A. N. Michel showed that the zero-state system equivalence and the zero-input system equivalence do not imply the system equivalence [8, p. 171]. They noted [8, p. 387] that different state-space realizations of the system transfer function matrix lead to the same zero-state system response, but the corresponding zero-input system response can be different. Their conclusions correspond to the real system environment and its history, which are the reasons to study the simultaneous influence of arbitrary initial conditions and arbitrary inputs. This is essential for the analysis of the system behavior, of the system equivalence, of the system realization and the system minimal realization, and of many system dynamic properties (e.g., *BIBO* and L-stability under arbitrary initial conditions, total system stability, system tracking and system optimality). This led to *Basic problem* 42 (Subsection 3.2.2). Another, more specific, form follows.

Problem 92 ***Main problem***

Is it possible to find in the complex domain $\mathfrak{C}$ a compact mathematical description of the temporal transfer of the influence of all inputs and of arbitrary initial conditions through the system on its internal behavior and/or on its output behavior such that it is invariant relative to the system input vector and to all initial conditions, i.e., that such description is fully independent of the system input vector and of all initial conditions?

The discovery of the existence of ***the full (complete) matrix transfer function*** $F(.)$ of the system, its definition, its determination and its complex matrix value that is ***the full (complete) transfer function matrix*** $F(s)$ solves the problem [81]. The senior undergraduate students in the linear system and control courses at the University of Natal in Durban, RSA in 1993 [81]; in the National Engineering School of Belfort (*Ecole Nationale d'Ingénieurs de Belfort-ENIB*), France 1994-1996, [78]; and at the University of Technology Belfort-Montbeliard, Belfort (*UTBM*), 1999-2003 [71], [150], [154], [168] had the opportunity to learn about $F(s)$ and to become able to realize its effective applications. It was presented also in 2003 [174]. We will show herein, additionally, for the first time, how the control system full transfer matrix function $F(.)$ can be used to study various tracking properties of the system and to solve the associated problems.

In order to present a precise study of the (full) transfer function matrices, it is necessary to explain the notion and features of *degenerate* and *nondegenerate matrix functions*, and to clarify their differences from the well known reducible and irreducible matrices.

This chapter is equivalent to the corresponding presentation in [148]. This reference shows the theoretical applications of $F(s)$ to the establishment of *the full block diagram technique* valid for arbitrary input vector and for arbitrary initial conditions, to the system equivalence under nonzero initial conditions, to refined Lyapunov stability study and to BI stability under nonzero initial conditions.

We will show in this book the theoretical applications of $F(s)$ to tracking studies, to trackability studies and to tracking control synthesis.

Chapter 6

Nondegenerate matrices

6.1 Nondegenerate and degenerate matrices

For what follows the reader will need the knowledge of definitions and properties of the greatest common (left, right) divisors of the matrix polynomials, of the unimodular matrix polynomials, as well as the knowledge of the conditions for their (left, right) coprimeness (see the books by J. P. Antsaklis and A. N. Michel [8, pp. 526-528, 535-540], C.-T. Chen [29, pp. 591-599] and T. Kailath [198, pp. 373-382]).

A rational matrix function $M(.) = M_D^{-1}(.)M_N(.)$ $[M(.) = M_N(.)M_D^{-1}(.)]$ is *irreducible* if, and only if, its polynomial matrices $M_D(.)$ and $M_N(.)$ are (left and/or right) coprime (see C.-T. Chen [29, pp. 591-599] and T. Kailath [198, pp. 373-382]). The greatest common (left $L(.)$, also right $R(.)$) divisor of $M_D(.)$ and of $M_N(.)$ cancels itself in $M(.)$, even though $L(.)$ and $R(.)$ are unimodular polynomial matrices,

$$M_D(s) = L(s)D(s),\ M_N(s) = L(s)N(s) \Longrightarrow$$
$$M(.) = M_D^{-1}(.)M_N(.) = D^{-1}(s)L^{-1}(s)L(s)N(s) = D^{-1}(s)N(s),$$

$$M_D(s) = D(s)R(s),\ M_N(s) = N(s)R(s) \Longrightarrow$$
$$M(.) = M_N(.)M_D^{-1}(.) = N(s)R(s)R^{-1}(s)D^{-1}(s) = N(s)D^{-1}(s).$$

In [81], [148] it was noted that such a definition of the irreducibility is not fully adequate for $MIMO$ systems. It was the reason to show that an irreducible complex matrix function can be *nondegenerate* or *degenerate* in the following sense [81], [148]:

Definition 93 *A rational matrix function* $M(.) = M_D^{-1}(.)M_N(.)$ *[respectively,* $M(.) = M_N(.)M_D^{-1}(.)$*] is*

a) ***row nondegenerate*** *if, and only if, respectively:*

(i) the greatest common left [right] divisor of $M_D(.)$ *and of* $M_N(.)$ *is a nonsingular constant matrix, and*

(ii) the greatest common scalar factors of $\det M_D(s)$ *and of all elements of every row of* $(\mathrm{adj} M_D(s))\, M_N(s)$ *[respectively, of all elements of every row of* $M_N(s)\,(\mathrm{adj} M_D(s))$*] are nonzero constants that can be mutually different.*

Otherwise, $M(.)$ *is* **row degenerate.**

b) **column nondegenerate** *if, and only if, respectively:*

(i) the greatest common left [right] divisor of $M_D(.)$ *and* $M_N(.)$ *is a nonsingular constant matrix, and*

(ii) the greatest common scalar factors of $\det M_D(s)$ *and of all elements of every column of* $(\mathrm{adj} M_D(s))\, M_N(s)$ *[respectively, of all elements of every column of* $M_N(s)\,(\mathrm{adj} M_D(s))$*] are nonzero constants that can be mutually different.*

Otherwise, $M(.)$ *is* **column degenerate.**

c) **nondegenerate** *if, and only if, respectively:*

(i) the greatest common left [right] divisor of $M_D(.)$ *and* $M_N(.)$ *is a nonsingular constant matrix, and*

(ii) the greatest common scalar factor of $\det M_D(s)$ *and of all elements of* $(\mathrm{adj} M_D(s))\, M_N(s)$ *[respectively, of all elements of* $M_N(s)\,(\mathrm{adj} F_D(s))$*] is a nonzero constant.*

Otherwise, $M(.)$ *is* **degenerate.**

This leads to the following notes [148].

Note 94 *If a rational matrix function* $M(.) = M_D^{-1}(.)M_N(.)$ *[respectively,* $M(.) = M_N(.)M_D^{-1}(.)$*] is either row nondegenerate or column nondegenerate, or both, then it is also nondegenerate.*

Example 95 *The matrix*

$$M(s) = M_D^{-1}(s)M_N(s) = [(s+7)(s-9)]^{-1} \begin{bmatrix} s+7 & s+7 \\ s-9 & s-9 \end{bmatrix}$$

is both column nondegenerate and nondegenerate even though it is row degenerate. The greatest common factor of $\det M_D(s) = (s+7)(s-9)$ *and of all elements of the first row of* $M_N(s)$,

$$M_N(s) = \begin{bmatrix} s+7 & s+7 \\ s-9 & s-9 \end{bmatrix},$$

is $s+7$. *The greatest common factor of* $\det M_D(s) = (s+7)(s-9)$ *and of all elements of the second row of* $M_N(s)$ *is* $s-9$ *that is different from* $s+7$. *The greatest common factor of* $\det M_D(s) = (s+7)(s-9)$ *and of all elements of the first column of* $M_N(s)$ *is* 1. *The greatest common factor of* $\det M_D(s)$ *and of all elements of the second column of* $M_N(s)$ *is also* 1, *as well as for the greatest common factor of* $\det M_D(s)$ *and of all elements of* $M_N(s)$. *The given* $M(s) = M_D^{-1}(s)M_N(s)$ *is both column nondegenerate and nondegenerate even though it is row degenerate.*

Example 96 *Let*

$$M(s) = \frac{1}{(s+7)(s-9)} \begin{bmatrix} s+7 & s-9 \\ s+7 & s-9 \end{bmatrix}.$$

It is both row nondegenerate and nondegenerate even though it is column degenerate.

Note 97 *If a rational matrix function* $M(.) = M_D^{-1}(.)M_N(.)$ *[respectively,* $M(.) = M_N(.)M_D^{-1}(.)$*] is degenerate then it is both row degenerate and column degenerate.*

Example 98 *Let*

$$M(s) = \frac{1}{(s+5)(s+3)} \begin{bmatrix} s\,(s+5) & (s+6)\,(s+5) \\ 12\,(s+5) & (s+5)\,(s+3) \end{bmatrix}.$$

It is degenerate, and both column degenerate and row degenerate. The binomial $(s+5)$ *cancels itself. The result of the cancellation is*

$$M_{nd}(s) = \frac{1}{s+3} \begin{bmatrix} s & s+6 \\ 12 & s+3 \end{bmatrix} = \begin{bmatrix} \frac{s}{s+3} & \frac{s+6}{s+3} \\ \frac{12}{s+3} & 1 \end{bmatrix}.$$

It is both column and row nondegenerate. It is the nondegenerate form $M_{nd}(s)$ *of* $M(s)$.

Note 99 *If a rational matrix function* $M(.) = M_D^{-1}(.)M_N(.)$ *[respectively,* $M(.) = M_N(.)M_D^{-1}(.)$*] is nondegenerate, then it is also irreducible, but the reverse does not hold in general (i.e., it can be irreducible but need not be nondegenerate).*

Example 100 *Let* $M(.) = M_D^{-1}(.)M_N(.)$,

$$M_D(s) = \begin{bmatrix} s^2 + s - 8 & 2s^2 + 3s - 2 \\ 2s^2 - 2s - 12 & 4s^2 - 2s \end{bmatrix} =$$
$$= \begin{bmatrix} s+2 & 6 \\ 2s & 12 \end{bmatrix} \begin{bmatrix} s-1 & 2s-1 \\ -1 & 0 \end{bmatrix},$$

$$M_N(s) = \begin{bmatrix} 2s^2 + 3\,s + 4 & 6s^2 - 3s & 4s^2 + 66s - 34 \\ 4s^2 - 2s + 12 & 12s^2 - 30s + 12 & 8s^2 + 116s - 60 \end{bmatrix} =$$
$$= \begin{bmatrix} s+2 & 6 \\ 2s & 12 \end{bmatrix} \begin{bmatrix} 2s-1 & 6s-3 & 4s-2 \\ 1 & -2s+1 & 10s-5 \end{bmatrix}.$$

The polynomial matrices $M_D(.)$ *and* $M_N(.)$ *are left coprime. Their greatest left common divisor* $L(.)$,

$$L(s) = \begin{bmatrix} s+2 & 6 \\ 2s & 12 \end{bmatrix}, \; L^{-1}(s) = \frac{1}{24} \begin{bmatrix} 12 & -6 \\ -2s & s+2 \end{bmatrix}, \; detL(s) = 24,$$

is unimodular and cancels itself in $M(.)$. The rational matrix function $M(.)$ is irreducible in the sense of the definition in [29, p. 605] and [198, p. 370]. However, it is really further reducible, i.e., it is degenerate. The reduced form of $M(s)$ obtained after the cancellation of $L(s)$ reads

$$M_{irr}(s) = \frac{1}{2s-1}\begin{bmatrix} 0 & -2s+1 \\ 1 & s-1 \end{bmatrix}\begin{bmatrix} 2s-1 & 6s-3 & 4s-2 \\ 0 & -2s+1 & 10s-5 \end{bmatrix} =$$

$$= \frac{1}{2s-1}\begin{bmatrix} 0 & (-2s+1)^2 & (-2s+1)(10s-5) \\ 2s-1 & (2s-1)(4-s) & (2s-1)(5s-3) \end{bmatrix}.$$

It is degenerate because the polynomial $2s-1$ is common to $\det M_{irrD}(s) = 2s-1$ and to all elements of $(adj M_{irrD}(s))\, M_{irrN}(s)$:

$$(adj M_{irrD}(s))\, M_{irrN}(s) = \begin{bmatrix} 0 & (-2s+1)^2 & (-2s+1)(10s-5) \\ 2s-1 & (2s-1)(4-s) & (2s-1)(5s-3) \end{bmatrix} =$$
$$= (2s-1)\begin{bmatrix} 0 & 2s-1 & -10s+5 \\ 1 & 4-s & 5s-3 \end{bmatrix}.$$

Evidently, the polynomial $2s-1$ is not constant. It cancels itself in the denominator and in all entries of the nominator matrix of $M_{irr}(s)$,

$$M_{irr}(s) = \frac{2s-1}{2s-1}\begin{bmatrix} 0 & 2s-1 & -5(2s-1) \\ 1 & 4-s & 5s-3 \end{bmatrix} \Longrightarrow$$
$$M_{irrnd}(s) = \begin{bmatrix} 0 & 2s-1 & -5(2s-1) \\ 1 & 4-s & 5s-3 \end{bmatrix}.$$

The final, completely reduced form, i.e., the nondegenerate form $M_{nd}(s)$, $M_{nd}(s) = M_{irrnd}(s)$, of both $M(s)$ and $M_{irr}(s)$, respectively, reads

$$M_{nd}(s) = \begin{bmatrix} 0 & 2s-1 & -5(2s-1) \\ 1 & 4-s & 5s-3 \end{bmatrix}.$$

It is different from the irreducible form $M_{irr}(s)$,

$$M_{nd}(s) = \begin{bmatrix} 0 & 2s-1 & -5(2s-1) \\ 1 & 4-s & 5s-3 \end{bmatrix} \neq$$
$$\neq \frac{1}{2s-1}\begin{bmatrix} 0 & (-2s+1)^2 & (-2s+1)(10s-5) \\ 2s-1 & (2s-1)(4-s) & (2s-1)(5s-3) \end{bmatrix} = M_{irr}(s).$$

Example 101 *Given 1x7 row matrix $M(s)$,*

$$M(s)=\left[(s-9)^2(s+2)(s+5)\right]^{-1}\bullet\begin{bmatrix}(s-9)^2(s+6)\,(7s-10)\,\vdots\\ \vdots\,-(s-9)(s-7)(17s+8)\,\vdots\\ \vdots\,-(s-9)(12s+11)\,\vdots\\ \vdots\,215(s-9)\,\vdots\\ \vdots\,-23(s-9)(2s-7)\,\vdots\\ \vdots\,(s-9)(2s^2+s-5)\,\vdots\\ \vdots\,(s-9)(s^2+34)\end{bmatrix}^T. \tag{6.1}$$

The binomial $(s-9)$ is a common factor to the denominator polynomial $(s-9)^2(s+2)(s+5)$ and to all entries of the numerator polynomial matrix that is row vector. It is (row) degenerate. Its row nondegenerate form $M_{rnd}(s)$ results after the cancellation of $(s-9)$:

$$M_{rnd}(s)=\left[(s-9)(s+2)(s+5)\right]^{-1}\bullet\begin{bmatrix}(s-1)(s+6)\,(7s-10)\,\vdots\\ \vdots\,-(s-7)(17s+8)\,\vdots\\ \vdots\,-(12s+11)\,\vdots\\ \vdots\,215\,\vdots\\ \vdots\,-23(2s-7)\,\vdots\\ \vdots\,(2s^2+s-5)\,\vdots\\ \vdots\,(s^2+34)\end{bmatrix}^T. \tag{6.2}$$

It is irreducible, too, $M_{irr}(s)=M_{rnd}(s)$.

6.2 Basic lemma

The following lemma is important for the linear tracking control synthesis.

Lemma 102 *[148]* ***Basic lemma***

Let $M(.)$ be a real rational proper matrix function of s. Let $\mathbf{Z}(.)$ and $\mathbf{W}(.)$ be real rational proper vector functions of s, which are interrelated via $M(.)$,

$$\mathbf{Z}(s)=M(s)\mathbf{W}(s),\ \mathbf{Z}(s)\in\mathfrak{C}^p,\ M(s)\in\mathfrak{C}^{pxq},\ \mathbf{W}(s)\in\mathfrak{C}^q. \tag{6.3}$$

1) Any equal pole and zero common to all elements of the same row of $M(s)$ do not influence the character of the original $\mathbf{z}(t)$ of $\mathbf{Z}(s)$ and may be cancelled.

2) Any equal pole and zero of any entry of $\mathbf{W}(s)$ do not influence the character of the corresponding entry of the original $\mathbf{z}(t)$ of $\mathbf{Z}(s)$ and may be cancelled.

3) Any equal pole and zero of any entry of $M(s)\mathbf{W}(s)$ do not influence the character of the corresponding entry of the original $\mathbf{z}(t)$ of $\mathbf{Z}(s)$ and may be cancelled.

4) The poles of the row nondegenerate form $[M(s)\mathbf{W}(s)]_{nd}$ of $M(s)\mathbf{W}(s)$ determine the character of the original $\mathbf{z}(t)$ of $\mathbf{Z}(s)$, where $\mathbf{z}(t)$ is inverse Laplace transform of $\mathbf{Z}(s)$,

$$\mathbf{z}(t) = \mathcal{L}^{-1}\left\{\mathbf{Z}(s)\right\} = \mathcal{L}^{-1}\left\{[M(s)\mathbf{W}(s)]_{rnd}\right\}. \tag{6.4}$$

5) If every zero of every element of every row of $M(s)$ is different from every pole of the corresponding entry of $\mathbf{W}(s)$ and every pole of every element of every row of $M(s)$ is different from every zero of the corresponding entry of $\mathbf{W}(s)$, then the row nondegenerate form $[M(s)\mathbf{W}(s)]_{rnd}$ of $M(s)\mathbf{W}(s)$ becomes the product of the row nondegenerate forms $M(s)_{rnd}$ and $\mathbf{W}(s)_{rnd}$ of $M(s)$ and $\mathbf{W}(s)$,

$$[M(s)\mathbf{W}(s)]_{rnd} = M(s)_{rnd}\mathbf{W}(s)_{rnd}. \tag{6.5}$$

Then (6.4) reduces to

$$\mathbf{z}(t) = \mathcal{L}^{-1}\left\{\mathbf{Z}(s)\right\} = \mathcal{L}^{-1}\left\{M(s)_{rnd}\mathbf{W}(s)_{rnd}\right\}. \tag{6.6}$$

The proof is in Appendix H.

Example 103 *[148] The IO system*

$$\begin{bmatrix} 1 & 1 \\ 1 & 2 \end{bmatrix}\mathbf{y}^{(2)}(t) - \begin{bmatrix} 1 & 1 \\ 1 & 2 \end{bmatrix}\mathbf{y}(t) =$$

$$= \begin{bmatrix} -1 & 2 \\ 0 & -3 \end{bmatrix}\mathbf{i}(t) + \begin{bmatrix} 1 & 1 \\ -1 & 1 \end{bmatrix}\mathbf{i}^{(1)}(t) + \begin{bmatrix} 0 & 0 \\ 1 & 0 \end{bmatrix}\mathbf{i}^{(2)}(t)$$

has the transfer function matrix

$$G_{IO}(s) = \frac{s^2-1}{(s^2-1)^2}\begin{bmatrix} -(s-1)(s-2) & s+7 \\ (s-1)^2 & -5 \end{bmatrix}.$$

Its nondegenerate form $G_{IOnd}(s)$ reads

$$G_{IOnd}(s) = \frac{1}{s^2-1}\begin{bmatrix} -(s-1(s-2) & s+7 \\ (s-1)^2 & -5 \end{bmatrix}.$$

It is also its row nondegenerate form $G_{IOrnd}(s)$,

$$G_{IOnd}(s) = G_{IOrnd}(s).$$

However, its column nondegenerate form $G_{IOcnd}(s)$ is different from them,

$$G_{IOcnd}(s) = \frac{1}{s+1}\begin{bmatrix} -(s-2) & s+7 \\ s-1 & -5 \end{bmatrix}.$$

If we wish to determine the system output response under all zero initial conditions, we should use the row nondegenerate form $G_{IOrnd}(s)$ of $G_{IO}(s)$, and we may not use its column nondegenerate form $G_{IOcnd}(s)$ because the pole $s^* = 1$ cannot be cancelled in the rows of $G_{IOrnd}(s)$ although it can be cancelled in its columns.

Chapter 7

Full transfer function matrix *F(s)*

7.1 General definitions of *F(S)*

This section develops further the concept of *the system full transfer function matrix* introduced in [81]. The general definitions presented in the sequel are equivalent to those in [148].

7.1.1 Definition of F(s): *IO* system

The full transfer function matrix $F_{IO}(s)$ of the *IO* system (2.9) (Section 2.1), repeated as

$$A^{(\nu)}\mathbf{Y}^{\nu}(t) = B^{(\mu)}\mathbf{I}^{\mu}(t),\ t \in \mathfrak{T}, \tag{7.1}$$

describes in the complex domain $\mathfrak{C}$ how the system temporally transfers a *simultaneous influence* of an arbitrary input vector $\mathbf{I}(t)$, of any input initial conditions $\mathbf{I}_{0\mp}$, $\mathbf{I}_{0\mp}^{(1)}$, ..., $\mathbf{I}_{0\mp}^{(\mu-1)}$, and of arbitrary output initial conditions $\mathbf{Y}_{0\mp}$, $\mathbf{Y}_{0\mp}^{(1)}$, ..., $\mathbf{Y}_{0\mp}^{(\nu-1)}$ on the system output response $\mathbf{Y}(t)$, Fig. 7.1 [148].

Int $\mathfrak{T}_0$ is *the interior of* $\mathfrak{T}_0$,

$$Int\ \mathfrak{T}_0 = \{t : t \in \mathfrak{T}_0,\ t > 0\} =]0, \infty[,\ Int\ \mathfrak{T}_0 \subset \mathfrak{T}_0,$$
$$inf\,(Int\ \mathfrak{T}_0) = 0 \in \mathfrak{T}_0,\ \ sup\,(Int\ \mathfrak{T}_0) = \infty. \tag{7.2}$$

Definition 104 *a)* ***The full (complete) (IO) matrix transfer function*** $F_{IO}(.)$, $F_{IO}(.) : \mathfrak{C} \rightarrow \mathfrak{C}^{N\times[(\mu+1)M+\nu N]}$, ***of the IO system (7.1)*** *is a matrix function of the complex variable s such that it determines uniquely (left, right) Laplace transform* $\mathbf{Y}^{(\mp)}(s)$ *of the system output* $\mathbf{Y}(t)$ *as a homogenous linear function of (left, right) Laplace transform* $\mathbf{I}^{(\mp)}(s)$ *of the system input vector* $\mathbf{I}(t)$ *for an arbitrary variation of* $\mathbf{I}(t)$, *of arbitrary initial vector values* $\mathbf{I}_{0(\mp)}^{\mu-1}$ *and* $\mathbf{Y}_{0(\mp)}^{\nu-1}$ *of the extended input vector* $\mathbf{I}^{\mu-1}(t)$ *and of the extended output vector* $\mathbf{Y}^{\nu-1}(t)$ *at* $t = 0^{(\mp)}$, *respectively, and its matrix value is* ***the system***

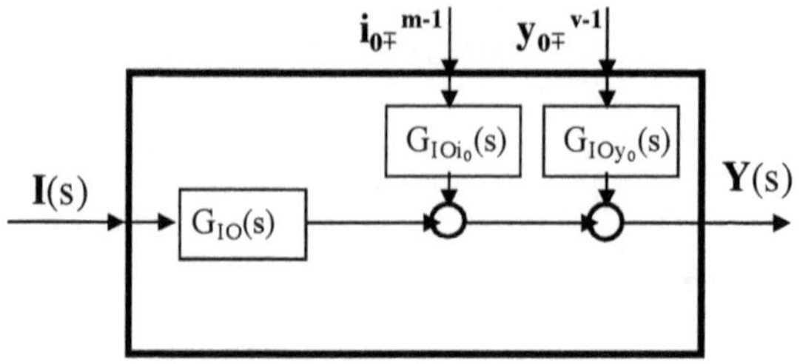

Figure 7.1: The full block diagram of the IO system shows the system transfer function matrices relative to the input vector and relative to all initial conditions in which m $= \mu$ and v $= \nu$.

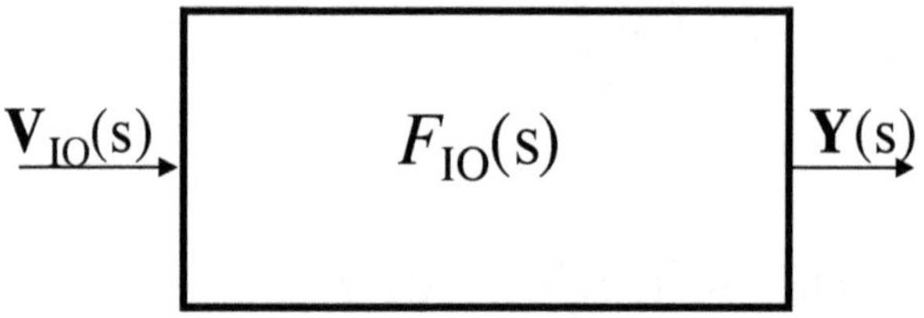

Figure 7.2: The full block of the IO system shows the dependence of $\mathbf{Y}(s)$ on the vector $\mathbf{V}_{IO}(s)$ through the full transfer function matrix $F_{IO}(s)$.

full (complete) input-output (IO) transfer function matrix *denoted by* $F_{IO}(s)$, $F_{IO}(s) \in \mathfrak{C}^{N\times[(\mu+1)M+\nu N]}$,

$$\mathbf{Y}^{(\mp)}(s) = F_{IO}(s)\mathbf{V}_{IO}^{(\mp)}(s),$$

$$\mathbf{V}_{IO}^{(\mp)}(s) = \left[\left(\mathbf{I}^{(\mp)}(s)\right)^T \vdots \left(\mathbf{I}_{0(\mp)}^{\mu-1}\right)^T \vdots \left(\mathbf{Y}_{0(\mp)}^{\nu-1}\right)^T\right]^T, \tag{7.3}$$

Fig. 7.2 [148].

b) ***The (IO) matrix transfer function*** $G_{IO}(.)$, $G_{IO}(.) : \mathfrak{C} \to \mathfrak{C}^{N\times M}$, ***of the IO system (7.1)*** *is a matrix function of the complex variable s such that it determines uniquely (left, right) Laplace transform* $\mathbf{Y}^{(\mp)}(s)$ *of the system output* $\mathbf{Y}(t)$ *as a homogenous linear function of (left, right) Laplace transform* $\mathbf{I}^{(\mp)}(s)$ *of the system input* $\mathbf{I}(t)$ *for an arbitrary variation of* $\mathbf{I}(t)$*, and under all zero initial conditions, that is, that the initial vector values* $\mathbf{I}_{0(\mp)}^{\mu-1}$ *and* $\mathbf{Y}_{0(\mp)}^{\nu-1}$ *of the extended input vector* $\mathbf{I}^{\mu-1}(t)$ *and of the extended output vector* $\mathbf{Y}^{\nu-1}(t)$ *at* $t = 0^{(\mp)}$ *are equal to zero vectors,* $\mathbf{I}_{0(\mp)}^{\mu-1} = \mathbf{0}_{\mu M}$ *and* $\mathbf{Y}_{0(\mp)}^{\nu-1} = \mathbf{0}_{\nu N}$*, and its matrix value is* ***the system input-output (IO) transfer function matrix*** $G_{IO}(s)$, $G_{IO}(s) \in \mathfrak{C}^{N\times M}$,

$$\mathbf{Y}^{(\mp)}(s) = G_{IO}(s)\mathbf{I}^{(\mp)}(s),\ \mathbf{I}_{0(\mp)}^{\mu-1} = \mathbf{0}_{\mu M},\ \mathbf{Y}_{0(\mp)}^{\nu-1} = \mathbf{0}_{\nu N}. \tag{7.4}$$

c) ***The (IICO) matrix transfer function*** $G_{IOi_0}(.)$ ***relative to*** $\mathbf{I}_{0(\mp)}^{\mu-1}$, $G_{IOi_0}(.) : \mathfrak{C} \to \mathfrak{C}^{N\times\mu M}$, ***of the IO system (7.1)*** *is a matrix function of the* *complex variable s such that it determines uniquely, respectively, (left, right) Laplace transform $\mathbf{Y}^{(\mp)}(s)$ of the system output $\mathbf{Y}(t)$ as a homogenous linear function of an arbitrary initial vector $\mathbf{I}_{0(\mp)}^{\mu-1}$ of the extended input vector $\mathbf{I}^{\mu-1}(t)$ at $t=0^{(\mp)}$ in the free regime on $Int\,\mathfrak{T}_0$ and for all zero output initial conditions, i.e., for $\mathbf{I}(t)=\mathbf{0}_M$ for every $t \in Int\,\mathfrak{T}_0$, and $\mathbf{Y}_{0(\mp)}^{\nu-1} \equiv \mathbf{0}_{\nu N}$, and its matrix value is* ***the system input initial condition-output (IICO) transfer function matrix relative to*** $\mathbf{I}_{0(\mp)}^{\mu-1}$ *, which is denoted by $G_{IOi_0}(s)$, $G_{IOi_0}(s) \in \mathfrak{C}^{N\times\mu M}$,*

$$\mathbf{Y}^{(\mp)}(s)=G_{IOi_0}(s)\mathbf{I}_{0(\mp)}^{\mu-1},\ \mathbf{I}(t)=\mathbf{0}_M,\ \forall t \in Int\mathfrak{T}_0,\ \mathbf{Y}_{0(\mp)}^{\nu-1} \equiv \mathbf{0}_{\nu N}. \tag{7.5}$$

d) ***The (OICO) matrix transfer function*** $G_{IOy_0}(.)$ ***relative to*** $\mathbf{Y}_{0(\mp)}^{\nu-1}$, $G_{IOy_0}(.): \mathfrak{C} \rightarrow \mathfrak{C}^{N\times\nu N}$, ***of the IO system (7.1)*** *is a matrix function of the complex variable s such that it determines uniquely, respectively, (left, right) Laplace transform $\mathbf{Y}^{(\mp)}(s)$ of the system output $\mathbf{Y}(t)$ as a homogenous linear function of an arbitrary initial vector $\mathbf{Y}_{0(\mp)}^{\nu-1}$ of the extended output vector $\mathbf{Y}^{\nu-1}(t)$ at $t=0^{(\mp)}$ for the system in a free regime and under all zero input initial conditions, i.e., for $\mathbf{I}(t) \equiv \mathbf{0}_M$ and $\mathbf{I}_{0(\mp)}^{\mu-1}=\mathbf{0}_{\mu M}$, and its matrix value is* ***the system output initial condition-output (OICO) transfer function matrix relative to*** $\mathbf{Y}_{0(\mp)}^{\nu-1}$*, which is denoted by $G_{IOy_0}(s)$, $G_{IOy_0}(s) \in \mathfrak{C}^{N\times\nu N}$,*

$$\mathbf{Y}^{(\mp)}(s)=G_{IOy_0}(s)\mathbf{Y}_{0(\mp)}^{\nu-1},\ \mathbf{I}(t) \equiv \mathbf{0}_M,\ \mathbf{I}_{0(\mp)}^{\mu-1}=\mathbf{0}_{\mu M}. \tag{7.6}$$

e) ***The (ICO) matrix transfer function*** $G_{IO_0}(.)$ ***relative to*** $[\mathbf{I}_{0(\mp)}^{\mu-1^T}\ \mathbf{Y}_{0(\mp)}^{\nu-1^T}]^T$, $G_{IO_0}(.): \mathfrak{C} \rightarrow \mathfrak{C}^{N\times(\mu M+\nu N)}$, ***of the IO system (7.1),*** *is a matrix function of the complex variable s such that it determines uniquely, respectively, (left, right) Laplace transform $\mathbf{Y}^{(\mp)}(s)$ of the system output $\mathbf{Y}(t)$ as a homogenous linear function of an arbitrary initial vector $[\mathbf{I}_{0(\mp)}^{\mu-1^T}\quad \mathbf{Y}_{0(\mp)}^{\nu-1^T}]^T \in \mathfrak{R}^{\mu M+\nu N}$ of the extended input vector $\mathbf{I}^{\mu-1}(t)$ and of the extended output vector $\mathbf{Y}^{\nu-1}(t)$ at $t=0^{(\mp)}$ for the system in a free regime on $Int\ \mathfrak{T}_0$, i.e., for $\mathbf{I}(t)=\mathbf{0}_M$ for every $t \in Int\mathfrak{T}_0$, and its matrix value is* ***the system initial conditions-output (ICO) transfer function matrix relative to all initial conditions,*** *which is denoted by $G_{IO_0}(s)$, $G_{IO_0}(s) \in \mathfrak{C}^{N\times(\mu M+\nu N)}$,*

$$\mathbf{Y}^{(\mp)}(s)=G_{IO_0}(s)\begin{bmatrix}\mathbf{I}_{0(\mp)}^{\mu-1}\\ \mathbf{Y}_{0(\mp)}^{\nu-1}\end{bmatrix},\ \mathbf{I}(t)=\mathbf{0}_M,\ \forall t \in Int\mathfrak{T}_0. \tag{7.7}$$

Note 105 *The system linearity enables us to conclude that $G_{IO}(s)$, $G_{IOi_0}(s)$, and $G_{IOy_0}(s)$, i.e., $G_{IO_0}(s)$, are submatrices of $F_{IO}(s)$, Fig. 7.1,*

$$F_{IO}(s)=\left[G_{IO}(s) \vdots G_{IOi_0}(s) \vdots G_{IOy_0}(s)\right]=$$

$$=\left[G_{IO}(s) \vdots G_{IO_0}(s)\right]. \tag{7.8}$$

$G_{IOi_0}(s)$ and $G_{IOy_0}(s)$ are submatrices of $G_{IO_0}(s)$,

$$G_{IO_0}(s)=\left[G_{IOi_0}(s) \vdots G_{IOy_0}(s)\right]. \tag{7.9}$$

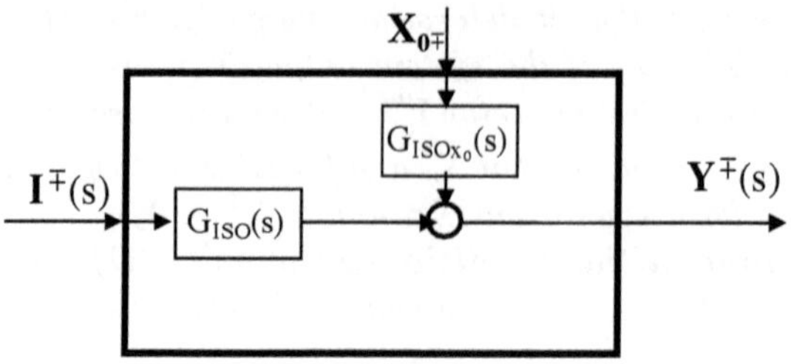

Figure 7.3: The full block diagram of the *ISO* system shows the system transfer function matrices and the influence of both the input vector $\mathbf{I}(t)$ (by its Laplace transform $\mathbf{I}^{\pm}(s)$) and the initial state vector $\mathbf{X}(0^{\mp})$ on the system output behavior expressed by $\mathbf{Y}^{\pm}(s)$.

It is easy to show that (7.4) through (7.6) follow from (7.3) and (7.8), and vice versa.

Note 106 *The book [148] generalizes and broadens the classical block diagram technique to the full block diagram technique. The system transfer function matrix $G_{IO}(s)$ is replaced by the full transfer function matrix $F_{IO}(s)$ and the vector $\mathbf{V}_{IO}^{(\mp)}(s)$,*

$$\mathbf{V}_{IO}^{(\mp)}(s) = \left[\left(\mathbf{I}^{(\mp)}(s)\right)^T \vdots \left(\mathbf{I}_{0(\mp)}^{\mu-1}\right)^T \vdots \left(\mathbf{Y}_{0(\mp)}^{\nu-1}\right)^T\right]^T$$

replaces $\mathbf{I}^{(\mp)}(s)$. Then, (7.3),

$$\mathbf{Y}^{(\mp)}(s) = F_{IO}(s)\mathbf{V}_{IO}^{(\mp)}(s),$$

Fig. 7.2.

7.1.2 Definition of $F(s)$: *ISO* system

The transfer of the influence of the input vector and of the initial conditions through the *ISO* system (2.29), (2.30) (Section 2.2) as

$$\frac{d\mathbf{X}(t)}{dt} = A\mathbf{X}(t) + B\mathbf{I}(t),\ \forall t \in \mathfrak{T}_0, \tag{7.10}$$

$$\mathbf{Y}(t) = C\mathbf{X}(t) + D\mathbf{I}(t),\ \forall t \in \mathfrak{T}_0. \tag{7.11}$$

is through two channels, Fig. 7.3 [148]. This figure presents *the full block diagram* of the *ISO* system (7.10), (7.11).

The *full transfer function matrix* $F_{ISO}(s)$ of the *ISO* system (7.10), (7.11) describes, in the complex domain $\mathfrak{C}$, how the system temporally transfers an influence of an arbitrary initial state $\mathbf{X}_{0\mp}$ and an arbitrary input vector $\mathbf{I}(t)$ on the system output response $\mathbf{Y}(t)$.

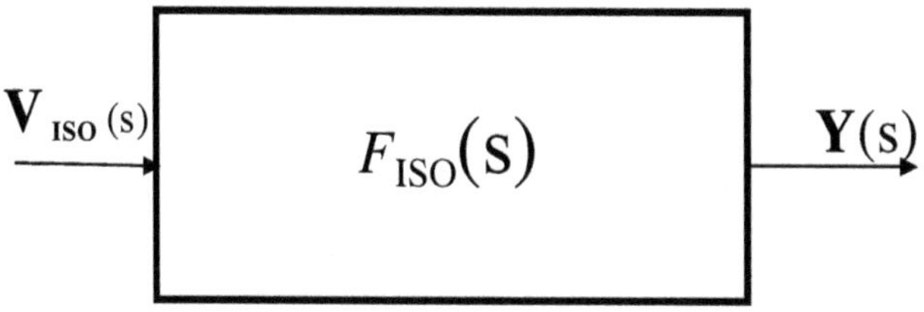

Figure 7.4: The full block of the ISO system shows the dependence of $\mathbf{Y}(s)$ on the vector $\mathbf{V}_{ISO}(s)$ through the system full transfer function matrix $F_{ISO}(s)$.

The form of a system mathematical model of a fixed physical system does not influence either the input vector or the output vector of the system. The system transfer function matrix $G(s)$ is independent of the form of the system mathematical model. However, initial conditions, although arbitrary, have in general different meanings, hence forms, for the input-output model (7.1) and for the input-state-output model (7.10), (7.11) of the same physical system. The consequence is that the form of the ISO system full transfer function matrix depends on the form of the ISO system mathematical model.

Definition 107 *a)* ***The full (complete) (ISO) matrix transfer function*** $F_{ISO}(.)$, $F_{ISO}(.) : \mathfrak{C} \to \mathfrak{C}^{N\times(M+n)}$, ***of the ISO system (7.10), (7.11)*** *is a matrix function of the complex variable s such that it determines uniquely (left, right) Laplace transform* $\mathbf{Y}^{(\mp)}(s)$ *of the system output* $\mathbf{Y}(t)$ *as a homogeneous linear function of (left, right) Laplace transform* $\mathbf{I}^{(\mp)}(s)$ *of the system input vector* $\mathbf{I}(t)$ *for an arbitrary variation of* $\mathbf{I}(t)$, *and of arbitrary initial vector values* $\mathbf{X}_{0^{(\mp)}}$ *of the state vector* $\mathbf{X}(t)$ *at* $t = 0^{(\mp)}$, *respectively; and its matrix value is* ***the system full (complete) input-(through state)-output (ISO) transfer function matrix*** $F_{ISO}(s)$, $F_{ISO}(s) \in \mathfrak{C}^{N\times(M+n)}$,

$$\mathbf{Y}^{(\mp)}(s) = F_{ISO}(s)\mathbf{V}_{ISO}^{(\mp)}(s),\ \mathbf{V}_{ISO}^{(\mp)}(s) = \left[\left(\mathbf{I}^{(\mp)}(s)\right)^T \vdots\, \mathbf{X}_{0^{(\mp)}}^T\right]^T, \tag{7.12}$$

Fig. 7.4 [148].

b) ***The (ISO) matrix transfer function*** $G_{ISO}(.)$, $G_{ISO}(.) : \mathfrak{C} \to \mathfrak{C}^{N\times M}$, ***of the ISO system (7.10), (7.11)*** *is a matrix function of the complex variable s such that it determines uniquely (left, right) Laplace transform* $\mathbf{Y}^{(\mp)}(s)$ *of the system output* $\mathbf{Y}(t)$ *as a homogeneous linear function of (left, right) Laplace transform* $\mathbf{I}^{(\mp)}(s)$ *of the system input vector* $\mathbf{I}(t)$ *for an arbitrary variation of* $\mathbf{I}(t)$, *and for zero initial state vector* $\mathbf{X}_{0^{(\mp)}}$ *of the state vector* $\mathbf{X}(t)$ *at* $t = 0^{(\mp)}$, *respectively; and its matrix value is* ***the system input-(through state)-output (ISO) transfer function matrix*** $G_{ISO}(s)$, $G_{ISO}(s) \in \mathfrak{C}^{N\times M}$,

$$\mathbf{Y}^{\mp}(s) = G_{ISO}(s)\mathbf{I}^{\mp}(s),\ \mathbf{X}_{0^{\mp}} = \mathbf{0}_n. \tag{7.13}$$

c) ***The (ISCO) matrix transfer function*** $G_{ISOx_0}(.)$ ***relative to*** $\mathbf{X}_{0^{\mp}}$, $G_{ISOx_0}(.) : \mathfrak{C} \to \mathfrak{C}^{N\times n}$, ***of the ISO system (7.10), (7.11)*** *is a matrix function* *of the complex variable s such that it determines uniquely (left, right) Laplace transform* $\mathbf{Y}^{\mp}(s)$ *of the system output* $\mathbf{Y}(t)$ *as a homogeneous linear function of an arbitrary initial vector value* $\mathbf{X}_{0\mp}$ *of the state vector* $\mathbf{X}(t)$ *at* $t = 0^{(\mp)}$ *for the system in the free regime (i.e., for* $\mathbf{I}(t) \equiv \mathbf{0}_M$*); and its matrix value is* ***the system initial state condition-output (ISCO) transfer function matrix relative to*** $\mathbf{X}_{0\mp}$, *which is denoted by* $G_{ISOx_0}(s)$, $G_{ISOx_0}(s) \in \mathfrak{C}^{Nxn}$,

$$\mathbf{Y}^{\mp}(s) = G_{ISOx_0}(s)\mathbf{X}_{0\mp},\ \mathbf{I}(t) \equiv \mathbf{0}_M. \tag{7.14}$$

Note 108 *The matrices* $G_{ISO}(s)$ *and* $G_{ISOx_0}(s)$ *are submatrices of the system full transfer function matrix* $F_{ISO}(s)$,

$$F_{ISO}(s) = \left[G_{ISO}(s) \vdots G_{ISOx_0}(s)\right]. \tag{7.15}$$

The full block diagram of the ISO system (7.10), (7.11) is in Fig. 7.3.

Note 109 *The use of the vector*

$$\mathbf{V}_{ISO}^{(\mp)}(s) = \left[\left(\mathbf{I}^{(\mp)}(s)\right)^T \vdots \mathbf{X}_{0(\mp)}^T\right]^T$$

instead of $\mathbf{I}^{(\mp)}(s)$ *and* $F_{ISO}(s)$ *instead of* $G_{ISO}(s)$ *permits [148] the formal application of the classical block diagram technique applied to the ISO system (7.10), (7.11), which results in the system full block shown in Fig. 7.4, due to (7.12),*

$$\mathbf{Y}^{\mp}(s) = F_{ISO}(s)\mathbf{V}_{ISO}^{(\mp)}(s).$$

Equations (7.12) and (7.15) yield (7.13) and (7.14), as well as vice versa, due to the system linearity.

Note 110 *The influences of the initial input and the initial output are contained in the influence of the initial state vector* $\mathbf{X}_{0\mp}$. *They do not appear explicitly in (7.12). The ISO system (7.10), (7.11) does not have the transfer function matrices relative to the initial input and output conditions.*

Note 111 *The full ISO transfer function matrix* $F_{ISO}(s)$ *of the ISO system is obtained from the given ISO model of the system. The full IO transfer function matrix* $F_{IOISO}(s)$ *of the ISO system is different from the full ISO transfer function matrix* $F_{ISO}(s)$ *of the system. The full IO transfer function matrix results from the IO system generated by the given ISO system. Its definition is given in Subsection 7.1.1. Its determination is the determination of the full transfer function matrix of the generated IO system (Subsection 7.2.1).*

Note 112 *In addition to the various system transfer function matrices related to the system output vector, there exist the system transfer function matrices relative to the system state vector. Their definitions follow.*

Definition 113 *a)* ***The full** (IS) **matrix transfer function*** $F_{ISOIS}(.)$, $F_{ISOIS}(.) : \mathfrak{C} \to \mathfrak{C}^{nx(M+n)}$, ***of the ISO system (7.10),(7.11)*** *is a matrix function of the complex variable s such that it determines uniquely (left, right) Laplace transform* $\mathbf{X}^{(\mp)}(s)$ *of the system state vector* $\mathbf{X}(t)$ *as a homogeneous linear function of (left, right) Laplace transform* $\mathbf{I}^{(\mp)}(s)$ *of the system input vector* $\mathbf{I}(t)$ *for an arbitrary variation of* $\mathbf{I}(t)$*, and of arbitrary initial vector value* $\mathbf{X}_{0(\mp)}$ *of the state vector* $\mathbf{X}(t)$ *at* $t = 0^{(\mp)}$*, respectively; and its matrix value is* ***the system full (complete) input-state** (IS) **transfer function matrix*** $F_{ISOIS}(s)$, $F_{ISOIS}(s) \in \mathfrak{C}^{nx(M+n)}$,

$$\mathbf{X}^{(\mp)}(s) = F_{ISOIS}(s) \left[\left(\mathbf{I}^{(\mp)}(s) \right)^T \vdots \mathbf{X}_{0(\mp)}^T \right]^T = F_{ISOIS}(s) \mathbf{V}_{ISO}^{(\mp)}(s). \quad (7.16)$$

b) ***The matrix** (IS) **transfer function*** $G_{ISOIS}(.)$, $G_{ISOIS}(.) : \mathfrak{C} \to \mathfrak{C}^{nxM}$, ***of the ISO system (7.10), (7.11)*** *is a matrix function of the complex variable s such that it determines uniquely (left, right) Laplace transform* $\mathbf{X}^{(\mp)}(s)$ *of the system state vector* $\mathbf{X}(t)$ *as a homogeneous linear function of (left, right) Laplace transform* $\mathbf{I}^{(\mp)}(s)$ *of the system input vector* $\mathbf{I}(t)$ *for an arbitrary variation of* $\mathbf{I}(t)$*, and for zero initial state vector* $\mathbf{X}_{0(\mp)}$ *of the state vector* $\mathbf{X}(t)$ *at* $t = 0^{(\mp)}$*, respectively; and its matrix value is* ***the** (IS) **transfer function matrix*** $G_{ISOIS}(s)$, $G_{ISOIS}(s) \in \mathfrak{C}^{nxM}$,

$$\mathbf{X}^{\mp}(s) = G_{ISOIS}(s) \mathbf{I}^{\mp}(s), \ \mathbf{X}_{0\mp} = \mathbf{0}_n. \quad (7.17)$$

c) ***The** (ISS) **matrix transfer function*** $G_{ISOSS}(.)$ ***relative to the initial state*** $\mathbf{X}_{0\mp}$, $G_{ISOSS}(.) : \mathfrak{C} \to \mathfrak{C}^{nxn}$, ***of the ISO system (7.10), (7.11)*** *is a matrix function of the complex variable s such that it determines uniquely (left, right) Laplace transform* $\mathbf{X}^{\mp}(s)$ *of the system state vector* $\mathbf{X}(t)$ *as a homogeneous linear function of an arbitrary initial vector value* $\mathbf{X}_{0\mp}$ *of the state vector* $\mathbf{X}(t)$ *at* $t = 0^{(\mp)}$ *for the system in the free regime (i.e., for* $\mathbf{I}(t) \equiv \mathbf{0}_M$*); and its matrix value is* ***the system initial state-state** (ISS) **transfer function matrix*** $G_{ISOSS}(s)$, $G_{ISOSS}(s) \in \mathfrak{C}^{nxn}$,

$$\mathbf{X}^{\mp}(s) = G_{ISOSS}(s) \mathbf{X}_{0\mp}, \ \mathbf{I}(t) \equiv \mathbf{0}_M. \quad (7.18)$$

Note 114 *The matrices* $G_{ISOIS}(s)$ *and* $G_{ISOSS}(s)$ *compose the system full IS transfer function matrix* $F_{ISOIS}(s)$,

$$F_{ISOIS}(s) = \left[G_{ISOIS}(s) \vdots G_{ISOSS}(s) \right]. \quad (7.19)$$

7.2 Determination of $F(s)$ in general

The determination of $F(s)$ of the *IO* systems, (Subsection 7.2.1), and of the *ISO* systems, (Subsection 7.2.2), in general is from [148]. It is the basis for the determination of $F(s)$ of *IO* and *ISO* plants, controllers, and control systems, (Sections 7.3 and 7.4).

7.2.1 $F(s)$ of the IO system

The transfer function matrix $G_{IO}(s)$ of the IO system (2.1), i.e., (2.9) (Section 2.1), which is repeated as

$$A^{(\nu)}\mathbf{Y}^{\nu}(t) = B^{(\mu)}\mathbf{I}^{\mu}(t),\ \forall t \in \mathfrak{T}_0. \tag{7.20}$$

can be set into the following *compact form* [71], [148], [154]:

$$G_{IO}(s) = \left(A^{(\nu)}S_N^{(\nu)}(s)\right)^{-1}\left(B^{(\mu)}S_M^{(\mu)}(s)\right) \tag{7.21}$$

by using the matrices $A^{(\nu)}$ and $B^{(\mu)}$ (2.4) (Section 2.1),

$$\begin{aligned} A^{(\nu)} &= \left[A_0 \vdots\ A_1 \vdots ... \vdots A_\nu\right] \in \mathfrak{R}^{N\mathrm{x}(\nu+1)N}, \\ B^{(\mu)} &= \left[B_0 \vdots\ B_1 \vdots ... \vdots B_\mu\right] \in \mathfrak{R}^{N\mathrm{x}(\mu+1)M}, \end{aligned} \tag{7.22}$$

and the matrix function $S_i^{(k)}(.) : \mathfrak{C} \longrightarrow \mathfrak{C}^{\,i(k+1)\mathrm{x}i}$ (3.7) (Subsection 3.3.2),

$$\begin{aligned} S_i^{(k)}(s) &= \left[s^0 I_i \vdots s^1 I_i \vdots s^2 I_i \vdots ... \vdots s^k I_i\right]^T \in \mathfrak{C}^{\,i(k+1)\mathrm{x}i}, \\ I_i &= diag\left\{1\ \ 1\ \ ...\ 1\right\} \in \mathfrak{R}^{i\mathrm{x}i},\ (k,i) \in \left\{(\mu, M),\ (\nu, N)\right\}. \end{aligned} \tag{7.23}$$

The compact form (7.21) of $G_{IO}(s)$ results from Laplace transform of (7.20) and from the fact that (7.22) and (7.23) imply

$$\left(\sum_{k=0}^{k=\nu} A_k s^k\right)^{-1}\left(\sum_{k=0}^{k=\mu\leq\nu} B_k s^k\right) = \left(A^{(\nu)}S_N^{(\nu)}(s)\right)^{-1}\left(B^{(\mu)}S_M^{(\mu)}(s)\right).$$

We use also the matrix function $Z_k^{(\varsigma-1)}(.) : \mathfrak{C} \rightarrow \mathfrak{C}^{(\varsigma+1)k\mathrm{x}\varsigma k}$ (3.8) (Subsection 3.3.2), in order to determine the compact form of the system full transfer function matrix $F_{IO}(s)$,

$$\begin{aligned} Z_k^{(\varsigma-1)}(s) &= \left\{ \begin{array}{c} \begin{bmatrix} O_k & O_k & O_k & ... & O_k \\ s^0 I_k & O_k & O_k & ... & O_k \\ ... & ... & ... & ... & ... \\ s^{\varsigma-1} I_k & s^{\varsigma-2} I_k & s^{\varsigma-3} I_k & ... & s^0 I_k \end{bmatrix},\ \varsigma \geq 1 \\ O_k,\ \varsigma < 1 \end{array} \right\}, \\ Z_k^{(\varsigma-1)}(s) &\in \mathfrak{C}^{(\varsigma+1)k\mathrm{x}\varsigma k},\ (\varsigma, k) \in \left\{(\mu, M),\ (\nu, N)\right\} \end{aligned} \tag{7.24}$$

We repeat the essence of Note 63 (Subsection 3.3.2) in order to avoid any ambiguity.

Note 115 *If $\varsigma = 0$ then the matrix $Z_k^{(\varsigma-1)}(s) = Z_k^{(-1)}(s)$ should be completely omitted rather than replaced by the zero matrix. The matrix $Z_k^{(\zeta-1)}(s)$ is not defined for $\zeta \leq 0$ and should be treated as the nonexisting one.*

Theorem 116 *a) The full (IO) transfer function matrix $F_{IO}(s)$ of the IO system (7.20) has the following form:*

$$If\ \mu \geq 1,\ then\ F_{IO}(s) = F_{IOD}^{-1}(s)F_{ION}(s) =$$
$$= \left(A^{(\nu)}S_N^{(\nu)}(s)\right)^{-1}\left[B^{(\mu)}S_M^{(\mu)}(s) \vdots -B^{(\mu)}Z_M^{(\mu-1)}(s) \vdots\ A^{(\nu)}Z_N^{(\nu-1)}(s)\right] =$$
$$= \left[G_{IO}(s) \vdots G_{IOi_0}(s) \vdots\ G_{IOy_0}(s)\right].$$
$$If\ \mu = 0,\ then\ F_{IO}(s) = F_{IOD}^{-1}(s)F_{ION}(s) =$$
$$= \left(A^{(\nu)}S_N^{(\nu)}(s)\right)^{-1}\left[B^{(\mu)}S_M^{(\mu)}(s) \vdots A^{(\nu)}Z_N^{(\nu-1)}(s)\right] =$$
$$= \left(A^{(\nu)}S_N^{(\nu)}(s)\right)^{-1}\left[B_0 \vdots A^{(\nu)}Z_N^{(\nu-1)}(s)\right]$$
$$= \left[G_{IO}(s) \vdots G_{IOy_0}(s)\right], \tag{7.25}$$

so that

$$\mathbf{Y}^{\mp}(s) = F_{IO}(s)\left\{\begin{array}{c} \left[(\mathbf{I}^{\mp}(s))^T \vdots \left(\mathbf{I}_{0\mp}^{\mu-1}\right)^T \vdots \left(\mathbf{Y}_{0\mp}^{\nu-1}\right)^T\right]^T\ if\ \mu \geq 1, \\ \left[(\mathbf{I}^{\mp}(s))^T \vdots \left(\mathbf{Y}_{0\mp}^{\nu-1}\right)^T\right]^T\ if\ \mu = 0 \end{array}\right\} =$$
$$= F_{IO}(s)\mathbf{V}_{IO}^{\mp}(s). \tag{7.26}$$

b) The (IO) transfer function matrix $G_{IO}(s)$ of the system (2.1) is determined by

$$G_{IO}(s) = \left(A^{(\nu)}S_N^{(\nu)}(s)\right)^{-1}\left(B^{(\mu)}S_M^{(\mu)}(s)\right). \tag{7.27}$$

c) The (IICO) transfer function matrix $G_{IOi_0}(s)$ of the system (2.1) is determined by

$$G_{IOi_0}(s) = \left(A^{(\nu)}S_N^{(\nu)}(s)\right)^{-1}\left\{\begin{array}{c} -B^{(\mu)}Z_M^{(\mu-1)}(s),\ if\ \mu \geq 1 \\ O,\ if\ \mu = 0, \end{array}\right\}. \tag{7.28}$$

d) The (OICO) transfer function matrix $G_{IOy_0}(s)$ of the system (2.1) is determined by

$$G_{IOy_0}(s) = \left(A^{(\nu)}S_N^{(\nu)}(s)\right)^{-1}A^{(\nu)}Z_N^{(\nu-1)}(s). \tag{7.29}$$

e) The (ICO) transfer function matrix $G_{IO_0}(s)$ of the system (2.1) is determined by

$$G_{IO_0}(s) = \left(A^{(\nu)}S_N^{(\nu)}(s)\right)^{-1}\left\{\begin{array}{c} \left[-B^{(\mu)}Z_M^{(\mu-1)}(s) \vdots A^{(\nu)}Z_N^{(\nu-1)}(s)\right],\ if\ \mu \geq 1 \\ A^{(\nu)}Z_N^{(\nu-1)}(s),\ if\ \mu = 0, \end{array}\right\}. \tag{7.30}$$

For the proof see Appendix I.

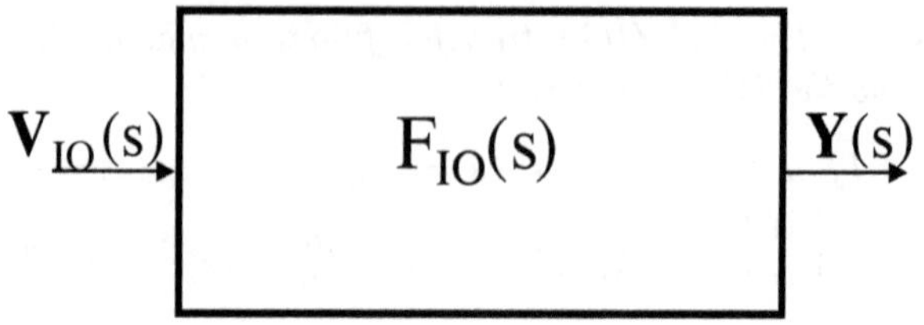

Figure 7.5: The full block of the *IO* system.

Note 117 *From* $A_k \in \mathfrak{R}^{N\times N}$, $B_k \in \mathfrak{R}^{N\times M}$, $k = 0,1,..,\nu$, $A_\nu \neq O_N$, *(2.1), and (7.24) follow*

$$\deg\left[A^{(\nu)}Z_N^{(\nu-1)}(s)\right] = \nu - 1 \text{ and } \mu \geq 1 \Longrightarrow \deg\left[B^{(\mu)}Z_M^{(\mu-1)}(s)\right] = \mu - 1. \tag{7.31}$$

Comment 118 *Equation (7.26) determines the overall action vector function* $\mathbf{V}_{IO}(.)$ *and its left Laplace transform* $\mathbf{V}_{IO}^-(s)$ *for the IO system,*

$$\mathbf{V}_{IO}(t) = \left\{ \begin{array}{c} \left[\mathbf{I}^T(t) \vdots \delta(t)\left(\mathbf{I}_{0^-}^{\mu-1}\right)^T \vdots \delta(t)\left(\mathbf{Y}_{0^-}^{\nu-1}\right)^T\right]^T, \mu \geq 1 \\ \left[\mathbf{I}^T(t) \vdots \delta(t)\left(\mathbf{Y}_{0^-}^{\nu-1}\right)^T\right]^T, \mu = 0 \end{array} \right\}, \tag{7.32}$$

$$\mathbf{V}_{IO}^-(s) = \left\{ \begin{array}{c} \left[\left(\mathbf{I}^-(s)\right)^T \vdots \left(\mathbf{I}_{0^-}^{\mu-1}\right)^T \vdots \left(\mathbf{Y}_{0^-}^{\nu-1}\right)^T\right]^T, \mu \geq 1 \\ \left[\left(\mathbf{I}^-(s)\right)^T \vdots \left(\mathbf{Y}_{0^-}^{\nu-1}\right)^T\right]^T, \mu = 0 \end{array} \right\} \equiv$$

$$= \left\{ \begin{array}{c} \mathbf{V}_{IO}^-(s;\mathbf{I}_{0^-}^{\mu-1},\mathbf{Y}_{0^-}^{\nu-1}),\ \mu \geq 1 \\ \mathbf{V}_{IO}^-(s;\mathbf{Y}_{0^-}^{\nu-1}),\ \mu = 0 \end{array} \right\}, \tag{7.33}$$

where $\delta(t)$ *is the unit Dirac impulse (for details see [148]). Equation (7.26) takes now formally the classical form of* $\mathbf{Y}^-(s)$ *obtained under zero initial conditions,*

$$\mathbf{Y}^-(s) = F_{IO}(s)\mathbf{V}_{IO}^-(s). \tag{7.34}$$

The system full block, which is valid now for arbitrary initial conditions, has the classical form established for zero initial conditions, Fig. 7.5 [148]. The vector functions $\mathbf{V}_{IO}^-(.) \equiv \mathbf{V}_{IO}^-(.;\mathbf{I}_{0^-}^{\mu-1},\mathbf{Y}_{0^-}^{\nu-1})$ *and* $\mathbf{V}_{IO}^-(.) = \mathbf{V}_{IO}^-(.;\mathbf{Y}_{0^-}^{\nu-1})$ *hide the nonzero initial conditions* $\mathbf{I}_{0^-}^{\mu-1}$ *and* $\mathbf{Y}_{0^-}^{\nu-1}$*, so that in fact*

$$\begin{aligned} \mathbf{Y}^-(s) &= \mathbf{Y}^-(s;\mathbf{I}_{0^-}^{\mu-1},\mathbf{Y}_{0^-}^{\nu-1}) = F_{IO}(s)\mathbf{V}_{IO}^-(s;\mathbf{I}_{0^-}^{\mu-1},\mathbf{Y}_{0^-}^{\nu-1}),\ \mu \geq 1, \\ \mathbf{Y}^-(s) &= \mathbf{Y}^-(s;\mathbf{Y}_{0^-}^{\nu-1}) = F_{IO}(s)\mathbf{V}_{IO}^-(s;\mathbf{Y}_{0^-}^{\nu-1}),\ \mu = 0. \end{aligned} \tag{7.35}$$

Comment 119 ***The generalization of the block diagram technique is the full block diagram technique***

The introduction of the generalized input vector function $\mathbf{V}_{IO}(.)$ *(7.32) permits us to use the classical block diagram technique with the full transfer function matrix* $F_{IO}(s)$ *of the system instead of its transfer function matrix* $G_{IO}(s)$*, and with* $\mathbf{V}_{IO}(s)$ *instead of* $\mathbf{I}(s)$*, Fig. 7.5 (for details see [148]).*

Comment 120 *The system full transfer function matrix* $F_{IO}(s)$ *is the system dynamic invariant. The order, the dimension, and the parameters of the system completely determine* $F_{IO}(s)$*. It is independent of the input vector and of all initial conditions; i.e., it is independent of the generalized input vector* $\mathbf{V}_{IO}(t)$*, i.e., of its Laplace transform* $\mathbf{V}_{IO}(s)$*. It has the same principal characteristics as the system transfer function matrix* $G_{IO}(s)$ *determined for zero initial conditions.*

Note 121 *The system full transfer function matrix* $F_{IO}(s)$ *incorporates the system transfer function matrix* $G_{IO}(s)$ *as a submatrix:*

$$F_{IO}(s) = F_{IOD}^{-1}(s)F_{ION}(s) = \underbrace{\left(A^{(\nu)}S_N^{(\nu)}(s)\right)^{-1}}_{F_{IOD}(s)=G_{IOD}(s)} \bullet$$

$$\bullet \underbrace{\left\{ \begin{array}{l} \left[\overbrace{B^{(\mu)}S_M^{(\mu)}(s)}^{G_{ION}(s)} \vdots -B^{(\mu)}Z_M^{(\mu-1)}(s) \vdots A^{(\nu)}Z_N^{(\nu-1)}(s) \right], \ \mu \geq 1 \\ \left[\overbrace{B^{(\mu)}S_M^{(\mu)}(s)}^{G_{ION}(s)} \vdots A^{(\nu)}Z_N^{(\nu-1)}(s) \right], \ \mu = 0 \end{array} \right\}}_{F_{ION}(s)} =$$

$$= \left\{ \begin{array}{l} \left[\begin{array}{c} G_{IO}(s) \vdots \left(A^{(\nu)}S_N^{(\nu)}(s)\right)^{-1} \left(-B^{(\mu)}Z_M^{(\mu-1)}(s)\right) \vdots \\ \vdots \left(A^{(\nu)}S_N^{(\nu)}(s)\right)^{-1} A^{(\nu)}Z_N^{(\nu-1)}(s) \end{array} \right], \mu \geq 1 \\ \left[G_{IO}(s) \vdots \left(A^{(\nu)}S_N^{(\nu)}(s)\right)^{-1} A^{(\nu)}Z_N^{(\nu-1)}(s) \right], \ \mu = 0 \end{array} \right\}. \quad (7.36)$$

$F_{IOD}(s)$ *and* $F_{ION}(s)$ *are the denominator and the numerator polynomial matrices of the system full transfer function matrix* $F_{IO}(s)$*,*

$$F_{IOD}(s) = A^{(\nu)}S_N^{(\nu)}(s),$$

$$F_{ION}(s) = \left[B^{(\mu)}S_M^{(\mu)}(s) \vdots -B^{(\mu)}Z_M^{(\mu-1)}(s) \vdots A^{(\nu)}Z_N^{(\nu-1)}(s) \right], \ \mu \geq 1$$

$$F_{ION}(s) = \left[B^{(\mu)}S_M^{(\mu)}(s) \vdots A^{(\nu)}Z_N^{(\nu-1)}(s) \right], \ \mu = 0,.$$

and

$$G_{IO}(s) = G_{IOD}^{-1}(s)G_{ION}(s), \ G_{IOD}(s) = F_{IOD}(s).$$

Example 122 *The given SISO IO system is described by*

$$Y^{(2)}(t) + Y^{(1)}(t) - 2Y(t) = 2I^{(2)}(t) - 14I^{(1)}(t) + 12Ii(t).$$

Its description and characteristics in the complex domain read

$$(s^2 + s - 2)Y^{\mp}(s) - (s+1)Y_{0\mp} - Y_{0\mp}^{(1)} =$$
$$= (2s^2 - 14s + 12)I^{\mp}(s) - (2s - 14)I_{0\mp} - 2I_{0\mp}^{(1)} \Longrightarrow$$

$$Y^{\mp}(s) = (s^2 + s - 2)^{-1} \bullet$$
$$\bullet \left[2s^2 - 14s + 12 \vdots -(2s-14) \vdots -2 \vdots s+1 \vdots 1 \right] \bullet$$
$$\bullet \left[I^{\mp}(s) \vdots \underbrace{I_{0\mp} \vdots I_{0\mp}^{(1)}}_{\mathbf{I}_{0\mp}^{1T}} \vdots \underbrace{Y_{0\mp} \vdots Y_{0\mp}^{(1)}}_{\mathbf{Y}_{0\mp}^{1T}} \right]^T =$$

$$= F_{IO}(s)\mathbf{V}^{\mp}(s), \ \mathbf{V}^{\mp}(s) = \left[I^{\mp}(s) \vdots \mathbf{I}_{0\mp}^{1} \vdots \mathbf{Y}_{0\mp}^{1} \right]^T \Longrightarrow$$
$$F_{IO}(s) = (s^2 + s - 2)^{-1} \bullet$$
$$\bullet \left[2s^2 - 14s + 12 \vdots -(2s-14) \vdots -2 \vdots s+1 \vdots 1 \right],$$

$$G_{IO}(s) = \frac{2s^2 - 14s + 12}{s^2 + s - 2} = \frac{(s-1)(s-6)}{(s-1)(s+2)} \Longrightarrow G_{IOnd}(s) = \frac{s-6}{s+2}.$$

The system transfer function matrix $G_{IO}(s)$ has the same zero $s^0 = 1$ and pole $s^* = 1$ so that they can be cancelled in $G_{IO}(s)$. However, the cancellation is not possible in the system full transfer function matrix $F_{IO}(s)$.

Example 123 *Let us consider the following example from [52, p. 58] of two ratios of polynomials,*

$$\frac{s+1}{s(s+2)} \text{ and } \frac{(s+1)(s-1)}{s(s+2)(s-1)}. \tag{7.37}$$

They correspond to the same rational function $f(.) : \mathfrak{T} \longrightarrow \mathfrak{R}$. They have two common zeros $s_1^0 = -1$ and $s_2^0 = \infty$, and two common poles $s_1^* = 0$ and $s_2^* = -2$. The second ratio has an additional positive real pole $s_3^* = 1$. However, if they represent the system transfer functions $G_1(.)$ and $G_2(.)$,

$$G_1(s) = \frac{s+1}{s(s+2)} \text{ and } G_2(s) = \frac{(s+1)(s-1)}{s(s+2)(s-1)}, \tag{7.38}$$

then they do not correspond to the same system. $G_1(s)$ is nondegenerate, while $G_2(s)$ is degenerate. The former is the nondegenerate form of the latter. $G_1(.)$ is the transfer function of the second-order SISO system described by

$$y^{(2)}(t) + 2y^{(1)}(t) = I(t) + I^{(1)}(t), \tag{7.39}$$

while $G_2(.)$ is the transfer function of the third-order SISO system determined by

$$y^{(3)}(t) + y^{(2)}(t) - 2y^{(1)}(t) = -I(t) + I^{(2)}(t). \tag{7.40}$$

The full transfer function matrix $F_1(s)$ of the former reads

$$F_1(s) = (s^2+s)^{-1}\left[s+1 \vdots -1 \vdots s+2 \vdots 1\right], \tag{7.41}$$

while the full transfer function matrix $F_2(s)$ of the latter is found as

$$F_2(s) = (s^3+s^2-2s)^{-1}\bullet$$
$$\bullet\left[s^2-1 \vdots -s \vdots -1 \vdots s^2+s-2 \vdots s+1 \vdots 1\right]. \tag{7.42}$$

Both $F_1(s)$ and $F_2(s)$ are row nondegenerate and nondegenerate.

Example 124 *A MIMO IO system is described by*

$$\mathbf{Y}^{(1)}(t) + \begin{bmatrix} 2 \vdots 0 \\ 0 \vdots 1 \end{bmatrix}\mathbf{Y}(t) = \begin{bmatrix} 2 \vdots 0 \\ 0 \vdots 0 \end{bmatrix}\mathbf{I}(t) + \mathbf{I}^{(1)}(t),$$
$$\mathbf{Y} = \begin{bmatrix} y_1 \\ y_2 \end{bmatrix}, \mathbf{I} = \begin{bmatrix} I_1 \\ I_2 \end{bmatrix}.$$

Its transfer function matrix $G_{IO}(s)$,

$$G_{IO}(s) = \begin{bmatrix} 1 & 0 \\ 0 & \frac{s}{s+1} \end{bmatrix}, \textit{ full rank } G_{IO}(s) = 2,$$

is rank defective for $s = 0$,

$$rank\ G_{IO}(0) = 1.$$

However, the system full transfer function matrix $F_{IO}(s)$,

$$F_{IO}(s) = \begin{bmatrix} 1 & 0 & -\frac{1}{s+2} & 0 & \frac{1}{s+2} & 0 \\ 0 & \frac{s}{s+1} & 0 & -\frac{1}{s+1} & 0 & \frac{1}{s+1} \end{bmatrix},$$

has the full rank over the field of complex numbers,

$$rankF_{IO}(s) \equiv \textit{full rank } F_{IO}(s) = 2.$$

In this example,

$$G_{IO}(s) = \begin{bmatrix} 1 & 0 \\ 0 & \frac{s}{s+1} \end{bmatrix},$$

$$G_{IOi_0}(s) = \begin{bmatrix} -\frac{1}{s+2} & 0 \\ 0 & -\frac{1}{s+1} \end{bmatrix},$$

$$G_{IOy_0}(s) = \begin{bmatrix} \frac{1}{s+2} & 0 \\ 0 & \frac{1}{s+1} \end{bmatrix},$$

$$G_{IO_0}(s) = \begin{bmatrix} -\frac{1}{s+2} & 0 & \frac{1}{s+2} & 0 \\ 0 & -\frac{1}{s+1} & 0 & \frac{1}{s+1} \end{bmatrix}.$$

An example of the determination of $F_{IO}(s)$ of a second order $MIMO$ system is worked out in [148].

Example 125 *The three-dimensional second-order IO system*

$$\begin{bmatrix} 3 \vdots 0 \vdots 0 \\ 0 \vdots 0 \vdots 0 \\ 0 \vdots 0 \vdots 0 \end{bmatrix} \mathbf{Y}^{(2)}(t) + \begin{bmatrix} 0 \vdots 0 \vdots 0 \\ 0 \vdots 1 \vdots 0 \\ 0 \vdots 0 \vdots 0 \end{bmatrix} \mathbf{Y}^{(1)}(t) + \begin{bmatrix} 0 \vdots 0 \vdots 0 \\ 0 \vdots 0 \vdots 0 \\ 0 \vdots 0 \vdots 1 \end{bmatrix} \mathbf{Y}(t) =$$

$$= \begin{bmatrix} 2 \vdots 0 \\ 0 \vdots 1 \\ 1 \vdots 0 \end{bmatrix} \mathbf{I}(t) + \begin{bmatrix} 1 \vdots 0 \\ 0 \vdots 1 \\ 1 \vdots 1 \end{bmatrix} \mathbf{I}^{(2)}(t)$$

yields

$$\nu = 2,\ \mu = 2,\ N = 3,\ M = 2,\ \det A_\nu = \det A_2 = \det \begin{bmatrix} 3 \vdots 0 \vdots 0 \\ 0 \vdots 0 \vdots 0 \\ 0 \vdots 0 \vdots 0 \end{bmatrix} = 0,$$

$$A^{(2)} = \begin{bmatrix} 0 \vdots 0 \vdots 0 & 0 \vdots 0 \vdots 0 & 3 \vdots 0 \vdots 0 \\ 0 \vdots 0 \vdots 0 & 0 \vdots 1 \vdots 0 & 0 \vdots 0 \vdots 0 \\ 0 \vdots 0 \vdots 1 & 0 \vdots 0 \vdots 0 & 0 \vdots 0 \vdots 0 \end{bmatrix},$$

$$S_3^{(2)}(s) = \begin{bmatrix} 1 & 0 & 0 \\ 0 & 1 & 0 \\ 0 & 0 & 1 \\ s & 0 & 0 \\ 0 & s & 0 \\ 0 & 0 & s \\ s^2 & 0 & 0 \\ 0 & s^2 & 0 \\ 0 & 0 & s^2 \end{bmatrix},$$

$$A^{(2)}S_3^{(2)}(s) = \begin{bmatrix} 3s^2 \vdots 0 \vdots 0 \\ 0 \vdots s \vdots 0 \\ 0 \vdots 0 \vdots 1 \end{bmatrix}, \ \deg\left[A^{(2)}S_3^{(2)}(s)\right] = 2,$$

$$adj\left(A^{(2)}S_3^{(2)}(s)\right) = \begin{bmatrix} s \vdots 0 \vdots 0 \\ 0 \vdots 3s^2 \vdots 0 \\ 0 \vdots 0 \vdots 3s^3 \end{bmatrix}, \ \deg\left[adj\left(A^{(2)}S_3^{(2)}(s)\right)\right] = 3,$$

$$\det\left[A^{(2)}S_3^{(2)}(s)\right] = 3s^3, \ \deg\left\{\det\left[A^{(2)}S_3^{(2)}(s)\right]\right\} = 3,$$

$$B^{(2)} = \begin{bmatrix} 2 \vdots 0 & 0 \vdots 0 & 1 \vdots 0 \\ 0 \vdots 1 & 0 \vdots 0 & 0 \vdots 1 \\ 1 \vdots 0 & 0 \vdots 0 & 1 \vdots 1 \end{bmatrix},$$

$$S_2^{(2)}(s) = \begin{bmatrix} 1 & 0 \\ 0 & 1 \\ s & 0 \\ 0 & s \\ s^2 & 0 \\ 0 & s^2 \end{bmatrix},$$

$$B^{(2)}S_2^{(2)}(s) = \begin{bmatrix} 2+s^2 \vdots & 0 \\ 0 & \vdots 1+s^2 \\ 1+s^2 \vdots & s^2 \end{bmatrix}, \ \deg\left[B^{(2)}S_2^{(2)}(s)\right] = 2,$$

$$B^{(2)}Z_2^{(2-1)}(s) = \begin{bmatrix} 2 \vdots 0 & 0 \vdots 0 & 1 \vdots 0 \\ 0 \vdots 1 & 0 \vdots 0 & 0 \vdots 1 \\ 1 \vdots 0 & 0 \vdots 0 & 1 \vdots 1 \end{bmatrix} \begin{bmatrix} 0 \vdots 0 \vdots 0 \vdots 0 \\ 0 \vdots 0 \vdots 0 \vdots 0 \\ 1 \vdots 0 \vdots 0 \vdots 0 \\ 0 \vdots 1 \vdots 0 \vdots 0 \\ s \vdots 0 \vdots 1 \vdots 0 \\ 0 \vdots s \vdots 0 \vdots 1 \end{bmatrix} =$$

$$= \begin{bmatrix} s \vdots 0 & 1 \vdots 0 \\ 0 \vdots s & 0 \vdots 1 \\ s \vdots s & 1 \vdots 1 \end{bmatrix}, \ \deg B^{(2)}Z_2^{(2-1)}(s) = deg(s) = 1,$$

$$A^{(2)} Z_3^{(2-1)}(s) = \begin{bmatrix} 0 \vdots 0 \vdots 0 & 0 \vdots 0 \vdots 0 & 3 \vdots 0 \vdots 0 \\ 0 \vdots 0 \vdots 0 & 0 \vdots 1 \vdots 0 & 0 \vdots 0 \vdots 0 \\ 0 \vdots 0 \vdots 1 & 0 \vdots 0 \vdots 0 & 0 \vdots 0 \vdots 0 \end{bmatrix} \begin{bmatrix} 0 \vdots 0 \vdots 0 \vdots 0 \vdots 0 \vdots 0 \\ 0 \vdots 0 \vdots 0 \vdots 0 \vdots 0 \vdots 0 \\ 0 \vdots 0 \vdots 0 \vdots 0 \vdots 0 \vdots 0 \\ 1 \vdots 0 \vdots 0 \vdots 0 \vdots 0 \vdots 0 \\ 0 \vdots 1 \vdots 0 \vdots 0 \vdots 0 \vdots 0 \\ 0 \vdots 0 \vdots 1 \vdots 0 \vdots 0 \vdots 0 \\ s \vdots 0 \vdots 0 \vdots 1 \vdots 0 \vdots 0 \\ 0 \vdots s \vdots 0 \vdots 0 \vdots 1 \vdots 0 \\ 0 \vdots 0 \vdots s \vdots 0 \vdots 0 \vdots 1 \end{bmatrix} =$$

$$= \begin{bmatrix} 3s \vdots 0 \vdots 0 & 3 \vdots 0 \vdots 0 \\ 0 \vdots 1 \vdots 0 & 0 \vdots 0 \vdots 0 \\ 0 \vdots 0 \vdots 0 & 0 \vdots 0 \vdots 0 \end{bmatrix}, \ \deg\left[A^{(2)} Z_3^{(2-1)}(s)\right] = \deg(3s) = 1,$$

and

$$F_{IO}(s) = \frac{\begin{bmatrix} s & \vdots & 0 & \vdots & 0 \\ 0 & \vdots & 3s^2 & \vdots & 0 \\ 0 & \vdots & 0 & \vdots & 3s^3 \end{bmatrix}}{3s^3} \bullet$$

$$\bullet \begin{bmatrix} 2+s^2 & 0 & -s & 0 & -1 & 0 & 3s & 0 & 0 & 3 & 0 & 0 \\ 0 & 1+s^2 & 0 & -s & 0 & -1 & 0 & 1 & 0 & 0 & 0 & 0 \\ 1+s^2 & s^2 & s & s & -1 & -1 & 0 & 0 & 0 & 0 & 0 & 0 \end{bmatrix}.$$

The full transfer function matrix $F_{IO}(s)$ is improper because the degree of its numerator matrix (which is equal to 5) exceeds the degree of its denominator polynomial (which is equal to 3).

Example 126 *The IO system*

$$\begin{bmatrix} 3 \vdots 0 \vdots 1 \\ 2 \vdots 0 \vdots 0 \\ 0 \vdots 1 \vdots 1 \end{bmatrix} \mathbf{Y}^{(2)}(t) + \begin{bmatrix} 0 \vdots 0 \vdots 0 \\ 0 \vdots 1 \vdots 0 \\ 0 \vdots 0 \vdots 0 \end{bmatrix} \mathbf{Y}^{(1)}(t) + \begin{bmatrix} 0 \vdots 0 \vdots 0 \\ 0 \vdots 0 \vdots 0 \\ 0 \vdots 0 \vdots 1 \end{bmatrix} \mathbf{Y}(t) =$$

$$= \begin{bmatrix} 2 \vdots 0 \\ 0 \vdots 1 \\ 1 \vdots 0 \end{bmatrix} \mathbf{I}(t) + \begin{bmatrix} 1 \vdots 0 \\ 0 \vdots 1 \\ 1 \vdots 1 \end{bmatrix} \mathbf{I}^{(2)}(t)$$

induces

$$\nu = 2,\ \mu = 2,\ N = 3,\ \ M = 2,\ \det A_\nu = \det A_2 = \det \begin{bmatrix} 3 \vdots 0 \vdots 1 \\ 2 \vdots 0 \vdots 0 \\ 0 \vdots 1 \vdots 1 \end{bmatrix} = 2 \neq 0,$$

$$A^{(2)} S_3^{(2)}(s) = \begin{bmatrix} 3s^2 & 0 & s^2 \\ 2s^2 & s & 0 \\ 0 & s^2 & s^2+1 \end{bmatrix},\ \deg\left[A^{(2)} S_3^{(2)}(s)\right] = 2,$$

$$adj\left(A^{(2)} S_3^{(2)}(s)\right) = \begin{bmatrix} s+s^3 & s^4 & -s^3 \\ -2s^4 - 2s2 & 3s^4+3s^2 & 2s^4 \\ 2s^4 & -3s^4 & 3s^3 \end{bmatrix},$$

$$\deg\left[adj\left(A^{(2)} S_3^{(2)}(s)\right)\right] = 4,$$

$$\det\left[A^{(2)} S_3^{(2)}(s)\right] = 2s^6 + 3s^5 + 3s^3,\ \deg\left\{\det\left[A^{(2)} S_3^{(2)}(s)\right]\right\} = 6,$$

$$B^{(2)} S_2^{(2)}(s) = \begin{bmatrix} 2+s^2 & 0 \\ 0 & 1+s^2 \\ 1+s^2 & s^2 \end{bmatrix},\ \deg\left[B^{(2)} S_2^{(2)}(s)\right] = 2,$$

$$B^{(2)} Z_2^{(2-1)}(s) = \begin{bmatrix} 2 \vdots 0 & 0 \vdots 0 & 1 \vdots 0 \\ 0 \vdots 1 & 0 \vdots 0 & 0 \vdots 1 \\ 1 \vdots 0 & 0 \vdots 0 & 1 \vdots 1 \end{bmatrix} \begin{bmatrix} 0 \vdots 0 \vdots 0 \vdots 0 \\ 0 \vdots 0 \vdots 0 \vdots 0 \\ 1 \vdots 0 \vdots 0 \vdots 0 \\ 0 \vdots 1 \vdots 0 \vdots 0 \\ s \vdots 0 \vdots 1 \vdots 0 \\ 0 \vdots s \vdots 0 \vdots 1 \end{bmatrix} =$$

$$= \begin{bmatrix} s \vdots 0 & 1 \vdots 0 \\ 0 \vdots s & 0 \vdots 1 \\ s \vdots s & 1 \vdots 1 \end{bmatrix},\ \deg B^{(2)} Z_2^{(2-1)}(s) = deg(s) = 1,$$

$$A^{(2)}Z_3^{(2-1)}(s) = \begin{bmatrix} 0 & 0 & 0 & 0 & 0 & 0 & 3 & 0 & 1 \\ 0 & 0 & 0 & 0 & 1 & 0 & 2 & 0 & 0 \\ 0 & 0 & 1 & 0 & 0 & 0 & 0 & 1 & 1 \end{bmatrix} \begin{bmatrix} 0 & \vdots & 0 & \vdots & 0 & \vdots & 0 & \vdots & 0 & \vdots & 0 \\ 0 & \vdots & 0 & \vdots & 0 & \vdots & 0 & \vdots & 0 & \vdots & 0 \\ 0 & \vdots & 0 & \vdots & 0 & \vdots & 0 & \vdots & 0 & \vdots & 0 \\ 1 & \vdots & 0 & \vdots & 0 & \vdots & 0 & \vdots & 0 & \vdots & 0 \\ 0 & \vdots & 1 & \vdots & 0 & \vdots & 0 & \vdots & 0 & \vdots & 0 \\ 0 & \vdots & 0 & \vdots & 1 & \vdots & 0 & \vdots & 0 & \vdots & 0 \\ s & \vdots & 0 & \vdots & 0 & \vdots & 1 & \vdots & 0 & \vdots & 0 \\ 0 & \vdots & s & \vdots & 0 & \vdots & 0 & \vdots & 1 & \vdots & 0 \\ 0 & \vdots & 0 & \vdots & s & \vdots & 0 & \vdots & 0 & \vdots & 1 \end{bmatrix} =$$

$$= \begin{bmatrix} 3s & 0 & 3+s & s & 3 & 1 \\ 2s & 1 & 0 & 2 & 0 & 0 \\ 0 & s & s & 0 & 1 & 1 \end{bmatrix}, \ \deg\left[A^{(2)}Z_3^{(2-1)}(s)\right] = \deg(3s) = 1,$$

and

$$F_{IO}(s) = \frac{\begin{bmatrix} s+s^3 & s^4 & -s^3 \\ -2s^2-2s^4 & 3s^2+3s^4 & 2s^4 \\ 2s^4 & -3s^4 & 3s^3 \end{bmatrix}}{2s^6+3s^5+3s^3} \bullet$$

$$\bullet \begin{bmatrix} 2+s^2 & 0 & -s & 0 & -1 & 0 & 3s & 0 & 3+s & s & 3 & 1 \\ 0 & 1+s^2 & 0 & -s & 0 & -1 & 2s & 1 & 0 & 2 & 0 & 0 \\ 1+s^2 & s^2 & -s & -s & -1 & -1 & 0 & s & s & 0 & 1 & 1 \end{bmatrix}.$$

The degree of the numerator matrix polynomial is equal to 6 that is also the degree of the denominator polynomial. The full transfer function matrix $F_{IO}(s)$ is proper in this case.

7.2.2 $F(s)$ of the *ISO* system

This subsection deals with the *ISO* system (2.29), (2.30) (Section 2.2) as

$$\frac{d\mathbf{X}(t)}{dt} = A\mathbf{X}(t) + B\mathbf{I}(t), \ \forall t \in \mathfrak{T}_0, \tag{7.43}$$

$$\mathbf{Y}(t) = C\mathbf{X}(t) + D\mathbf{I}(t), \ \forall t \in \mathfrak{T}_0. \tag{7.44}$$

The rank of the matrix C in (7.44) can be arbitrary.

The *ISO* transfer function $G_{ISO}(s)$ has the well known form,

$$G_{ISO}(s) = C(sI_n - A)^{-1}B + D. \tag{7.45}$$

Theorem 127 *a) The full (ISO) transfer function matrix $F_{ISO}(s)$ of the ISO system (7.43), (7.44) has the following form in general:*

$$F_{ISO}(s) = \left[C(sI_n - A)^{-1}B + D \vdots C(sI_n - A)^{-1} \right] =$$
$$= F_{ISOD}^{-1}(s)F_{ISON}(s), \tag{7.46}$$

so that

$$\mathbf{Y}^{\mp}(s) = F_{ISO}(s) \left[\left(\mathbf{I}^{\mp}(s)\right)^T \vdots \mathbf{X}_{0\mp}^T \right]^T = F_{ISO}(s)\mathbf{V}_{ISO}^{\mp}(s;\mathbf{X}_{0\mp}),$$
$$\mathbf{V}_{ISO}^{\mp}(s;\mathbf{X}_{0\mp}) = \left[\left(\mathbf{I}^{\mp}(s)\right)^T \vdots \mathbf{X}_{0\mp}^T \right]^T. \tag{7.47}$$

b) The (ISCO) transfer function matrix $G_{ISOx_0}(s)$ relative to $\mathbf{X}_0$ of the ISO system (7.43), (7.44) has the following form in general:

$$G_{ISOx_0}(s) = C(sI_n - A)^{-1}. \tag{7.48}$$

c) The full (complete) (IS) transfer function matrix $F_{ISOIS}(s)$ of the ISO system (7.43), (7.44) reads

$$F_{ISOIS}(s) = \left[(sI_n - A)^{-1}B \vdots (sI_n - A)^{-1} \right] =$$
$$= F_{ISOISD}^{-1}(s)F_{ISOISN}(s). \tag{7.49}$$

d) The (IS) transfer function matrix $G_{ISOIS}(s)$ of the ISO system (7.43), (7.44) obeys

$$G_{ISOIS}(s) = (sI_n - A)^{-1}B \;. \tag{7.50}$$

e) The (ISS) transfer function matrix $G_{ISOSS}(s)$ relative to $X_{0\mp}$ of the ISO system (7.43), (7.44) fulfills

$$G_{ISOSS}(s) = (sI_n - A)^{-1}. \tag{7.51}$$

Although the proof of Theorem 127 is almost evident, we present it immediately in order to illustrate the procedure of the determination of the system full transfer function matrix $F_{ISO}(s)$ and to show the origin of the difference between it and the system transfer function matrix $G_{ISO}(s)$.

Proof. a) Left, right Laplace transform of (7.43), (7.44) yields

$$\mathbf{X}^{\mp}(s) = (sI_n - A)^{-1} \left(\mathbf{X}_{0\mp} + B\mathbf{I}^{(\mp)}(s) \right) \Longrightarrow$$
$$\mathbf{X}^{\mp}(s) = \left[\begin{array}{cc} (sI_n - A)^{-1}B & (sI_n - A)^{-1} \end{array} \right] \left[\begin{array}{c} \mathbf{I}^{(\mp)}(s) \\ \mathbf{X}_{0\mp} \end{array} \right],$$

$$\mathbf{Y}^{\mp}(s) = C\mathbf{X}^{\mp}(s) + D\mathbf{I}^{\mp}(s) =$$

$$\mathbf{Y}^{\mp}(s) = C\left[(sI_n - A)^{-1}\left(\mathbf{X}_{0\mp} + B\mathbf{I}^{(\mp)}(s)\right)\right] + D\mathbf{I}^{\mp}(s) =$$

$$= \underbrace{\left[\overbrace{C(sI_n - A)^{-1}B + D}^{G_{ISO}(s)} \vdots \overbrace{C(sI_n - A)^{-1}}^{G_{ISOx_0}(s)}\right]}_{F_{ISO}(s)} \underbrace{\begin{bmatrix} \mathbf{I}^{\mp}(s) \\ \mathbf{X}_{0\mp} \end{bmatrix}}_{\mathbf{V}_{ISO}^{\mp}(s;\mathbf{X}_{0\mp})} =$$

$$= [det(sI_n - A)]^{-1}\left[Cadj(sI_n - A)B + Ddet(sI_n - A) \vdots Cadj(sI_n - A)\right] \bullet$$

$$\bullet \begin{bmatrix} \mathbf{I}^{\mp}(s) \\ \mathbf{X}_{0\mp} \end{bmatrix} = F_{ISOD}^{-1}(s)F_{ISON}(s)\mathbf{V}_{ISO}^{\mp}(s;\mathbf{X}_{0\mp}),$$

$$F_{ISOD}(s) = det(sI_n - A),$$

$$F_{ISON}(s) = \left[Cadj(sI_n - A)B + Ddet(sI_n - A) \vdots Cadj(sI_n - A)\right],$$

$$F_{ISO}(s) = \left[C(sI_n - A)^{-1}B + D \vdots C(sI_n - A)^{-1}\right] =$$

$$= [det(sI_n - A)]^{-1}\left[Cadj(sI_n - A)B + Ddet(sI_n - A) \vdots Cadj(sI_n - A)\right]. \tag{7.52}$$

These results and (7.12) (Definition 107, Subsection 7.1.2) prove the statement under a) of the theorem, i.e., the equations (7.46) and (7.47).

b) (7.48) results directly from a) and from the definition of $G_{ISOx_0}(s)$ (7.14) (Definition 107, Subsection 7.1.2).

c) The first two equations in (7.52) and a) of Definition 113 (Subsection 7.1.2) imply (7.49).

d) Equation (7.50) results from (7.52) and b) of Definition 113 (Subsection 7.1.2).

e) The combination of (7.52) and c) of Definition 113 (Subsection 7.1.2) gives (7.51) ■

Comment 128 *The input vector function* $\mathbf{I}(.)$ *and the initial state vector* $\mathbf{X}_{0\mp}$ *are all the actions on the system. This inspires us to introduce the action vector function* $\mathbf{V}_{ISO}(.)$ *for the ISO system by*

$$\mathbf{V}_{ISO}(t;\mathbf{X}_0) = \begin{bmatrix} \mathbf{I}(t) \\ \delta(t)\mathbf{X}_{0^-} \end{bmatrix} \in \mathfrak{R}^{M+n},\ \mathbf{V}_{ISO}^{-}(s;\mathbf{X}_0) = \begin{bmatrix} \mathbf{I}^{-}(s) \\ \mathbf{X}_{0^-} \end{bmatrix} \in \mathfrak{C}^{M+n}. \tag{7.53}$$

This permits us to set (7.47) into the compact form, i.e.,

$$\mathbf{Y}^{-}(s) = \mathbf{Y}^{-}(s;\mathbf{X}_0) = F_{ISO}(s)\mathbf{V}_{ISO}^{-}(s;\mathbf{X}_0). \tag{7.54}$$

This is the well known form of the classical relationship between $\mathbf{Y}^{-}(s)$ *and* $\mathbf{I}^{-}(s)$ *for the zero initial state vector,* $\mathbf{X}_{0^-} = \mathbf{0}_n$*, which is expressed via the*

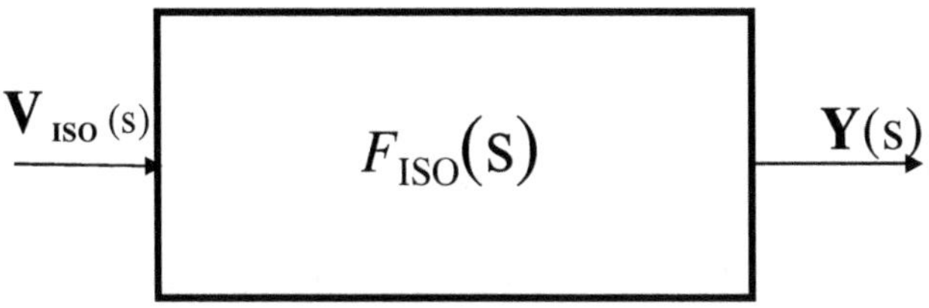

Figure 7.6: The full block of the ISO system.

system transfer function matrix $G_{ISO}(s)$,

$$\mathbf{Y}^{-}(s) = \mathbf{Y}^{-}(s;\mathbf{0}_n) = G_{ISO}(s)\mathbf{I}^{-}(s),\ \mathbf{X}_{0^-} = \mathbf{0}_n. \tag{7.55}$$

We emphasize once more that (7.55) is valid only for the zero initial state vector, $\mathbf{X}_{0^-} = \mathbf{0}_n$, *while (7.47), i.e., (7.54), holds for arbitrary initial conditions.*

Comment 129 *The system full transfer function matrix* $F_{ISO}(s)$ *incorporates the system transfer function matrix* $G_{ISO}(s)$:

$$F_{ISO}(s) = \left[G_{ISO}(s) \vdots G_{ISOx_0}(s)\right].$$

Note 130 ***The full block diagram technique is the generalization of the classical block diagram technique***

The action vector $\mathbf{V}_{ISO}(t;\ \mathbf{X}_0)$ *permits us to apply directly the block diagram technique if* $\mathbf{I}(t)$ *is replaced by* $\mathbf{V}_{ISO}(t;\ \mathbf{X}_0)$*; i.e.,* $\mathbf{I}(s)$ *is replaced by* $\mathbf{V}_{ISO}^{-}(s;\ \mathbf{X}_0)$, *and* $F_{ISO}(s)$ *replaces* $G_{ISO}(s)$, *Fig. 7.6 [148].*

Note 131 *The system transfer function matrix* $G_{ISO}(s)$ *is invariant relative to a mathematical description of a fixed physical system. This does not apply to the system full transfer function matrix* $F_{ISO}(s)$ *in general, which depends on the choice of the state variables, i.e., on the choice of the state vector* $\mathbf{X}$. *This is clear from the equation (7.47). It shows the influence of the initial state vector* $\mathbf{X}_0$, *rather than the initial output vector* $\mathbf{Y}_0$, *on the output response. Therefore, the transfer function matrix* $G_{ISOx_0}(s)$ *relative to* $\mathbf{X}_0$ *shows how* $\mathbf{X}_0$ *influences the output vector* $\mathbf{Y}$. *The form of* $G_{ISOx_0}(s)$ *depends on the choice of* $\mathbf{X}$. *Let* $F_{IOISO}(s)$ *and* $G_{IOISO}(s)$ *be the full transfer function matrix and the transfer function obtained from the IO mathematical model of the given ISO system. Then*

$$F_{IOISO}(s) \neq F_{ISO}(s) \text{ and } G_{IOISO}(s) = G_{ISO}(s) \neq G_{ISOx_0}(s) \text{ in general.}$$

For a special class of the systems we can express $\mathbf{X}_0$ in terms of $\mathbf{I}_0$ and $\mathbf{Y}_0$. Recall $n \geq N$ so that the full rank of $C \in \mathfrak{R}^{Nxn}$ equals N. Hence, the full rank of $C^T C \in \mathfrak{R}^{nxn}$ is n. These facts imply that $det C^T C \neq 0$ is possible if, and only if, $N = n$. Consequently, we can solve uniquely (7.44) for $\mathbf{X}$ if, and only if, both $N = n$ and $det C \neq 0$.

Theorem 132 *If in a special case (subscript sp) $N = n$ and the matrix C is nonsingular, then the following statements hold:*

a) The full transfer function matrix $F_{ISOsp}(s)$ of the ISO system (7.43), (7.44) has the following form in this special case (the subscript sp):

$$F_{ISOsp}(s) = \left[C(sI_n - A)^{-1}B + D \vdots -C(sI_n - A)^{-1}C^{-1}D \vdots C(sI_n - A)^{-1}C^{-1} \right], \tag{7.56}$$

so that

$$\mathbf{Y}^{\mp}(s) = F_{ISOsp}(s) \begin{bmatrix} \mathbf{I}^{\mp}(s) \\ \mathbf{I}_{0\mp} \\ \mathbf{Y}_{0\mp} \end{bmatrix} = F_{ISOsp}(s)\mathbf{V}^{\mp}_{ISOsp}(s). \tag{7.57}$$

b) The transfer function matrix $G_{ISOi_0sp}(s)$ relative to $\mathbf{I}_0$ of the system (7.43), (7.44) has the following form:

$$G_{ISOi_0sp}(s) = -C(sI_n - A)^{-1}C^{-1}D = -(sI_n - CAC^{-1})^{-1}D. \tag{7.58}$$

c) The transfer function matrix $G_{ISOy_0}(s)$ relative to $\mathbf{Y}_0$ of the system (7.43), (7.44) has the following special form:

$$G_{ISOy_0sp}(s) = C(sI_n - A)^{-1}C^{-1} = (sI_n - CAC^{-1})^{-1}. \tag{7.59}$$

Proof. Let $N = n$ and let C be nonsingular. This permits us to resolve (7.44) for $\mathbf{X}(t)$ at the initial moment $t_0 = 0^{\mp}$, i.e., to solve $\mathbf{Y}_{0\mp} = C\mathbf{X}_{0\mp} + D\mathbf{I}_{0\mp}$ for $\mathbf{X}_{0\mp}$,

$$\mathbf{X}_{0\mp} = -C^{-1}D\mathbf{I}_{0\mp} + C^{-1}\mathbf{Y}_{0\mp}.$$

This, (7.47) and (7.52) imply (7.56) through (7.59) ■

Comment 133 *We determined in general the action vector $\mathbf{V}_{ISO}(t) = \mathbf{V}_{ISO}(t; \mathbf{X}_0)$ and its left/right Laplace transform $\mathbf{V}^{\mp}_{ISO}(s; \mathbf{X}_0)$ in (7.53) for the ISO system so that (7.47) becomes in the special case for* $\det C \neq 0$,

$$\mathbf{Y}^{\mp}(s) = F_{ISOsp}(s)\mathbf{V}^{\mp}_{ISOsp}(s; \mathbf{I}_0; \mathbf{Y}_0). \tag{7.60}$$

From (7.57) it follows that the action vector function

$$\mathbf{V}_{ISOsp}(.) = \mathbf{V}_{ISOsp}(.; \mathbf{I}_0; \mathbf{Y}_0)$$

and its left/right Laplace transform $\mathbf{V}^{\mp}_{ISOsp}(s;\ \mathbf{I}_0;\ \mathbf{Y}_0)$ are determined for the ISO system in the special case by

$$\mathbf{V}_{ISOsp}(t; \mathbf{I}_0; \mathbf{Y}_0) = \left[\mathbf{I}^T(t) \vdots \delta(t)\mathbf{I}^T_{0-} \vdots \delta(t)\mathbf{Y}^T_{0-} \right]^T,$$

$$\mathbf{V}^{-}_{ISOsp}(s; \mathbf{I}_0; \mathbf{Y}_0) = \left[\left(\mathbf{I}^{-}(s)\right)^T \vdots \mathbf{I}^T_{0-} \vdots \mathbf{Y}^T_{0-} \right]^T. \tag{7.61}$$

This permits us to set (7.57) into the classical form (7.55),

$$\mathbf{Y}^{\mp}(s) = F_{ISOsp}(s)\mathbf{V}^{\mp}_{ISOsp}(s; \mathbf{I}_0; \mathbf{Y}_0). \tag{7.62}$$

Example 134 *Given* $G_{ISO}(s) = (s^2-1)^{-1}(s-1) = (s+1)^{-1}$. *Four different (state space, i.e., ISO) realizations* (A,B,C,D) *of* $G_{ISO}(s)$ *are determined in [8, p. 395]. We show first how to determine the full transfer function matrix and other transfer function matrices for each.*

$$1)\left\{A_1=\begin{bmatrix}0&1\\1&0\end{bmatrix},\ B_1=\begin{bmatrix}0\\1\end{bmatrix},\ C_1=\begin{bmatrix}-1\ \vdots\ 1\end{bmatrix},\ D_1=0\right\}$$
$$\Longrightarrow$$

$$Y_1^{\mp}(s)=\begin{bmatrix}-1\vdots 1\end{bmatrix}\begin{bmatrix}s&-1\\-1&s\end{bmatrix}^{-1}\left\{\begin{bmatrix}0\\1\end{bmatrix}I^{\mp}(s)+\mathbf{X}_{0\mp}\right\}=$$
$$=\left(s^2-1\right)^{-1}\begin{bmatrix}s-1\ \vdots\ 1-s\ \vdots\ s-1\end{bmatrix}\begin{bmatrix}I^{\mp}(s)\\X_{10\mp}\\X_{20\mp}\end{bmatrix},\ \mathbf{X}_{0\mp}=\begin{bmatrix}X_{10\mp}\\X_{20\mp}\end{bmatrix},$$

$$Y_1^{\mp}(s)=F_{ISO1}(s)\mathbf{V}_{ISO1}^{\mp}(s;\mathbf{X}_0),\ \mathbf{V}_{ISO1}^{\mp}(s;\mathbf{X}_0)=\begin{bmatrix}I^{\mp}(s)\\\mathbf{X}_{0\mp}\end{bmatrix}\Longrightarrow$$
$$F_{ISO1}(s)=\left(s^2-1\right)^{-1}\begin{bmatrix}s-1\ \vdots\ 1-s\ \vdots\ s-1\end{bmatrix}=$$
$$=\begin{bmatrix}C_1(sI_2-A_1)^{-1}B_1+D_1\ \vdots\ C_1(sI_2-A_1)^{-1}\end{bmatrix},$$
$$\Longrightarrow$$
$$F_{ISO1nd}(s)=\begin{bmatrix}\dfrac{1}{s+1}\ \vdots\ -\dfrac{1}{s+1}\ \vdots\ \dfrac{1}{s+1}\end{bmatrix},$$

$$G_{ISO1}(s)=\left[C_1(sI_2-A_1)^{-1}B_1+D_1\right]=\frac{s-1}{s^2-1}\Longrightarrow G_{ISO1nd}(s)=\frac{1}{s+1},$$
$$G_{ISOx_01}(s)=C_1(sI_2-A_1)^{-1}=\frac{\begin{bmatrix}1-s\ \vdots\ -1+s\end{bmatrix}}{s^2-1},$$
$$F_{ISOIS1}(s)=\begin{bmatrix}(sI_2-A_1)^{-1}B_1\ \vdots\ (sI_2-A_1)^{-1}\end{bmatrix}=\frac{\begin{bmatrix}1&s&1\\s&1&s\end{bmatrix}}{s^2-1},$$
$$G_{ISOIS1}(s)=(sI_2-A_1)^{-1}B_1=\frac{\begin{bmatrix}1\\s\end{bmatrix}}{s^2-1},$$
$$G_{ISOSS1}(s)=(sI_2-A_1)^{-1}=\frac{\begin{bmatrix}s&1\\1&s\end{bmatrix}}{s^2-1}.$$

We cannot show the influence of the initial output value $Y_{0\mp}$ *on the system response because we cannot express the state variables* X_1 *and* X_2 *in terms of the output* Y_1 *due to* $Y_1=-X_1+X_2$, *i.e., due to rank* $C_1=1=M_1<2=n_1$.

Notice that the given ISO system description, i.e.,

$$\begin{bmatrix} \frac{dX_1}{dt} \\ \frac{dX_2}{dt} \end{bmatrix} = \begin{bmatrix} 0 & 1 \\ 1 & 0 \end{bmatrix} \begin{bmatrix} X_1 \\ X_2 \end{bmatrix} + \begin{bmatrix} 0 \\ 1 \end{bmatrix} I,\ Y_1 = \begin{bmatrix} -1 \vdots 1 \end{bmatrix} \begin{bmatrix} X_1 \\ X_2 \end{bmatrix},$$

allows

$$X_1^{(1)} = X_2,\ X_2^{(1)} = X_1 + I,$$
$$Y_1 = -X_1 + X_2 \Longrightarrow$$
$$Y_1^{(1)} = -X_1^{(1)} + X_2^{(1)} = -X_2 + X_1 + I = -Y_1 + I \Longrightarrow$$
$$Y_1^{(1)} + Y_1 = I.$$

This IO mathematical model of the given ISO system yields the IO full transfer function matrix $F_{IOISO1}(s)$ of the given ISO system,

$$\begin{aligned} sY_1^{\mp}(s) - Y_{10\mp} + Y_1^{\mp}(s) &= I^{\mp}(s) \Longrightarrow Y_1^{\mp}(s) = \frac{1}{s+1} \begin{bmatrix} 1 \vdots 1 \end{bmatrix} \begin{bmatrix} I^{\mp}(s) \\ Y_{10\mp} \end{bmatrix} = \\ &= F_{IOISO1}(s) \mathbf{V}_{IOISO1}^{\mp}(s; Y_{10\mp}) \Longrightarrow \end{aligned}$$

$$F_{IOISO1}(s) = \begin{bmatrix} \frac{1}{s+1} \vdots \frac{1}{s+1} \end{bmatrix} \neq F_{ISO1}(s),\ G_{IOISO1}(s) = \frac{1}{s+1} = G_{ISO1nd}(s),$$

$$\mathbf{V}_{IOISO1}^{\mp}(s; Y_{10\mp}) = \begin{bmatrix} I^{\mp}(s) \\ Y_{10\mp} \end{bmatrix} \neq \begin{bmatrix} I^{\mp}(s) \\ \mathbf{X}_{0\mp}^{T} \end{bmatrix} = \mathbf{V}_{ISO1}^{\mp}(s; \mathbf{X}_0).$$

$G_{IOISO1}(s)$ is the nondegenerate form $G_{ISO1nd}(s)$ of $G_{ISO1}(s)$. $F_{IOISO1}(s)$ is different from $F_{ISO1}(s)$ and from the nondegenerate form $F_{ISO1nd}(s)$ of $F_{ISO1}(s)$ that is

$$F_{ISO1nd}(s) = \begin{bmatrix} \frac{1}{s+1} \vdots -\frac{1}{s+1} \vdots \frac{1}{s+1} \end{bmatrix}.$$

Notice that

$$\begin{aligned} Y_1^{\mp}(s) = F_{ISO1}(s) \mathbf{V}_{ISO1}^{\mp}(s; \mathbf{X}_0) &= \begin{bmatrix} \frac{s-1}{s^2-1} \vdots -\frac{s-1}{s^2-1} \vdots \frac{s-1}{s^2-1} \end{bmatrix} \begin{bmatrix} I^{\mp}(s) \\ \mathbf{X}_{0\mp} \end{bmatrix} = \\ &= \begin{bmatrix} \frac{1}{s+1} \vdots \frac{1}{s+1} \vdots \frac{1}{s+1} \end{bmatrix} \begin{bmatrix} I^{\mp}(s) \\ -X_{10\mp} \\ X_{20\mp} \end{bmatrix}, \end{aligned}$$

and

$$Y_1 = -X_1 + X_2 \Longrightarrow Y_{10} = -X_{10} + X_{20}$$

imply

$$Y_1^{\mp}(s) = \begin{bmatrix} \frac{1}{s+1} \vdots \frac{1}{s+1} \end{bmatrix} \begin{bmatrix} I^{\mp}(s) \\ Y_{10\mp} \end{bmatrix} = F_{IOISO1}(s) \mathbf{V}_{IOISO1}^{\mp}(s; Y_{10\mp}).$$

This shows the equivalence between $F_{ISO1}(s)$ and $F_{IOISO1}(s)$ in this example,

$$F_{ISO1}(s)\mathbf{V}_{ISO1}^{\mp}(s;\mathbf{X}_0) = F_{IOISO1}(s)\mathbf{V}_{IOISO1}^{\mp}(s;Y_{10\mp}) = Y_1^{\mp}(s).$$

2) $\left\{A_2 = \begin{bmatrix} 0 & 1 \\ 1 & 0 \end{bmatrix},\ B_2 = \begin{bmatrix} -1 \\ 1 \end{bmatrix},\ C_2 = \begin{bmatrix} 0 \vdots 1 \end{bmatrix},\ D_2 = 0\right\}$
$\Longrightarrow$

$$Y_2^{\mp}(s) = \begin{bmatrix} 0 \vdots 1 \end{bmatrix} \begin{bmatrix} s & -1 \\ -1 & s \end{bmatrix}^{-1} \left\{ \begin{bmatrix} -1 \\ 1 \end{bmatrix} I^{\mp}(s) + \mathbf{X}_{0\mp} \right\} =$$

$$= \underbrace{\left(s^2-1\right)^{-1} \begin{bmatrix} s-1 \vdots 1 \vdots s \end{bmatrix}}_{F_{ISO2}(s)} \begin{bmatrix} I^{\mp}(s) \\ X_{10\mp} \\ X_{20\mp} \end{bmatrix},$$

$$Y_2^{\mp}(s) = F_{ISO2}(s)\mathbf{V}_{ISO2}^{\mp}(s;\mathbf{X}_0),\ \mathbf{V}_{ISO2}^{\mp}(s;\mathbf{X}_0) = \begin{bmatrix} I^{\mp}(s) \\ \mathbf{X}_{0\mp} \end{bmatrix} \Longrightarrow$$

$$F_{ISO2}(s) = \left(s^2-1\right)^{-1} \begin{bmatrix} s-1 \vdots 1 \vdots s \end{bmatrix} = F_{ISO2nd}(s),$$

$$G_{ISO2}(s) = \frac{s-1}{s^2-1},\ G_{ISO2nd}(s) = \frac{1}{s+1},$$

$$G_{ISOx_02}(s) = \frac{\begin{bmatrix} 1 \vdots s \end{bmatrix}}{s^2-1},\ F_{ISOIS2}(s) = \frac{\begin{bmatrix} 1-s & s & 1 \\ -1+s & 1 & s \end{bmatrix}}{s^2-1},$$

$$G_{ISOIS2}(s) = \frac{\begin{bmatrix} 1-s \\ -1+s \end{bmatrix}}{s^2-1},\ G_{ISOSS2}(s) = \frac{\begin{bmatrix} s & 1 \\ 1 & s \end{bmatrix}}{s^2-1}.$$

We find the IO system model as follows:

$$\frac{dX_1}{dt} = X_2 - I,\ \frac{dX_2}{dt} = X_1 + I,\ Y_2 = X_2 \Longrightarrow$$

$$\frac{dy_2}{dt} = X_1 + I \Longrightarrow \frac{d^2Y_2}{dt^2} = X_2 - I + I^{(1)} = Y_2 - I + I^{(1)} \Longrightarrow$$

$$Y_2^{(2)} - Y_2 = I^{(1)} - I.$$

This IO system model implies

$$s^2Y_2^{\mp}(s) - sY_{20\mp} - Y_{20\mp}^{(1)} - Y_2^{\mp}(s) = sI^{\mp}(s) - I_{0\mp} - I^{\mp}(s) \Longrightarrow$$

$$Y_2^{\mp}(s) = \frac{1}{s^2-1} \begin{bmatrix} s-1 \vdots -1 \vdots s \vdots 1 \end{bmatrix} \begin{bmatrix} I^{\mp}(s) \\ I_{0\mp} \\ Y_{20\mp} \\ Y_{20\mp}^{(1)} \end{bmatrix} =$$

$$= F_{IOISO2}(s)\mathbf{V}_{IOISO2}^{\mp}(s;I_0;\mathbf{Y}_{20\mp}^1) \Longrightarrow$$

$$F_{IOISO2}(s)=\left[\frac{s-1}{s^2-1}\,\vdots\,-\frac{1}{s^2-1}\,\vdots\,\frac{s}{s^2-1}\,\vdots\,\frac{1}{s^2-1}\right]\neq F_{ISO2}(s),$$

$$\mathbf{V}^{\mp}_{IOISO2}(s;I_0;\mathbf{Y}^1_{20\mp})=\begin{bmatrix} I^{\mp}(s)\\ I_{0\mp}\\ Y_{20\mp}\\ Y^{(1)}_{20\mp}\end{bmatrix}\neq\begin{bmatrix} I^{\mp}(s)\\ \mathbf{X}_{0\mp}\end{bmatrix}=\mathbf{V}^{\mp}_{ISO2}(s;\mathbf{X}_0),$$

$$G_{IOISO2}(s)=\frac{s-1}{s^2-1}=G_{ISO2}(s).$$

Besides,

$$Y_2^{\mp}(s)=\frac{1}{s^2-1}\left[s-1\,\vdots\,-1\,\vdots\,s\,\vdots\,1\right]\begin{bmatrix} I^{\mp}(s)\\ I_{0\mp}\\ Y_{20\mp}\\ Y^{(1)}_{20\mp}\end{bmatrix}=$$

$$=F_{IOISO2}(s)\mathbf{V}^{\mp}_{IOISO2}(s;I_0;\mathbf{Y}^1_{20\mp})=$$

$$=\frac{1}{s^2-1}\left[(s-1)\,I^{\mp}(s)\ -I_{0\mp}\ +\ sX_{20}\ +\ X_{10}+I_{0\mp}\right]=$$

$$=\frac{1}{s^2-1}\left[(s-1)\,I^{\mp}(s)\ +\ X_{10}\ +\ sX_{20}\ \right]=$$

$$=\left[\frac{s-1}{s^2-1}\,\vdots\,1\,\vdots\,s\right]\begin{bmatrix} I^{\mp}(s)\\ X_{10}\\ X_{20}\end{bmatrix}=F_{ISO2}(s)\mathbf{V}^{\mp}_{ISO2}(s;\mathbf{X}_0).$$

This shows the equivalence between $F_{ISO2}(s)$ and $F_{IOISO2}(s)$ in this case.

3) $\left\{A_3=\begin{bmatrix}1&0\\0&-1\end{bmatrix},\ B_3=\begin{bmatrix}0\\1\end{bmatrix},\ C_3=\left[0\vdots 1\right],\ D_3=0\right\}$ $\Longrightarrow$

$$Y_3^{\mp}(s)=\left[0\vdots 1\right]\begin{bmatrix}s-1&0\\0&s+1\end{bmatrix}^{-1}\left\{\begin{bmatrix}0\\1\end{bmatrix}I^{\mp}(s)+\mathbf{X}_{0\mp}\right\}=$$

$$=\left(s^2-1\right)^{-1}\left[s-1\vdots 0\vdots s-1\right]\begin{bmatrix} I^{\mp}(s)\\ X_{10\mp}\\ X_{20\mp}\end{bmatrix}_{X_{20\mp}=Y_{30\mp}}$$

$$=\underbrace{\left(s^2-1\right)^{-1}\left[s-1\,\vdots\,0\,\vdots\,s-1\right]}_{F_{ISO3}(s)}\begin{bmatrix} I^{\mp}(s)\\ X_{10\mp}\\ Y_{30\mp}\end{bmatrix},$$

$$F_{ISO3}(s)=\left(s^2-1\right)^{-1}\left[s-1\,\vdots\,0\,\vdots\,s-1\right]\Longrightarrow$$

$$F_{ISO3nd}(s)=\left[\frac{1}{s+1}\,\vdots\,0\,\vdots\,\frac{1}{s+1}\right].$$

$$Y_3^{\mp}(s) = F_{ISO3}(s)\mathbf{V}_{ISO3}^{\mp}(s;\mathbf{X}_0),\ \mathbf{V}_{ISO3}^{\mp}(s;\mathbf{X}_0) = \begin{bmatrix} I^{\mp}(s) \\ \mathbf{X}_{0\mp} \end{bmatrix} \Longrightarrow$$

$$F_{ISO3}(s) = \left(s^2-1\right)^{-1}\left[s-1 \vdots 0 \vdots s-1\right] \Longrightarrow$$

$$G_{ISO3}(s) = \frac{s-1}{s^2-1},\ G_{ISOx_03}(s) = \frac{\left[0 \vdots s-1\right]}{s^2-1},$$

$$F_{ISOIS3}(s) = \frac{\begin{bmatrix} 0 & s+1 & 0 \\ -1+s & 0 & s-1 \end{bmatrix}}{s^2-1} \Longrightarrow$$

$$G_{ISOIS3}(s) = \frac{\begin{bmatrix} 0 \\ -1+s \end{bmatrix}}{s^2-1},\ G_{ISOSS3}(s) = \frac{\begin{bmatrix} s+1 & 0 \\ 0 & s-1 \end{bmatrix}}{s^2-1}.$$

The nondegenerate form $F_{ISO3nd}(s)$ of $F_{ISO3}(s)$ reads

$$F_{ISO3nd}(s) = \frac{1}{s+1}\left[1 \vdots 0 \vdots 1\right].$$

We determine now the IO model of the system,

$$\frac{dX_1}{dt} = X_1,\ \frac{dX_2}{dt} = -X_2 + I,\ Y_3 = X_2 \Longrightarrow$$

$$Y_3^{(1)} + Y_3 = I \Longrightarrow$$

$$Y_3^{\mp}(s) = \frac{1}{s+1}\left[1 \vdots 1\right]\begin{bmatrix} I^{\mp}(s) \\ Y_{30\mp} \end{bmatrix} = F_{IOISO3}(s)\mathbf{V}_{IOISO3}^{\mp}(s;Y_{30\mp}) \Longrightarrow$$

$$F_{IOISO3}(s) = \left[\frac{1}{s+1} \vdots \frac{1}{s+1}\right],\ \mathbf{V}_{IOISO3}^{\mp}(s;Y_{30\mp}) = \begin{bmatrix} I^{\mp}(s) \\ Y_{30\mp} \end{bmatrix}.$$

Notice that we can write

$$Y_3^{\mp}(s) = \underbrace{\frac{1}{s+1}\left[1 \vdots 1\right]}_{F_{IOISO3}(s)}\underbrace{\begin{bmatrix} I^{\mp}(s) \\ Y_{30\mp} \end{bmatrix}}_{\mathbf{V}_{IOISO3}^{\mp}(s;Y_{30\mp})} = \underbrace{\frac{1}{s+1}\left[1 \vdots 0 \vdots 1\right]}_{F_{ISO3nd}(s)}\begin{bmatrix} I^{\mp}(s) \\ X_{10\mp} \\ Y_{30\mp} \end{bmatrix} =$$

$$= F_{ISO3nd}(s)\begin{bmatrix} I^{\mp}(s) \\ X_{10\mp} \\ X_{20\mp} \end{bmatrix}.$$

We may conclude the equivalence between $F_{IOISO3}(s)$ and the nondegenerate form $F_{ISO3nd}(s)$ of $F_{ISO3}(s)$ in this example.

4) $\{A_4 = [-1],\ B_4 = [1],\ C_4 = [1],\ D_4 = 0\} \Longrightarrow$

$$Y_4^{\mp}(s) = [1]\,[s+1]^{-1}\left\{[1]\,I^{\mp}(s) + \mathbf{X}_{0\mp}\right\} =$$
$$= (s+1)^{-1}\begin{bmatrix} 1 \vdots 1 \end{bmatrix}\begin{bmatrix} I^{\mp}(s) \\ X_{10\mp} \end{bmatrix}_{X_{10\mp}=Y_{40\mp}} =$$
$$= (s+1)^{-1}\begin{bmatrix} 1 \vdots 1 \end{bmatrix}\begin{bmatrix} I^{\mp}(s) \\ Y_{40\mp} \end{bmatrix} =$$
$$= F_{ISO4}(s)\mathbf{V}_{ISO4}^{\mp}(s;\mathbf{X}_0),\ \mathbf{V}_{ISO4}^{\mp}(s;\mathbf{X}_0) = \begin{bmatrix} I^{\mp}(s) \\ X_{10\mp} \end{bmatrix} \Longrightarrow$$

$$F_{ISO4}(s) = (s+1)^{-1}\begin{bmatrix} 1 \vdots 1 \end{bmatrix} = F_{IO4}(s),$$
$$G_{ISO4}(s) = \frac{1}{s+1},$$

$$G_{ISOx_04}(s) = \frac{1}{s+1},\ F_{ISOIS4}(s) = \frac{\begin{bmatrix} 1 \vdots 1 \end{bmatrix}}{s+1},$$
$$G_{ISOIS4}(s) = \frac{1}{s+1},\ G_{ISOSS4}(s) = \frac{1}{s+1}.$$

When we replace $\mathbf{X}_{0\mp} = (X_{0\mp}) = (X_{10\mp})$ *by* $Y_{40\mp}$ *due to* $X_{10\mp} = Y_{40\mp}$*, then formally, there is not an explicit influence of the initial state variable on the system output response.*

The state space model under (4) corresponds to the following first-order IO differential equation and the full transfer function matrix $F_{IO4}(s)$*:*

$$Y_4^{(1)}(t) + Y_4(t) = I(t) \Longrightarrow$$
$$(s+1)\,Y_4^{\mp}(s) - Y_{40\mp} = I^{\mp}(s) \Longrightarrow$$
$$Y_4^{\mp}(s) = \underbrace{(s+1)^{-1}\begin{bmatrix} 1 \vdots 1 \end{bmatrix}}_{F_{IOISO4}(s)}\begin{bmatrix} I^{\mp}(s) \\ Y_{40\mp} \end{bmatrix},\ F_{IOISO4}(s) = \frac{1}{s+1}\begin{bmatrix} 1 \vdots 1 \end{bmatrix},$$

$$Y_4^{\mp}(s) = F_{IOISO4}(s)\mathbf{V}_{IOISO4}^{\mp}(s;Y_0),\ \mathbf{V}_{IOISO4}^{\mp}(s;Y_{40}) = \begin{bmatrix} I^{\mp}(s) \\ Y_{40\mp} \end{bmatrix} \Longrightarrow$$
$$F_{IOISO4}(s) = (s+1)^{-1}\begin{bmatrix} 1 \vdots 1 \end{bmatrix} = F_{ISO4}(s),\ G_{IOISO4} = \frac{1}{s+1} = G_{ISO4}.$$

Example 135 *R. E. Kalman considered a time-varying LC network in [201, Example 1, pp. 163-165], which is neither completely controllable nor observable. Without losing these properties we accept that all network parameters are* *constant, i.e.,* $C(t) \equiv C$ *and* $L(t) \equiv L$*, so that the system description reads*

$$\frac{dX_1}{dt} = -\frac{1}{L}X_1 + U \Longrightarrow X_1(s) = (s + \frac{1}{L})^{-1}\left[X_{10} + U(s)\right],$$
$$\frac{dX_2}{dt} = -\frac{1}{L}X_2 \Longrightarrow X_2(s) = (s + \frac{1}{L})^{-1}X_{20},$$

$$Y = \frac{2}{L}X_2 + U \Longrightarrow Y(s) = \frac{2}{L}(s + \frac{1}{L})^{-1}X_{20} + U(s) \Longrightarrow$$
$$Y(s) = \left[1 \vdots \frac{2}{L}(s + \frac{1}{L})^{-1}\right]\begin{bmatrix} U(s) \\ X_{20} \end{bmatrix} \Longrightarrow$$

$$G_{ISO}(s) \equiv 1,\; F_{ISO}(s) = \left[1 \vdots \frac{2}{L}\left(s + \frac{1}{L}\right)^{-1}\right].$$

The transfer function $G_{ISO}(s)$ *leads to the conclusion that the system is static. However, the full transfer function matrix* $F_{ISO}(s)$ *shows that the system is dynamic. If we write* $F_{ISO}(s)$ *in the form*

$$F_{ISO}(s) = (s + \frac{1}{L})^{-1}\left[s + \frac{1}{L} \vdots \frac{2}{L}\right],$$

then

$$G_{ISO}(s) = \frac{s + \frac{1}{L}}{s + \frac{1}{L}} \equiv 1 = G_{ISOnd}(s)$$

shows also that the system is dynamic, and that it is not completely controllable and observable. $F_{ISO}(s)$ *is not either degenerate or reducible, while* $G_{ISO}(s)$ *is both degenerate and reducible.*

Example 136 *We use the following ISO system from [201, Example 8, pp. 188,189]:*

$$\frac{d\mathbf{X}}{dt} = \begin{bmatrix} 0 & 1 & 0 \\ 5 & 0 & 2 \\ -2 & 0 & -2 \end{bmatrix}\mathbf{X} + \begin{bmatrix} 0 \\ 0 \\ 0.5 \end{bmatrix} I,$$
$$Y = \left[-2 \vdots 1 \vdots 0\right]\mathbf{X}.$$

We apply the last equation of (7.52)

$$F_{ISO}(s) = \left[det(sI_n - A)\right]^{-1} \bullet$$
$$\bullet \left[Cadj(sI_n - A)B + Ddet(sI_n - A) \vdots Cadj(sI_n - A)\right] =$$

$$= \left(s^3 + 2s^2 - 5s - 6\right)^{-1} \bullet$$
$$\bullet \left[s-2 \vdots -2s^2 + s + 6 \vdots s^2 - 4 \vdots 2s - 4 \right] =$$
$$= \left[(s+1)(s-2)(s+3) \right]^{-1} \bullet$$
$$\bullet \left[s-2 \vdots (s-2)(-2s-3) \vdots (s-2)(s+2) \vdots 2(s-2) \right] =$$
$$= \frac{s-2}{s-2} \left[(s+1)(s+3) \right]^{-1} \left[1 \vdots -2s-3 \vdots s+2 \vdots 2 \right] \Longrightarrow$$

$$G_{ISO}(s) = \left[det(sI_n - A) \right]^{-1} \left[Cadj(sI_n - A)B + Ddet(sI_n - A) \right] =$$
$$= \left(s^3 + 2s^2 - 5s - 6\right)^{-1} (s-2) = \frac{s-2}{s-2} \left[(s+1)(s+3) \right]^{-1}.$$

Since $G_{ISO}(s)$ is reducible, i.e., since it degenerates to

$$G_{ISOird}(s) = \frac{1}{(s+1)(s+3)} = G_{ISOnd}(s),$$

it follows that the system is not completely controllable and observable. In this example $F_{ISO}(s)$ is also both reducible and degenerate. After cancelling the same zero and pole at $s = 2$, we determine the nondegenerate form $F_{ISOnd}(s)$ of $F_{ISO}(s)$,

$$F_{ISOnd}(s) = \frac{\left[1 \vdots -2s-3 \vdots s+2 \vdots 2 \right]}{(s+1)(s+3)}.$$

It is also the irreducible form $F_{ISOird}(s)$ of $F_{ISO}(s)$,

$$F_{ISOird}(s) = F_{ISOnd}(s).$$

7.3 *F(s)* of the *IO* control system

We will broaden the concept of the system full transfer function matrix to the control systems in what follows.

7.3.1 *F(s)* of the *IO* plant

The *IO* plant $\mathcal{P}$ is described by (2.15) (Subsection 2.1.2),

$$A_P^{(\nu)} \mathbf{Y}^{\nu}(t) = C_{Pu}^{(\mu_{Pu})} \mathbf{U}^{\mu_{Pu}}(t) + D_{Pd}^{(\mu_{Pd})} \mathbf{D}^{\mu_{Pd}}(t), \ \det A_{P\nu} \neq 0, \ \forall t \in \mathfrak{T}_0,$$
$$A_P^{(\nu)} \in \mathfrak{R}^{N \times (\nu+1)N}, \ C_{Pu}^{(\mu_{Pu})} \in \mathfrak{R}^{N \times (\mu_{Pu}+1)r}, \ D_{Pd}^{(\mu_{Pd})} \in \mathfrak{R}^{N \times (\mu_{Pd}+1)d},$$
$$\nu \geq \max\left\{\mu_{Pd}, \ \mu_{Pu}\right\}. \quad (7.63)$$

Let us determine its full transfer function matrix. We apply Laplace transform to (7.63) and use the definitions of $S_i^{(k)}(s)$ (7.23) and $Z_k^{(\varsigma-1)}(s)$ (7.24) (Subsection 7.2.1):

$$\mathfrak{L}\left\{A_P^{(\nu)}\mathbf{Y}^{\nu}(t)\right\} = \mathfrak{L}\left\{C_{Pu}^{(\mu_{Pu})}\mathbf{U}^{\mu_{Pu}}(t) + D_{Pd}^{(\mu_{Pd})}\mathbf{D}^{\mu_{Pd}}(t)\right\} \Longrightarrow$$

$$\begin{gathered} A_P^{(\nu)}S_N^{(\nu)}(s)\mathbf{Y}(s) - A_P^{(\nu)}Z_N^{(\nu-1)}(s)\mathbf{Y}_0^{\nu-1} = \\ = C_{Pu}^{(\mu_{Pu})}S_r^{(\mu_{Pu})}(s)\mathbf{U}(s) - C_{Pu}^{(\mu_{Pu})}Z_r^{(\mu_{Pu}-1)}(s)\mathbf{U}_0^{\mu_{Pu}-1} + \\ +D_{Pd}^{(\mu_{Pd})}S_d^{(\mu_{Pd})}(s)\mathbf{D}(s) - D_{Pd}^{(\mu_{Pd})}Z_d^{(\mu_{Pd}-1)}(s)\mathbf{D}_0^{\mu_{Pd}-1} \Longrightarrow \end{gathered} \tag{7.64}$$

$$\mathbf{Y}^{\mp}(s) = F_{IOP}(s)\mathbf{V}_{IOP}(s) = \Bigg(\underbrace{A_P^{(\nu)}S_N^{(\nu)}(s)}_{F_{IOPD}(s)}\Bigg)^{-1} \bullet$$

$$\bullet \underbrace{\begin{bmatrix} D_{Pd}^{(\mu_{Pd})}S_d^{(\mu_{Pd})}(s) \vdots C_{Pu}^{(\mu_{Pu})}S_r^{(\mu_{Pu})}(s) \vdots -D_{Pd}^{(\mu_{Pd})}Z_d^{(\mu_{Pd}-1)} \vdots \\ \vdots -C_{Pu}^{(\mu_{Pu})}Z_r^{(\mu_{Pu}-1)}(s) \vdots A_P^{(\nu)}Z_N^{(\nu-1)}(s) \end{bmatrix}}_{F_{IOPN}(s)} \underbrace{\begin{bmatrix} \mathbf{D}(s) \\ \mathbf{U}(s) \\ \mathbf{D}_0^{\mu_{Pd}-1} \\ \mathbf{U}_0^{\mu_{Pu}-1} \\ \mathbf{Y}_0^{\nu-1} \end{bmatrix}}_{\mathbf{V}_{IOP}(s)}. \tag{7.65}$$

$F_{IOPD}(s)$ is the denominator matrix polynomial of the IO plant full transfer function matrix $F_{IOP}(s)$,

$$F_{IOPD}(s) = A_P^{(\nu)}S_N^{(\nu)}(s).$$

$F_{IOPN}(s)$ is the numerator matrix polynomial of the IO plant full transfer function matrix $F_{IOP}(s)$,

$$F_{IOPN}(s) = \begin{bmatrix} \left(D_{Pd}^{(\mu_{Pd})}S_d^{(\mu_{Pd})}(s)\right)^T \\ \left(C_{Pu}^{(\mu_{Pu})}S_r^{(\mu_{Pu})}(s)\right)^T \\ -\left(D_{Pd}^{(\mu_{Pd})}Z_d^{(\mu_{Pd}-1)}\right)^T \\ -\left(C_{Pu}^{(\mu_{Pu})}Z_r^{(\mu_{Pu}-1)}(s)\right)^T \\ \left(A_P^{(\nu)}Z_N^{(\nu-1)}(s)\right)^T \end{bmatrix}^T$$

The IO plant full transfer function matrix $F_{IOP}(s)$ reads

$$F_{IOP}(s) = \left(A_P^{(\nu)} S_N^{(\nu)}(s)\right)^{-1} \bullet \begin{bmatrix} \left(D_{Pd}^{(\mu_{Pd})} S_d^{(\mu_{Pd})}(s)\right)^T \\ \left(C_{Pu}^{(\mu_{Pu})} S_r^{(\mu_{Pu})}(s)\right)^T \\ -\left(D_{Pd}^{(\mu_{Pd})} Z_d^{(\mu_{Pd}-1)}\right)^T \\ -\left(C_{Pu}^{(\mu_{Pu})} Z_r^{(\mu_{Pu}-1)}(s)\right)^T \\ \left(A_P^{(\nu)} Z_N^{(\nu-1)}(s)\right)^T \end{bmatrix}^T . \tag{7.66}$$

7.3.2 $F(s)$ of the IO controller

The IO controller (CR), (2.22), i.e., (2.23) (Subsection 2.1.3), can be described by

$$A_{CR}^{(\nu_C)} \mathbf{U}^{\nu_C}(t) = P_{CR}^{(\mu_{Cy})} \left(\mathbf{Y}_d^{\mu_{Cy}}(t) - \mathbf{Y}^{\mu_{Cy}}(t)\right), \ \det A_{CR\nu} \neq 0, \ t \in \mathfrak{T}$$
$$P_{CR}^{(\mu_{Cy})} \in \mathfrak{R}^{r\text{x}(\mu_{Cy}+1)N}, \ \mathbf{Y}_d - \mathbf{Y} = \boldsymbol{\varepsilon} = -\mathbf{y}. \tag{7.67}$$

Laplace transform of (7.67) yields

$$\mathfrak{L}\left\{A_{CR}^{(\nu_C)} \mathbf{U}^{\nu_C}(t)\right\} = \mathfrak{L}\left\{P_{CR}^{(\mu_{Cy})} \left(\mathbf{Y}_d^{\mu_{Cy}}(t) - \mathbf{Y}^{\mu_{Cy}}(t)\right)\right\} \Longrightarrow$$

$$A_{CR}^{(\nu_C)} S_r^{(\nu_C)}(s)\mathbf{U}(s) - A_{CR}^{(\nu_C)} Z_r^{(\nu_C-1)}(s)\mathbf{U}_0^{\nu_C-1} =$$
$$= P_{CR}^{(\mu_{Cy})} S_N^{(\mu_{Cy})}(s)\left[\mathbf{Y}_d(s) - \mathbf{Y}(s)\right] - P_{CR}^{(\mu_{Cy})} Z_N^{(\mu_{Cy}-1)}(s)\left(\mathbf{Y}_{d0}^{\nu-1} - \mathbf{Y}_0^{\nu-1}\right) \Longrightarrow$$

$$\mathbf{U}(s) = F_{IOCR}(s)\mathbf{V}_{IOCR}(s) = \left(\underbrace{A_{CR}^{(\nu_C)} S_r^{(\nu_C)}(s)}_{F_{IOCRD}(s)}\right)^{-1} \bullet$$
$$\bullet \underbrace{\left[P_{CR}^{(\mu_{Cy})} S_N^{(\mu_{Cy})}(s) \vdots A_{CR}^{(\nu_C)} Z_r^{(\nu_C-1)}(s) \vdots -P_{CR}^{(\mu_{Cy})} Z_N^{(\mu_{Cy}-1)}(s)\right]}_{F_{IOCRN}(s)} \bullet$$
$$\bullet \underbrace{\begin{bmatrix} \mathbf{Y}_d(s) - \mathbf{Y}(s) \\ \mathbf{U}_0^{\nu-1} \\ \mathbf{Y}_{d0}^{\nu-1} - \mathbf{Y}_0^{\nu-1} \end{bmatrix}}_{\mathbf{V}_{IOCR}(s)} . \tag{7.68}$$

$F_{IOCRD}(s)$ is the denominator matrix polynomial of the IO controller full transfer function matrix $F_{IOCR}(s)$,

$$F_{IOPD}(s) = A_{CR}^{(\nu_C)} S_r^{(\nu_C)}(s).$$

$F_{IOCRN}(s)$ is the numerator matrix polynomial of the IO controller full transfer function matrix $F_{IOCR}(s)$,

$$F_{IOCRN}(s) = \left[P_{CR}^{(\mu_{Cy})} S_N^{(\mu_{Cy})}(s) \vdots A_{CR}^{(\nu_C)} Z_r^{(\nu_C - 1)}(s) \vdots - P_{CR}^{(\mu_{Cy})} Z_N^{(\mu_{Cy}-1)}(s)\right]$$

The full transfer function matrix of the IO controller follows:

$$F_{IOCR}(s) = \left(A_{CR}^{(\nu_C)} S_r^{(\nu_C)}(s)\right)^{-1} \bullet$$
$$\bullet \left[P_{CR}^{(\mu_{Cy})} S_N^{(\mu_{Cy})}(s) \vdots A_{CR}^{(\nu_C)} Z_r^{(\nu_C - 1)}(s) \vdots - P_{CR}^{(\mu_{Cy})} Z_N^{(\mu_{Cy}-1)}(s)\right].$$

7.3.3 $F(s)$ of the IO control system

We close the loop by replacing $\mathbf{U}(s)$ from the right-hand side of (7.68) into the right-hand side of (7.64). Detailed calculation follows step-by-step in order to enable an easy checking and understanding the form of the result:

$$A_P^{(\nu)} S_N^{(\nu)}(s)\mathbf{Y}(s) - A_P^{(\nu)} Z_N^{(\nu-1)}(s)\mathbf{Y}_0^{\nu-1} =$$
$$= C_{Pu}^{(\mu_{Pu})} S_r^{(\mu_{Pu})}(s) \left(A_{CR}^{(\nu_C)} S_r^{(\nu_C)}(s)\right)^{-1} \bullet$$
$$\bullet \left[P_{CR}^{(\mu_{Cy})} S_N^{(\mu_{Cy})}(s) \vdots A_{CR}^{(\nu_C)} Z_r^{(\nu_C - 1)}(s) \vdots - P_{CR}^{(\mu_{Cy})} Z_N^{(\mu_{Cy}-1)}(s)\right] \bullet$$
$$\bullet \begin{bmatrix} \mathbf{Y}_d(s) - \mathbf{Y}(s) \\ \mathbf{U}_0^{\nu-1} \\ \mathbf{Y}_{d0}^{\nu-1} - \mathbf{Y}_0^{\nu-1} \end{bmatrix} -$$
$$-C_{Pu}^{(\mu_{Pu})} Z_r^{(\mu_{Pu}-1)}(s)\mathbf{U}_0^{\mu_{Pu}-1} +$$
$$+D_{Pd}^{(\mu_{Pd})} S_d^{(\mu_{Pd})}(s)\mathbf{D}(s) - D_{Pd}^{(\mu_{Pd})} Z_d^{(\mu_{Pd}-1)}(s)\mathbf{D}_0^{\mu_{Pd}-1} \Longrightarrow$$

$$A_P^{(\nu)} S_N^{(\nu)}(s)\mathbf{Y}(s) - A_P^{(\nu)} Z_N^{(\nu-1)}(s)\mathbf{Y}_0^{\nu-1} =$$
$$= C_{Pu}^{(\mu_{Pu})} S_r^{(\mu_{Pu})}(s) \left(A_{CR}^{(\nu_C)} S_r^{(\nu_C)}(s)\right)^{-1} \bullet$$
$$\bullet \left\{ \begin{array}{c} P_{CR}^{(\mu_{Cy})} S_N^{(\mu_{Cy})}(s)\left[\mathbf{Y}_d(s) - \mathbf{Y}(s)\right] \\ + A_{CR}^{(\nu_C)} Z_r^{(\nu_C-1)}(s)\mathbf{U}_0^{\nu-1} - \\ -P_{CR}^{(\mu_{Cy})} Z_N^{(\mu_{Cy}-1)}(s)\left[\mathbf{Y}_{d0}^{\nu-1} - \mathbf{Y}_0^{\nu-1}\right] \end{array} \right\} -$$
$$-C_{Pu}^{(\mu_{Pu})} Z_r^{(\mu_{Pu}-1)}(s)\mathbf{U}_0^{\mu_{Pu}-1} +$$
$$+D_{Pd}^{(\mu_{Pd})} S_d^{(\mu_{Pd})}(s)\mathbf{D}(s) - D_{Pd}^{(\mu_{Pd})} Z_d^{(\mu_{Pd}-1)}(s)\mathbf{D}_0^{\mu_{Pd}-1} \Longrightarrow$$

$$\left(A_P^{(\nu)} S_N^{(\nu)}(s) + C_{Pu}^{(\mu_{Pu})} S_r^{(\mu_{Pu})}(s) \left(A_{CR}^{(\nu_C)} S_r^{(\nu_C)}(s)\right)^{-1} P_{CR}^{(\mu_{Cy})} S_N^{(\mu_{Cy})}(s)\right) \mathbf{Y}(s) =$$

$$= D_{Pd}^{(\mu_{Pd})} S_d^{(\mu_{Pd})}(s)\mathbf{D}(s) - D_{Pd}^{(\mu_{Pd})} Z_d^{(\mu_{Pd}-1)}(s)\mathbf{D}_0^{\mu_{Pd}-1} +$$

$$+ C_{Pu}^{(\mu_{Pu})} S_r^{(\mu_{Pu})}(s) \left(A_{CR}^{(\nu_C)} S_r^{(\nu_C)}(s)\right)^{-1} P_{CR}^{(\mu_{Cy})} S_N^{(\mu_{Cy})}(s)\mathbf{Y}_d(s) +$$

$$+ \left\{ \begin{array}{c} C_{Pu}^{(\mu_{Pu})} S_r^{(\mu_{Pu})}(s) \left(A_{CR}^{(\nu_C)} S_r^{(\nu_C)}(s)\right)^{-1} A_{CR}^{(\nu_C)} Z_r^{(\nu_C-1)}(s) - \\ - \left[C_{Pu}^{(\mu_{Pu})} Z_r^{(\mu_{Pu}-1)}(s) \vdots O_{N,\nu-\mu_{Pu}} \right] \end{array} \right\} \mathbf{U}_0^{\nu-1} -$$

$$- C_{Pu}^{(\mu_{Pu})} S_r^{(\mu_{Pu})}(s) \left(A_{CR}^{(\nu_C)} S_r^{(\nu_C)}(s)\right)^{-1} P_{CR}^{(\mu_{Cy})} Z_N^{(\mu_{Cy}-1)}(s) \left(\mathbf{Y}_{d0}^{\nu-1} - \mathbf{Y}_0^{\nu-1}\right) +$$

$$+ A_P^{(\nu)} Z_N^{(\nu-1)}(s)\mathbf{Y}_0^{\nu-1} \Longrightarrow$$

$$\mathbf{Y}(s) = \left\langle \begin{array}{c} A_P^{(\nu)} S_N^{(\nu)}(s) + \\ + C_{Pu}^{(\mu_{Pu})} S_r^{(\mu_{Pu})}(s) \left(A_{CR}^{(\nu_C)} S_r^{(\nu_C)}(s)\right)^{-1} P_{CR}^{(\mu_{Cy})} S_N^{(\mu_{Cy})}(s) \end{array} \right\rangle^{-1} \bullet$$

$$\bullet \left(\begin{array}{c} D_{Pd}^{(\mu_{Pd})} S_d^{(\mu_{Pd})}(s)\mathbf{D}(s) - D_{Pd}^{(\mu_{Pd})} Z_d^{(\mu_{Pd}-1)}(s)\mathbf{D}_0^{\mu_{Pd}-1} + \\ + C_{Pu}^{(\mu_{Pu})} S_r^{(\mu_{Pu})}(s) \left(A_{CR}^{(\nu_C)} S_r^{(\nu_C)}(s)\right)^{-1} P_{CR}^{(\mu_{Cy})} S_N^{(\mu_{Cy})}(s)\mathbf{Y}_d(s) + \\ + \left\{ \begin{array}{c} \left\{ \begin{array}{c} C_{Pu}^{(\mu_{Pu})} S_r^{(\mu_{Pu})}(s) \left(A_{CR}^{(\nu_C)} S_r^{(\nu_C)}(s)\right)^{-1} \bullet \\ \bullet A_{CR}^{(\nu_C)} Z_r^{(\nu_C-1)}(s) \end{array} \right\} - \\ - \left[C_{Pu}^{(\mu_{Pu})} Z_r^{(\mu_{Pu}-1)}(s) \vdots O_{N,\nu-\mu_{Pu}} \right] \end{array} \right\} \mathbf{U}_0^{\nu-1} - \\ - C_{Pu}^{(\mu_{Pu})} S_r^{(\mu_{Pu})}(s) \left(A_{CR}^{(\nu_C)} S_r^{(\nu_C)}(s)\right)^{-1} P_{CR}^{(\mu_{Cy})} Z_N^{(\mu_{Cy}-1)}(s)\varepsilon_0^{\nu-1} + \\ + \left(A_P^{(\nu)} S_N^{(\nu)}(s)\right)^{-1} A_P^{(\nu)} Z_N^{(\nu-1)}(s)\mathbf{Y}_0^{\nu-1} \end{array} \right),$$

$$\Longrightarrow$$

$$\mathbf{Y}(s) = F_{IOCSy}(s)\mathbf{V}_{IOCS}(s) =$$

$$= \underbrace{\left[G_{CSd}(s) \vdots G_{CSYd}(s) \vdots G_{CSdo}(s) \vdots G_{CSuo}(s) \vdots G_{CS\varepsilon o}(s) \vdots G_{CSyo}(s) \right]}_{F_{IOCSy}(s)} \bullet$$

$$\bullet \underbrace{\begin{bmatrix} \mathbf{D}(s) \\ \mathbf{Y}_d(s) \\ \mathbf{D}_0^{\mu_{Pd}-1} \\ \mathbf{U}_0^{\nu-1} \\ \varepsilon_0^{\nu-1} \\ \mathbf{Y}_0^{\nu-1} \end{bmatrix}}_{\mathbf{V}_{IOCS}(s)}. \tag{7.69}$$

The full transfer function matrix $F_{CSy}(s)$ of the IO control system, with $\mathbf{Y}$ considered as its output vector, has the following form:

$$F_{IOCSy}(s) =$$
$$= \left[G_{IOCSd}(s) \vdots G_{IOCSyd}(s) \vdots G_{IOCSdo}(s) \vdots G_{IOCSuo}(s) \vdots G_{IOCS\varepsilon o}(s) \vdots G_{IOCSyo}(s) \right], \tag{7.70}$$

where

$$G_{IOCSd}(s) =$$
$$= \left(A_P^{(\nu)} S_N^{(\nu)}(s) + C_{Pu}^{(\mu_{Pu})} S_r^{(\mu_{Pu})}(s) \left(A_{CR}^{(\nu_C)} S_r^{(\nu_C)}(s) \right)^{-1} P_{CR}^{(\mu_{Cy})} S_N^{(\mu_{Cy})}(s) \right)^{-1} \bullet$$
$$\bullet D_{Pd}^{(\mu_{Pd})} S_d^{(\mu_{Pd})}(s), \tag{7.71}$$

$$G_{IOCSyd}(s) =$$
$$= \left(A_P^{(\nu)} S_N^{(\nu)}(s) + C_{Pu}^{(\mu_{Pu})} S_r^{(\mu_{Pu})}(s) \left(A_{CR}^{(\nu_C)} S_r^{(\nu_C)}(s) \right)^{-1} P_{CR}^{(\mu_{Cy})} S_N^{(\mu_{Cy})}(s) \right)^{-1} \bullet$$
$$\bullet C_{Pu}^{(\mu_{Pu})} S_r^{(\mu_{Pu})}(s) \left(A_{CR}^{(\nu_C)} S_r^{(\nu_C)}(s) \right)^{-1} P_{CR}^{(\mu_{Cy})} S_N^{(\mu_{Cy})}(s), \tag{7.72}$$

$$G_{IOCSdo}(s) =$$
$$= - \left(A_P^{(\nu)} S_N^{(\nu)}(s) + C_{Pu}^{(\mu_{Pu})} S_r^{(\mu_{Pu})}(s) \left(A_{CR}^{(\nu_C)} S_r^{(\nu_C)}(s) \right)^{-1} P_{CR}^{(\mu_{Cy})} S_N^{(\mu_{Cy})}(s) \right)^{-1} \bullet$$
$$\bullet D_{Pd}^{(\mu_{Pd})} Z_d^{(\mu_{Pd}-1)}(s), \tag{7.73}$$

$$G_{IOCSuo}(s) =$$
$$= \left(A_P^{(\nu)} S_N^{(\nu)}(s) + C_{Pu}^{(\mu_{Pu})} S_r^{(\mu_{Pu})}(s) \left(A_{CR}^{(\nu_C)} S_r^{(\nu_C)}(s) \right)^{-1} P_{CR}^{(\mu_{Cy})} S_N^{(\mu_{Cy})}(s) \right)^{-1} \bullet$$
$$\bullet \left\{ \begin{array}{c} C_{Pu}^{(\mu_{Pu})} S_r^{(\mu_{Pu})}(s) \left(A_{CR}^{(\nu_C)} S_r^{(\nu_C)}(s) \right)^{-1} A_{CR}^{(\nu_C)} Z_r^{(\nu_C-1)}(s) - \\ - \left[C_{Pu}^{(\mu_{Pu})} Z_r^{(\mu_{Pu}-1)}(s) \vdots O_{N,\nu-\mu_{Pu}} \right] \end{array} \right\}, \tag{7.74}$$

$$G_{IOCS\varepsilon o}(s) =$$
$$= - \left(A_P^{(\nu)} S_N^{(\nu)}(s) + C_{Pu}^{(\mu_{Pu})} S_r^{(\mu_{Pu})}(s) \left(A_{CR}^{(\nu_C)} S_r^{(\nu_C)}(s) \right)^{-1} P_{CR}^{(\mu_{Cy})} S_N^{(\mu_{Cy})}(s) \right)^{-1} \bullet$$
$$\bullet C_{Pu}^{(\mu_{Pu})} S_r^{(\mu_{Pu})}(s) \left(A_{CR}^{(\nu_C)} S_r^{(\nu_C)}(s) \right)^{-1} P_{CR}^{(\mu_{Cy})} Z_N^{(\mu_{Cy}-1)}(s), \tag{7.75}$$

$$G_{IOCSyo}(s) =$$
$$= \left(A_P^{(\nu)} S_N^{(\nu)}(s) + C_{Pu}^{(\mu_{Pu})} S_r^{(\mu_{Pu})}(s) \left(A_{CR}^{(\nu_C)} S_r^{(\nu_C)}(s)\right)^{-1} P_{CR}^{(\mu_{Cy})} S_N^{(\mu_{Cy})}(s)\right)^{-1} \bullet$$
$$\bullet \left(A_P^{(\nu)} S_N^{(\nu)}(s)\right)^{-1} A_P^{(\nu)} Z_N^{(\nu-1)}(s). \tag{7.76}$$

For the IO feedback control synthesis it is useful to know the full transfer function matrix $F_{IOCS\varepsilon}(s)$ of the control system, with the error vector ε considered as its output vector. We find it by starting with

$$\varepsilon(\mathbf{s}) = \mathbf{Y}_d(s) - \mathbf{Y}(s).$$

This and (7.69) imply $G_{IOCS\varepsilon yd}(s)$

$$\varepsilon(s) = \mathbf{Y}_d(s) - F_{IOCSy}(s)\mathbf{V}_{IOCS}(s) = F_{IOCS\varepsilon}(s)\mathbf{V}_{IOCS}(s) =$$
$$= \underbrace{\left[G_{IOCSd} \vdots \underbrace{\begin{array}{c} I_N- \\ -G_{IOCSyd} \end{array}}_{G_{IOCS\varepsilon yd}} \vdots G_{IOCSdo} \vdots G_{IOCSuo} \vdots G_{IOCS\varepsilon o} \vdots G_{IOCSyo} \right]}_{F_{IOCS\varepsilon}(s)} \bullet$$
$$\bullet \underbrace{\begin{bmatrix} \mathbf{D}(s) \\ \mathbf{Y}_d(s) \\ \mathbf{D}_0^{\mu_{Pd}-1} \\ \mathbf{U}_0^{\nu-1} \\ \varepsilon_0^{\nu-1} \\ \mathbf{Y}_0^{\nu-1} \end{bmatrix}}_{\mathbf{V}_{IOCS}(s)} \tag{7.77}$$

so that

$$F_{IOCS\varepsilon}(s) =$$
$$= \left[G_{IOCSd}(s) \vdots G_{IOCS\varepsilon yd}(s) \vdots G_{IOCSdo}(s) \vdots G_{IOCSuo}(s) \vdots G_{IOCS\varepsilon o}(s) \vdots G_{IOCSyo}(s) \right], \tag{7.78}$$

where

$$G_{IOCS\varepsilon yd}(s) = I_N - G_{IOCSyd}(s), \tag{7.79}$$

i.e.,

$$G_{IOCS\varepsilon yd}(s) = I_N -$$
$$- \left(A_P^{(\nu)} S_N^{(\nu)}(s) + C_{Pu}^{(\mu_{Pu})} S_r^{(\mu_{Pu})}(s) \left(A_{CR}^{(\nu_C)} S_r^{(\nu_C)}(s)\right)^{-1} P_{CR}^{(\mu_{Cy})} S_N^{(\mu_{Cy})}(s)\right)^{-1} \bullet$$
$$\bullet C_{Pu}^{(\mu_{Pu})} S_r^{(\mu_{Pu})}(s) \left(A_{CR}^{(\nu_C)} S_r^{(\nu_C)}(s)\right)^{-1} P_{CR}^{(\mu_{Cy})} S_N^{(\mu_{Cy})}(s). \tag{7.80}$$

We can put this into another form:

$$G_{IOCS\varepsilon yd}(s) =$$
$$= \left\langle A_P^{(\nu)} S_N^{(\nu)}(s) + C_{Pu}^{(\mu_{Pu})} S_r^{(\mu_{Pu})}(s) \left(A_{CR}^{(\nu_C)} S_r^{(\nu_C)}(s) \right)^{-1} P_{CR}^{(\mu_{Cy})} S_N^{(\mu_{Cy})}(s) \right\rangle^{-1} \bullet$$
$$\bullet \left\langle A_P^{(\nu)} S_N^{(\nu)}(s) + C_{Pu}^{(\mu_{Pu})} S_r^{(\mu_{Pu})}(s) \left(A_{CR}^{(\nu_C)} S_r^{(\nu_C)}(s) \right)^{-1} P_{CR}^{(\mu_{Cy})} S_N^{(\mu_{Cy})}(s) \right\rangle -$$
$$- \left\langle A_P^{(\nu)} S_N^{(\nu)}(s) + C_{Pu}^{(\mu_{Pu})} S_r^{(\mu_{Pu})}(s) \left(A_{CR}^{(\nu_C)} S_r^{(\nu_C)}(s) \right)^{-1} P_{CR}^{(\mu_{Cy})} S_N^{(\mu_{Cy})}(s) \right\rangle^{-1} \bullet$$
$$\bullet \left\langle C_{Pu}^{(\mu_{Pu})} S_r^{(\mu_{Pu})}(s) \left(A_{CR}^{(\nu_C)} S_r^{(\nu_C)}(s) \right)^{-1} P_{CR}^{(\mu_{Cy})} S_N^{(\mu_{Cy})}(s) \right\rangle,$$

$$G_{IOCS\varepsilon yd}(s) =$$
$$= \left\langle A_P^{(\nu)} S_N^{(\nu)}(s) + C_{Pu}^{(\mu_{Pu})} S_r^{(\mu_{Pu})}(s) \left(A_{CR}^{(\nu_C)} S_r^{(\nu_C)}(s) \right)^{-1} P_{CR}^{(\mu_{Cy})} S_N^{(\mu_{Cy})}(s) \right\rangle^{-1} \bullet$$
$$\bullet \left\{ \begin{array}{c} \left\langle A_P^{(\nu)} S_N^{(\nu)}(s) + C_{Pu}^{(\mu_{Pu})} S_r^{(\mu_{Pu})}(s) \left(A_{CR}^{(\nu_C)} S_r^{(\nu_C)}(s) \right)^{-1} P_{CR}^{(\mu_{Cy})} S_N^{(\mu_{Cy})}(s) \right\rangle - \\ - \left\langle C_{Pu}^{(\mu_{Pu})} S_r^{(\mu_{Pu})}(s) \left(A_{CR}^{(\nu_C)} S_r^{(\nu_C)}(s) \right)^{-1} P_{CR}^{(\mu_{Cy})} S_N^{(\mu_{Cy})}(s) \right\rangle \end{array} \right\}$$

$$G_{IOCS\varepsilon yd}(s) =$$
$$= \left\langle A_P^{(\nu)} S_N^{(\nu)}(s) + C_{Pu}^{(\mu_{Pu})} S_r^{(\mu_{Pu})}(s) \left(A_{CR}^{(\nu_C)} S_r^{(\nu_C)}(s) \right)^{-1} P_{CR}^{(\mu_{Cy})} S_N^{(\mu_{Cy})}(s) \right\rangle^{-1} \bullet$$
$$\bullet A_P^{(\nu)} S_N^{(\nu)}(s) \tag{7.81}$$

Exercise 137 *We leave for the exercise the determination of the full transfer function matrix of the control system of the IO plant controlled by the ISO controller.*

7.4 *F(s)* of the *ISO* control system

7.4.1 *F(s)* of the *ISO* plant

The *ISO* description (2.29), (2.30) of the plant (2.33), (2.34) is given in Subsection 2.2.2. It comprises the state differential equation and the algebraic output equation,

$$\frac{d\mathbf{X}_P(t)}{dt} = A_P\mathbf{X}_P(t) + B_P\mathbf{U}(t) + L_P\mathbf{D}(t),\ \mathbf{X}_P \in R^{n_P},\ \forall t \in \mathfrak{T}_0, \tag{7.82}$$
$$\mathbf{Y}(t) = C_P\mathbf{X}_P(t) + H_P\mathbf{U}(t) + D_P\mathbf{D}(t),\ C_P \in R^{N x n_P},\ \forall t \in \mathfrak{T}_0. \tag{7.83}$$

Their Laplace transforms follow. We apply Laplace transform to (7.82), (7.83):

$$\begin{aligned}\mathfrak{L}\left\{\frac{d\mathbf{X}_P(t)}{dt}\right\} &= \mathfrak{L}\left\{A_P\mathbf{X}_P(t)+B_P\mathbf{U}(t)+L_P\mathbf{D}(t)\right\},\\ \mathfrak{L}\left\{\mathbf{Y}(t)\right\} &= \mathfrak{L}\left\{C_P\mathbf{X}_P(t)+H_P\mathbf{U}(t)+D_P\mathbf{D}(t)\right\}\Longrightarrow\end{aligned}$$

$$\begin{aligned}s\mathbf{X}_P(s)-\mathbf{X}_{P0} &= A_P\mathbf{X}_P(s)+B_P\mathbf{U}(s)+L_P\mathbf{D}(s),\\ \mathbf{Y}(s) &= C_P\mathbf{X}_P(s)+H_P\mathbf{U}(s)+D_P\mathbf{D}(s)\Longrightarrow\end{aligned}$$

$$\begin{gathered}\mathbf{X}_P(s)=\left(sI_{n_P}-A_P\right)^{-1}\left[B_P\mathbf{U}(s)+L_P\mathbf{D}(s)+\mathbf{X}_{P0}\right],\\ \mathbf{Y}(s)=C_P\left(sI_{n_P}-A_P\right)^{-1}\left[B_P\mathbf{U}(s)+L_P\mathbf{D}(s)+\mathbf{X}_{P0}\right]+\\ H_P\mathbf{U}(s)+D_P\mathbf{D}(s)\Longrightarrow\end{gathered}$$

$$\begin{gathered}\mathbf{Y}(s)=\left[C_P\left(sI_{n_P}-A_P\right)^{-1}B_P+H_P\right]\mathbf{U}(s)\\ +\left[C_P\left(sI_{n_P}-A_P\right)^{-1}L_P+D_P\right]\mathbf{D}(s)+C_P\left(sI_{n_P}-A_P\right)^{-1}\mathbf{X}_{P0},\end{gathered}\tag{7.84}$$

i.e.,

$$\mathbf{Y}(s)=F_{ISOP}(s)\underbrace{\begin{bmatrix}\mathbf{D}(s)\\ \mathbf{U}(s)\\ \mathbf{X}_{P0}\end{bmatrix}}_{\mathbf{V}_{ISOP}(s)},\tag{7.85}$$

where the *ISO* plant full transfer function matrix reads

$$F_{ISOP}(s)=\left[G_{ISOPD}(s)\ \vdots\ G_{ISOPU}(s)\ \vdots\ G_{ISOPXo}(s)\right],\tag{7.86}$$

i.e.,

$$F_{ISOP}(s)=\begin{bmatrix}\left(\underbrace{C_P\left(sI_{n_P}-A_P\right)^{-1}L_P+D_P}_{G_{ISOPD}(s)}\right)^T\\ \left(\underbrace{C_P\left(sI_{n_P}-A_P\right)^{-1}B_P+H_P}_{G_{ISOPU}(s)}\right)^T\\ \left(\underbrace{C_P\left(sI_{n_P}-A_P\right)^{-1}}_{G_{ISOPXo}(s)}\right)^T\end{bmatrix}^T.\tag{7.87}$$

The denominator (scalar) polynomial $F_{ISOPD}(s)$ and the numerator matrix polynomial $F_{ISOPN}(s)$ of $F_{ISOP}(s)$ follow from (7.87):

$$F_{ISOP}(s) = F_{ISOPD}^{-1}(s)\, F_{ISOPN}(s)\,,\;\; F_{ISOPD}(s) = \det(sI_{n_P} - A_P)\,,$$

$$F_{ISOPN}(s) = \begin{bmatrix} \left(C_P adj\,(sI_{n_P} - A_P)\, L_P + \det(sI_{n_P} - A_P)\, D_P\right)^T \\ \left(C_P adj\,(sI_{n_P} - A_P)\, B_P + \det(sI_{n_P} - A_P)\, H_P\right)^T \\ \left(C_P adj\,(sI_{n_P} - A_P)\right)^T \end{bmatrix}^T .$$

7.4.2 *F(s)* of the *ISO* controller

From Subsection 2.2.3 follows the *ISO* description of the controller

$$\frac{d\mathbf{X}_{CR}(t)}{dt} = A_{CR}\mathbf{X}_{CR}(t) + B_{CR}\varepsilon(t), \tag{7.88}$$

$$\mathbf{Y}_{CR}(t) = C_{CR}\mathbf{X}_{CR}(t) + H_{CR}\varepsilon(t) = \mathbf{U}(t). \tag{7.89}$$

Laplace transform of (7.88), (7.89) provides $\mathfrak{L}\{\mathbf{U}(t)\} = \mathbf{U}(s)$:

$$\mathfrak{L}\left\{\frac{d\mathbf{X}_{CR}(t)}{dt}\right\} = \mathfrak{L}\{A_{CR}\mathbf{X}_{CR}(t) + B_{CR}\varepsilon(t)\}\,,$$
$$\mathfrak{L}\{\mathbf{Y}_{CR}(t)\} = \mathfrak{L}\{C_{CR}\mathbf{X}_{CR}(t) + H_{CR}\varepsilon(t)\} = \mathfrak{L}\{\mathbf{U}(t)\} \Longrightarrow$$

$$s\mathbf{X}_{CR}(s) - \mathbf{X}_{CR0} = A_{CR}\mathbf{X}_{CR}(s) + B_{CR}\varepsilon(s),$$
$$\mathbf{Y}_{CR}(s) = C_{CR}\mathbf{X}_{CR}(s) + H_{CR}\varepsilon(s) = \mathbf{U}(s) \Longrightarrow$$

$$\mathbf{X}_{CR}(s) = (sI_{n_c} - A_{CR})^{-1}\left[B_{CR}\varepsilon(s) + \mathbf{X}_{CR0}\right],$$
$$\mathbf{U}(s) = C_{CR}\mathbf{X}_{CR}(s) + H_{CR}\varepsilon(s) \Longrightarrow$$

$$\mathbf{X}_{CR}(s) = (sI_{n_c} - A_{CR})^{-1}\left[B_{CR}\varepsilon(s) + \mathbf{X}_{CR0}\right],$$
$$\mathbf{U}(s) = \left[C_{CR}\,(sI_{n_c} - A_{CR})^{-1}\, B_{CR} + H_{CR}\right]\varepsilon(s) +$$
$$+ C_{CR}\,(sI_{n_c} - A_{CR})^{-1}\,\mathbf{X}_{CR0}. \tag{7.90}$$

We can set the second equation in (7.90) in the following form:

$$\mathbf{U}(s) = \underbrace{\left[\; C_{CR}\,(sI_{n_c} - A_{CR})^{-1}\, B_{CR} + H_{CR} \quad C_{CR}\,(sI_{n_c} - A_{CR})^{-1} \;\right]}_{F_{ISOCR}(s)} \underbrace{\begin{bmatrix} \varepsilon(s) \\ \mathbf{X}_{CR0} \end{bmatrix}}_{\mathbf{V}_{ISOCR}(s)}.$$

The *ISO* controller full transfer function matrix $F_{ISOCR}(s)$ reads

$$F_{ISOCR}(s) = \left[\; C_{CR}\,(sI_{n_c} - A_{CR})^{-1}\, B_{CR} + H_{CR} \quad C_{CR}\,(sI_{n_c} - A_{CR})^{-1} \;\right],$$

i.e.,

$$F_{ISOCR}(s) = F_{ISOCRD}^{-1}(s)\, F_{ISOCRN}(s) = \left(\underbrace{\det(sI_{n_c} - A_{CR})}_{F_{ISOCRD}(s)} \right)^{-1} \bullet$$

$$\bullet \underbrace{\left[\begin{array}{c} \left(C_{CR} adj(sI_{n_c} - A_{CR}) B_{CR} + \det(sI_{n_c} - A_{CR}) H_{CR}\right)^T \\ \left(C_{CR} adj(sI_{n_c} - A_{CR})\right)^T \end{array} \right]^T}_{F_{ISOCRN}(s)}.$$

7.4.3 *F(s)* of the *ISO* control system

We close the loop by replacing $\mathbf{U}(s)$ from (7.90) into (7.84), and by noting that $\mathbf{Y}_P = \mathbf{Y}_{CS} = \mathbf{Y}$ and $\mathbf{Y}_{Pd} = \mathbf{Y}_{CSd} = \mathbf{Y}_d$ so that $\boldsymbol{\varepsilon} = \mathbf{Y}_{Pd} - \mathbf{Y} = \mathbf{Y}_d - \mathbf{Y}$:

$$\begin{gathered} \mathbf{Y}(s) = \left[C_P (sI_{n_P} - A_P)^{-1} B_P + H_P \right] \bullet \\ \bullet \left\{ \left[C_{CR} (sI_{n_c} - A_{CR})^{-1} B_{CR} + H_{CR} \right] \boldsymbol{\varepsilon}(s) + C_{CR} (sI_{n_c} - A_{CR})^{-1} \mathbf{X}_{CR0} \right\} + \\ + \left[C_P (sI_{n_P} - A_P)^{-1} L_P + D_P \right] \mathbf{D}(s) + C_P (sI_{n_P} - A_P)^{-1} \mathbf{X}_{P0} \Longrightarrow \end{gathered} \tag{7.91}$$

$$\begin{gathered} \left\langle \begin{array}{c} I_N + \left[C_P (sI_{n_P} - A_P)^{-1} B_P + H_P \right] \bullet \\ \bullet \left[C_{CR} (sI_{n_c} - A_{CR})^{-1} B_{CR} + H_{CR} \right] \end{array} \right\rangle \mathbf{Y}(s) = \\ = \left[C_P (sI_{n_P} - A_P)^{-1} B_P + H_P \right] \bullet \\ \bullet \left\{ \left[C_{CR} (sI_{n_c} - A_{CR})^{-1} B_{CR} + H_{CR} \right] \mathbf{Y}_d(s) + C_{CR} (sI_{n_c} - A_{CR})^{-1} \mathbf{X}_{CR0} \right\} + \\ + \left[C_P (sI_{n_P} - A_P)^{-1} L_P + D_P \right] \mathbf{D}(s) + C_P (sI_{n_P} - A_P)^{-1} \mathbf{X}_{P0} \Longrightarrow \end{gathered}$$

$$\begin{gathered} \mathbf{Y}(s) = \left\langle \begin{array}{c} I_N + \left[C_P (sI_{n_P} - A_P)^{-1} B_P + H_P \right] \bullet \\ \bullet \left[C_{CR} (sI_{n_c} - A_{CR})^{-1} B_{CR} + H_{CR} \right] \end{array} \right\rangle^{-1} \bullet \\ \bullet \left\langle \begin{array}{c} \bullet \left[C_P (sI_{n_P} - A_P)^{-1} B_P + H_P \right] \bullet \\ \bullet \left\{ \begin{array}{c} \left[C_{CR} (sI_{n_c} - A_{CR})^{-1} B_{CR} + H_{CR} \right] \mathbf{Y}_d(s) + \\ + C_{CR} (sI_{n_c} - A_{CR})^{-1} \mathbf{X}_{CR0} \end{array} \right\} + \\ + \left[C_P (sI_{n_P} - A_P)^{-1} L_P + D_P \right] \mathbf{D}(s) + C_P (sI_{n_P} - A_P)^{-1} \mathbf{X}_{P0} \end{array} \right\rangle \\ \Longrightarrow \end{gathered}$$

$$\mathbf{Y}(s) = \underbrace{\left[G_{ISOCSyd}(s) \vdots G_{ISOCSyyd}(s) \vdots G_{ISOCSyxcro}(s) \vdots G_{ISOCSyxpo}(s) \right]}_{F_{ISOCSy}(s)} \bullet \bullet \underbrace{\begin{bmatrix} \mathbf{D}(s) \\ \mathbf{Y}_d(s) \\ \mathbf{X}_{CR0} \\ \mathbf{X}_{P0} \end{bmatrix}}_{\mathbf{V}_{ISOCS}(s)}. \tag{7.92}$$

We have determined the full transfer function matrix $F_{ISOCSy}(s)$ of the overall control system with $\mathbf{Y}$ as the output vector,

$$F_{ISOCSy}(s) = = \left[G_{ISOCSyd}(s) \vdots G_{ISOCSyyd}(s) \vdots G_{ISOCSyxcro}(s) \vdots G_{ISOCSyxpo}(s) \right], \tag{7.93}$$

with the transfer function submatrices:

$$G_{ISOCSyd}(s) = \left\langle \begin{array}{c} I_N + \left[C_P \left(sI_{n_P} - A_P \right)^{-1} B_P + H_P \right] \bullet \\ \bullet \left[C_{CR} \left(sI_{n_c} - A_{CR} \right)^{-1} B_{CR} + H_{CR} \right] \end{array} \right\rangle^{-1} \bullet \bullet \left[C_P \left(sI_{n_P} - A_P \right)^{-1} L_P + D_P \right], \tag{7.94}$$

$$G_{ISOCSyyd}(s) = \left\langle \begin{array}{c} I_N + \left[C_P \left(sI_{n_P} - A_P \right)^{-1} B_P + H_P \right] \bullet \\ \bullet \left[C_{CR} \left(sI_{n_c} - A_{CR} \right)^{-1} B_{CR} + H_{CR} \right] \end{array} \right\rangle^{-1} \bullet \bullet \left[C_P \left(sI_{n_P} - A_P \right)^{-1} B_P + H_P \right] \bullet \bullet \left[C_{CR} \left(sI_{n_c} - A_{CR} \right)^{-1} B_{CR} + H_{CR} \right], \tag{7.95}$$

$$G_{ISOCSyxcro}(s) = = \left\langle \begin{array}{c} I_N + \left[C_P \left(sI_{n_P} - A_P \right)^{-1} B_P + H_P \right] \bullet \\ \bullet \left[C_{CR} \left(sI_{n_c} - A_{CR} \right)^{-1} B_{CR} + H_{CR} \right] \end{array} \right\rangle^{-1} \bullet \bullet \left[C_P \left(sI_{n_P} - A_P \right)^{-1} B_P + H_P \right] C_{CR} \left(sI_{n_c} - A_{CR} \right)^{-1}, \tag{7.96}$$

$$G_{ISOCSyxpo}(s) = \left\langle \begin{array}{c} I_N + \left[C_P \left(sI_{n_P} - A_P \right)^{-1} B_P + H_P \right] \bullet \\ \bullet \left[C_{CR} \left(sI_{n_c} - A_{CR} \right)^{-1} B_{CR} + H_{CR} \right] \end{array} \right\rangle^{-1} \bullet \bullet C_P \left(sI_{n_P} - A_P \right)^{-1}. \tag{7.97}$$

We calculate now Laplace transform of the output error vector from the preceding results and in view of $\boldsymbol{\varepsilon} = \mathbf{Y}_d - \mathbf{Y}$:

$$\boldsymbol{\varepsilon}(s) = \mathbf{Y}_d(s) - \mathbf{Y}(s) = \mathbf{Y}_d(s) - F_{ISOCSy}(s)\mathbf{V}_{ISOCS}(s) = \\ = \mathbf{Y}_d(s) - \\ - \left[G_{ISOCSyd}(s) \vdots G_{ISOCSyyd}(s) \vdots G_{ISOCSyxcro}(s) \vdots G_{ISOCSyxpo}(s) \right] \bullet \\ \bullet \begin{bmatrix} \mathbf{D}(s) \\ \mathbf{Y}_d(s) \\ \mathbf{X}_{CR0} \\ \mathbf{X}_{P0} \end{bmatrix} \Longrightarrow$$

$$\boldsymbol{\varepsilon}(s) = F_{ISOCS\varepsilon}(s)\mathbf{V}_{ISOCS}(s) \\ = \left[\underbrace{-G_{ISOCSyd}(s)}_{G_{ISOCS\varepsilon d}(s)} \vdots \underbrace{I_N - G_{ISOCSyyd}(s)}_{G_{ISOCS\varepsilon yd}} \vdots \underbrace{-G_{ISOCSyxcro}(s)}_{G_{ISOCS\varepsilon xcro}(s)} \vdots \underbrace{-G_{ISOCSyxpo}(s)}_{G_{ISOCS\varepsilon xpo}(s)} \right] \bullet \\ \bullet \begin{bmatrix} \mathbf{D}(s) \\ \mathbf{Y}_d(s) \\ \mathbf{X}_{CR0} \\ \mathbf{X}_{P0} \end{bmatrix} \Longrightarrow \tag{7.98}$$

$$G_{ISOCS\varepsilon d}(s) = -G_{ISOCSyd}(s),\ G_{ISOCS\varepsilon yd} = I_N - G_{ISOCSyyd}(s), \\ G_{ISOCS\varepsilon xcro}(s) = -G_{ISOCSyxcro}(s),\ G_{ISOCS\varepsilon xpo}(s) = -G_{ISOCSyxpo}(s), \tag{7.99}$$

$$F_{ISOCS\varepsilon}(s) = \\ = \left[G_{ISOCS\varepsilon d}(s) \vdots G_{ISOCS\varepsilon yd} \vdots G_{ISOCS\varepsilon xcro}(s) \vdots G_{ISOCS\varepsilon xpo}(s) \right]. \tag{7.100}$$

$$G_{ISOCS\varepsilon yd} = I_N - \\ - \left\langle \begin{array}{c} I_N + \left[C_P \left(sI_{n_P} - A_P \right)^{-1} B_P + H_P \right] \bullet \\ \bullet \left[C_{CR} \left(sI_{n_c} - A_{CR} \right)^{-1} B_{CR} + H_{CR} \right] \end{array} \right\rangle^{-1} \bullet \\ \bullet \left[C_P \left(sI_{n_P} - A_P \right)^{-1} B_P + H_P \right] \bullet \\ \bullet \left[C_{CR} \left(sI_{n_c} - A_{CR} \right)^{-1} B_{CR} + H_{CR} \right] \Longrightarrow$$

$$G_{ISOCS\varepsilon yd} = \left\langle \begin{array}{c} I_N + \left[C_P \left(sI_{n_P} - A_P \right)^{-1} B_P + H_P \right] \bullet \\ \bullet \left[C_{CR} \left(sI_{n_c} - A_{CR} \right)^{-1} B_{CR} + H_{CR} \right] \end{array} \right\rangle^{-1} \bullet$$

$$\bullet \left\{ \begin{array}{c} I_N + \left[C_P \left(sI_{n_P} - A_P \right)^{-1} B_P + H_P \right] \bullet \\ \bullet \left[C_{CR} \left(sI_{n_c} - A_{CR} \right)^{-1} B_{CR} + H_{CR} \right] - \\ - \left[C_P \left(sI_{n_P} - A_P \right)^{-1} B_P + H_P \right] \bullet \\ \bullet \left[C_{CR} \left(sI_{n_c} - A_{CR} \right)^{-1} B_{CR} + H_{CR} \right] \Longrightarrow \end{array} \right\} \Longrightarrow$$

$$G_{ISOCS\varepsilon yd} = \left\langle \begin{array}{c} I_N + \left[C_P \left(sI_{n_P} - A_P \right)^{-1} B_P + H_P \right] \bullet \\ \bullet \left[C_{CR} \left(sI_{n_c} - A_{CR} \right)^{-1} B_{CR} + H_{CR} \right] \end{array} \right\rangle^{-1} . \tag{7.101}$$

We can verify this as follows by starting with (7.91):

$$\varepsilon(s) = \mathbf{Y}_d(s) - \mathbf{Y}(s) = \mathbf{Y}_d(s) - \left[C_P \left(sI_{n_P} - A_P \right)^{-1} B_P + H_P \right] \bullet$$
$$\bullet \left\{ \begin{array}{c} \left[C_{CR} \left(sI_{n_c} - A_{CR} \right)^{-1} B_{CR} + H_{CR} \right] \varepsilon(s) + \\ C_{CR} \left(sI_{n_c} - A_{CR} \right)^{-1} \mathbf{X}_{CR0} \end{array} \right\} -$$
$$- \left[C_P \left(sI_{n_P} - A_P \right)^{-1} L_P + D_P \right] \mathbf{D}(s) - C_P \left(sI_{n_P} - A_P \right)^{-1} \mathbf{X}_{P0} \Longrightarrow$$

$$\left\langle \begin{array}{c} I_N + \left[C_P \left(sI_{n_P} - A_P \right)^{-1} B_P + H_P \right] \bullet \\ \bullet \left[C_{CR} \left(sI_{n_c} - A_{CR} \right)^{-1} B_{CR} + H_{CR} \right] \end{array} \right\rangle \varepsilon(s) =$$
$$= \mathbf{Y}_d(s) - \left[C_P \left(sI_{n_P} - A_P \right)^{-1} B_P + H_P \right] \left[C_{CR} \left(sI_{n_c} - A_{CR} \right)^{-1} \mathbf{X}_{CR0} \right] -$$
$$- \left[C_P \left(sI_{n_P} - A_P \right)^{-1} L_P + D_P \right] \mathbf{D}(s) - C_P \left(sI_{n_P} - A_P \right)^{-1} \mathbf{X}_{P0} \Longrightarrow$$

$$\varepsilon(s) = \left\langle \begin{array}{c} I_N + \left[C_P \left(sI_{n_P} - A_P \right)^{-1} B_P + H_P \right] \bullet \\ \bullet \left[C_{CR} \left(sI_{n_c} - A_{CR} \right)^{-1} B_{CR} + H_{CR} \right] \end{array} \right\rangle^{-1} \bullet$$
$$\bullet \left\langle \begin{array}{c} \mathbf{Y}_d(s) - \left[C_P \left(sI_{n_P} - A_P \right)^{-1} B_P + H_P \right] \left[C_{CR} \left(sI_{n_c} - A_{CR} \right)^{-1} \mathbf{X}_{CR0} \right] - \\ - \left[C_P \left(sI_{n_P} - A_P \right)^{-1} L_P + D_P \right] \mathbf{D}(s) - C_P \left(sI_{n_P} - A_P \right)^{-1} \mathbf{X}_{P0} \end{array} \right\rangle \tag{7.102}$$

$$\Longrightarrow$$

From this we deduce $G_{ISOCS\varepsilon yd}$:

$$G_{ISOCS\varepsilon yd} = \left\langle \begin{array}{c} I_N + \left[C_P \left(sI_{n_P} - A_P \right)^{-1} B_P + H_P \right] \bullet \\ \bullet \left[C_{CR} \left(sI_{n_c} - A_{CR} \right)^{-1} B_{CR} + H_{CR} \right] \end{array} \right\rangle^{-1} . \tag{7.103}$$

This verifies (7.101).

Exercise 138 *We leave for the exercise the determination of the full transfer function matrix of the control system of the ISO plant controlled by the IO controller.*

7.5 Conclusion: general form of *F(s)*

We refer to [148] for what follows.

For all considered systems, the following relationship holds between $\mathbf{Y}^{-}(s)$ and $\mathbf{V}^{-}(s)$ due to (7.25), (7.26), (7.34), (7.46), (7.47) and (7.54),

$$\mathbf{Y}^{-}(s) = F(s)\mathbf{V}^{-}(s),$$
$$F(s) \in \mathfrak{C}^{N\times W},\ \mathbf{V}^{-}(s) \in \mathfrak{C}^{W},\ (L+1)\,U = W. \tag{7.104}$$

The system full transfer function matrix $F(s)$ can be decomposed in the inverse $F_D^{-1}(s)$ of its denominator polynomial matrix $F_D(s)$ and in its numerator polynomial matrix $F_N(s)$,

$$F(s) = F_D^{-1}(s)F_N(s),$$
$$F_D(s) = A_D^{(J)}S_N^{(J)}(s) = \sum_{k=0}^{k=J} A_{Dk}s^k \in \mathfrak{C}^{N\times N},$$
$$A_{Dk} \in \mathfrak{R}^{N\times N},\ A_D^{(J)} \in \mathfrak{R}^{N\times(J+1)N},$$
$$F_N(s) = B_N^{(L)}S_W^{(L)}(s) = \sum_{k=0}^{k=L} B_{Nk}s^k \in \mathfrak{C}^{N\times W},$$
$$B_{Nk} \in \mathfrak{R}^{N\times U},\ B_N^{(L)} \in \mathfrak{R}^{N\times(L+1)U},\ (L+1)\,U = W, \tag{7.105}$$

where

$$F_D^{-1}(s) = [detF_D(s)]^{-1}\, adjF_D(s),$$
$$\Delta(s) = detF_D(s) = det\left[A_D^{(J)}S_N^{(J)}(s)\right] = det\left[\sum_{k=0}^{k=J} A_{Dk}s^k\right]. \tag{7.106}$$

$\Delta(s) = detF_D(s)$ is the characteristic polynomial of the system in general.

Conclusion 139 ***The full block diagram technique generalizes and extends the classical block diagram technique***

Laplace transform $\mathbf{V}^{-}(s)$ *of the generalized action vector* $\mathbf{V}(t)$ *incorporates both Laplace transform of the input vector and the vector of all initial conditions. It enables the generalization and the extension of the classical block diagram technique to incorporate all initial conditions.*

The system full transfer function matrix $F(s)$ *replaces the system transfer function matrix* $G(s)$*, and Laplace transform* $\mathbf{V}(s)$ *of the action vector* $\mathbf{V}(t)$ *replaces Laplace transform* $\mathbf{I}(s)$ *of the input vector* $\mathbf{I}(t)$*, Fig. 7.7 [148].*

The rules of the classic block diagram technique compose the algebra of the block diagrams. The book [148] establishes the analogous ***algebra of the full block diagrams.*** *It induces* ***the full block diagram technique.***

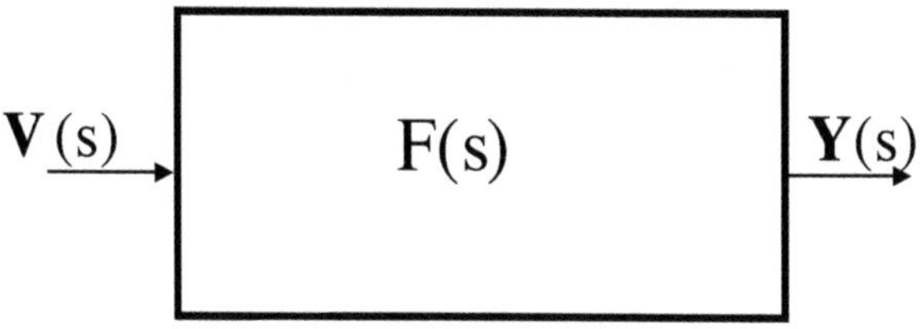

Figure 7.7: The full block of the *time*-invariant continuous-*time* system.

7.6 Physical meaning of *F(s)*

This section is from [148].

7.6.1 The *IO* system

The full transfer function matrix $F(s)$ has an important physical meaning, additional to that expressed in its definition. Let $\mathbf{1} = [1 \;\; 1 \; ... \; 1]^T$ be of the appropriate dimension. For the definition of Dirac unit impulse $\delta(.)$ see [8], [148].

Definition 140 *A matrix function $\Phi_{IO}(.) : \mathfrak{T} \longrightarrow \mathfrak{R}^{N\mathrm{x}[(\mu+1)M+\nu N]}$ is **the fundamental matrix function** of the IO system (2.9) (Section 2.1) if, and only if, it obeys both (i) and (ii) for an arbitrary input vector function $\mathbf{I}(.)$, and for arbitrary initial conditions $\mathbf{I}_{0^-}^{\mu-1}$ and $\mathbf{Y}_{0^-}^{\nu-1}$,*

$$(i)\ \mathbf{Y}(t;\mathbf{Y}_{0^-}^{\nu-1};\mathbf{I}) = \int_{0^-}^{t} \left\{ \Phi_{IO}(\tau) \begin{bmatrix} \mathbf{I}(t-\tau) \\ \delta(t-\tau)\mathbf{I}_{0^-}^{\mu-1} \\ \delta(t-\tau)\mathbf{Y}_{0^-}^{\nu-1} \end{bmatrix} \right\} d\tau,$$

$$\Phi_{IO}(t) = \left[\Gamma_{IO}(t) \vdots \Gamma_{IOi_0}(t) \vdots \Gamma_{IOy_0}(t) \right],$$

$$\Gamma_{IO}(t) \in \mathfrak{R}^{N\mathrm{x}M},\ \Gamma_{IOi_0}(t) \in \mathfrak{R}^{N\mathrm{x}\mu M},\ \Gamma_{IOy_0}(t) \in \mathfrak{R}^{N\mathrm{x}\nu N},$$

$$(ii)\ \Gamma_{IOi_0}(0^-)\mathbf{I}_{0^-}^{\mu-1} = \left[\int_{0^-}^{t=0^-} \Gamma_{IO}(\tau)\mathbf{I}(t-\tau)d\tau \vdots O_{N,(\mu-1)M} \right],$$

$$\Gamma_{IOy_0}(0^-) \equiv \left[I_N \vdots O_{N,(\nu-1)N} \right].$$

Note 141 *The equations under (i) of Definition 140 and the properties of $\delta(.)$ (for details see [8], [148]) yield*

$$\mathbf{Y}(t;\mathbf{Y}_{0^-}^{\nu-1};\mathbf{I}) = \left(\int_{0^-}^{t} \Gamma_{IO}(\tau)\mathbf{I}(t-\tau)d\tau \right) + \Gamma_{IOi_0}(t)\mathbf{I}_{0^-}^{\mu-1} + \Gamma_{IOy_0}(t)\mathbf{Y}_{0^-}^{\nu-1},\ t \in \mathfrak{T}_0. \tag{7.107}$$

Theorem 142 *(i) The fundamental matrix function $\Gamma_{IO}(.)$ of the IO system (2.9) is the inverse of the left Laplace transform of the system full transfer function matrix $F_{IO}(s)$,*

$$\Phi_{IO}(t) = \mathcal{L}^{-1}\left\{F_{IO}(s)\right\}. \tag{7.108}$$

(ii) The full transfer function matrix $F_{IO}(s)$ of the IO system (2.9) is the left Laplace transform of the system fundamental matrix $\Phi_{IO}(t)$,

$$F_{IO}(s) = \mathcal{L}^{-}\left\{\Phi_{IO}(t)\right\}. \tag{7.109}$$

Appendix J contains the proof.

This theorem expresses a physical meaning of the full transfer function matrix of the IO system (2.9).

Example 143 *Let us refer to Example 122 (Subsection 7.2.1). The given second-order $SISO$ IO system is described by*

$$Y^{(2)}(t) + Y^{(1)}(t) - 2Y(t) = 2I^{(2)}(t) - 14I^{(1)}(t) + 12I(t).$$

Its full transfer function matrix was found to read

$$F_{IO}(s) = \left[G_{IO}(s) \vdots G_{IOio}(s) \vdots G_{IOyo}(s)\right] =$$

$$= \begin{bmatrix} \underbrace{\dfrac{2(s-1)(s-6)}{(s-1)(s+2)}}_{G_{IO}(s)} \vdots \\ \left(\underbrace{\dfrac{-2(s-7)}{(s-1)(s+2)} \vdots \dfrac{-2}{(s-1)(s+2)}}_{G_{IOio}(s)}\right)^T \vdots \\ \left(\underbrace{\dfrac{s+1}{(s-1)(s+2)} \vdots \dfrac{1}{(s-1)(s+2)}}_{G_{IOyo}(s)}\right)^T \end{bmatrix}^T .$$

Notice that $F_{IO}(.)$ is only proper, not strictly proper, real rational matrix func-

tion. Its inverse Laplace transform $\mathcal{L}^{-1}\{F_{IO}(s)\}$,

$$\mathcal{L}^{-1}\{F_{IO}(s)\} = \mathcal{L}^{-1}\left\{\left[G_{IO}(s) \vdots G_{IOio}(s) \vdots G_{IOyo}(s)\right]\right\} =$$

$$= \begin{bmatrix} \underbrace{\mathcal{L}^{-1}\left\{\frac{2(s-1)(s-6)}{(s-1)(s+2)}\right\}}_{\mathcal{L}^{-1}\{G_{IO}(s)\}} \vdots \\ \left(\underbrace{\mathcal{L}^{-1}\left\{\frac{-2(s-7)}{(s-1)(s+2)}\right\} \vdots \mathcal{L}^{-1}\left\{\frac{-2}{(s-1)(s+2)}\right\}}_{\mathcal{L}^{-1}\{G_{IOio}(s)\}}\right)^T \vdots \\ \left(\underbrace{\mathcal{L}^{-1}\left\{\frac{s+1}{(s-1)(s+2)}\right\} \vdots \mathcal{L}^{-1}\left\{\frac{1}{(s-1)(s+2)}\right\}}_{\mathcal{L}^{-1}\{G_{IOyo}(s)\}}\right)^T \end{bmatrix},$$

determines the fundamental matrix function $\Phi_{IO}(.)$ *of the system in view of (7.107):*

$$\Phi_{IO}(t) = \left[\Gamma_{IO}(t) \vdots \Gamma_{IOi_0}(t) \vdots \Gamma_{IOy_0}(t)\right] =$$

$$= \begin{bmatrix} \underbrace{2\left[\delta(t) - 8e^{-2t}\right]}_{\Gamma_{IOi}(t)} \vdots \\ \left(\underbrace{\left(4e^t - 6e^{-2t}\right) \vdots -\frac{2}{3}\left(e^t - e^{-2t}\right)}_{\Gamma_{IOi_0}(t)}\right)^T \vdots \\ \left(\underbrace{\frac{1}{3}\left(2e^t + e^{-2t}\right) \vdots \frac{1}{3}\left(e^t - e^{-2t}\right)}_{\Gamma_{IOy_0}(t)}\right)^T \end{bmatrix}^T.$$

Notice that

$$\int_{0^-}^{t=0^-} \Gamma_{IO}(\tau)i(t-\tau)d\tau = \int_{0^-}^{t=0^-} 2\left[\delta(\tau) - 8e^{-2\tau}\right] i(t-\tau)d\tau =$$

$$= 2\int_{0^-}^{t=0^-} \delta(t)i(t-\tau)dt = 2i_{0^-},$$

$$\Gamma_{IOi_0}(0^-)\mathbf{I}^1(0^-) = \left[4e^t - 6e^{-2t} \vdots -\frac{2}{3}\left(e^t - e^{-2t}\right)\right]_{t=0^-} \mathbf{I}^1(0^-) =$$
$$= \left[-2 \vdots 0\right] \mathbf{I}^1(0^-) = -2i_{0^-},$$

$$\Gamma_{IOy_0}(0^-)\mathbf{Y}^1(0^-) = \left[\frac{1}{3}\left(2e^t + e^{-2t}\right) \vdots \frac{1}{3}\left(e^t - e^{-2t}\right)\right]_{t=0^-} \mathbf{Y}^1(0^-) =$$
$$= \left[\frac{1}{3}(3) \vdots \frac{1}{3}(0)\right] \mathbf{Y}^1(0^-) = \left[1 \vdots 0\right] \mathbf{Y}^1(0^-) = y_{0^-} \Longrightarrow$$

$$y(0^-; \mathbf{Y}^1_{0^-}; i) = \int_{0^-}^{t=0^-} \Gamma_{IO}(\tau)i(t-\tau)d\tau + \Gamma_{IOi_0}(0^-)\mathbf{I}^1(0^-) +$$
$$+ \Gamma_{IOy_0}(0^-)\mathbf{Y}^1(0^-) = 2i_{0^-} - 2i_{0^-} + y_{0^-} = y_{0^-}.$$

$\Gamma_{IO}(t) = 2\left[\delta(t) - 8e^{-2t}\right]$ *contains Dirac unit impulse* $\delta(t)$ *because it is inverse Laplace transform of the proper ratio*

$$\frac{2(s-1)(s-6)}{(s-1)(s+2)},$$

which implies that $F_{IO}(.)$ *is only proper, not strictly proper, real rational matrix function. However,* $\Gamma_{IOi_0}(t)$ *and* $\Gamma_{IOy_0}(t)$ *do not contain Dirac impulse. They are matrix exponential functions. Their Laplace transforms are strictly proper rational functions* $G_{IOio}(s)$ *and* $G_{IOyo}(s)$*, respectively.*

7.6.2 The *ISO* system

Definition 144 *A matrix function* $\Phi_{ISOIS}(.) : \mathfrak{T} \longrightarrow \mathfrak{R}^{nx(M+n)}$ *is* ***the IS fundamental matrix function*** *of the ISO system (2.29), (2.30) (Section (2.2)) if, and only if, it obeys both (i) and (ii):*

(i) $$\mathbf{X}(t; \mathbf{X}_{0^-}; \mathbf{I}) = \int_{0^-}^{t} \left\{\Phi_{ISOIS}(\tau)\begin{bmatrix} \mathbf{I}(t-\tau) \\ \delta(t-\tau)\mathbf{X}_{0^-} \end{bmatrix}\right\} d\tau =$$

$$= \int_{0^-}^{t} \Gamma_{ISOIS}(\tau)\mathbf{I}(t-\tau)d\tau + \Gamma_{ISOSS}(t)\mathbf{X}_{0^-},$$
$$\Gamma_{ISOIS}(t) \in \mathfrak{R}^{nxM}, \Gamma_{ISOSS}(t) \in \mathfrak{R}^{nxn}, \tag{7.110}$$

(ii) $\Gamma_{ISOIS}(0^-) = O_{n,M}, \Gamma_{ISOSS}(0^-) = I_n.$

It is well known [8], [29], [52], [198], [214], [215], [246], [268], [269] that

$$\Gamma_{ISOIS}(t) = e^{At}B,\ \Gamma_{ISOSS}(t) = e^{At}. \tag{7.111}$$

e^{At} can be computed from **the resolvent matrix** $(sI - A)^{-1}$ [240] of the ISO system (2.29), (2.30),

$$e^{At} = \mathcal{L}^{-1}\left\{(sI - A)^{-1}\right\}, \tag{7.112}$$

and vice versa,

$$(sI - A)^{-1} = \mathcal{L}\left\{e^{At}\right\}. \tag{7.113}$$

Theorem 145 *(i) The IS fundamental matrix function $\Phi_{ISOIS}(.)$ of the ISO system (2.29), (2.30) is the inverse of left Laplace transform of the system full IS transfer function matrix $F_{ISOIS}(s)$,*

$$\Phi_{ISOIS}(t) = \mathcal{L}^{-1}\left\{F_{ISOIS}(s)\right\}. \tag{7.114}$$

(ii) The full transfer function matrix $F_{ISOIS}(s)$ of the ISO system (2.29), (2.30) is left Laplace transform of the system fundamental matrix $\Phi_{ISOIS}(t)$,

$$F_{ISOIS}(s) = \mathcal{L}^{-}\left\{\Phi_{ISOIS}(t)\right\}. \tag{7.115}$$

The proof is in Appendix K.

Definition 146 *A matrix function $\Phi_{ISO}(.) : \mathfrak{T} \longrightarrow \mathfrak{R}^{N\times(M+n)}$ is **the ISO fundamental matrix function** of the ISO system (2.29), (2.30) if, and only if, it obeys both (i) and (ii) for an arbitrary input $\mathbf{I}(.)$ and for an arbitrary initial state vector $\mathbf{X}_{0^-}$:*

(i) $\mathbf{Y}(t;\mathbf{X}_{0^-};\mathbf{I}) = \int_{0^-}^{t}\left\{\Phi_{ISO}(\tau)\begin{bmatrix}\mathbf{I}(t-\tau)\\ \delta(t-\tau)\mathbf{X}_{0^-}\end{bmatrix}\right\}d\tau,$

(ii) $\Phi_{ISO}(t) = \left[\Gamma_{ISO}(t) \vdots \Gamma_{ISOx_o}(t)\right], \Gamma_{ISO}(t) \in \mathfrak{R}^{N\times M}, \Gamma_{ISOx_o}(t) \in \mathfrak{R}^{N\times n}.$

Note 147 *The equations of Definition 146 and the properties of $\delta(.)$ (for details see [8], [148]) permit*

$$\mathbf{Y}(t;\mathbf{X}_{0^-};\mathbf{I}) = \left(\int_{0^-}^{t}\Gamma_{ISO}(\tau)\mathbf{I}(t-\tau)d\tau\right) + \Gamma_{ISOx_o}(t)\mathbf{X}_{0^-},\ t \in \mathfrak{T}_0. \tag{7.116}$$

Note 148 *From the linear systems theory [8], [29], [52], [198], [214], [215], [246], [268], [269] and from (7.116) follow*

$$\Gamma_{ISO}(t) = Ce^{At}B + \delta(t)D,\ \Gamma_{ISOx_o}(t) = Ce^{At}. \tag{7.117}$$

Theorem 149 *(i) The fundamental matrix function $\Phi_{ISO}(.)$ of the ISO system (2.29), (2.30) is the inverse of left Laplace transform of the system full transfer function matrix $F_{ISO}(s)$,*

$$\Phi_{ISO}(t) = \mathcal{L}^{-1}\left\{F_{ISO}(s)\right\}. \tag{7.118}$$

(ii) The full transfer function matrix $F_{ISO}(s)$ of the ISO system (2.29), (2.30) is left Laplace transform of the system fundamental matrix $\Phi_{ISO}(t)$,

$$F_{ISO}(s) = \mathcal{L}^{-}\left\{\Phi_{ISO}(t)\right\}. \tag{7.119}$$

The proof is in Appendix L.

Theorem 149 shows a physical meaning of the full transfer function matrix of the ISO system (2.29), (2.30).

Note 150 *The equations (2.30) and (7.116), written for* $t = 0$, *and (7.117) establish the relationship between* $\mathbf{Y}_{0^-} = \mathbf{Y}(0^-)$, $\mathbf{I}_{0^-} = \mathbf{I}(0^-)$ *and* $\mathbf{X}_{0^-} = \mathbf{X}(0^-)$,

$$\mathbf{Y}_{0^-} = \int_{0^-}^{0^-} \Gamma_{ISO}(\tau)\mathbf{I}(t-\tau)d\tau + \Gamma_{ISOx_0}(0^-)\mathbf{X}_{0^-} = C\mathbf{X}_{0^-} + D\mathbf{I}_{0^-} \Longrightarrow$$

$$\int_{0^-}^{0^-} \Gamma_{ISO}(t)dt = \int_{0^-}^{0^-} \left(Ce^{At}B + \delta(t)D\right) dt = D,\ \Gamma_{ISOx_0}(0^-) = C. \quad (7.120)$$

Note 151 *In a special case,* $N = n$ *and* $det\Gamma_{ISOx_0}(0) = detC \neq 0$. *This permits us to resolve (7.120) for* $\mathbf{X}_0$ *in terms of* $\mathbf{Y}_0$,

$$\mathbf{X}_0 = C^{-1}\left(\mathbf{Y}_0 - D\mathbf{I}_0\right). \quad (7.121)$$

This equation transforms (7.46) and (7.47) (Subsection 7.2.2) into

$$\mathbf{Y}^-(s) = F_{ISOsp}(s)\left[\left(\mathbf{I}^-(s)\right)^T \vdots\ \mathbf{I}_0^T \vdots\ \mathbf{Y}_0^T\right]^T,$$

$$F_{ISOsp}(s) =$$

$$= \left[C(sI_n - A)^{-1}B + D \vdots (sI_n - CAC^{-1})^{-1}D \vdots (sI_n - CAC^{-1})^{-1}\right], \quad (7.122)$$

and

$$\mathbf{Y}(s) = F_{ISOsp}(s)\mathbf{V}_{ISOsp}(s),\ \mathbf{V}_{ISOsp}(s) = \left[\mathbf{I}^T(s) \vdots \mathbf{I}_0^T \vdots \mathbf{Y}_0^T\right]^T. \quad (7.123)$$

Equation (7.122) agrees with Equation (7.56) (Theorem 132, Subsection 7.2.2).

Part III

NOVEL CONTROL THEORIES: TRACKING AND TRACKABILITY

Chapter 8

Tracking theory

8.1 Tracking generally

The purpose of control of a dynamic system called *object* or *plant* is to force the plant *to behave sufficiently closely* to (*to follow* sufficiently precisely, *to track* sufficiently accurately) its *desired output behavior* over some, usually prespecified, *time* interval and under real (usually unpredictable and unknown) both external (input) actions and initial conditions [216, pp. 121-127].

The very, the primary, goal of control is to assure that the controlled plant exhibits a requested kind of **output tracking** that we will call, for short, **tracking** in this special sense. However, the term *tracking* in the wide sense concerns all kinds of the plant real output vector *following* its desired output vector.

Historically considered, tracking studies started as the *servomechanism* or *servosystem* theory in the general sense. Among its pioneers are L. A. MacColl, 1945 [233]; H. Lauer, R. Lesnick and L. E. Matson, 1947 [219]; J. C. West, 1953 [317]; H. Chestnut and R. W. Mayer, 1955 [33]; I. Flügge-Lotz and C. F. Taylor, 1956 [60]; J. C. Lozier, 1956 [231]; and G. S. Brown and D. P. Campbell, 1948 [22]. A. I. Talkin used the term "servo tracking" in 1961 [309]. The name *servomechanism* or *servosystem* signifies the controller that should force the plant output, or forces the controlled plant output, *to follow*, i.e., *to track*, its, in general, *time*-varying desired output rather than to track only a constant desired output. The latter is the purpose of the feedback controller (the controller in the closed-loop control system) called classically *regulator*.

In the control literature tracking has been mainly and largely studied as the problem of the zero steady state error problem. This means that the control synthesis should assure that the control forces the plant real output (or state) to approach asymptotically the plant desired output (or state) as *time* t tends to infinity, respectively. The tracking studies started in this sense, then the control studies began. They have been known under different names such as studies of servomechanisms/servosystems, or of regulation systems, or of control systems in general comprising the preceding ones, i.e., as studies on the servomechanism/servosystem problem, or studies on the regulation problem, or on the control problem in general incorporating the preceding ones. The problem of the zero steady state error has been commonly treated as a part of stability and stabilization studies, which might be a reason for which the control need for the tracking theory in its own right was not recognized until 1980 [145], [146], [280], [281]. During his first visit to Belgrade B. Porter presented the problem of tracking of nonlinear Lurie-Postnikov systems to the author and proposed the common research that resulted in those papers and in [278].

The notion, the sense, and the meaning of **tracking** herein signify in general that the real plant output *follows*, i.e., *tracks*, every plant desired output belonging to a family $\mathfrak{Y}_d^k$, $k \in \{-, 0, 1, ..., n, ..\}$,

- *regardless of whether the desired output is constant* (in a part of the control literature this is related to the regulation systems) *or time-varying* (in another part of the control literature it is associated with the servomechanisms/servosystems),

- *under the actions of arbitrary external disturbances* belonging to a set $\mathfrak{D}^k$, $k \in \{-, 0, 1, ..., n, ..\}$,

and

- *under arbitrary (input and output) initial conditions.*

Therefore, tracking incorporates both the servomechanism/servosystem problematic and the regulation issues; i.e., it spans the whole control thematic.

Subsection 2.1.2 presents the characterization of *the disturbance family* $\mathfrak{D}^k$ and of *the desired output family* $\mathfrak{Y}_d^k$.

Tracking is ***perfect*** (***ideal***) if, and only if, the plant real output behavior is always equal to the plant desired output behavior. If the initial real output is different from the initial desired output then tracking is only ***imperfect***. We will consider both perfect and imperfect tracking.

The definition of any tracking property should clarify the following:

- *the characterization of the plant behavior we are interested in*, whether we are interested in the internal dynamical behavior of the plant, or in the plant output dynamical behavior;

- *the space in which the demanded closeness is to be achieved*, which means that, although originally tracking concerns the output behavior, we can consider the output tracking either via *the output space* or via *the state space*;

- *the definition of the distance between the real behavior and the desired behavior* of the plant;

- *the definition of the demanded closeness of the real behavior to the desired behavior* of the plant;

- *the nonempty sets of the initial conditions of all plant variables* under which the demanded closeness is to be achieved;

- *the nonempty set* $\mathfrak{D}^{(.)}$ *of permitted external disturbances* acting on the plant, under which the demanded closeness is to be realized;
- *the nonempty set* $\mathfrak{Y}_d^{(.)}$ *of realizable desired plant behaviors* that can be demanded;
- *the time interval over which the demanded closeness is to be guaranteed*; and
- *the requested quality with which the real behavior is to follow the desired behavior* of the plant.

We can differently specify the preceding items. Their different specifications lead to numerous various tracking concepts, each containing a number of different tracking properties.

The references [71]-[77], [90], [92], [94], [96], [98]-[103], [108]-[112], [118], [120]-[123], [125], [128], [129], [132]-[135], [137], [138], [140], [142], [149], [159], [163], [174]-[177] deal with various types of tracking in Lyapunov sense.

The references [98], [100], [106], [111], [120], [134], [136], [139], [141], [143], [166], [178], [180], [267] established the theory of control synthesis for tracking with finite (scalar or vector) reachability *time.*

We will present the fundamentals of *the concept of tracking in Lyapunov sense* and of *the concept of tracking with finite (scalar or vector) reachability time.*

The concept of *practical tracking (with finite settling and/or finite reachability time)* was introduced and studied in [98], [111], [128], A. Kökösy [206] - [210], D. V. Lazić (Lazitch) [220] - [224], M. J. Stojčić (M. Y. Stoychitch) [307], and M. R. Jovanović (M. R. Yovanovitch) [325, Definition 5.1, p. 45] developed it further theoretically and applied it effectively to the control of technical objects.

The concept of *tracking with a prespecified performance index* occupies [73], [105]-[107], [126], [128], [132]-[141], [149], [159]-[161], [163], [164], [166], [167], [174]-[180]. The further development is due to A. Kökösy [206], D. V. Lazić (Lazitch) [220], N. N. Nedić (Neditch) and D. Pršić (Prshitch) [261]-[263], [285], Z. B. Ribar et al. [288], [289], M. J. Stojčić (M. Y. Stoychitch) [307], and M. R. Jovanović (Yovanovitch) [325].

Every tracking concept comprises a number of various tracking properties We will define only basic ones of the concept of tracking in Lyapunov sense, and of the concept of tracking with (scalar or vector) *finite reachability time (FRT)*.

All the preceding tracking concepts open new directions in the control theory. They are open for further research and development.

Before we start control synthesis to ensure a requested tracking property, we should examine whether the plant is able to track under an appropriate action of control, i.e., whether the plant is **trackable**.

We will consider ***tracking*** and ***trackability*** of *time*-invariant continuous-*time* linear systems, as well as ***tracking control synthesis*** for them. Various tracking properties and trackability kinds defined in the sequel will illustrate richness of the tracking and trackability phenomena as well as their greater complexity than that of the related stability or controllability properties, respectively.

8.2 Tracking versus stability

There is the great variety of stability concepts, e.g., Lyapunov stability concept, Bounded-Input-Bounded-Output (*BIBO*) stability concept, $L_{(.)}$-stability concept, Lagrange stability concept, Poisson stability concept, and Practical stability concept. Lyapunov stability concept, *BIBO* stability concept, and Practical stability concept contain various stability properties of an equilibrium vector, and/or of a motion, and/or of a set, and/or of a system.

Some of the stability concepts concern the internal dynamic behavior (Lyapunov stability concept, Lagrange stability concept, Poisson stability concept, and Practical stability concept), and others treat the output system behavior (*BIBO* stability concept, and $L_{(.)}$ stability concept). Practical stability concept treats system behavior over a bounded time interval $[0, \tau[\subset \mathfrak{T}_0$ and over the unbounded infinite time interval $\mathfrak{T}_0$, while other mentioned stability concepts take into account only system behavior over the unbounded infinite time interval $\mathfrak{T}_0$.

The important characteristics of stability concepts express how they define the distance between two system behaviors and which kind of closeness they demand. Another significant feature shows the influence of initial conditions and/or of external (input) actions on the system. Their common characteristic is their validity for dynamic systems in general. They do not reflect originally the primary control goal.

Lyapunov stability concept does not permit originally nonnominal total input actions on the system and it does not concern the system output behavior. It does not correspond, conceptually and originally, directly to the control goal. This holds also for Lagrange and Poisson stability concepts.

BIBO stability concept and $L_{(.)}$ stability concept do not treat originally the influence of nonzero initial conditions, and they do not demand that system real behaviors follow the system desired output behavior. They do not satisfy, conceptually and originally, directly the primary control purpose.

Practical stability concept does not treat the system output behavior. It does not satisfy, conceptually and originally, directly the control goal.

It is well known that a linear system can be stable (its equilibrium vector can be globally asymptotically stable), but it need not exhibit tracking. The steady state error of its unit step response can be nonzero. System stability is not sufficient for system tracking.

A linear system can exhibit a kind of tracking in even though it is unstable. The following simple example illustrates this statement.

Example 152 *[99, p. 11, Example 4] Let*

$$\frac{d\mathbf{X}(t)}{dt} = \frac{1}{7}\begin{bmatrix} 9 & -4 \\ 8 & -9 \end{bmatrix}\mathbf{X}(t) + \begin{bmatrix} 3 \\ 4 \end{bmatrix} D(t) + \begin{bmatrix} 2 \\ 6 \end{bmatrix} Y_D(t),$$

$$Y(t) = \begin{bmatrix} -\frac{1}{2} \vdots 1 \end{bmatrix} \mathbf{X}(t)$$

be the ISO description of a closed-loop feedback control system. Its motions in the free regime are determined by

$$\mathbf{X}(t;\mathbf{X}_0;0;0) = \frac{1}{7}\begin{bmatrix} 8e^t - e^{-t} & -2e^t + 2e^{-t} \\ 4e^t - 4e^{-t} & -e^t + 8e^{-t} \end{bmatrix} \mathbf{X}_0.$$

The zero equilibrium state $X_e = \mathbf{0}_2$ is unstable; i.e., the system is unstable.

Let the desired output be $Y_D(t) = 0$. The real output

$$Y(t) = Y_0 e^{-t}$$

asymptotically tracks the desired output for every $Y_0 \in \mathfrak{R}$ and for $D(t) \equiv 0$. The system exhibits asymptotic (output) tracking of the desired output $Y_D(t) = 0$ for every initial state $X_0 \in \mathfrak{R}^2$, i.e., for every initial output $Y_0 \in \mathfrak{R}$ and for $D(t) \equiv 0$. The system achieves global tracking in the free regime despite being unstable.

System motions converge asymptotically to the set

$$\Upsilon(Y_D) = \left\{ \mathbf{X} \in \mathfrak{R}^2 : Y(\mathbf{X}) = -\frac{1}{2}X_1 + X_2 = Y_D = 0 \right\},$$

which we call ***the target set*** *of the system. It is time-invariant in this case. It is the set of all system states for which the system real output is equal to its desired output.*

However, if we change $Y_D(t)$, for example, to $Y_D(t) = h(t)$, ($h(t) = 0$ for all $t < 0$, $h(t) \in [0,1]$ for $t = 0$, $h(t) = 1$ for all $t > 0$), then the system real output $Y(t)$ does not track the new $Y_D(t)$. The real output diverges to infinity as time t goes to infinity,

$$Y(t) \to \infty, \text{ as } t \to \infty.$$

This example illustrates the need for a study of tracking both as a self-contained issue and as the phenomenon related to stability because in many cases an appropriate stability property is necessary (but not sufficient) for the demanded tracking property.

Conclusion 153 ***Stability and tracking are in general mutually independent concepts***

Stability properties do not guarantee tracking properties in general, and tracking properties do not imply stability properties in general. They can be mutually independent. However, system stability appears often necessary for tracking. Besides, in some cases tracking is sufficient for system stability.

Note 154 ***Control vector partitioning***

If the plant to be controlled is not stable, then we will partition the full total control vector $\mathbf{U}_F$ *into*

- *the stabilizing total control vector* $\mathbf{U}_S$,

and

- *the tracking total control vector* $\mathbf{U}_T$,

$$\mathbf{U_F} = \mathbf{U}_S + \mathbf{U}_T.$$

We should start the control synthesis by applying any of the stabilization methods to synthesize the stabilizing control vector $\mathbf{U}_S$*. This should be the first step. The stabilizing controller and so controlled plant constitute then a stable system that will represent a stable plant for the synthesis of the tracking control* $\mathbf{U}_T$*. The next, the second, step is to synthesize the tracking control* $\mathbf{U}_T$*.*

We will denote in the sequel $\mathbf{U}_T$ *simply by* $\mathbf{U}$,

$$\mathbf{U}_T = \mathbf{U}.$$

If the plant is stable, then we set simply

$$\mathbf{U_F} = \mathbf{U_T} = \mathbf{U}.$$

We will study only the tracking control synthesis in the sequel.

Claim 155 ***Plant stability***

The system called plant (or object) to be controlled by the tracking control $\mathbf{U}$ *is stable (i.e., its zero error vector, equivalently its total desired motion, is globally asymptotically stable).*

If not stated otherwise, we accept this claim to hold.

8.3 Perfect tracking: characterization

8.3.1 On perfect tracking generally

The concept of *perfect (ideal) tracking* helps us to discover what is theoretically the best possible real dynamic total output behavior $\mathbf{Y}(t)$ of a plant *relative to its desired total output behavior* $\mathbf{Y}_d(t)$ (Definition 43, Subsection 3.3.1). The precise meaning of perfect tracking reads as follows:

Definition 156 ***The k-th order perfect tracking of the plant***

The plant exhibits ***the*** *k-th* ***order perfect tracking of its desired k-th order output vector response*** $\mathbf{Y}_d^k(t)$ *if, and only if, its real k-th order output vector response* $\mathbf{Y}^k(t)$ *is always equal to its desired k-th order output vector response* $\mathbf{Y}_d^k(t)$,

$$\mathbf{Y}^k(t) = \mathbf{Y}_d^k(t),\ \forall t \in \mathfrak{T}_0. \tag{8.1}$$

If, and only if, $k = 0$*, then the zero order perfect tracking is simply called* ***perfect tracking.***

We discover now the relationship between perfect tracking and the *k-th* order perfect tracking.

Theorem 157 ***Perfect tracking and the*** k***-th order perfect tracking***
If the real output vector function $\mathbf{Y}(.)$ *and the desired output vector function* $\mathbf{Y}_d(.)$ *are k-times continuously differentiable on* $\mathfrak{T}_0$*, then for the validity of (8.1) it is necessary and sufficient that*

$$\mathbf{Y}(t) = \mathbf{Y}_d(t),\ \forall t \in \mathfrak{T}_0, \tag{8.2}$$

holds, i.e.,

$$\mathbf{Y}(t) \in \mathfrak{C}^k \text{ and } \mathbf{Y}_d(t) \in \mathfrak{C}^k \Longrightarrow$$
$$\left\langle \mathbf{Y}^k(t) = \mathbf{Y}_d^k(t),\ \forall t \in \mathfrak{T}_0 \right\rangle \Longleftrightarrow \left\langle \mathbf{Y}(t) = \mathbf{Y}_d(t),\ \forall t \in \mathfrak{T}_0 \right\rangle. \tag{8.3}$$

Proof. Let

$$\mathbf{Y}(t) \in \mathfrak{C}^k \text{ and } \mathbf{Y}_d(t) \in \mathfrak{C}^k$$

hold.

Necessity. Let (8.1) be valid. The definition of $\mathbf{Y}^k$,

$$\mathbf{Y}^k = \begin{bmatrix} \mathbf{Y} \\ \mathbf{Y}^{(1)} \\ \mathbf{Y}^{(2)} \\ \dots\dots \\ \mathbf{Y}^{(k)} \end{bmatrix} = \begin{bmatrix} \mathbf{Y}^{(0)} \\ \mathbf{Y}^{(1)} \\ \mathbf{Y}^{(2)} \\ \dots\dots \\ \mathbf{Y}^{(k)} \end{bmatrix},$$

and (8.1) prove the validity of (8.2).

Sufficiency. Let (8.2) be satisfied. By the definition, the first derivative of $\mathbf{Y}(t)$ on $\mathfrak{T}_0$ reads

$$\mathbf{Y}^{(1)}(t) = \lim \left[\frac{\mathbf{Y}(t+\theta) - \mathbf{Y}(t)}{\theta} : \theta \to 0 \right],\ \forall t \in \mathfrak{T}_0.$$

This and (8.2) yield

$$\mathbf{Y}^{(1)}(t) = \lim \left[\frac{\mathbf{Y}_d(t+\theta) - \mathbf{Y}_d(t)}{\theta} : \theta \to 0 \right] = \mathbf{Y}_d^{(1)}(t),\ \forall t \in \mathfrak{T}_0. \tag{8.4}$$

Hence,

$$\mathbf{Y}^{(i)}(t) = \mathbf{Y}_d^{(i)}(t),\ \forall t \in \mathfrak{T}_0, \tag{8.5}$$

holds for $i = 0, 1$. Let it be fulfilled also for $i = 2, 3, .., j$,

$$\mathbf{Y}^{(i)}(t) = \mathbf{Y}_d^{(i)}(t),\ \forall t \in \mathfrak{T}_0,\ \forall i = 0, 1, 2, ..., j. \tag{8.6}$$

By the definition,

$$\mathbf{Y}^{(j+1)}(t) = \lim \left[\frac{\mathbf{Y}^{(j)}(t+\theta) - \mathbf{Y}^{(j)}(t)}{\theta} : \theta \to 0 \right],\ \forall t \in \mathfrak{T}_0.$$

This and (8.6) lead to

$$\mathbf{Y}^{(j+1)}(t) = \lim \left[\frac{\mathbf{Y}_d^{(j)}(t+\theta) - \mathbf{Y}_d^{(j)}(t)}{\theta} : \theta \to 0 \right] = \mathbf{Y}_d^{(j+1)}(t), \ \forall t \in \mathfrak{T}_0.$$

We have proved that (8.6) holds for $i = 0, 1$, and since it holds for $i = j$, then it holds also for $i = j + 1$. By mathematical induction, we have proved that (8.2) guarantees (8.1) ∎

Conclusion 158 *This theorem is general. Its proof is valid regardless of the form of the mathematical model of the system, which can be either linear or nonlinear, time-invariant or time-varying. It enables us to reduce the study of the k-th order perfect tracking to the perfect tracking.*

8.3.2 The *IO* systems

Perfect tracking and the target set of the IO system

By following the general Definition 43 of the desired regime (Subsection 3.3.1), Definition 156 (Subsection 8.3.1) for the *IO* plant (2.15) (Subsection 2.1.2) we present

Definition 159 ***The k-th order perfect tracking of the IO plant (2.15)***
The IO plant (2.15) exhibits ***the k-th order perfect tracking of its desired k-th order output vector response*** $\mathbf{Y}_d^k(t)$ *if, and only if, its real k-th order output vector response* $\mathbf{Y}^k(t)$ *is always equal to its desired k-th order output vector response* $\mathbf{Y}_d^k(t)$,

$$\mathbf{Y}^k(t) = \mathbf{Y}_d^k(t), \ \forall t \in \mathfrak{T}_0, \ k \in \{0, 1, ..., \nu - 1\} \tag{8.7}$$

If, and only if, $k = 0$, *then the zero order perfect tracking is simply called* ***perfect tracking.***

The condition (8.7), which is in fact (8.1) (Subsection 8.3.1), determines a particular set important for tracking of the *IO* plant (2.15) (Subsection 2.1.2). It is the set of all vectors $\mathbf{Y}^{\nu-1} \in \mathfrak{R}^{\nu N}$ such that their subvectors $\mathbf{Y}^k$ are equal to $\mathbf{Y}_d^k(t)$ at the moment $t \in \mathfrak{T}_0$.

Definition 160 ***The k-th order target set of the IO plant (2.15)***
The set $\Upsilon_{IO}^k(t; \mathbf{Y}_d)$, $\Upsilon_{IO}^k(t; \mathbf{Y}_d) \subset \mathfrak{R}^{\nu N}$, *of all vectors* $\mathbf{Y}^{\nu-1} \in \mathfrak{R}^{\nu N}$ *such that* $\mathbf{Y}^k$ *is equal to* $\mathbf{Y}_d^k(t)$ *is* ***the k-th order target set of the IO plant (2.15) relative to its desired output vector response*** $\mathbf{Y}_d(t)$ ***at a moment*** $t \in \mathfrak{T}_0$,

$$\Upsilon_{IO}^k(t; \mathbf{Y}_d) \equiv \left\{ \mathbf{Y}^{\nu-1} : \left(\begin{array}{c} \mathbf{Y}^k = \mathbf{Y}_d^k(t), \ \mathbf{Y}^{(j)} \in \mathfrak{R}^N, \\ \forall j = k+1, \ k+2, \ ... \ , \nu - 1 \end{array} \right) \right\},$$
$$\forall t \in \mathfrak{T}_0, \ 0 \leq k \leq \nu - 1. \tag{8.8}$$

If, and only if, $k = 0$, *then the zero order target set* $\Upsilon^0_{IO}(t; \mathbf{Y}_d)$ *at a moment* t *is called* ***the target set*** $\Upsilon(t; \mathbf{Y}_d)$ ***at the moment*** t,

$$\Upsilon^0_{IO}(t; \mathbf{Y}_d) = \Upsilon_{IO}(t; \mathbf{Y}_d) \equiv$$
$$\equiv \left\{ \mathbf{Y}^{\nu-1} : \left(\begin{array}{c} \mathbf{Y} = \mathbf{Y}_d(t),\ \mathbf{Y}^{(j)} \in \mathfrak{R}^N, \\ \forall j = 1,\ 2,\ \dots\ ,\nu - 1 \end{array} \right) \right\},\ \forall t \in \mathfrak{T}_0.$$

The instantaneous target set $\Upsilon^k_{IO}(t; \mathbf{Y}_d)$ is time-varying as soon as $\mathbf{Y}_d(t)$ is variable; i.e., $\mathbf{Y}_d(t) \neq \mathbf{const}$. For example, if

$$\mathbf{Y}_d(t) = e^{-t}\mathbf{1}_N,\ \mathbf{1}_N = [1\ \ 1\ \ .\ .\ .\ \ 1]^T \in \mathfrak{R}^N,$$

then the target set $\Upsilon^k_{IOIO}(t; \mathbf{Y}_d)$ is time-varying,

$$\Upsilon^k_{IO}(t; \mathbf{Y}_d) \equiv \left\{ \mathbf{Y}^{\nu-1} : \left(\begin{array}{c} \mathbf{Y}^k = \mathbf{Y}^k_d(t),\ \mathbf{Y}^{(j)} \in \mathfrak{R}^N, \\ \forall j = k+1,\ k+2,\ \dots\ ,\nu - 1, \\ \mathbf{Y}^{(i)} = \left\langle \begin{array}{c} e^{-t}\mathbf{1}_N,\ if\ i\ is\ even, \\ -e^{-t}\mathbf{1}_N,\ if\ i\ is\ odd \end{array} \right\rangle, \\ \forall i = 0, 1, \dots, k \end{array} \right) \right\},$$
$$\forall t \in \mathfrak{T}_0,\ 0 \leq k \leq \nu - 1.$$

The target set $\Upsilon^k_{IO}(t; \mathbf{Y}_d)$ is a hyperplane in the space $\mathfrak{R}^{\nu N}$.

Note 161 *The IO plant (2.15) exhibits perfect tracking relative to its desired output vector response* $\mathbf{Y}_d(t)$ *if, and only if,* $\mathbf{Y}^{\nu-1}(t)$ *is always in the target set* $\Upsilon_{IO}(t; \mathbf{Y}_d)$ *(8.8),*

$$\mathbf{Y}^{\nu-1}(t) \in \Upsilon_{IO}(t; \mathbf{Y}_d),\ \forall t \in \mathfrak{T}_0.$$

Note 162 ***Necessary condition for perfect tracking***

From Definition 156 (Subsection 8.3.1), i.e., from Definition 159, follows the necessary, but insufficient, condition for the k-th order perfect tracking of the plant desired output response $\mathbf{Y}_d(t)$. *The initial k-th order output vector* $\mathbf{Y}^k_0$ *of the plant should be equal to its initial k-th order desired output vector* $\mathbf{Y}^k_{d0}$,

$$\mathbf{Y}^k_0 = \mathbf{Y}^k_{d0}.$$

Comment 163 ***Realizability of perfect tracking***

The above necessary condition for the k-th order perfect tracking is rarely realizable. The real initial k-th order output $\mathbf{Y}^k_0$ *of the object is most often unpredictable, hence different from* $\mathbf{Y}^k_{d0}$. *Perfect tracking occurs rarely. We should look for other forms of tracking, which represent* ***imperfect tracking****, i.e.,* ***realistic tracking****, with at least satisfactory, or with a very good, quality according to an accepted criterion.*

Perfect tracking and the desired regime of the IO system

We will specify Definition 43 (Subsection 3.3.1) of the desired regime from the tracking point of view for the plant (2.15).

Definition 164 ***The k-th order desired (nominal) regime of the plant***
The IO plant (2.15) is ***in the k-th order desired (nominal) regime*** *if, and only if, it exhibits the k-th order perfect tracking, i.e., if, and only if,*

$$\mathbf{Y}^k(t) = \mathbf{Y}_d^k(t),\ \forall t \in \mathfrak{T}_0. \tag{8.9}$$

Perfect tracking and the nominal control of the IO system

We will accommodate Definition 48 (Subsection 3.3.1) of the nominal control to the plant (2.15).

Definition 165 ***The k-th order nominal control***
A control $\mathbf{U}^*(.)$ *is* ***a nominal k-th order control for the plant (2.15) relative to the (disturbance, desired output) pair*** $[\mathbf{D}^i(.),\ \mathbf{Y}_d(.)]$*, which is denoted by* $\mathbf{U}_N(.)$*,*

$$\mathbf{U}^*(.) = \mathbf{U}_N(.;\mathbf{D},\mathbf{Y}_d), \tag{8.10}$$

if, and only if, it guarantees (8.9), i.e.,

$$\mathbf{U}^*(t) = \mathbf{U}_N(t) \Longleftrightarrow \mathbf{Y}^k(t) = \mathbf{Y}_d^k(t),\ \forall t \in \mathfrak{T}_0. \tag{8.11}$$

Theorem 66 (Subsection 3.3.2), Definition 159 and Definition 165 imply the following.

Theorem 166 ***Necessary condition for the existence of the nominal control***
The initial k-th order nominal control vector $\mathbf{U}_{N0}^{\mu_{Pu}}$ *of (2.15) satisfies*

$$C_{Pu}^{(\mu_{Pu})}\mathbf{U}_{N0}^{\mu_{Pu}} = A_P^{(\nu)}\mathbf{Y}_0^{\nu} - D_{Pd}^{(\mu_{Pd})}\mathbf{D}^{\mu_{Pd}},\ \mathbf{Y}_0^k = \mathbf{Y}_{d0}^k. \tag{8.12}$$

It depends explicitly on both $\mathbf{Y}_{d0}^k$ *and* $\mathbf{D}^{\mu_{Pd}}$ *in general.*

Note 167 *Theorem 68 (Subsection 3.3.2) shows that the equation (8.12) is uniquely solvable in* $\mathbf{U}_{N0}^{\mu_{Pu}}$ *if, and only if,*

$$rank\ C_{Pu}^{(\mu_{Pu})} = N \leq r. \tag{8.13}$$

If we accept that the controller is at rest in the nominal regime at the initial moment,

$$\mathbf{U}_{N0}^{\mu_{Pu}} = [\mathbf{U}_{N0}^T\ \ \mathbf{U}_{N0}^{(1)T}\ \ \mathbf{U}_{N0}^{(2)T}\ \cdots\ \mathbf{U}_{N0}^{(\mu_{Pu})T}]^T = [\mathbf{U}_{N0}^T\ \ \mathbf{0}_r^T\ \ \mathbf{0}_r^T\ \cdots\ \mathbf{0}_r^T\ ]^T, \tag{8.14}$$

then Theorem 68 shows that the equation (8.12) is uniquely solvable in $\mathbf{U}_{N0}^{\mu_{Pu}}$ *if, and only if,* $rankC_{Pu0} = N \leq r$*, where* C_{Pu0} *is the first submatrix of* $C_{Pu}^{(\mu_{Pu})}$ *(Subsection 2.1.2).*

Theorem 168 ***Theorem on the nominal control***

In order for a control $\mathbf{U}^*(.)$ *to be the k-th order nominal control for the IO plant (2.15) relative to the (disturbance-desired output) pair* $[\mathbf{D}^{\mu_{Pd}}(.),\ \mathbf{Y}_d(.)]$ *it is necessary and sufficient that*

$$C_{Pu}^{(\mu_{Pu})}\mathbf{U}^{*\mu_{Pu}}(t) = A_P^{(\nu)}\mathbf{Y}^{\nu}(t) - D_{Pd}^{(\mu_{Pd})}\mathbf{D}^{\mu_{Pd}}(t),\ \mathbf{Y}^k(t) = \mathbf{Y}_d^k(t), \forall t \in \mathfrak{T}_0. \quad (8.15)$$

This theorem is a special form of Theorem 66 (Subsection 3.3.2) applied to the plant (2.15).

Comment 169 *Equation (8.15) for* $k = \nu$*-1 can be set into the complex form,*

$$\left(C_{Pu}^{(\mu_{Pu})}S_r^{(\mu_{Pu})}(s)\right)\mathbf{U}^*(s) = -\left(D_{Pd}^{(\mu_{Pd})}S_d^{(\mu_{Pd})}(s)\right)\mathbf{D}(s) + \left(A_P^{(\nu)}S_N^{(\nu)}(s)\right)\mathbf{Y}_d(s) +$$
$$+\left(D_{Pd}^{(\mu_{Pd})}Z_d^{(\mu_{Pd}-1)}(s)\right)\mathbf{D}_0^{\nu_{Pd}-1} + \left(C_{Pu}^{(\mu_{Pu})}Z_r^{(\mu_{Pu}-1)}(s)\right)\mathbf{U}_0^{*\nu_{Pu}-1} -$$
$$-\left(A_P^{(\nu)}Z_N^{(\nu-1)}(s)\right)\mathbf{Y}_{d0}^{\nu-1}.$$

This equation has the unique solution for arbitrary initial conditions if, and only if,

$$det\left[\left(C_{Pu}^{(\mu_{Pu})}S_r^{(\mu_{Pu})}(s)\right)\left(C_{Pu}^{(\mu_{Pu})}S_r^{(\mu_{Pu})}(s)\right)^T\right] \neq 0, \quad (8.16)$$

or equivalently if, and only if, (8.13) holds. Then the solution reads in the complex domain

$$\mathbf{U}^*(s) = \left(C_{Pu}^{(\mu_{Pu})}S_r^{(\mu_{Pu})}(s)\right)^T \bullet$$
$$\bullet\left[\left(C_{Pu}^{(\mu_{Pu})}S_r^{(\mu_{Pu})}(s)\right)\left(C_{Pu}^{(\mu_{Pu})}S_r^{(\mu_{Pu})}(s)\right)^T\right]^{-1} \bullet$$
$$\bullet\left[\begin{array}{c} -\left(D_{Pd}^{(\mu_{Pd})}S_d^{(\mu_{Pd})}(s)\right)\mathbf{D}(s) + \left(A_P^{(\nu)}S_N^{(\nu)}(s)\right)\mathbf{Y}_d(s) + \\ +\left(D_{Pd}^{(\mu_{Pd})}Z_d^{(\mu_{Pd}-1)}(s)\right)\mathbf{D}_0 + \left(C_{Pu}^{(\mu_{Pu})}Z_r^{(\mu_{Pu}-1)}(s)\right)\mathbf{U}_0^* - \\ -\left(A_P^{(\nu)}Z_N^{(\nu-1)}(s)\right)\mathbf{Y}_{d0}. \end{array}\right].$$

Theorem 170 *If control* $\mathbf{U}(.)$ *of the IO plant (2.15) is the nominal control* $\mathbf{U}_N(.)$ *of the order* ν*-1 relative to* $[\mathbf{D}^{\mu_{Pd}}(.),\ \mathbf{Y}_d(.)]$ *and the initial real output* $\mathbf{Y}_0^{\nu-1}$ *is the desired initial output* $\mathbf{Y}_{d0}^{\nu-1}$, $\mathbf{Y}_0^{\nu-1} = \mathbf{Y}_{d0}^{\nu-1}$, *then the plant exhibits perfect tracking of* $\mathbf{Y}_d(.)$ *of the order* ν*-1 under the action of* $\mathbf{D}^{\mu_{Pd}}(t)$.

Proof. Let all the conditions of the theorem statement hold. Since $\mathbf{U}(t) \equiv \mathbf{U}_N(t)$, then (8.15) is valid. It and (2.15) yield

$$A_P^{(\nu)}\left[\mathbf{Y}_d^{\nu}(t) - \mathbf{Y}^{\nu}(t)\right] = \mathbf{0}_N,\ i.e.,\ A_P^{(\nu)}\varepsilon^{\nu}(t) = \mathbf{0}_N,\ \forall t \in \mathcal{T}_0.$$

These homogeneous differential linear equations have the trivial solution for the zero initial conditions:

$$\mathbf{Y}_{d0}^{\nu-1} - \mathbf{Y}_0^{\nu-1} = \varepsilon_o^{\nu-1} = \mathbf{0}_N \Longrightarrow$$
$$\mathbf{Y}_d^{\nu-1}(t) - \mathbf{Y}^{\nu-1}(t) = \varepsilon^{\nu-1}(t) = \mathbf{0}_N,\ \forall t \in \mathcal{T}_0.$$

This proves perfect tracking of the order ν-1 ■

8.3.3 The *ISO* systems

Perfect tracking and the target set of the *ISO* plant

We consider the *ISO* plant (2.33), (2.34) (Subsection 2.2.2).

By following the general Definition 43 (Subsection 3.3.1) of the desired regime, we present Definition 156, (Subsection 8.3.1) for the *ISO* plant (2.33), (2.34).

Definition 171 ***The k-th order perfect tracking of the ISO plant (2.33), (2.34)***

The plant (2.33), (2.34) exhibits ***the k-th order perfect tracking of its desired output vector response*** $\mathbf{Y}_d^k(t)$ *if, and only if, its real k-th order output vector response* $\mathbf{Y}^k(t)$ *is always equal to its desired k-th order output vector response* $\mathbf{Y}_d^k(t)$,

$$\mathbf{Y}^k(t) = \mathbf{Y}_d^k(t),\ \forall t \in \mathfrak{T}_0,\ \ k \in \{0, 1, ..., n-1\}. \tag{8.17}$$

If, and only if, $k = 0$, then the zero order perfect tracking is simply called ***perfect tracking.***

Definition 172 ***The k-th order target set of the ISO plant (2.33), (2.34)***

The set $\Upsilon^k(t; \mathbf{D}; \mathbf{U}; \mathbf{Y}_d^k)$, $\Upsilon^k(t; \mathbf{D}; \mathbf{U}; \mathbf{Y}_d^k) \subset \mathfrak{R}^n$, *of all vectors* $\mathbf{X}_P$ *such that* $\mathbf{Y}^k$ *is equal to* $\mathbf{Y}_d^k(t)$ *is* ***the k-th order target set of the ISO plant (2.33), (2.34) relative to its desired k-th order output vector response*** $\mathbf{Y}_d^k(t)$ ***at a moment*** $t \in \mathfrak{T}_0$,

$$\Upsilon^k(t; \mathbf{D}; \mathbf{U}; \mathbf{Y}_d^k) = \left\{ \mathbf{X}_P:\ \mathbf{Y}^k(\mathbf{X}_P) = \mathbf{Y}_d^k(t) \right\} =$$
$$= \left\{ \mathbf{X}_P: \left(\begin{array}{c} C_P \mathbf{X}_P + H_P \mathbf{U}(t) + D_P \mathbf{D}(t) = \mathbf{Y}_d(t), \\ C_P A_p^j \mathbf{X}_P + C_P \left[\sum_{i=1}^{i=j} A_p B_p \mathbf{U}^{(i-1)}(t) \right] + \\ + H_P \mathbf{U}^{(j)}(t) + C_P \left[\sum_{i=1}^{i=j} A_p L_P \mathbf{D}^{(i-1)}(t) \right] + \\ + D_P \mathbf{D}^{(j)}(t) = \mathbf{Y}_d^{(j)}(t),\ \forall j = 1, ..., k \end{array} \right) \right\}. \tag{8.18}$$

If, and only if, $k = 0$, then the zero order target set $\Upsilon^0(t; \mathbf{D}; \mathbf{U}; \mathbf{Y}_d)$ *at a moment t is called* ***the target set*** $\Upsilon(t; \mathbf{D}; \mathbf{U}; \mathbf{Y}_d)$ ***at the moment*** *t*,

$$\Upsilon^0(t; \mathbf{D}; \mathbf{U}; \mathbf{Y}_d) \equiv \Upsilon(t; \mathbf{D}; \mathbf{U}; \mathbf{Y}_d) =$$
$$= \{ \mathbf{X}_P:\ C_P \mathbf{X}_P + H_P \mathbf{U}(t) + D_P \mathbf{D}(t) = \mathbf{Y}_d(t) \}.$$

Note 173 *The k-th order target set* $\Upsilon^k(t; \mathbf{D}; \mathbf{U}; \mathbf{Y}_d^k)$ *at a moment t of the ISO plant (2.33), (2.34) is a subset of the plant state space* $\mathfrak{R}^n$, $\Upsilon^k(t; \mathbf{D}; \mathbf{U}; \mathbf{Y}_d^k) \subset \mathfrak{R}^n$, *which corresponds to the internal dynamics space of the IO plant (2.21) (Subsection 2.1.2), rather than being a subset of the plant output space* $\mathfrak{R}^N$. *It depends not only on* $\mathbf{Y}_d^k$ *but also on* $\mathbf{D}^k$ *and* $\mathbf{U}^k$, *which means that it depends on derivatives of* $\mathbf{Y}_d$, $\mathbf{D}$ *and* $\mathbf{U}$, *but it is independent of all derivatives of the state vector* $\mathbf{X}_P$.

Note 174 *In order for the ISO plant described by (2.33), (2.34) to exhibit the k-th order perfect tracking relative to its desired k-th order output vector response* $\mathbf{Y}_d^k(.)$ *it is necessary and sufficient that its state* $\mathbf{X}_P(t;\mathbf{X}_{P0};\mathbf{D};\mathbf{U})$ *is always in its k-th order target set* $\Upsilon^k(t;\mathbf{D};\mathbf{U};\mathbf{Y}_d^k)$,

$$\mathbf{X}_P(t;\mathbf{X}_{P0};\mathbf{D};\mathbf{U}) \in \Upsilon^k(t;\mathbf{D};\mathbf{U};\mathbf{Y}_d^k),\ \forall t \in \mathfrak{T}_0.$$

Note 175 ***Realizability of perfect tracking***

The preceding definitions imply that the initial desired state vector $\mathbf{X}_{Pd0}$ *and the initial nominal control vector* $\mathbf{U}_{0N}$ *of (2.33), (2.34) obey*

$$C_P\mathbf{X}_{Pd0} + H_P\mathbf{U}_{0N} = \mathbf{Y}_{d0} - D_P\mathbf{D}_0. \tag{8.19}$$

They depend explicitly not only on $\mathbf{Y}_{d0}$, *but also on* $\mathbf{D}_0$.

The real initial conditions $\mathbf{X}_{P0}$ *and* $\mathbf{U}_0$ *satisfy (8.19) only in very special cases. Perfect tracking is possible only in such special cases. A kind of an* ***imperfect tracking*** *is possible in the most real cases.*

Perfect tracking and the desired regime of the ISO plant

We will refine Definition 43, (Subsection 3.3.1), for the plant (2.33), (2.34).

Definition 176 ***The k-th order desired (nominal) regime of the plant***

The plant (2.33), (2.34) is ***in the k-th order desired (nominal) regime*** *if, and only if, it exhibits the k-th order perfect tracking (8.17),*

$$\mathbf{Y}^k(t) = \mathbf{Y}_d^k(t),\ \forall t \in \mathfrak{T}_0. \tag{8.20}$$

Perfect tracking and the nominal control of the ISO plant

We will adjust Definition 50 (Subsection 3.3.1) to the plant (2.33), (2.34).

Definition 177 ***The nominal motion and the k-th order nominal control***

A pair $[\mathbf{X}_P^*(.),\ \mathbf{U}^*(.)]$ *is a* [***nominal motion (nominal state)-k-th order control***] ***pair for the plant (2.33), (2.34) relative to*** (***the disturbance-k-th order desired output***) ***pair*** $[\mathbf{D}(.),\ \mathbf{Y}_d^k(.)]$, *which is denoted by* $[\mathbf{X}_{PN}(.),\ \mathbf{U}_N(.)]$,

$$[\mathbf{X}_P^*(.), \mathbf{U}^*(.)] = [\mathbf{X}_{PN}(.;\mathbf{D},\mathbf{Y}_d), \mathbf{U}_N(.;\mathbf{D},\mathbf{Y}_d)] = [\mathbf{X}_{PN}(.), \mathbf{U}_N(.)] \tag{8.21}$$

if, and only if, it guarantees (8.20)

$$[\mathbf{X}_P^*(t), \mathbf{U}^*(t)] = [\mathbf{X}_{PN}(t), \mathbf{U}_N(t)] \Longleftrightarrow \mathbf{Y}^k(t) = \mathbf{Y}_d^k(t),\ \forall t \in \mathfrak{T}_0. \tag{8.22}$$

Theorem 178 ***The nominal (state-control) pair***

In order for the state-control pair $[\mathbf{X}_P^*(.), \mathbf{U}^*(.)]$ *to be a nominal state-control pair for the plant (2.33), (2.34) relative to the (disturbance-desired output) pair* $[\mathbf{D}(.), \mathbf{Y}_d(.)]$ *it is necessary and sufficient that*

$$\frac{d\mathbf{X}_P^*(t)}{dt} - A_P\mathbf{X}_P^*(t) - B_P\mathbf{U}^*(t) = D_P\mathbf{D}_\mathbf{P}(t),\ \forall t \in \mathfrak{T}_0,$$
$$C_P\mathbf{X}_P^*(t) + H_P\mathbf{U}^*(t) = \mathbf{Y}_d(t) - L_P\mathbf{D}_\mathbf{P}(t),\ \forall t \in \mathfrak{T}_0. \tag{8.23}$$

This theorem is a special form of Theorem 72 (Subsection 3.3.3) applied to the plant (2.33), (2.34).

Comment 179 *Equations (8.23) can be set into the complex form, respectively,*

$$\begin{bmatrix} sI_n - A_P & -B_P \\ C_P & H_P \end{bmatrix} \begin{bmatrix} \mathbf{X}_P^*(s) \\ \mathbf{U}^*(s) \end{bmatrix} = \begin{bmatrix} D_P & I_n & O_{n,N} \\ -L_P & O_{N,n} & I_N \end{bmatrix} \begin{bmatrix} \mathbf{D}(s) \\ \mathbf{X}_{P0}^* \\ \mathbf{Y}_d(s) \end{bmatrix}. \tag{8.24}$$

This equation shows that, if $r = N$*, then the full rank,*

$$rank\ P_{ISOR}(s;\mathbf{0}_n) = (n+N) = (n+r) \tag{8.25}$$

of the ISO system matrix $P_{ISOR}(s;\mathbf{0}_n)$ *[148],*

$$P_{ISOR}(s;\mathbf{0}_n) = \begin{bmatrix} sI_n - A_P & -B_P \\ C_P & H_P \end{bmatrix},$$

is necessary and sufficient for the existence of the unique solution $[\mathbf{X}_P^*(s), \mathbf{U}^*(s)]$ *of Equation (8.24) for arbitrary* $\mathbf{X}_{P0}^* \in \mathfrak{R}^n$*. The solution is then determined by*

$$\begin{bmatrix} \mathbf{X}_P^*(s) \\ \mathbf{U}^*(s) \end{bmatrix} = \begin{bmatrix} sI_n - A_P & -B_P \\ C_P & H_P \end{bmatrix}^{-1} \begin{bmatrix} D_P & I_n & O_{n,r} \\ -L_P & O_{N,n} & I_N \end{bmatrix} \begin{bmatrix} \mathbf{D}(s) \\ \mathbf{X}_{P0}^* \\ \mathbf{Y}_d(s) \end{bmatrix}.$$

If $N < r$ *and*

$$rankP_{ISOR}(s;\mathbf{0}_n) = (n+N) < (n+r),$$

then Equation (8.24) has the following solution for arbitrary $\mathbf{X}_{P0}^* \in \mathfrak{R}^n$*,*

$$\begin{bmatrix} \mathbf{X}_P^*(s) \\ \mathbf{U}^*(s) \end{bmatrix} = \begin{bmatrix} sI_n - A_P & -B_P \\ C_P & H_P \end{bmatrix}^T \bullet$$
$$\bullet \left\langle \begin{bmatrix} sI_n - A_P & -B_P \\ C_P & H_P \end{bmatrix} \begin{bmatrix} sI_n - A_P & -B_P \\ C_P & H_P \end{bmatrix}^T \right\rangle^{-1} \bullet$$
$$\bullet \begin{bmatrix} D_P & I_n & O_{n,r} \\ -L_P & O_{N,n} & I_N \end{bmatrix} \begin{bmatrix} \mathbf{D}(s) \\ \mathbf{X}_{P0}^* \\ \mathbf{Y}_d(s) \end{bmatrix}.$$

If

$$rankP_{ISOR}(s;\mathbf{0}_n) < (n+N),$$

then Equation (8.24) does not have a solution.

8.4 Imperfect tracking: characterization

8.4.1 Output space: tracking in Lyapunov sense

In the sequel $m \in \{1, \nu\}$.We will define **in the output space** several typical tracking properties in ***Lyapunov sense***. This means the following:

1. A real behavior of a plant should track a desired plant behavior so that Lyapunov closeness among them is achieved on the infinite and unbounded time interval $\mathfrak{T}_0$.

2. Lyapunov closeness means that for every closeness over the time interval $\mathfrak{T}_0$, which is specified by the corresponding positive real number ε, there exists an initial closeness determined by another positive real number δ such that every real initial plant behavior in the δ closeness of the plant initial desired behavior at the initial moment $t_0 = 0$ rests in the ε-closeness of the plant desired behavior forever, i.e., for all $t \in \mathfrak{T}_0$. If this closeness holds for the whole internal dynamics vector $\mathbf{Y}^{m-1}$, then, and only then, it ensures simultaneously Lyapunov stability to the plant desired output behavior $\mathbf{Y}_d^{m-1}(.)$.

3. The initial closeness of the real initial plant behavior to the plant initial desired behavior is arbitrary in its δ closeness, which permits δ-arbitrariness of the initial conditions.

4. The real plant behaviors starting initially in some Δ closeness of the plant initial desired behavior converge asymptotically to the plant desired behavior as time t goes to infinity. If this closeness holds for the whole internal dynamics vector $\mathbf{Y}^{m-1}$, then, and only then, it ensures the attraction property to the plant desired output behavior $\mathbf{Y}_d^{m-1}(.)$.

When all above requirements hold for the whole internal dynamics vector $\mathbf{Y}^{m-1}$, then, and only then, the desired output behavior $\mathbf{Y}_d^{m-1}(.)$ possesses the asymptotic (hence, the exponential) stability property. However, tracking in Lyapunov sense demands, hence ensures, more than Lyapunov stability for the following reasons:

The requested closeness is to be realized:

i) for the *k-th* order *output behaviors* expressed by the temporal evolution of $\mathbf{Y}^k(t) = \mathbf{Y}^k(t; \mathbf{Y}_0^{m-1}; \mathbf{D}; \mathbf{U})$, $k \in \{0, 1, 2, ...\}$, in general, rather than only for the whole internal dynamics behavior described by the temporal evolution of $\mathbf{Y}^{m-1}(t) = \mathbf{Y}^{m-1}(t; \mathbf{Y}_0^{m-1}; \mathbf{D}; \mathbf{U})$, which is strictly demanded in Lyapunov stability theory (which appears as a special case of Lyuapunov tracking theory);

ii) *for every desired plant behavior from a (given, or to be determined) family* $\mathfrak{Y}_d^k$ of possibly demanded realizable plant desired behaviors $\mathbf{Y}_d(.)$, i.e., *tracking should hold over the desired output family* $\mathfrak{Y}_d^k$, $k \in \{0, 1, 2, ...\}$;

iii) *for every external disturbance* $\mathbf{D}(.)$ *from a (given, or to be determined) family* $\mathfrak{D}^k$ of permitted external disturbances, rather than only if $\mathbf{D}(.)$ is nominal. Lyapunov stability theory does not permit nonnominal disturbances. Therefore, Lyapunov stability properties represent special cases of the corresponding tracking properties in Lyapunov sense.

The above consideration explains why the following definitions represent extensions and generalizations of the corresponding definitions of Lyapunov stability properties.

We assume at least continuity of the desired plant output vector function $\mathbf{Y}_d(.)$ and of the disturbance vector function $\mathbf{D}(.)$ in order to be adequate descriptions of the corresponding physical variables that are continuous in *time* [for details see new *Physical Continuity and Uniqueness Principle* 5-7, and *Time Continuity and Uniqueness Principle* 9 (Introduction 1) [170], [171]].

The following definitions hold for both the *IO* plant (2.21) (Subsection 2.1.2) and the *ISO* plant (2.33), (2.34) (Subsection 2.2.2). Subsection 8.4.3 contains the specific definitions related to the *IO* plant (2.21). Its order m is $\nu \geq 1$, i.e., $m = \nu \geq 1$. Subsection 8.4.4 presents definitions of stable state tracking properties. They are characteristic for the *ISO* plant (2.33), (2.34), the order m of which is one, $m = 1$.

Definition 180 ***The k-th order tracking in Lyapunov sense of the desired output of the IO plant (2.21) (Subsection 2.1.2) or of the ISO plant (2.33), (2.34)***

The m-th order plant exhibits ***the k-th order asymptotic output tracking****, for short* ***the k-th order tracking, over*** $\mathfrak{D}^k x \mathfrak{Y}_d^k$ *if, and only if, for every* $[\mathbf{D}(.), \mathbf{Y}_d(.)] \in \mathfrak{D}^k x \mathfrak{Y}_d^k$ *there exists a positive real number* Δ, $\Delta = \Delta(\mathbf{D}, \mathbf{Y}_d)$, *or* $\Delta = \infty$, *such that* $\left\|\mathbf{Y}_0^{m-1} - \mathbf{Y}_{do}^{m-1}\right\| < \Delta$ *guarantees that* $\mathbf{Y}^k(t)$ *approaches asymptotically* $\mathbf{Y}_d^k(t)$ *as time t goes to infinity, i.e.,*

$$\forall [\mathbf{D}(.), \mathbf{Y}_d(.)] \in \mathfrak{D}^k x \mathfrak{Y}_d^k, \ \exists \Delta \in \mathfrak{R}^+ \cup \{\infty\},$$
$$\left\|\mathbf{Y}_0^{m-1} - \mathbf{Y}_{do}^{m-1}\right\| < \Delta \Longrightarrow \lim \left[\left\| \mathbf{Y}^k(t) - \mathbf{Y}_d^k(t) \right\| : t \rightarrow \infty \right] = 0,$$
$$k \in \{0, 1, 2, ...\}. \tag{8.26}$$

The zero, $(k = 0)$, *order tracking is simply called* ***tracking****.*

If, and only if, the value of Δ *depends at most on* $\mathfrak{D}^k x \mathfrak{Y}_d^k$ *but not on a particular choice of* $[\mathbf{D}(.), \mathbf{Y}_d(.)]$ *from* $\mathfrak{D}^k x \mathfrak{Y}_d^k$, *then the k-th order tracking is* ***uniform over*** $\mathfrak{D}^k x \mathfrak{Y}_d^k$.

The k-th order tracking over $\mathfrak{D}^k x \mathfrak{Y}_d^k$ *is* ***global (in the whole)*** *if, and only if,* $\Delta = \infty$ *for every* $[\mathbf{D}(.), \mathbf{Y}_d(.)] \in \mathfrak{D}^k x \mathfrak{Y}_d^k$. *It is also uniform over* $\mathfrak{D}^k x \mathfrak{Y}_d^k$.

Note 181 ***Tracking and realizability of*** $\mathbf{Y}_d(t)$

The desired plant output $\mathbf{Y}_d(t)$ *can be unrealizable even though the plant can mathematically exhibit tracking. This is due to the asymptotic convergence of the real output behavior to the desired one only as* $t \longrightarrow \infty$. *This essentially means that* $\mathbf{Y}(t)$ *converges to* $\mathbf{Y}_d(\infty)$ *as* $t \longrightarrow \infty$. *We do not require for tracking any closeness of* $\mathbf{Y}(t)$ *to* $\mathbf{Y}_d(t)$ *at any* $t < \infty$, $t \in \mathfrak{T}_0$.

Tracking does not give any information about the real output behavior relative to the desired one at any finite moment, after the initial one. The deviation of the former from the latter can be arbitrarily large at any finite instant. The following definition eliminates this drawback.

Definition 182 ***The k-th order stablewise tracking in Lyapunov sense of the desired output of the IO plant (2.21) or of the ISO plant (2.33), (2.34)***

The m-th order plant exhibits **the k-th order stablewise asymptotic output tracking,** *for short* **the k-th order stablewise tracking,** *over* $\mathfrak{D}^k x \mathfrak{Y}_d^k$ *if, and only if, it exhibits* **the k-th order tracking over** $\mathfrak{D}^k x \mathfrak{Y}_d^k$, *and for every positive real number* ε *and for every* $[\mathbf{D}(.), \mathbf{Y}_d(.)] \in \mathfrak{D}^k x \mathfrak{Y}_d^k$ *there is a positive real number* δ, $\delta = \delta(\varepsilon, \mathbf{D}, \mathbf{Y}_d)$, *such that* $\left\|\mathbf{Y}_0^{m-1} - \mathbf{Y}_{d0}^{m-1}\right\| < \delta$ *guarantees* $\left\|\mathbf{Y}^k(t) - \mathbf{Y}_d^k(t)\right\| < \varepsilon$ *for all* $t \in \mathfrak{T}_0$, *i.e.,*

$$\forall \varepsilon \in \mathfrak{R}^+, \forall [\mathbf{D}(.), \mathbf{Y}_d(.)] \in \mathfrak{D}^k x \mathfrak{Y}_d^k, \exists \delta \in \mathfrak{R}^+,$$
$$\left\|\mathbf{Y}_0^{m-1} - \mathbf{Y}_{d0}^{m-1}\right\| < \delta \Longrightarrow \left\|\mathbf{Y}^k(t) - \mathbf{Y}_d^k(t)\right\| < \varepsilon, \ \forall t \in \mathfrak{T}_0,$$
$$k \in \{0, 1, 2, ...\}. \tag{8.27}$$

The zero, $(k = 0)$, *order stablewise tracking is simply called* ***stablewise tracking.***

If, and only if, the values of both δ *and* Δ *depend at most on* $\mathfrak{D}^k x \mathfrak{Y}_d^k$, *but not on a particular choice of* $[\mathbf{D}(.), \mathbf{Y}_d(.)]$ *from* $\mathfrak{D}^k x \mathfrak{Y}_d^k$, *then, and only then, the k-th order stablewise tracking over* $\mathfrak{D}^k x \mathfrak{Y}_d^k$ *is* **uniform over** $\mathfrak{D}^k x \mathfrak{Y}_d^k$.

The k-th order stablewise tracking over $\mathfrak{D}^k x \mathfrak{Y}_d^k$ *is* **global (in the whole)** *if, and only if, it is both the k-th order global tracking over* $\mathfrak{D}^k x \mathfrak{Y}_d^k$ *and the k-th order stablewise tracking over* $\mathfrak{D}^k x \mathfrak{Y}_d^k$ *with* $\delta = \delta(\varepsilon, \mathfrak{D}^k, \mathfrak{Y}_d^k) \to \infty$ *as* $\varepsilon \to \infty$ *for every* $[\mathbf{D}(.), \mathbf{Y}_d(.)] \in \mathfrak{D}^k x \mathfrak{Y}_d^k$. *It is also uniform over* $\mathfrak{D}^k x \mathfrak{Y}_d^k$.

Theorem 183 ***Stablewise tracking and realizability of*** $\mathbf{Y}_d(t)$

If a plant exhibits stablewise tracking of the desired output $\mathbf{Y}_d(t)$, *then* $\mathbf{Y}_d(t)$ *is realizable for the desired output initial conditions, i.e.,*

$$\mathbf{Y}(0) = \mathbf{Y}_d(0) \Longrightarrow \mathbf{Y}(t) = \mathbf{Y}_d(t), \ \forall t \in \mathfrak{T}_0.$$

Proof. Let the *m-th* order plant exhibit stablewise tracking of the desired output $\mathbf{Y}_d(t)$. Conditions of Definition 182 hold, i.e., (8.27) is valid. Let us assume that $\mathbf{Y}_d(t)$ is not realizable. We will disprove this assumption by showing that it leads to a contradiction. Let $\mathbf{Y}_0^{m-1} = \mathbf{Y}_{d0}^{m-1}$. Hence, $\left\|\mathbf{Y}_0^{m-1} - \mathbf{Y}_{d0}^{m-1}\right\| = 0 < \delta$ for every $\delta \in \mathfrak{R}^+$, which implies $\|\mathbf{Y}(t; \mathbf{Y}_{d0}) - \mathbf{Y}_d(t; \mathbf{Y}_{d0})\| < \varepsilon$, $\forall t \in \mathfrak{T}_0$, $\forall \varepsilon \in \mathfrak{R}^+$, due to (8.27). If $\mathbf{Y}_d(t)$ were unrealizable, then there would be a moment $\tau \in \mathfrak{T}_0$ and $\xi \in \mathfrak{R}^+$ such that $\|\mathbf{Y}(\tau; \mathbf{Y}_{d0}) - \mathbf{Y}_d(\tau; \mathbf{Y}_{d0})\| = \xi$. This would contradict $\|\mathbf{Y}(t; \mathbf{Y}_{d0}) - \mathbf{Y}_d(t; \mathbf{Y}_{d0})\| < \varepsilon, \forall t \in \mathfrak{T}_0, \forall \varepsilon \in \mathfrak{R}^+$. Hence, in view of arbitrarily small $\varepsilon > 0$, we can accept $\varepsilon < \xi$. According to the assumption for such ε there does not exist $\delta > 0$ satisfying Definition 182 because for $\mathbf{Y}_0^{m-1} = \mathbf{Y}_{d0}^{m-1}$, $\left\|\mathbf{Y}_0^{m-1} - \mathbf{Y}_{d0}^{m-1}\right\| = 0 < \delta$ for every $\delta \in \mathfrak{R}^+$, implies $\|\mathbf{Y}(\tau; \mathbf{Y}_{d0}) - \mathbf{Y}_d(\tau; \mathbf{Y}_{d0})\| = \xi > \varepsilon$. This means that the plant does not exhibit stablewise tracking of the desired output $\mathbf{Y}_d(t)$, which is the contradiction. Hence, the assumption that $\mathbf{Y}_d(t)$ is not realizable is invalid. Therefore, $\mathbf{Y}_d(t)$ is realizable ■

The $(m-1)\,th$ order stablewise tracking expresses stability of the desired output behavior, in addition to its tracking. It does not allow arbitrarily large output error for bounded initial conditions and for the bounded input vector function. However, it does not show the rate of the convergence of the real output behavior to the desired one.

Definition 184 ***The k-th order exponential output tracking***

The m-th order plant exhibits ***the k-th order exponential output tracking,*** *for short* ***the k-th order exponential tracking, over*** $\mathfrak{D}^k x\mathfrak{Y}_d^k$ *if, and only if, for every* $[\mathbf{D}(.), \mathbf{Y}_d(.)] \in \mathfrak{D}^k x\mathfrak{Y}_d^k$ *there exist positive real numbers* $\alpha \geq 1$, β, *and* Δ, *or* $\Delta = \infty$, $\alpha = \alpha(\mathbf{D}, \mathbf{Y}_d)$, $\beta = \beta(\mathbf{D}, \mathbf{Y}_d)$, $\Delta = \Delta(\mathbf{D}, \mathbf{Y}_d)$, *such that* $\left\| \mathbf{Y}_0^{m-1} - \mathbf{Y}_{d0}^{m-1} \right\| < \Delta$ *guarantees that* $\mathbf{Y}^k(t)$ *approaches exponentially* $\mathbf{Y}_d^k(t)$ *all the time, i.e.,*

$$\forall [\mathbf{D}(.), \mathbf{Y}_d(.)] \in \mathfrak{D}^k x\mathfrak{Y}_d^k, \ \exists \Delta \in \mathfrak{R}^+ \cup \{\infty\},$$
$$\left\| \mathbf{Y}_0^{m-1} - \mathbf{Y}_{d0}^{m-1} \right\| < \Delta \Longrightarrow \left\| \mathbf{Y}^k(t) - \mathbf{Y}_d^k(t) \right\| \leq \alpha \left\| \mathbf{Y}_0^{m-1} - \mathbf{Y}_{d0}^{m-1} \right\| \exp(-\beta t),$$
$$k \in \{0, 1, 2, ...\}. \tag{8.28}$$

The zero, $(k = 0)$, *order exponential tracking is simply called* ***exponential tracking.***

The k-th order exponential tracking over $\mathfrak{D}^k x\mathfrak{Y}_d^k$ *is* ***global (in the whole)*** *if, and only if,* $\Delta = \Delta(\mathbf{D}, \mathbf{Y}_d) = \infty$ *for every* $[\mathbf{D}(.), \mathbf{Y}_d(.)] \in \mathfrak{D}^k x\mathfrak{Y}_d^k$.

If, and only if, the values of α, β, *and* Δ *depend at most on* $\mathfrak{D}^k x\mathfrak{Y}_d^k$ *but not on a particular choice of* $[\mathbf{D}(.), \mathbf{Y}_d(.)]$ *from* $\mathfrak{D}^k x\mathfrak{Y}_d^k$, *then the k-th order exponential tracking is* ***uniform over*** $\mathfrak{D}^k x\mathfrak{Y}_d^k$.

Theorem 185 ***Exponential tracking and stablewise tracking***

If the m-th order plant exhibits exponential tracking of the desired output $\mathbf{Y}_d(t)$, *then the tracking is also stablewise tracking.*

Proof. Let a plant exhibit exponential tracking. The conditions of Definition 184 hold, i.e., (8.28) is valid. Let $\varepsilon \in \mathfrak{R}^+$ be arbitrarily small so that $\alpha^{-1}\varepsilon < \Delta$, and let $\delta(\varepsilon) = \alpha^{-1}\varepsilon \in \mathfrak{R}^+$. Hence, (8.28) guarantees

$$\left\| \mathbf{Y}_0^{m-1} - \mathbf{Y}_{d0}^{m-1} \right\| < \delta(\varepsilon) = \alpha^{-1}\varepsilon < \Delta \Longrightarrow$$
$$\left\| \mathbf{Y}(t) - \mathbf{Y}_d(t) \right\| \leq$$
$$\leq \alpha \left\| \mathbf{Y}_0 - \mathbf{Y}_{d0} \right\| \exp(-\beta t) < \alpha\alpha^{-1}\varepsilon \exp(-\beta t) \leq \varepsilon, \forall t \in \mathfrak{T}_0.$$

For every $\varepsilon \geq \alpha\Delta$ we accept $\delta(\varepsilon) = \alpha^{-1}\Delta$. This shows that Definition 182 is satisfied for $k = 0$, i.e., that exponential tracking is also stablewise tracking ■

The preceding theorems imply directly the following results.

Corollary 186 ***Exponential tracking and realizability of*** $\mathbf{Y}_d(t)$

If the plant exhibits exponential tracking of the desired output $\mathbf{Y}_d(t)$, *then* $\mathbf{Y}_d(t)$ *is realizable for the desired output initial conditions.*

Corollary 187 ***Necessity of realizability of*** $\mathbf{Y}_d(t)$

Realizability of $\mathbf{Y}_d(t)$ *is necessary for stablewise tracking, hence also for exponential tracking.*

Note 188 *Tracking allows arbitrary big error overshoot for arbitrary small initial output error. Stablewise tracking eliminates this drawback. Both tracking and stablewise tracking permit very slow error convergence to the zero error. Exponential tracking eliminates this drawback.*

Note 189 ***Tracking and the k-th order tracking***

Tracking is necessary for the k-th order tracking. It is also necessary for global tracking, for the k-th order (global) stablewise tracking and for the k-th order (global) exponential tracking.

Note 190 ***Tracking in Lyapunov sense and Lyapunov stability***

The specifications under 2. and 4. at the beginning of this subsection, together with the preceding definitions, show that, and why, the k-th order tracking properties in Lyapunov sense and Lyapunov stability properties are mutually different. They are also mutually independent for $0 < k < \nu$-1. *If* $k = \nu$-1 ≥ 0, *then*

- *(global) attraction of the desired internal behavior of the IO plant is necessary, but not sufficient, for its (global) tracking over* $\mathfrak{D}^k x \mathfrak{Y}_d^k$;
- *the* $(\nu - 1)$ *th order (global) tracking of the desired output behavior over* $\mathfrak{D}^k x \mathfrak{Y}_d^k$ *is sufficient (but not necessary) for the (global) attraction of the desired internal behavior of the IO plant, i.e., the former ensures the latter;*
- *(global) asymptotic stability of the desired internal behavior of the IO plant is necessary, but not sufficient, for the (global) stablewise tracking over* $\mathfrak{D}^k x \mathfrak{Y}_d^k$;
- *the* $(\nu - 1)$ *th order (global) stablewise tracking of the desired output behavior over* $\mathfrak{D}^k x \mathfrak{Y}_d^k$ *is sufficient (but not necessary) for the (global) asymptotic stability of the desired internal behavior of the IO plant, i.e., the former ensures the latter;*
- *(global) exponential stability of the desired internal behavior of the IO plant is necessary, but not sufficient, for the (global) exponential tracking over* $\mathfrak{D}^k x \mathfrak{Y}_d^k$;
- *the* $(\nu - 1)$ *th order (global) exponential tracking of the desired output behavior over* $\mathfrak{D}^k x \mathfrak{Y}_d^k$ *is sufficient (but not necessary) for the (global) exponential stability of the desired internal behavior of the IO plant, i.e., the former ensures the latter.*

Lyapunov stability properties do not guarantee the tracking properties in Lyapunov sense. The former are valid only for the nominal disturbance, while the latter hold for any disturbance from the family $\mathfrak{D}^k$. *The former concern stability properties of a particular, single, desired internal dynamic behavior, while the latter are valid for every desired (output or internal dynamic) behavior from* $\mathfrak{Y}_d^k$.

Lyapunov stability concept concerns all dynamic systems, not only plants and their control systems.

The concept of tracking in Lyapunov sense concerns the noncontrolled plant, or the controlled plant, hence its control system. It is the original qualitative dynamic concept of the control theory. It has significance only in the framework of control issues.

Note 191 *Tracking is necessary for all other, above defined, tracking properties.*

8.4.2 Output space: tracking with FRT

Tracking with the finite scalar reachability time

A higher tracking quality is *tracking with the finite scalar reachability time.*

A reachability time is a moment when a real output variable, or the whole real output vector, becomes, respectively, equal to a desired output variable, or to the whole desired output vector, and after that moment they stay equal forever.

The tracking properties in Lyapunov sense guarantee asymptotic convergence of the real output response to the desired one only for infinite *time* (as $t \to \infty$). They do not ensure that the real output response reaches the desired one in a finite *time* and stays equal since then forever. In order to overcome this essential drawback from the engineering and control system customer points of view, we will present definitions of some tracking properties with the finite reachability *time.*

We can demand that the reachability *time* is the same for all output variables (and their derivatives). It is then scalar valued denoted by τ_R, $\tau_R \in Int\,\mathfrak{T}_0$. It induces the *time* sets $\mathfrak{T}_R$ and $\mathfrak{T}_{R\infty}$ as the subsets of $\mathfrak{T}_0$,

$$\mathfrak{T}_R = \{t:\ 0 \le t \le \tau_R\} \subset \mathfrak{T}_0,\ \mathfrak{T}_{R\infty} = \{t: \tau_R \le t < \infty\} \subset \mathfrak{T}_0,\ \mathfrak{T}_R \cup \mathfrak{T}_{R\infty} = \mathfrak{T}_0. \tag{8.29}$$

The following definitions determine **in the output space** various types of ***tracking with the finite scalar reachability time.***

Definition 192 ***The k-th order tracking with the finite reachability time of the desired output of the IO plant (2.21) (Subsection 2.1.2) or of the ISO plant (2.33), (2.34) (Subsection 2.2.2)***

The m-th order plant exhibits ***the k-th order output tracking with the finite scalar reachability time*** τ_R*,* $\tau_R \in Int\,\mathfrak{T}_0$*, for short* ***the k-th order tracking with the finite reachability time*** τ_R*,* ***over*** $\mathfrak{D}^k \mathrm{x} \mathfrak{Y}_d^k$ *if, and only if, for every* $(\mathbf{D}(.), \mathbf{Y}_d(.)) \in \mathfrak{D}^k \mathrm{x} \mathfrak{Y}_d^k$ *there exists a positive real number* Δ*,* $\Delta = \Delta(\mathbf{D}, \mathbf{Y}_d)$*, or* $\Delta = \infty$*, such that* $\left\| \mathbf{Y}_{d0}^{m-1} - \mathbf{Y}_0^{m-1} \right\| < \Delta$ *guarantees that* $\mathbf{Y}^k(t)$ *reaches* $\mathbf{Y}_d^k(t)$ *at latest at the moment* τ_R*, after which they rest equal forever, i.e.,*

$$\begin{gathered} \forall\, [\mathbf{D}(.), \mathbf{Y}_d(.)] \in \mathfrak{D}^k \mathrm{x} \mathfrak{Y}_d^k,\ \exists \Delta \in \mathfrak{R}^+ \cup \{\infty\}, \Delta = \Delta(\mathbf{D}, \mathbf{Y}_d), \\ \left\| \mathbf{Y}_{d0}^{m-1} - \mathbf{Y}_0^{m-1} \right\| < \Delta \Longrightarrow \mathbf{Y}^k(t) = \mathbf{Y}_d^k(t),\ \forall t \in \mathfrak{T}_{R\infty}, \\ k \in \{0, 1, 2, ...\}. \end{gathered} \tag{8.30}$$

The zero, $(k = 0)$*, order tracking with the finite reachability time* τ_R *is simply called* ***tracking with the finite reachability time*** τ_R*.*

If, and only if, the value of Δ *depends at most on* $\mathfrak{D}^k \mathrm{x} \mathfrak{Y}_d^k$*, but not on a particular choice of* $(\mathbf{D}(.), \mathbf{Y}_d(.))$ *from* $\mathfrak{D}^k \mathrm{x} \mathfrak{Y}_d^k$*, then the k-th order tracking is* ***uniform over*** $\mathfrak{D}^k \mathrm{x} \mathfrak{Y}_d^k$*.*

The k-th order tracking with the finite reachability time τ_R is ***global (in the whole)*** *if, and only if, $\Delta = \infty$. It is also uniform.*

The overshoot can be very big on the *time* set $\mathfrak{T}_R$. In order to prevent it we introduce the following.

Definition 193 ***The k-th order stablewise tracking with the finite reachability time of the desired output of the IO plant (2.21) or of the ISO plant (2.33), (2.34)***

The m-th order plant exhibits ***the k-th order stablewise output tracking with the finite scalar reachability time*** *τ_R, $\tau_R \in Int\, \mathfrak{T}_0$, for short* ***the k-th order stablewise tracking with the finite reachability time τ_R, over $\mathfrak{D}^k \mathrm{x} \mathfrak{Y}_d^k$*** *if, and only if, it exhibits the k-th order tracking with the finite reachability time τ_R over $\mathfrak{D}^k \mathrm{x} \mathfrak{Y}_d^k$, and for every positive real number ε and for every $(\mathbf{D}(.), \mathbf{Y}_d(.)) \in \mathfrak{D}^k \mathrm{x} \mathfrak{Y}_d^k$ there exists a positive real number δ, $\delta = \delta(\varepsilon, \mathbf{D}, \mathbf{Y}_d)$, such that $\left\| \mathbf{Y}_{d0}^{m-1} - \mathbf{Y}_0^{m-1} \right\| < \delta$ guarantees $\left\| \mathbf{Y}_d^k(t) - \mathbf{Y}^k(t) \right\| < \varepsilon$ for all $t \in \mathfrak{T}_R = [0, \tau_R]$, i.e.,*

$$\begin{gathered} \forall \left[\mathbf{D}(.), \mathbf{Y}_d(.)\right] \in \mathfrak{D}^k \mathrm{x} \mathfrak{Y}_d^k,\ \forall \varepsilon \in \mathfrak{R}^+,\ \exists \delta \in \mathfrak{R}^+, \\ \left\| \mathbf{Y}_{d0}^{m-1} - \mathbf{Y}_0^{m-1} \right\| < \delta \Longrightarrow \left\| \mathbf{Y}^k(t) - \mathbf{Y}_d^k(t) \right\| < \varepsilon,\ \forall t \in \mathfrak{T}_R, \\ k \in \{0, 1, 2, ...\}. \end{gathered} \tag{8.31}$$

The zero, $(k = 0)$, order stablewise tracking with the finite reachability time τ_R, is simply called ***stablewise tracking with the finite reachability time τ_R.***

The k-th order stablewise tracking with the finite reachability time τ_R is ***global (in the whole)*** *if, and only if, it is both the k-th order global tracking with the finite reachability time τ_R, and the k-th order stablewise tracking with the finite reachability time τ_R with $\delta = \delta(\varepsilon, \mathbf{D}, \mathbf{Y}_d) \to \infty$ as $\varepsilon \to \infty$.*

If, and only if, the values of both δ and Δ depend at most on $\mathfrak{D}^k \mathrm{x} \mathfrak{Y}_d^k$, but not on a particular choice of $(\mathbf{D}(.), \mathbf{Y}_d(.))$ from $\mathfrak{D}^k \mathrm{x} \mathfrak{Y}_d^k$, then the k-th order stablewise tracking with the finite reachability time τ_R is ***uniform over $\mathfrak{D}^k \mathrm{x} \mathfrak{Y}_d^k$.***

Tracking with the finite vector reachability time

We will define **in the output space** various types of ***tracking with the finite vector reachability time.***

Elementwise tracking with the finite vector reachability time represents better tracking than the preceding tracking types.

We can associate with every output variable Y_i its own scalar reachability *time* $\tau_{Ri} \in Int\, \mathfrak{T}_0$, and with its derivatives $Y_i^{(1)}$, $Y_i^{(2)}$, ..., $Y_i^{(k)}$ their own scalar reachability times $\tau_{Ri(1)} \in Int\, \mathfrak{T}_0$, $\tau_{Ri(2)} \in Int\, \mathfrak{T}_0$, ..., $\tau_{Ri(k)} \in Int\, \mathfrak{T}_0$, respectively. They compose the following vector reachability times:

$$\boldsymbol{\tau}_R^N = \boldsymbol{\tau}_{R(0)}^N = \begin{bmatrix} \tau_{R1} \\ \tau_{R2} \\ ... \\ \tau_{RN} \end{bmatrix} = \begin{bmatrix} \tau_{R1,(0)} \\ \tau_{R2,(0)} \\ ... \\ \tau_{RN,(0)} \end{bmatrix} \in (Int\ \mathfrak{T}_0)^N, \tag{8.32}$$

where

$$(Int\mathfrak{T}_0)^i = \underbrace{Int\mathfrak{T}_0 \text{x} Int\mathfrak{T}_0 \text{x} ... \text{x} Int\mathfrak{T}_0}_{i-times} \tag{8.33}$$

and

$$\boldsymbol{\tau}_{R(j)}^N = \begin{bmatrix} \tau_{R1,(j)} \\ \tau_{R2,(j)} \\ ... \\ \tau_{RN,(j)} \end{bmatrix} \in (Int\ \mathfrak{T}_0)^N, \ j \in \{0, 1, 2, ..\}. \tag{8.34}$$

In order to treat mathematically effectively and simply such cases, we define *the vector reachability time* $\boldsymbol{\tau}_{R[k]}^{(j+1)N} \in (Int\mathfrak{T}_0)^{(k+1)N}$,

$$\boldsymbol{\tau}_{R[k]}^{(k+1)N} = \begin{bmatrix} \boldsymbol{\tau}_R^N \\ \boldsymbol{\tau}_{R(1)}^N \\ \boldsymbol{\tau}_{R(2)}^N \\ ... \\ \boldsymbol{\tau}_{R(j)}^N \end{bmatrix} = \begin{bmatrix} \boldsymbol{\tau}_{R(0)}^N \\ \boldsymbol{\tau}_{R(1)}^N \\ \boldsymbol{\tau}_{R(2)}^N \\ ... \\ \boldsymbol{\tau}_{R(k)}^N \end{bmatrix} \in (Int\mathfrak{T}_0)^{(k+1)N}, \ \boldsymbol{\tau}_{R[0]}^N = \boldsymbol{\tau}_R^N, \tag{8.35}$$

i.e.,

$$\boldsymbol{\tau}_{R[k]}^{(k+1)N} = \left[\tau_{R1,(0)} \ \cdots \ \tau_{RN,(0)} \ \ \tau_{R1,(1)} \cdots \ \tau_{RN,(1)} \ \cdots \ \tau_{R1,(k)} \ \cdots \ \tau_{RN,(k)}\right]^T,$$
$$\forall k = 0, \ 1, ..., m-1, \tag{8.36}$$

where $\tau_{Ri} = \tau_{Ri(0)}$ is the reachability *time* of the *i-th* output variable, $i = 1, 2, .., N$, and $\tau_{Ri,(j)}$ is the reachability *time* of the *j-th* derivative of the *i-th* output variable, $j = 0, 1, 2, .., m$-1.

We relate $\boldsymbol{\tau}_{R[k]}^{(k+1)N}$ to tracking treated via *the extended output space* $\mathfrak{R}^{(k+1)N}$, which for $k = m$-1, $m = \nu$, becomes *the internal dynamics space* $\mathfrak{R}^{mN} = \mathfrak{R}^{\nu N}$ if the plant is the *IO* plant. However, $\mathfrak{R}^{(k+1)N}$ becomes the ordinary output space $\mathfrak{R}^N$ for $k = m\text{-}1 = 0$ if the plant is the *ISO* plant since then $m = 1$.

We will use *the elementwise unit* $(k+1)N$ *vector* $\mathbf{1}_{(k+1)N}$, all elements of which are equal to one,

$$\mathbf{1}_{(k+1)N} = \begin{bmatrix} \underbrace{1\ 1...1}_{(k+1)-times} \end{bmatrix}^T \in \mathfrak{R}^{(k+1)N}, \ k \in \{0, 1, 2, ...\}, \tag{8.37}$$

and *the time vector* $\mathbf{t}^{(k+1)N}$ [170, p. 387], all elements of which are the same temporal variable-*time* t,

$$\mathbf{t}^{(k+1)N} = t\mathbf{1}_{(k+1)N} = [t\ t...t]^T \in \mathfrak{T}_0^{(k+1)N} \cup \{\infty\}^{(k+1)N}, \ k \in \{0, 1, 2, ...\}, \tag{8.38}$$

where

$$\mathfrak{T}_0^i = \underbrace{\mathfrak{T}_0 \text{x} \mathfrak{T}_0 \text{x} ... \text{x} \mathfrak{T}_0}_{i-times}. \tag{8.39}$$

The above notation leads to

$$\mathfrak{T}_R^{(k+1)N} = \left\{ \mathbf{t}^{(k+1)N} : \mathbf{0}_{(k+1)N} \leq \mathbf{t}^{(k+1)N} \leq \boldsymbol{\tau}_{R[k]}^{(k+1)N} \right\},$$
$$\mathfrak{T}_{R\infty}^{(k+1)N} = \left\{ \mathbf{t}^{(k+1)N} : \boldsymbol{\tau}_{R[k]}^{(k+1)N} \leq \mathbf{t}^{(k+1)N} < \infty \mathbf{1}_{(k+1)N} \right\}. \tag{8.40}$$

The symbolic vector notation

$$\mathbf{Y}^k(\mathbf{t}^{(k+1)N}) = \mathbf{Y}_d^k(\mathbf{t}^{(k+1)N}),\ \forall \mathbf{t}^{(k+1)N} \in [\boldsymbol{\tau}_{R[k]}^{(k+1)N},\ \infty \mathbf{1}_{(k+1)N}[,$$
$$k \in \{0, 1, 2, ...\}$$

means in the scalar form

$$Y_i^{(j)}(t) = Y_{di}^{(j)}(t),\ \forall t \in [\tau_{Ri(j)},\ \infty[,\ \forall i = 1, 2, ..., N,\ \forall j \in \{0, 1, 2, .., k\}.$$

Besides

$$\left| \mathbf{Y}^{(j)}(\mathbf{t}^N) - \mathbf{Y}_d^{(j)}(\mathbf{t}^N) \right| = \left| \begin{array}{c} Y_1^{(j)}(t) - Y_{d1}^{(j)}(t) \\ Y_2^{(j)}(t) - Y_{d2}^{(j)}(t) \\ ... \\ Y_N^{(j)}(t) - Y_{dN}^{(j)}(t) \end{array} \right| \in \mathfrak{R}_+^N,\ \forall j = 0, 1, 2, ..., k,$$

and

$$\left| \mathbf{Y}^k(\mathbf{t}^{(k+1)N}) - \mathbf{Y}_d^k(\mathbf{t}^{(k+1)N}) \right| = \left| \begin{array}{c} \mathbf{Y}(\mathbf{t}^N) - \mathbf{Y}_d(\mathbf{t}^N) \\ \mathbf{Y}^{(1)}(\mathbf{t}^N) - \mathbf{Y}_d^{(1)}(\mathbf{t}^N) \\ ... \\ \mathbf{Y}^{(k)}(\mathbf{t}^N) - \mathbf{Y}_d^{(k)}(\mathbf{t}^N) \end{array} \right| \in \mathfrak{R}_+^{(k+1)N},$$
$$k \in \{0, 1, 2, ...\}.$$

Let positive real numbers $\Delta_{i(j)}$, or $\Delta_{i(j)} = \infty$, be associated with the *j-th* derivative of Y_i and of Y_{di}, and be taken for the entries of the positive N vector $\boldsymbol{\Delta}_{(j)}$, i.e., of the positive $(k+1)N$ vector $\boldsymbol{\Delta}^{kN}$,

$$\boldsymbol{\Delta}_{(j)}^N = \left[\begin{array}{c} \Delta_{1,(j)} \\ \Delta_{2,(j)} \\ ... \\ \Delta_{N,(j)} \end{array} \right] \in \mathfrak{R}^{+^N} \cup \{\infty\}^N,\ \forall j = 0, 1, 2, ..., k,\ \Delta_{i,(0)} \equiv \Delta_i,\ \boldsymbol{\Delta}_{(0)}^N \equiv \boldsymbol{\Delta}^N, \tag{8.41}$$

$$\boldsymbol{\Delta}^{(k+1)N} = \left[\begin{array}{c} \boldsymbol{\Delta}_{(0)}^N \\ \boldsymbol{\Delta}_{(1)}^N \\ ... \\ \boldsymbol{\Delta}_{(k)}^N \end{array} \right] = \left[\begin{array}{c} \boldsymbol{\Delta}^N \\ \boldsymbol{\Delta}_{(1)}^N \\ ... \\ \boldsymbol{\Delta}_{(k)}^N \end{array} \right] \in \mathfrak{R}^{+^{(k+1)N}} \cup \{\infty\}^{(k+1)N},\ k \in \{1, 2, .., m-1\}, \tag{8.42}$$

so that

$$\left|\mathbf{Y}_0^k - \mathbf{Y}_{d0}^k\right| < \mathbf{\Delta}^{(k+1)N},\ \forall k = 0,1,2,...,m-1, \tag{8.43}$$

signifies

$$\left|Y_{i0}^{(j)} - Y_{di0}^{(j)}\right| < \Delta_{i,(j)},\ \forall i = 1,2,...,N,\ \forall j = 0,1,2,...,\ k. \tag{8.44}$$

Definition 194 ***The k-th order elementwise tracking with the finite vector reachability time*** $\tau_{R[k]}^{(k+1)N}$ ***of the desired output of the IO plant (2.21) or of the ISO plant (2.33), (2.34)***

The m-th order plant exhibits ***the k-th order elementwise output tracking with the finite vector reachability time*** $\boldsymbol{\tau}_{R[k]}^{(k+1)N}$, *for short* ***the k-th order elementwise tracking with the finite vector reachability time*** $\boldsymbol{\tau}_{R[k]}^{(k+1)N}$, ***over*** $\mathfrak{D}^k \mathrm{x} \mathfrak{Y}_d^k$ *if, and only if, for every* $(\mathbf{D}(.), \mathbf{Y}_d(.)) \in \mathfrak{D}^k \mathrm{x} \mathfrak{Y}_d^k$ *there exist positive real numbers* $\Delta_{i(j)}$, *or* $\Delta_{i(j)} = \infty$, *which are the entries of the vector* $\mathbf{\Delta}^{mN}$, $\mathbf{\Delta}^{mN} = \mathbf{\Delta}^{mN}(\mathbf{D}, \mathbf{Y}_d)$, *such that* $\left|\mathbf{Y}_{d0}^{m-1} - \mathbf{Y}_0^{m-1}\right| < \mathbf{\Delta}^{mN}$ *guarantees that* $\mathbf{Y}^k(\mathbf{t}^{(k+1)N})$ *becomes equal to* $\mathbf{Y}_d^k(\mathbf{t}^{(k+1)N})$ *at latest at the vector moment* $\boldsymbol{\tau}_{R[k]}^{(k+1)N}$, *after which they stay equal forever, i.e.,*

$$\begin{gathered}\forall\left[\mathbf{D}(.), \mathbf{Y}_d(.)\right] \in \mathfrak{D}^k \mathrm{x} \mathfrak{Y}_d^k,\ \exists \mathbf{\Delta}^{mN} \in \mathfrak{R}^{+^{\nu N}} \cup \left\{\infty \mathbf{1}^{mN}\right\}, \\ \left|\mathbf{Y}_{d0}^{m-1} - \mathbf{Y}_0^{m-1}\right| < \mathbf{\Delta}^{mN} \Longrightarrow \mathbf{Y}^k(\mathbf{t}^{(k+1)N}) = \mathbf{Y}_d^k(\mathbf{t}^{(k+1)N}), \\ \forall \mathbf{t}^{(k+1)N} \in \mathfrak{T}_{R\infty}^{(k+1)N},\ k \in \{0,1,2,...\}.\end{gathered} \tag{8.45}$$

The zero order, $(k = 0)$, *elementwise tracking with the finite vector reachability time* $\boldsymbol{\tau}_{R[k]}^{(k+1)N}$ *is simply called* ***elementwise tracking with the finite vector reachability time*** $\boldsymbol{\tau}_{R[k]}^{(k+1)N}$.

If, and only if, the value of $\mathbf{\Delta}^{mN}$ *depends at most on* $\mathfrak{D}^k \mathrm{x} \mathfrak{Y}_d^k$, *but not on a particular choice of* $(\mathbf{D}(.), \mathbf{Y}_d(.))$ *from* $\mathfrak{D}^k \mathrm{x} \mathfrak{Y}_d^k$, *then the k-th order elementwise tracking is* ***uniform over*** $\mathfrak{D}^k \mathrm{x} \mathfrak{Y}_d^k$.

The k-th order elementwise tracking with the finite vector reachability time $\boldsymbol{\tau}_{R[k]}^{(k+1)N}$ *is* ***global (in the whole)*** *if, and only if,* $\mathbf{\Delta}^{mN} = \infty \mathbf{1}^{mN}$. *It is also uniform.*

Definition 195 ***The k-th order stablewise elementwise tracking with the finite vector reachability time*** $\tau_{R[k]}^{(k+1)N}$ ***of the desired output of the IO plant (2.21) or of the ISO plant (2.33), (2.34)***

The m-th order plant exhibits ***the k-th order stablewise elementwise output tracking with the finite vector reachability time*** $\boldsymbol{\tau}_{R[k]}^{(k+1)N}$, *for short* ***the k-th order stablewise elementwise tracking with the finite vector reachability time*** $\boldsymbol{\tau}_{R[k]}^{(k+1)N}$, ***over*** $\mathfrak{D}^k \mathrm{x} \mathfrak{Y}_d^k$ *if, and only if, it exhibits the k-th order elementwise tracking with the finite vector reachability time* $\boldsymbol{\tau}_{R[k]}^{(k+1)N}$ *over* $\mathfrak{D}^k \mathrm{x} \mathfrak{Y}_d^k$, *and for every positive real* $(k+1)N$ *vector* $\boldsymbol{\varepsilon}^{(k+1)N}$ *and for every*

$(\mathbf{D}(.), \mathbf{Y}_d(.)) \in \mathfrak{D}^k \text{x} \mathfrak{Y}_d^k$ *there exists a positive real* mN *vector* $\boldsymbol{\delta}^{mN}$, $\boldsymbol{\delta}^{mN} = \boldsymbol{\delta}^{mN}(\boldsymbol{\varepsilon}^{(k+1)N}, \mathbf{D}, \mathbf{Y}_d)$, *such that* $\left|\mathbf{Y}_{d0}^{m-1} - \mathbf{Y}_0^{m-1}\right| < \boldsymbol{\delta}^{mN}$ *guarantees*

$$\left|\mathbf{Y}^k(\mathbf{t}^{(k+1)N}) - \mathbf{Y}_d^k(\mathbf{t}^{(k+1)N})\right| < \boldsymbol{\varepsilon}^{(k+1)N}$$

for all $\mathbf{t}^{(k+1)N} \in \mathfrak{T}_R^{(k+1)N}$, *i.e.,*

$$\begin{gathered}
\forall\left[\mathbf{D}(.), \mathbf{Y}_d(.)\right] \in \mathfrak{D}^k \text{x} \mathfrak{Y}_d^{(k+1)N}, \forall \boldsymbol{\varepsilon}^{(k+1)N} \in \mathfrak{R}^{+(k+1)N}, \exists \boldsymbol{\delta}^{mN} \in \mathfrak{R}^{+mN}, \\
\left|\mathbf{Y}_{d0}^{m-1} - \mathbf{Y}_0^{m-1}\right| < \boldsymbol{\delta}^{mN} \Longrightarrow \\
\left|\mathbf{Y}^k(\mathbf{t}^{(k+1)N}) - \mathbf{Y}_d^k(\mathbf{t}^{(k+1)N})\right| < \boldsymbol{\varepsilon}^{(k+1)N}, \ \forall \mathbf{t}^{(k+1)N} \in \mathfrak{T}_R^{(k+1)N} \\
k \in \{0, 1, 2, ...\}. \qquad (8.46)
\end{gathered}$$

The zero, $(k = 0)$, *order stablewise elementwise tracking with the finite vector reachability time* $\boldsymbol{\tau}_{R[k]}^{(k+1)N}$ *is simply called* ***stablewise elementwise tracking with the finite vector reachability time*** $\boldsymbol{\tau}_{R[k]}^{(k+1)N}$.

The k*-th order stablewise elementwise tracking with the finite vector reachability time* $\boldsymbol{\tau}_{R[k]}^{(k+1)N}$ *is* ***global (in the whole)*** *if, and only if, it is both the* k*-th order global elementwise tracking with the finite vector reachability time* $\boldsymbol{\tau}_{R[k]}^{(k+1)N}$ *and stablewise elementwise tracking with the finite vector reachability time* $\boldsymbol{\tau}_{R[k]}^{(k+1)N}$ *with* $\boldsymbol{\delta}^{mN} = \boldsymbol{\delta}^{mN}(\boldsymbol{\varepsilon}^{(k+1)N}, \mathbf{D}, \mathbf{Y}_d) \rightarrow \infty \mathbf{1}^{mN}$ *as* $\boldsymbol{\varepsilon}^{(k+1)N} \rightarrow \infty \mathbf{1}_{(k+1)N}$.

If, and only if, the values of both $\boldsymbol{\delta}^{mN}$ *and* $\boldsymbol{\Delta}^{mN}$ *depend at most on* $\mathfrak{D}^k \text{x} \mathfrak{Y}_d^k$, *but not on a particular choice of* $(\mathbf{D}(.), \mathbf{Y}_d(.))$ *from* $\mathfrak{D}^k \text{x} \mathfrak{Y}_d^k$, *then the* k*-th order stablewise elementwise tracking with the finite vector reachability time* $\boldsymbol{\tau}_{R[k]}^{(k+1)N}$ *is* ***uniform over*** $\mathfrak{D}^k \text{x} \mathfrak{Y}_d^k$.

Comment 196 *Every tracking with the finite (scalar or vector) reachability time implies perfect tracking that starts at the (scalar or vector) reachability instant and continues forever. It expresses very high tracking quality.*

8.4.3 Internal dynamics space: the *IO* plant tracking

IO plant tracking in Lyapunov sense

In what follows we will define various tracking properties via **the target set** $\Upsilon_{IO}^k(t; \mathbf{Y}_d^k)$ as a subset of **the internal dynamics space** $\mathfrak{R}^{\nu N}$ of the system, rather than via the system output space (Subsections 8.4.1 and 8.4.2), in order to consider *how the real output tracks the desired output.*

Let $d\left[\mathbf{v}, \Upsilon_{IO}^k(t; \mathbf{Y}_d^k)\right]$ be the distance of a vector $\mathbf{v}$ from the *IO* target set $\Upsilon_{IO}^k(t; \mathbf{Y}_d^k)$, [Equation (8.8) in Definition 160, Subsection 8.3.2], $\Upsilon_{IO}^k(t; \mathbf{Y}_d^k) \subset$

$\mathfrak{R}^{\nu N}$,

$$\Upsilon_{IO}^{k}(t;\mathbf{Y}_d^k)=\left\{\mathbf{Y}^{\nu-1}:\ \left(\begin{array}{c}\mathbf{Y}^k=\mathbf{Y}_d^k(t),\ \mathbf{Y}^{(j)}\in\mathfrak{R}^N,\\ \forall j=k+1,k+2,...,\nu-1\end{array}\right)\right\},$$
$$k\in\{0,1,...,\nu-1\}, \tag{8.47}$$

$$d\left[\mathbf{v},\Upsilon_{IO}^{k}(t;\mathbf{Y}_d^k)\right]=\inf[\|\mathbf{v}-\mathbf{w}\|:\mathbf{w}\in\Upsilon_{IO}^{k}(t;\mathbf{Y}_d^k)]. \tag{8.48}$$

The target set $\Upsilon_{IO}^{k}(t;\mathbf{Y}_d^k)$ is a subset of *the extended output space* $\mathfrak{R}^{\nu N}$ that is simultaneously *the internal dynamics space* of the *IO* plant (2.21) (Subsection 2.1.2). The target set $\Upsilon_{IO}^{k}(t;\mathbf{Y}_d^k)$ of the *IO* plant (2.21) does not depend on the disturbance vector $\mathbf{D}(.)$. It is time-varying as soon as $\mathbf{Y}_d^k(t)$ is variable. It is the instantaneous set at the instant t of all the extended real output vectors $\mathbf{Y}^{\nu-1}$,

$$\mathbf{Y}^{\nu-1}=[\mathbf{Y}^T\ \ \mathbf{Y}^{(1)^T}\ \ ...\ \mathbf{Y}^{(\nu-1)^T}]^T\in\mathfrak{R}^{\nu N},$$

such that (only) $\mathbf{Y}^k$, (i.e., the real output $\mathbf{Y}$ and its first k derivatives),

$$\mathbf{Y}^k=[\mathbf{Y}^T\ \ \mathbf{Y}^{(1)^T}\ \ ...\ \mathbf{Y}^{(k)^T}]^T\in\mathfrak{R}^{(k+1)N},$$

is (are) equal to $\mathbf{Y}_d^k(t)$ (to the desired output $\mathbf{Y}_d(t)$ and to its first k derivatives) at every moment $t\in\mathfrak{T}_0$, respectively. The higher derivatives of $\mathbf{Y}$, i.e.,

$$\mathbf{Y}^{(k+1)},\ \ ...\ \mathbf{Y}^{(\nu)},$$

can be different, respectively, from

$$\mathbf{Y}_d^{(k+1)}(t),\ \ ...\ \mathbf{Y}_d^{(\nu)}(t)$$

at any instant $t\in\mathfrak{T}_0$. However, they obey

$$A_P^{(\nu)}\mathbf{Y}^\nu=C_{Pu}^{(\mu_{Pu})}\mathbf{U}^{\mu_{Pu}}(t)+D_{Pd}^{(\mu_{Pd})}\mathbf{D}^{\mu_{Pd}}(t)\ \textit{with}\ \mathbf{Y}^k\equiv\mathbf{Y}_d^k(t).$$

They depend on both the disturbance vector $\mathbf{D}(t)$ and the control vector $\mathbf{U}(t)$.

Definition 197 ***Tracking in Lyapunov sense of the desired output of the IO plant (2.21)***

The ν-th order IO plant (2.21) exhibits ***the k-th order asymptotic output tracking,*** *for short* ***the k-th order tracking, over*** $\mathfrak{D}^k\text{x}\mathfrak{Y}_d^k$ *if, and only if, for every* $(\mathbf{D}(.),\mathbf{Y}_d(.))\in\mathfrak{D}^k\text{x}\mathfrak{Y}_d^k$ *there exists a positive real number* Δ, $\Delta=\Delta(\mathbf{D},\mathbf{Y}_d)$, *or* $\Delta=\infty$, *such that* $d\left(\mathbf{Y}_0^{\nu-1},\Upsilon_{IO}^{k}(0;\mathbf{Y}_d^k)\right)<\Delta$ *guarantees that* $\mathbf{Y}^{\nu-1}(t)$ *approaches asymptotically* $\Upsilon_{IO}^{k}(t;\mathbf{Y}_d^k)$ *as time t goes to infinity, i.e.,*

$$\forall\left[\mathbf{D}(.),\mathbf{Y}_d(.)\right]\in\mathfrak{D}^k\text{x}\mathfrak{Y}_d^k,\ \exists\Delta\in\mathfrak{R}^+\cup\{\infty\},$$
$$d\left[\mathbf{Y}_0^{\nu-1},\Upsilon_{IO}^{k}(0;\mathbf{Y}_d^k)\right]<\Delta\Longrightarrow\lim\left\langle d\left[\mathbf{Y}^{\nu-1}(t),\Upsilon_{IO}^{k}(t;\mathbf{Y}_d^k)\right]:t\to\infty\right\rangle=0,$$
$$k\in\{0,1,2,...,\nu-1\}. \tag{8.49}$$

The zero, $(k=0)$, order tracking is simply called ***tracking.***

If, and only if, the value of Δ *depends at most on* $\mathfrak{D}^k \mathrm{x} \mathfrak{Y}_d^k$*, but not on a particular choice of* $(\mathbf{D}(.), \mathbf{Y}_d(.))$ *from* $\mathfrak{D}^k \mathrm{x} \mathfrak{Y}_d^k$*, then the k-th order tracking is* ***uniform over*** $\mathfrak{D}^k \mathrm{x} \mathfrak{Y}_d^k$*.*

The k-th order tracking is ***global (in the whole)*** *if, and only if,* $\Delta = \infty$*. It is also uniform over* $\mathfrak{D}^k \mathrm{x} \mathfrak{Y}_d^k$*.*

Tracking guarantees the asymptotic convergence of the real output behaviors to the desired one. The asymptotic convergence means that the error vector tends to the zero vector as time t diverges to infinity. In other words, it signifies that the output steady state deviation (hence, steady state error) is equal to the zero vector.

Definition 198 ***The k-th order stablewise tracking of the desired output of the IO plant (2.21)***

The ν*-th order IO plant (2.21) exhibits* ***the k-th order stablewise asymptotic output tracking****, for short* ***the k-th order stablewise tracking, over*** $\mathfrak{D}^k \mathrm{x} \mathfrak{Y}_d^k$ *if, and only if, it exhibits the k-th order tracking over* $\mathfrak{D}^k \mathrm{x} \mathfrak{Y}_d^k$*, and for every positive real number* ε *and for every* $(\mathbf{D}(.), \mathbf{Y}_d(.)) \in \mathfrak{D}^k \mathrm{x} \mathfrak{Y}_d^k$ *there exists a positive real number* δ*,* $\delta = \delta(\varepsilon, \mathbf{D}, \mathbf{Y}_d)$*, such that* $d\left[\mathbf{Y}_0^{\nu-1}, \Upsilon_{IO}^k(0; \mathbf{Y}_d^k)\right] < \delta$ *guarantees* $d\left[\mathbf{Y}^{\nu-1}(t), \Upsilon_{IO}^k(t; \mathbf{Y}_d^k)\right] < \varepsilon$ *for all* $t \in \mathfrak{T}_0$*, i.e.,*

$$\forall (\mathbf{D}(.), \mathbf{Y}_d(.)) \in \mathfrak{D}^k \mathrm{x} \mathfrak{Y}_d^k,\ \forall \varepsilon \in \mathfrak{R}^+,\ \exists \delta \in \mathfrak{R}^+,\ \delta = \delta(\varepsilon, \mathbf{D}, \mathbf{Y}_d),$$
$$d\left[\mathbf{Y}_0^{\nu-1}, \Upsilon_{IO}^k(0; \mathbf{Y}_d^k)\right] < \delta \Longrightarrow d\left[\mathbf{Y}^{\nu-1}(t), \Upsilon_{IO}^k(t; \mathbf{Y}_d^k)\right] < \varepsilon,\ \forall t \in \mathfrak{T}_0,$$
$$k \in \{0, 1, 2, ..., \nu - 1\}. \tag{8.50}$$

The zero, $(k = 0)$*, order stablewise tracking is simply called* ***stablewise tracking.***

The k-th order stablewise tracking is ***global (in the whole)*** *if, and only if, it is both the k-th order global tracking, and the k-th order stablewise tracking with* $\delta = \delta(\varepsilon, \mathbf{D}, \mathbf{Y}_d) \to \infty$ *as* $\varepsilon \to \infty$ *for every* $(\mathbf{D}(.), \mathbf{Y}_d(.)) \in \mathfrak{D}^k \mathrm{x} \mathfrak{Y}_d^k$*.*

If, and only if, the values of both δ *and* Δ *depend at most on* $\mathfrak{D}^k \mathrm{x} \mathfrak{Y}_d^k$*, but not on a particular choice of* $(\mathbf{D}(.), \mathbf{Y}_d(.))$ *from* $\mathfrak{D}^k \mathrm{x} \mathfrak{Y}_d^k$*, then the k-th order stablewise tracking is* ***uniform over*** $\mathfrak{D}^k \mathrm{x} \mathfrak{Y}_d^k$*.*

Stablewise tracking incorporates tracking and ensures boundedness of the output deviations (i.e., of the output errors), but it does not provide any information about the rate of the real output vector convergence to the desired output vector.

The plant can exhibit tracking over one product set $\mathfrak{D}_1^k \mathrm{x} \mathfrak{Y}_{d1}^k$, but need not over another one, $\mathfrak{D}_2^k \mathrm{x} \mathfrak{Y}_{d2}^k$.

Definition 199 ***The k-th order exponential tracking of the desired output of the IO plant (2.21)***

The ν*-th order IO plant (2.21) exhibits* ***the k-th order exponential output tracking****, for short* ***the k-th order exponential tracking, over*** $\mathfrak{D}^k \mathrm{x} \mathfrak{Y}_d^k$ *if, and only if, for every* $(\mathbf{D}(.), \mathbf{Y}_d(.)) \in \mathfrak{D}^k \mathrm{x} \mathfrak{Y}_d^k$ *there exist positive real numbers*

$\alpha \geq 1$, β, and Δ, or $\Delta = \infty$, $\alpha = \alpha(\mathbf{D}, \mathbf{Y}_d)$, $\beta = \beta(\mathbf{D}, \mathbf{Y}_d)$, $\Delta = \Delta(\mathbf{D}, \mathbf{Y}_d)$, such that $d\left[\mathbf{Y}_0^{\nu-1}, \Upsilon_{IO}^k(0;\mathbf{Y}_d^k)\right] < \Delta$ guarantees that $\mathbf{Y}^{\nu-1}(t)$ approaches exponentially $\Upsilon_{IO}^k(t;\mathbf{Y}_d^k)$ all the time, i.e.,

$$\forall\left(\mathbf{D}(.), \mathbf{Y}_d(.)\right) \in \mathfrak{D}^k \mathrm{x} \mathfrak{Y}_d^k,\ \exists\alpha \in \mathfrak{R}^+,\ \alpha \geq 1,\ \exists\beta \in \mathfrak{R}^+,\ \exists\Delta \in \mathfrak{R}^+ \cup \{\infty\},$$
$$d\left[\mathbf{Y}_0^{\nu-1}, \Upsilon_{IO}^k(0;\mathbf{Y}_d^k)\right] < \Delta \Longrightarrow$$
$$d\left[\mathbf{Y}^{\nu-1}(t), \Upsilon_{IO}^k(t;\mathbf{Y}_d^k)\right] \leq \alpha d\left[\mathbf{Y}_0^{\nu-1}, \Upsilon_{IO}^k(0;\mathbf{Y}_d^k)\right] \exp\left(-\beta t\right),\ \forall t \in \mathfrak{T}_0,$$
$$k \in \{0, 1, 2, ..., \nu - 1\}. \tag{8.51}$$

The zero, $(k = 0)$, order exponential tracking is simply called ***exponential tracking.***

The k-th order exponential tracking is ***global (in the whole)*** *if, and only if, $\Delta = \infty$.*

If, and only if, the values of α, β, and Δ depend at most on $\mathfrak{D}^k \mathrm{x} \mathfrak{Y}_d^k$, but not on a particular choice of $\left(\mathbf{D}(.), \mathbf{Y}_d(.)\right)$ from $\mathfrak{D}^k \mathrm{x} \mathfrak{Y}_d^k$, then the k-th order exponential tracking is ***uniform over*** *$\mathfrak{D}^k \mathrm{x} \mathfrak{Y}_d^k$.*

The exponential tracking shows the rate of the real output vector convergence to the desired output vector, and simultaneously incorporates stablewise tracking. The rate of the convergence is exponential.

IO plant tracking with the finite scalar reachability time

Designers of technical plants predict that the plants will work only over bounded time intervals. The asymptotic convergence for them can be without any practical use, without any technical significance. Technical needs demand usually that the real behavior of the plant *reaches* the desired plant behavior in finite time that can be prespecified or should be determined. Such *finite reachability time* can be the same for all output variables, which means that it is the latest instant when the real value of every output variable should become equal to its desired value. After the reachability time, the real and the desired output should stay equal as long as the plant works. Although the lifetime of the plant is bounded, which can lead us to *the practical tracking concept* [98], [99], [111], [112], [127], [128], [220] - [224], we sometimes consider it as unbounded. The justification is that we do not know in advance the real lifetime of the plant. Besides, if we guarantee a satisfactory, or even excellent, tracking quality over an unbounded time interval $\mathfrak{T}_0 = [0, \infty[$, then we ensure simultaneously the same tracking quality over every subinterval of $\mathfrak{T}_0$ that incorporates the finite reachability time, and which is a bounded subinterval of $\mathfrak{T}_0$.

Definition 200 ***The k-th order tracking with the finite scalar reachability time of the desired output of the IO plant (2.21)***

The ν-th order IO plant (2.21) exhibits ***the k-th order output tracking with the finite reachability time*** *τ_R, $\tau_R \in Int\ \mathfrak{T}_0$, for short* ***the k-th order tracking with the finite reachability time*** *τ_R,* ***over*** *$\mathfrak{D}^k \mathrm{x} \mathfrak{Y}_d^k$ if, and* *only if, for every* $[\mathbf{D}(.), \mathbf{Y}_d(.)] \in \mathfrak{D}^k \text{x} \mathfrak{Y}_d^k$ *there exists a positive real number* Δ, $\Delta = \Delta(\mathbf{D}, \mathbf{Y}_d)$, *or* $\Delta = \infty$, *such that* $d\left[\mathbf{Y}_0^{\nu-1}, \Upsilon_{IO}^k(0; \mathbf{Y}_d^k)\right] < \Delta$ *guarantees that* $\mathbf{Y}^{\nu-1}(t)$ *reaches* $\Upsilon_{IO}^k(t; \mathbf{Y}_d^k)$ *at latest at the moment* τ_R, *after which it rests in* $\Upsilon_{IO}^k(t; \mathbf{Y}_d^k)$ *for ever, i.e.,*

$$\forall [\mathbf{D}(.), \mathbf{Y}_d(.)] \in \mathfrak{D}^k \text{x} \mathfrak{Y}_d^k,\ \exists \Delta \in \mathfrak{R}^+ \cup \{\infty\},$$
$$d\left[\mathbf{Y}_0^{\nu-1}, \Upsilon_{IO}^k(0; \mathbf{Y}_d^k)\right] < \Delta \Longrightarrow \mathbf{Y}^{\nu-1}(t) \in \Upsilon_{IO}^k(t; \mathbf{Y}_d^k),\ \forall t \in \mathfrak{T}_R,$$
$$k \in \{0, 1, 2, ..., \nu - 1\}. \tag{8.52}$$

The zero, $(k = 0)$, *order tracking with the finite reachability time* τ_R *is simply called* ***tracking with the finite reachability time*** τ_R.

If, and only if, the value of Δ *depends at most on* $\mathfrak{D}^k \text{x} \mathfrak{Y}_d^k$, *but not on a particular choice of* $(\mathbf{D}(.), \mathbf{Y}_d(.))$ *from* $\mathfrak{D}^k \text{x} \mathfrak{Y}_d^k$, *then the k-th order tracking with the finite reachability time* τ_R *is* ***uniform over*** $\mathfrak{D}^k \text{x} \mathfrak{Y}_d^k$.

The k-th order tracking with the finite reachability time τ_R *is* ***global*** *(**in the whole**) if, and only if,* $\Delta = \infty$. *It is also uniform.*

The following tracking property combines the stablewise tracking and the tracking with the finite reachability time.

Definition 201 ***The k-th order stablewise tracking with the finite reachability time of the desired output of the IO plant (2.21)***

The ν*-th order IO plant (2.21) exhibits* ***the k-th order stablewise output tracking with the finite reachability time*** τ_R, $\tau_R \in Int\, \mathfrak{T}_0$, *for short* ***the k-th order stablewise tracking with the finite reachability time*** τ_R, ***over*** $\mathfrak{D}^k \text{x} \mathfrak{Y}_d^k$ *if, and only if, it exhibits the k-th order tracking with the finite reachability time* τ_R *over* $\mathfrak{D}^k \text{x} \mathfrak{Y}_d^k$, *and for every positive real number* ε *and for every* $[\mathbf{D}(.), \mathbf{Y}_d(.)] \in \mathfrak{D}^k \text{x} \mathfrak{Y}_d^k$ *there exists a positive real number* δ, $\delta = \delta(\varepsilon, \mathbf{D}, \mathbf{Y}_d)$, *such that* $d\left[\mathbf{Y}_0^{\nu-1}, \Upsilon_{IO}^k(0; \mathbf{Y}_d^k)\right] < \delta$ *guarantees* $d\left[\mathbf{Y}^{\nu-1}(t), \Upsilon_{IO}^k(t; \mathbf{Y}_d^k)\right] < \varepsilon$ *for all* $t \in \mathfrak{T}_R$, *i.e.,*

$$\forall [\mathbf{D}(.), \mathbf{Y}_d(.)] \in \mathfrak{D}^k \text{x} \mathfrak{Y}_d^k,\ \forall \varepsilon \in \mathfrak{R}^+,\ \exists \delta \in \mathfrak{R}^+,\ \delta = \delta(\varepsilon, \mathbf{D}, \mathbf{Y}_d),$$
$$d\left[\mathbf{Y}_0^{\nu-1}, \Upsilon_{IO}^k(0; \mathbf{Y}_d^k)\right] < \delta \Longrightarrow d\left[\mathbf{Y}^{\nu-1}(t), \Upsilon_{IO}^k(t; \mathbf{Y}_d^k)\right] < \varepsilon,\ \forall t \in \mathfrak{T}_R,$$
$$k \in \{0, 1, 2, ..., \nu - 1\}. \tag{8.53}$$

The zero, $(k = 0)$, *order stablewise tracking with the finite reachability time* τ_R *is simply called* ***stablewise tracking with the finite reachability time*** τ_R.

The k-th order stablewise tracking with the finite reachability time τ_R *is* ***global*** *(**in the whole**) if, and only if, it is both the k-th order global tracking with the finite reachability time* τ_R, *and the k-th order stablewise tracking with the finite reachability time* τ_R *with* $\delta = \delta(\varepsilon, \mathbf{D}, \mathbf{Y}_d) \to \infty$ *as* $\varepsilon \to \infty$ *for every* $[\mathbf{D}(.), \mathbf{Y}_d(.)] \in \mathfrak{D}^k \text{x} \mathfrak{Y}_d^k$.

If, and only if, the values of both δ *and* Δ *depend at most on* $\mathfrak{D}^k \text{x} \mathfrak{Y}_d^k$, *but not on a particular choice of* $(\mathbf{D}(.), \mathbf{Y}_d(.))$ *from* $\mathfrak{D}^k \text{x} \mathfrak{Y}_d^k$, *then the k-th order stablewise tracking with the finite reachability time* τ_R *is* ***uniform over*** $\mathfrak{D}^k \text{x} \mathfrak{Y}_d^k$.

IO plant tracking with the finite vector reachability time

We should introduce some simple new symbolic vector notation in order to treat the following complex task.

We demand different reachability times for different output variables and for their derivatives. This opened the need for the formal introduction of the *time* vector notation in Subsection 8.4.2

Let $Dist\left[\mathbf{Y}^{\nu-1}(\mathbf{t}^{\nu N}), \Upsilon_{IO}^{k}(t;\mathbf{Y}_{d}^{k})\right]$ be **the vector** (i.e., **the elementwise**) **distance of the vector** $\mathbf{Y}^{\nu-1}(\mathbf{t}^{\nu N})$ **from the target set** $\Upsilon_{IO}^{k}(t;\mathbf{Y}_{d}^{k})$,

$$Dist\left[\mathbf{Y}^{\nu-1}(\mathbf{t}^{\nu N}), \Upsilon_{IO}^{k}(\mathbf{t}^{\nu N};\mathbf{Y}_{d}^{k})\right] = \inf[\left|\mathbf{Y}^{\nu-1}(\mathbf{t}^{\nu N})-\mathbf{w}\right| : \mathbf{w} \in \Upsilon_{IO}^{k}(\mathbf{t}^{\nu N};\mathbf{Y}_{d}^{k})], \tag{8.54}$$

where the infimum holds elementwise. The notation

$$\mathbf{Y}^{\nu-1}(\mathbf{t}^{\nu N}) \in \Upsilon_{IO}^{k}(\mathbf{t}^{\nu N};\mathbf{Y}_{d}^{k}),\ \forall \mathbf{t}^{\nu N} \in \mathfrak{T}_{R[k]\infty}^{\nu N}$$

means that the real output vector $\mathbf{Y}(\mathbf{t}^{N})$ and only its first k derivatives become equal to their desired values at latest at the reachability time, and thereafter rest equal,

$$\mathbf{Y}^{\nu-1}(\mathbf{t}^{\nu N}) \in \Upsilon_{IO}^{k}(\mathbf{t}^{\nu N};\mathbf{Y}_{d}^{k}),\ \forall \mathbf{t}^{\nu N} \in \mathfrak{T}_{R[k]\infty}^{\nu N} = [\boldsymbol{\tau}_{R[k]}^{\nu N},\ \infty \mathbf{1}_{\nu N}[,$$

$$\Longleftrightarrow$$

$$Y_{i}^{(j)}(t) = Y_{di}^{(j)}(t),\ \forall t \in [\tau_{Ri(j)},\ \infty[,\ \forall i = 1,2,...,N,\ \forall j = 0,1,2,...,k,$$
$$k \in \{0,1,2,...,\nu-1\}.$$

The derivatives of the order higher than k are left free, $Y_{i}^{(j)}(t) \in \mathfrak{R}$, $\forall j = k+1$, $k+2$, ..., ν-1, so that formally we let $\tau_{Ri(j)} = \infty$, $\forall j = k+1,\ k+2,\ ...,\ \nu$-1. This leads to define $\boldsymbol{\tau}_{R[k]}^{\nu N}$ by

$$\boldsymbol{\tau}_{R[k]}^{\nu N} = \begin{bmatrix} \boldsymbol{\tau}_{R[k]}^{(k+1)N} \\ \boldsymbol{\tau}_{\mathfrak{R}k}^{\nu} \end{bmatrix} \Longrightarrow \boldsymbol{\tau}_{R[\nu-1]}^{\nu N} = \begin{bmatrix} \boldsymbol{\tau}_{R(0)}^{N} \\ \boldsymbol{\tau}_{R(1)}^{N} \\ \boldsymbol{\tau}_{R(2)}^{N} \\ \cdots \\ \boldsymbol{\tau}_{R(\nu-1)}^{N} \end{bmatrix} = \begin{bmatrix} \boldsymbol{\tau}_{R}^{N} \\ \boldsymbol{\tau}_{R(1)}^{N} \\ \boldsymbol{\tau}_{R(2)}^{N} \\ \cdots \\ \boldsymbol{\tau}_{R(\nu-1)}^{N} \end{bmatrix},$$

where

$$\boldsymbol{\tau}_{\mathfrak{R}k}^{\nu} \in \mathfrak{R}^{\nu-k-1},\ k < \nu - 1.$$

Besides,

$$\boldsymbol{\Delta}_{IO[k]}^{\nu N} = [\Delta_{1}\ ...\ \Delta_{N}\ \ \Delta_{1(1)}\ ...\ \Delta_{N(1)}\ ......\Delta_{1(\nu-1)}\ ...\ \Delta_{N(\nu-1)}]^{T},$$
$$\Delta_{i(j)} \left\{ \begin{array}{c} \in \mathfrak{R}^{+} \cup \{\infty\},\ \forall i = 1,2,...,N,\ \forall j = 0,1,...,k, \\ = \infty,\ \forall i = 1,2,...,N,\ \forall j = k+1,k+2,...,\nu-1 \end{array} \right\}.$$

Definition 202 ***The elementwise tracking with the finite vector reachability time*** $\boldsymbol{\tau}_{R[k]}^{\nu N}$ ***of the desired output of the IO plant (2.21)***

The ν-th order IO plant (2.21) exhibits ***the k-th order output elementwise tracking with the finite vector reachability time*** $\boldsymbol{\tau}_{R[k]}^{\nu N}$, *for short* ***the k-th order elementwise tracking with the finite vector reachability time*** $\boldsymbol{\tau}_{R[k]}^{\nu N}$, ***over*** $\mathfrak{D}^k \mathrm{x} \mathfrak{Y}_d^k$ *if, and only if, for every* $[\mathbf{D}(.), \mathbf{Y}_d(.)] \in \mathfrak{D}^k \mathrm{x} \mathfrak{Y}_d^k$ *there exist positive real numbers* $\Delta_{i(j)}$, *or* $\Delta_{i(j)} = \infty$, *which are the entries of the vector* $\boldsymbol{\Delta}_{IO[k]}^{\nu N}$, $\boldsymbol{\Delta}_{IO[k]}^{\nu N} = \boldsymbol{\Delta}_{IO[k]}^{\nu N}(\mathbf{D}, \mathbf{Y}_d)$, *such that*

$$Dist\left[\mathbf{Y}_0^{\nu-1}, \Upsilon_{IO}^k(0; \mathbf{D}, \mathbf{Y}_d^k]\right) < \boldsymbol{\Delta}_{IO[k]}^{\nu N}$$

guarantees that $\mathbf{Y}^{\nu-1}(\mathbf{t}^{\nu N})$ *reaches* $\Upsilon_{IO}^k(\mathbf{t}^{\nu N}; \mathbf{D}, \mathbf{Y}_d^k)$ *at latest at the vector moment* $\boldsymbol{\tau}_{R[k]}^{\nu N}$, *after which it rests in* $\Upsilon_{IO}^k(t; \mathbf{D}, \mathbf{Y}_d^k)$ *forever, i.e.,*

$$\begin{gathered}\forall [\mathbf{D}(.), \mathbf{Y}_d(.)] \in \mathfrak{D}^k \mathrm{x} \mathfrak{Y}_d^k,\ \exists \boldsymbol{\Delta}_{IO[k]}^{\nu N} \in \mathfrak{R}^{+^{\nu N}} \cup \{\infty \mathbf{1}_{\nu N}\}, \\ Dist\left[\mathbf{Y}_0^{\nu-1}, \Upsilon_{IO}^k(0; \mathbf{D}, \mathbf{Y}_d^k)\right] < \boldsymbol{\Delta}_{IO[k]}^{\nu N} \Longrightarrow \\ \mathbf{Y}^{\nu-1}(\mathbf{t}^{\nu N}) \in \Upsilon_{IO}^k(\mathbf{t}^{\nu N}; \mathbf{D}, \mathbf{Y}_d^k),\ \forall \mathbf{t}^{\nu N} \in \mathfrak{T}_{R\infty}^{\nu N}, \\ k \in \{0, 1, 2, ..., \nu - 1\}. \end{gathered} \tag{8.55}$$

The zero, $(k = 0)$, order elementwise tracking with the finite vector reachability time $\boldsymbol{\tau}_{R[k]}^{\nu N}$ *is simply called* ***the elementwise tracking with the finite vector reachability time*** $\boldsymbol{\tau}_{R[k]}^{\nu N}$.

If, and only if, the value of $\boldsymbol{\Delta}_{IO[k]}^{\nu N}$ *depends at most on* $\mathfrak{D}^k \mathrm{x} \mathfrak{Y}_d^k$, *but not on a particular choice of* $[\mathbf{D}(.), \mathbf{Y}_d(.)]$ *from* $\mathfrak{D}^k \mathrm{x} \mathfrak{Y}_d^k$, *then the k-th order elementwise tracking with the finite vector reachability time* $\boldsymbol{\tau}_{R[k]}^{\nu N}$ *is* ***uniform over*** $\mathfrak{D}^k \mathrm{x} \mathfrak{Y}_d^k$.

The k-th order elementwise tracking with the finite vector reachability time $\boldsymbol{\tau}_{R[k]}^{\nu N}$ *is* ***global (in the whole)*** *if, and only if,* $\boldsymbol{\Delta}_{IO[k]}^{\nu N} = \infty \mathbf{1}_{\nu N}$. *It is also uniform.*

This tracking property permits arbitrarily big overshoots to the output variables before their reachability times elapse. The following property ensures boundedness of the overshoots of the difference $\left|\mathbf{Y}^k - \mathbf{Y}_d^k\right|$. It does not mind the values of the differences $\left|Y^{(j)} - Y_d^{(j)}\right|$ of the order higher than k, i.e., of $\left|Y^{(j)} - Y_d^{(j)}\right|$ for $j = k+1,\ k+2,\ ...\ ,\ \nu$. Hence, we define the νN vector $\boldsymbol{\varepsilon}_{[k]}^{\nu N} \in \mathfrak{R}^{+^{\nu N}}$ by

$$\begin{gathered}\boldsymbol{\varepsilon}_{[k]}^{\nu N} = \left[\varepsilon_1\ \varepsilon_2\ ...\ \varepsilon_N\ .....\ \varepsilon_{1(k)}\ \varepsilon_{2(k)}\ ...\ \varepsilon_{N(k)}\ \infty\ \ \infty\ ...\ \infty\right]^T, \\ k \in \{0, 1, 2, ..., \nu - 1\}, \\ \varepsilon_{i(j)} \left\{ \begin{array}{c} \in \mathfrak{R}^+,\ i = 1,\ 2, ...,\ N,\quad j = 0,\ 1, ...,\ k, \\ = \infty,\ i = 1, 2, ..., N,\ j = k+1,\ k+2, ..., \nu \end{array} \right\}. \end{gathered} \tag{8.56}$$

Definition 203 ***The k-th order stablewise elementwise tracking with the finite vector reachability time*** $\boldsymbol{\tau}$ ***of the desired output of the IO plant (2.21)***

The ν-th order IO plant (2.21) exhibits **the k-th order stablewise elementwise output tracking with the finite vector reachability time $\boldsymbol{\tau}_{R[k]}^{\nu N}$**, *for short* **the k-th order stablewise elementwise tracking with the finite vector reachability time $\boldsymbol{\tau}_{R[k]}^{\nu N}$, over $\mathfrak{D}^k x \mathfrak{Y}_d^k$** *if, and only if, it exhibits the k-th order elementwise tracking with the finite vector reachability time $\boldsymbol{\tau}_{R[k]}^{\nu N}$ over $\mathfrak{D}^k x \mathfrak{Y}_d^k$, and for every νN vector $\boldsymbol{\varepsilon}_{[k]}^{\nu N}$ (8.56) and for every $[\mathbf{D}(.), \mathbf{Y}_d(.)] \in \mathfrak{D}^k x \mathfrak{Y}_d^k$ there exists a positive real νN vector $\boldsymbol{\delta}^{\nu N}$, $\boldsymbol{\delta}^{\nu N} = \boldsymbol{\delta}^{\nu N}(\boldsymbol{\varepsilon}_{[k]}^{\nu N}, \mathbf{D}, \mathbf{Y}_d)$, such that $Dist[\mathbf{Y}_0^{\nu-1}, \Upsilon_{IO}^k(0; \mathbf{D}, \mathbf{Y}_d^k)] < \boldsymbol{\delta}^{\nu N}$ guarantees*

$$Dist\left[\mathbf{Y}^{\nu-1}(\mathbf{t}^{\nu N}), \Upsilon_{IO}^k(\mathbf{t}^{\nu N}; \mathbf{D}, \mathbf{Y}_d^k)\right] < \boldsymbol{\varepsilon}_{[k]}^{\nu N}$$

for every $\mathbf{t}^{\nu N} \in \mathfrak{T}_{R[k]}^{\nu N} = [\mathbf{0}_{\nu N}, \boldsymbol{\tau}_{R[k]}^{\nu N}]$, i.e.,

$$\begin{gathered}\forall [\mathbf{D}(.), \mathbf{Y}_d(.)] \in \mathfrak{D}^k x \mathfrak{Y}_d^k,\ \forall \boldsymbol{\varepsilon}_{[k]}^{\nu N} \in \mathfrak{R}^{+^{\nu N}},\ \exists \boldsymbol{\delta}^{\nu N} \in \mathfrak{R}^{+^{\nu N}}, \\ \boldsymbol{\delta}^{\nu N} = \boldsymbol{\delta}^{\nu N}(\boldsymbol{\varepsilon}_{[k]}^{\nu N}, \mathbf{D}, \mathbf{Y}_d),\ Dist\left[\mathbf{Y}_0^{\nu-1}, \Upsilon_{IO}^k(0; \mathbf{D}, \mathbf{Y}_d^k)\right] < \boldsymbol{\delta}^{\nu N} \Longrightarrow \\ Dist\left[\mathbf{Y}^{\nu-1}(\mathbf{t}^{\nu N}), \Upsilon_{IO}^k(\mathbf{t}^{\nu N}; \mathbf{D}, \mathbf{Y}_d^k)\right] < \boldsymbol{\varepsilon}_{[k]}^{\nu N},\ \forall \mathbf{t}^{\nu N} \in \mathfrak{T}_R^{\nu N}, \\ k \in \{0, 1, 2, ..., \nu - 1\}. \end{gathered} \tag{8.57}$$

The zero, ($k = 0$), order stablewise elementwise tracking with the finite vector reachability time $\boldsymbol{\tau}_{R[k]}^{\nu N}$ is simply called **stablewise elementwise tracking with the finite vector reachability time $\boldsymbol{\tau}_{R[k]}^{\nu N}$**.

The k-th order stablewise elementwise tracking with the finite vector reachability time $\boldsymbol{\tau}_{R[k]}^{\nu N}$ is **global (in the whole)** *if, and only if, it is both the k-th order global elementwise tracking with the finite vector reachability time $\boldsymbol{\tau}_{R[k]}^{\nu N}$ and the stablewise elementwise tracking with the finite vector reachability time $\boldsymbol{\tau}_{R[k]}^{\nu N}$ with $\boldsymbol{\delta}^{\nu N} = \boldsymbol{\delta}^{\nu N}(\boldsymbol{\varepsilon}_{[k]}^{\nu N}, \mathbf{D}, \mathbf{Y}_d) \to \infty \mathbf{1}_{\nu N}$ as $\boldsymbol{\varepsilon}_{[k]}^{\nu N} \to \infty \mathbf{1}_{\nu N}$ for every $[\mathbf{D}(.), \mathbf{Y}_d(.)] \in \mathfrak{D}^k x \mathfrak{Y}_d^k$.*

If, and only if, the values of both $\boldsymbol{\delta}^{\nu N}$ and $\boldsymbol{\Delta}_{IO[k]}^{\nu N}$ depend at most on $\mathfrak{D}^k x \mathfrak{Y}_d^k$, but not on a particular choice of $(\mathbf{D}(.), \mathbf{Y}_d(.))$ from $\mathfrak{D}^k x \mathfrak{Y}_d^k$, then the k-th order stablewise elementwise tracking with the finite vector reachability time $\boldsymbol{\tau}_{R[k]}^{\nu N}$ is **uniform over $\mathfrak{D}^k x \mathfrak{Y}_d^k$**.

Note 204 *Every (ν-1) th order stablewise tracking property ensures simultaneously global asymptotic stability of the desired output behavior of the IO plant (2.21).*

However, the lower order stablewise tracking properties do not ensure global asymptotic stability of the desired output behavior of the IO plant (2.21).

It is smart to test stability of the plant itself. If it is stable then we can continue with the tracking control synthesis. If it is not, then we should first stabilize the plant so that Claim 155 (Section 8.2) holds. The tracking control should be designed for the stabilized plant.

8.4.4 The *ISO* plant tracking in Lyapunov sense

We will present in the sequel the definitions of various tracking properties via **the state space** $\mathfrak{R}^n$ of the *ISO* plant (2.33), (2.34) (Subsection 2.2.2) in order to consider *how the plant real output tracks the plant desired output*. Subsection 8.4.1 presents the definitions of various tracking properties via **the output space** $\mathfrak{R}^{kN}$ of the system in general. They hold unchanged for both the *IO* plant (2.15) (Subsection 2.1.2) and for the *ISO* plant (2.33), (2.34).

A better tracking behavior than state (stabilizing or exponential) tracking is **stable** *state (stabilizing or exponential) tracking*, which guarantees also asymptotic or exponential stability of the desired state $\mathbf{X}_{Pd}(t;\mathbf{X}_{Pd0};\mathbf{D}_N;\mathbf{U}_N;\mathbf{Y}_d^k)$, (3.16) (Subsection 3.3.3), for short $\mathbf{X}_{Pd}(t;\mathbf{X}_{Pd0})$,

$$\mathbf{X}_{Pd}(t;\mathbf{X}_{Pd0}) \equiv \mathbf{X}_{Pd}(t;\mathbf{X}_{Pd0};\mathbf{D}_N;\mathbf{U}_N;\mathbf{Y}_d^k). \tag{8.58}$$

The applications of the following definitions of various *stable state (stabilizing) tracking properties* demand the knowledge of the desired motion $\mathbf{X}_{Pd}(t;\mathbf{X}_{Pd0})$. Therefore, we accept its knowledge, i.e., the following assumption to hold:

Assumption 205 *The plant desired motion* $\mathbf{X}_{Pd}(.;\mathbf{X}_{Pd0};\mathbf{D}_N;\mathbf{U}_N;\mathbf{Y}_d^k)$ *is well defined on* $\mathfrak{D}^k \mathrm{x} \mathfrak{Y}_d^k \mathrm{x} \mathfrak{X}_0$.

This means our knowledge of the solution $\mathbf{X}_{Pd}(t;\mathbf{X}_{Pd0};\mathbf{D}_N;\mathbf{U}_N;\mathbf{Y}_d^k)$ to (3.17)-(3.19) (Subsection 3.3.3) for every $\mathbf{Y}_d^k \in \mathfrak{Y}_d^k$ and for some $\mathbf{D}_N \in \mathfrak{D}^k$.

Definition 206 ***The state tracking in Lyapunov sense of the desired output of the** ISO **plant (2.33), (2.34)***

The ISO plant (2.33), (2.34) exhibits ***the state asymptotic output tracking,*** *for short* ***the state tracking, over*** $\mathfrak{D}^k \mathrm{x} \mathfrak{Y}_d^k$ *if, and only if, for every* $[\mathbf{D}(.), \mathbf{Y}_d(.)] \in \mathfrak{D}^k \mathrm{x} \mathfrak{Y}_d^k$ *there exists a positive real number* Δ, $\Delta = \Delta(\mathbf{D}, \mathbf{U}, \mathbf{Y}_d)$, *or* $\Delta = \infty$, *such that* $\|\mathbf{X}_{Pd}(t;\mathbf{X}_{Pd0}) - \mathbf{X}_P(t;\mathbf{X}_{P0})\|$ *tends to zero as time* t *goes to infinity as soon as* $\|\mathbf{X}_{Pd0} - \mathbf{X}_{P0}\| < \Delta$, *i.e.,*

$$\begin{gathered}\forall [\mathbf{D}(.), \mathbf{Y}_d(.)] \in \mathfrak{D}^k \mathrm{x} \mathfrak{Y}_d^k,\ \exists \Delta \in \mathfrak{R}^+ \cup \{\infty\}, \\ \|\mathbf{X}_{Pd0} - \mathbf{X}_{P0}\| < \Delta \Longrightarrow \\ \lim \{\|\mathbf{X}_{Pd}(t;\mathbf{X}_{Pd0}) - \mathbf{X}_P(t;\mathbf{X}_{P0})\| : t \to \infty\} = 0, \\ k \in \{0,1,2,...\}. \end{gathered} \tag{8.59}$$

If, and only if, the value of Δ *depends at most on* $\mathfrak{D}^k \mathrm{x} \mathfrak{Y}_d^k$ *and* $\mathbf{U}$, *but not on a particular choice of* $[\mathbf{D}(.), \mathbf{Y}_d(.)]$ *from* $\mathfrak{D}^k \mathrm{x} \mathfrak{Y}_d^k$, *then the state tracking is* ***uniform over*** $\mathfrak{D}^k \mathrm{x} \mathfrak{Y}_d^k$.

The state tracking is ***global (in the whole)*** *if, and only if,* $\Delta = \infty$. *It is also uniform over* $\mathfrak{D}^k \mathrm{x} \mathfrak{Y}_d^k$.

The state tracking implies tracking of the plant desired motion $\mathbf{X}_{Pd}(.;\mathbf{X}_{Pd0})$ and of its desired output response $\mathbf{Y}_d(.;\mathbf{Y}_{d0})$. It does not take into account the behavior of the plant real motion $\mathbf{X}_P(.;\mathbf{X}_{P0})$ relative to the plant desired motion at any finite moment $t \in Int\,\mathfrak{T}_0$. We overcome this drawback with the following state tracking property.

The notation $\min(\delta, \Delta)$ denotes the smaller between δ and Δ,

$$\min(\delta, \Delta) = \left\{ \begin{array}{c} \delta,\ \delta \leq \Delta, \\ \Delta,\ \Delta \leq \delta \end{array} \right\}.$$

Definition 207 ***The k-th order stable state tracking in Lyapunov sense of the desired output of the ISO plant (2.33), (2.34)***

The ISO plant (2.33), (2.34) exhibits ***the k-th order stable state asymptotic output tracking,*** *for short* ***the k-th order stable state tracking, over*** $\mathfrak{D}^k x \mathfrak{Y}_d^k$ *if, and only if, for every* $[\varepsilon, \mathbf{D}(.), \mathbf{Y}_d(.)] \in \mathfrak{R}^+ x \mathfrak{D}^k x \mathfrak{Y}_d^k$ *there exist positive real numbers* δ *and* Δ, $\delta = \delta(\varepsilon, \mathbf{D}, \mathbf{U}, \mathbf{Y}_d)$, $\Delta = \Delta(\mathbf{D}, \mathbf{U}, \mathbf{Y}_d)$, *or* $\Delta = \infty$, *such that* $\|\mathbf{X}_{Pd0} - \mathbf{X}_{P0}\| < \min(\delta, \Delta)$ *guarantees both* $\|\mathbf{X}_{Pd}(t; \mathbf{X}_{Pd0}) - \mathbf{X}_P(t; \mathbf{X}_{P0})\| < \varepsilon$ *for all* $t \in \mathfrak{T}_0$ *and that* $\mathbf{Y}^k(t)$ *approaches asymptotically* $\mathbf{Y}_d^k(t)$ *as time* t *goes to infinity, i.e.,*

$$\begin{gathered}\forall [\mathbf{D}(.), \mathbf{Y}_d(.)] \in \mathfrak{D}^k x \mathfrak{Y}_d^k,\ \exists \delta \in \mathfrak{R}^+,\ \exists \Delta \in \mathfrak{R}^+ \cup \{\infty\}, \\ \|\mathbf{X}_{Pd0} - \mathbf{X}_{P0}\| < \min(\delta, \Delta) \implies \\ \|\mathbf{X}_{Pd}(t; \mathbf{X}_{Pd0}) - \mathbf{X}_P(t; \mathbf{X}_{P0})\| < \varepsilon,\ \forall t \in \mathfrak{T}_0,\ and \\ \lim \left\{ \left\| \mathbf{Y}_d^k(t) - \mathbf{Y}^k(t) \right\| : t \to \infty \right\} = 0,\ k \in \{0, 1, 2, ...\}. \end{gathered} \tag{8.60}$$

The zero, ($k = 0$*), order stable state tracking is simply called* ***stable state tracking.***

If, and only if, the values of δ *and* Δ *depend at most on* $\mathfrak{D}^k x \mathfrak{Y}_d^k$ *and* $\mathbf{U}$, *but not on a particular choice of* $[\mathbf{D}(.), \mathbf{Y}_d(.)]$ *from* $\mathfrak{D}^k x \mathfrak{Y}_d^k$, *then the k-th order stable state tracking is* ***uniform over*** $\mathfrak{D}^k x \mathfrak{Y}_d^k$.

The k-th order stable state tracking is ***global (in the whole)*** *if, and only if,* $\Delta = \infty$. *It is* ***strictly global (strictly in the whole)*** *if, and only if, both* $\varepsilon \longrightarrow \infty$ *implies* $\delta_{\max}(\varepsilon) \longrightarrow \infty$ *and* $\Delta = \infty$, *where* $\delta_{\max}(\varepsilon)$ *is the maximal* $\delta(\varepsilon)$ *that satisfies (8.60). It is also uniform over* $\mathfrak{D}^k x \mathfrak{Y}_d^k$.

This definition does not ensure stability of the desired output $\mathbf{Y}_d(t)$.

Definition 208 ***The k-th order stable state stabilizing tracking in Lyapunov sense of the desired output of the ISO plant (2.33), (2.34)***

The ISO plant (2.33), (2.34) exhibits ***the k-th order stable state stabilizing output tracking,*** *for short* ***the k-th order stable state stabilizing tracking, over*** $\mathfrak{D}^k x \mathfrak{Y}_d^k$ *if, and only if, it exhibits* ***the k-th order stable state tracking over*** $\mathfrak{D}^k x \mathfrak{Y}_d^k$, *and for every positive real number* ε *and for every* $[\mathbf{D}(.), \mathbf{Y}_d(.)] \in \mathfrak{D}^k x \mathfrak{Y}_d^k$ *there exists a positive real number* δ, $\delta = \delta(\varepsilon, \mathbf{D}, \mathbf{U}, \mathbf{Y}_d)$, *such that* $\left\| \mathbf{Y}_{d0}^k - \mathbf{Y}_0^k \right\| < \delta$ *guarantees* $\left\| \mathbf{Y}_d^k(t) - \mathbf{Y}^k(t; \mathbf{Y}_0; \mathbf{D}, \mathbf{U}) \right\| < \varepsilon$ *for all* $t \in \mathfrak{T}_0$, *i.e.,*

$$\begin{gathered}\forall [\mathbf{D}(.), \mathbf{Y}_d(.)] \in \mathfrak{D}^k x \mathfrak{Y}_d^k,\ \forall \varepsilon \in \mathfrak{R}^+,\ \exists \delta \in \mathfrak{R}^+, \delta = \delta(\varepsilon, \mathbf{D}, \mathbf{U}, \mathbf{Y}_d) \\ \left\| \mathbf{Y}_{d0}^k - \mathbf{Y}_0^k \right\| < \delta \implies \left\| \mathbf{Y}_d^k(t) - \mathbf{Y}^k(t; \mathbf{Y}_0; \mathbf{D}, \mathbf{U}) \right\| < \varepsilon,\ \forall t \in \mathfrak{T}_0, \\ k \in \{0, 1, 2, ...\}. \end{gathered} \tag{8.61}$$

The zero, $(k = 0)$, order stable state stabilizing tracking is simply called **the stable state stabilizing tracking.**

The k-th order stable state stabilizing tracking is **global (in the whole)** *if, and only if, it is both the k-th order stable state tracking, and the k-th order stable state stabilizing tracking with* $\delta = \delta(\varepsilon, \mathbf{D}, \mathbf{U}, \mathbf{Y}_d) \rightarrow \infty$ *as* $\varepsilon \rightarrow \infty$ *for every* $[\mathbf{D}(.), \mathbf{Y}_d(.)] \in \mathfrak{D}^k \mathrm{x} \mathfrak{Y}_d^k$. *It is* **strictly global (strictly in the whole)** *if, and only if, additionally it is* **the k-th order global stable state tracking over** $\mathfrak{D}^k \mathrm{x} \mathfrak{Y}_d^k$.

If, and only if, the values of both δ *and* Δ *depend at most on* $\mathfrak{D}^k \mathrm{x} \mathfrak{Y}_d^k$ *and* $\mathbf{U}$, *but not on a particular choice of* $[\mathbf{D}(.), \mathbf{Y}_d(.)]$ *from* $\mathfrak{D}^k \mathrm{x} \mathfrak{Y}_d^k$, *then the k-th order stable state stabilizing tracking is* **uniform over** $\mathfrak{D}^k \mathrm{x} \mathfrak{Y}_d^k$.

The *k-th* order stable state stabilizing tracking concerns both the internal dynamic behavior of the plant in the state space $\mathfrak{R}^n$ and its output dynamic behavior in the extended output space $\mathfrak{R}^{kN}$.

The asymptotic convergence of the real output vector $\mathbf{Y}(t)$ to the desired output vector, as well as of the real motions to the desired motion, does not provide any information about the rate of the convergence.

Definition 209 ***The k-th order exponentially stable state stabilizing tracking in Lyapunov sense of the desired output of the ISO plant (2.33), (2.34)***

The ISO plant (2.33), (2.34) exhibits **the k-th order exponentially stable state stabilizing output tracking,** *for short* **the k-th order exponentially stable state stabilizing tracking, over** $\mathfrak{D}^k \mathrm{x} \mathfrak{Y}_d^k$ *if, and only if, for every positive real number* ε *and for every* $[\mathbf{D}(.), \mathbf{Y}_d(.)] \in \mathfrak{D}^k \mathrm{x} \mathfrak{Y}_d^k$ *there exists a positive real number* δ, $\delta = \delta(\varepsilon, \mathbf{D}, \mathbf{U}, \mathbf{Y}_d)$, *such that* $\left\|\mathbf{Y}_{d0}^k - \mathbf{Y}_0^k\right\| < \delta$ *guarantees* $\left\|\mathbf{Y}_d^k(t) - \mathbf{Y}^k(t; \mathbf{Y}_0; \mathbf{D}, \mathbf{U})\right\| < \varepsilon$ *for all* $t \in \mathfrak{T}_0$, *i.e., (8.61) holds and for every* $[\mathbf{D}(.), \mathbf{Y}_d(.)] \in \mathfrak{D}^k \mathrm{x} \mathfrak{Y}_d^k$ *there exist positive real numbers* $\xi \geq 1$, ψ, *and* θ, *or* $\theta = \infty$, $\xi = \xi(\mathbf{D}, \mathbf{U}, \mathbf{Y}_d)$, $\psi = \psi(\mathbf{D}, \mathbf{U}, \mathbf{Y}_d)$, $\theta = \theta(\mathbf{D}, \mathbf{U}, \mathbf{Y}_d)$, *such that* $\|\mathbf{X}_{d0} - \mathbf{X}_0\| < \theta$ *guarantees that* $\mathbf{X}(t)$ *approaches exponentially* $\mathbf{X}_d(t)$ *all the time, i.e.,*

$$\forall [\mathbf{D}(.), \mathbf{Y}_d(.)] \in \mathfrak{D}^k \mathrm{x} \mathfrak{Y}_d^k,\ \exists \xi \in \mathfrak{R}^+,\ \xi \geq 1,\ \exists \psi \in \mathfrak{R}^+,\ \exists \theta \in \mathfrak{R}^+ \cup \{\infty\},$$
$$\|\mathbf{X}_{d0} - \mathbf{X}_0\| < \theta \Longrightarrow$$
$$\|\mathbf{X}(t) - \mathbf{X}_d(t)\| \leq \xi \|\mathbf{X}_{d0} - \mathbf{X}_0\| \exp(-\psi t),\ \forall t \in \mathfrak{T}_0. \tag{8.62}$$

The zero, $(k = 0)$, order exponentially stable state stabilizing tracking is simply called **the exponentially stable state stabilizing tracking.**

The k-th order exponentially stable state stabilizing tracking is **global (in the whole)** *if, and only if,* $\theta = \infty$. *It is* **strictly global** *if, and only if, the k-th order stable state stabilizing tracking is strict global and* $\theta = \infty$.

If, and only if, the values of ξ, ψ, *and* θ *depend at most on* $\mathfrak{D}^k \mathrm{x} \mathfrak{Y}_d^k$ *and* $\mathbf{U}$, *but not on a particular choice of* $[\mathbf{D}(.), \mathbf{Y}_d(.)]$ *from* $\mathfrak{D}^k \mathrm{x} \mathfrak{Y}_d^k$, *then the k-th order exponentially stable state stabilizing tracking is* **uniform over** $\mathfrak{D}^k \mathrm{x} \mathfrak{Y}_d^k$.

Note 210 *Exponentially stable state stablewise tracking signifies that the state vector converges exponentially to the desired motion* $\mathbf{X}_d(t)$ *in the state space, which does not ensure the exponential convergence of the real output* $\mathbf{Y}(t)$ *to the desired* $\mathbf{Y}_d(t)$ *in the output space. In order to ensure the exponential tracking it is necessary to deal additionally with the output space, or with the output error space, rather than only with the state space.*

Definition 211 ***The k-th order exponentially stable state exponential tracking in Lyapunov sense of the desired output of the*** *ISO* ***plant (2.33), (2.34)***

The ISO plant (2.33), (2.34) exhibits ***the k-th order exponentially stable state exponential output tracking****, for short* ***the k-th order exponentially stable state exponential tracking, over*** $\mathfrak{D}^k x \mathfrak{Y}_d^k$ *if, and only if, it exhibits the k-th order exponentially stable state stabilizing tracking over* $\mathfrak{D}^k x \mathfrak{Y}_d^k$ *and, additionally, it exhibits the k-th order exponential tracking over* $\mathfrak{D}^k x \mathfrak{Y}_d^k$*, i.e., for every* $[\mathbf{D}(.), \mathbf{Y}_d(.)] \in \mathfrak{D}^k x \mathfrak{Y}_d^k$ *there exist positive real numbers* $\alpha \geq 1$*,* β*, and* Δ*, or* $\Delta = \infty$*,* $\alpha = \alpha(\mathbf{D}, \mathbf{U}, \mathbf{Y}_d)$*,* $\beta = \beta(\mathbf{D}, \mathbf{U}, \mathbf{Y}_d)$*,* $\Delta = \Delta(\mathbf{D}, \mathbf{U}, \mathbf{Y}_d)$*, such that* $\|\mathbf{Y}_{d0} - \mathbf{Y}_0\| < \Delta$ *guarantees that* $\mathbf{Y}(t)$ *approaches exponentially* $\mathbf{Y}_d(t)$ *all the time, i.e.,*

$$\forall [\mathbf{D}(.), \mathbf{Y}_d(.)] \in \mathfrak{D}^k x \mathfrak{Y}_d^k, \; \exists \alpha \in \mathfrak{R}^+, \; \alpha \geq 1, \; \exists \beta \in \mathfrak{R}^+, \; \exists \Delta \in \mathfrak{R}^+ \cup \{\infty\},$$
$$\left\|\mathbf{Y}_{d0}^k - \mathbf{Y}_0^k\right\| < \Delta \Longrightarrow$$
$$\left\|\mathbf{Y}^k(t) - \mathbf{Y}_d^k(t)\right\| \leq \alpha \left\|\mathbf{Y}_{d0}^k - \mathbf{Y}_0^k\right\| \exp(-\beta t), \; \forall t \in \mathfrak{T}_0. \tag{8.63}$$

The zero , ($k = 0$*), order exponentially stable state exponential tracking is simply called* ***the exponentially stable state exponential tracking****.*

The k-th order exponentially stable state exponential tracking is ***global (in the whole)*** *if, and only if, the k-th order exponentially stable state stabilizing tracking is global and the k-th order exponential tracking is global. It is* ***strictly global*** *if, and only, if the k-th order exponentially stable state stabilizing tracking is strictly global and* $\Delta = \infty$*.*

If, and only if, the values of ξ*,* ψ*,* θ*,* α*,* β*, and* Δ *depend at most on* $\mathfrak{D}^k x \mathfrak{Y}_d^k$ *and* $\mathbf{U}$*, but not on a particular choice of* $[\mathbf{D}(.), \mathbf{Y}_d(.)]$ *from* $\mathfrak{D}^k x \mathfrak{Y}_d^k$*, then the k-th order exponentially stable state exponential tracking is* ***uniform over*** $\mathfrak{D}^k x \mathfrak{Y}_d^k$*.*

This definition guarantees the exponential stability of the desired motion $\mathbf{X}_d(t)$ and of the desired output behavior $\mathbf{Y}_d^k(t)$.

Note 212 *Tracking is necessary for all other above specified tracking properties.*

8.4.5 State space: the *ISO* plant tracking with FRT

ISO plant tracking with the finite scalar reachability time

The exponential convergence of the plant state vector $\mathbf{X}_P(t)$ to the plant desired motion $\mathbf{X}_{Pd}(t)$ does not imply the exponential convergence of the real output behavior $\mathbf{Y}(t)$ to the desired output vector $\mathbf{Y}_d(t)$.

The scalar state reachability time τ_R, $\tau_R \in Int\ \mathfrak{T}_0$, i.e., the reachability time related to the state space means that at the moment $t = \tau_R$ the real plant state $\mathbf{X}_P(\tau_R)$ becomes desired $\mathbf{X}_{Pd}(\tau_R)$, $\mathbf{X}_P(\tau_R) = \mathbf{X}_{Pd}(\tau_R)$, and they rest equal forever. This assures that the real output vector $\mathbf{Y}(t)$ becomes equal to the desired output vector $\mathbf{Y}_d(t)$ at the state reachability time τ_R that is simultaneously the scalar output reachability time τ_R. This comes out from Definition 71 (Subsection 3.3.3) of the desired state (i.e., of the desired motion). We accept the validity of Assumption 205 (Subsection 8.4.4) in what follows.

Definition 213 ***The tracking with the finite scalar reachability time of the desired output of the ISO plant (2.29), (2.30)***

The ISO plant (2.29), (2.30) exhibits ***the output tracking with the finite scalar reachability time*** τ_R, $\tau_R \in Int\ \mathfrak{T}_0$, *for short* ***the tracking with the finite reachability time*** τ_R, ***over*** $\mathfrak{D}^k \mathrm{x} \mathfrak{Y}^k$ *if, and only if, for every* $[\mathbf{D}(.), \mathbf{Y}_d(.)] \in \mathfrak{D}^k \mathrm{x} \mathfrak{Y}^k$ *there exists a positive real number* Δ, $\Delta = \Delta(\mathbf{D}, \mathbf{U}, \mathbf{Y}_d)$, *or* $\Delta = \infty$, *such that* $\|\mathbf{X}_{Pd0} - \mathbf{X}_{P0}\| < \Delta$ *guarantees that* $\mathbf{X}_P(t)$ *reaches* $\mathbf{X}_{Pd}(t)$ *at latest at the moment* τ_R, *after which they rest equal forever, i.e.,*

$$\forall [\mathbf{D}(.), \mathbf{Y}_d(.)] \in \mathfrak{D}^k \mathrm{x} \mathfrak{Y}^k,\ \exists \Delta \in \mathfrak{R}^+ \cup \{\infty\},$$
$$\|\mathbf{X}_{Pd0} - \mathbf{X}_{P0}\| < \Delta \Longrightarrow \mathbf{X}_P(t) = \mathbf{X}_{Pd}(t),\ \forall t \in [\tau_R,\ \infty[,$$
$$k \in \{0, 1, 2, ...\}. \tag{8.64}$$

If, and only if, the value of Δ *depends at most on* $\mathfrak{D}^k \mathrm{x} \mathfrak{Y}^k$ *and* $\mathbf{U}$, *but not on a particular choice of* $[\mathbf{D}(.), \mathbf{Y}_d(.)]$ *from* $\mathfrak{D}^k \mathrm{x} \mathfrak{Y}^k$, *then the tracking is* ***uniform over*** $\mathfrak{D}^k \mathrm{x} \mathfrak{Y}^k$.

The tracking with the finite reachability time τ_R *is* ***global (in the whole) over*** $\mathfrak{D}^k \mathrm{x} \mathfrak{Y}^k$ *if, and only if,* $\Delta = \infty$. *It is also uniform.*

Definition 214 ***The k-th order stable state stabilizing tracking with the finite scalar reachability time*** τ_R ***of the desired output of the ISO plant (2.29), (2.30)***

The ISO plant (2.29), (2.30) exhibits ***the k-th order stable state stabilizing output tracking with the finite scalar reachability time*** τ_R, $\tau_R \in Int\ \mathfrak{T}_0$, *for short* ***the k-th order stable state stabilizing tracking with the finite reachability time*** τ_R, ***over*** $\mathfrak{D}^k \mathrm{x} \mathfrak{Y}^k$ *if, and only if, it exhibits the tracking with the finite reachability time* τ_R *over* $\mathfrak{D}^k \mathrm{x} \mathfrak{Y}^k$, *and for every positive real number* ε *and for every* $[\mathbf{D}(.), \mathbf{Y}_d(.)] \in \mathfrak{D}^k \mathrm{x} \mathfrak{Y}^k$ *there exists a positive real number* δ, $\delta = \delta(\varepsilon, \mathbf{D}, \mathbf{U}, \mathbf{Y}_d)$, *such that* $\|\mathbf{X}_{P0} - \mathbf{X}_{Pd0}\| < \delta$ *and* $\|\mathbf{Y}_{d0} - \mathbf{Y}_0\| < \delta$ *guarantee both* $\|\mathbf{X}_{Pd}(t) - \mathbf{X}_P(t)\| < \varepsilon$ *for all* $t \in \mathfrak{T}_R = [0, \tau_R]$, *and* $\left\|\mathbf{Y}_d^k(t) - \mathbf{Y}^k(t)\right\| < \varepsilon$ *for all* $t \in \mathfrak{T}_0$, *i.e.,*

$$\forall [\mathbf{D}(.), \mathbf{Y}_d(.)] \in \mathfrak{D}^k \mathrm{x} \mathfrak{Y}^k,\ \forall \varepsilon \in \mathfrak{R}^+,\ \exists \delta \in \mathfrak{R}^+, \delta = \delta(\varepsilon, \mathbf{D}, \mathbf{U}, \mathbf{Y}_d),$$
$$\|\mathbf{X}_{Pd0} - \mathbf{X}_{P0}\| < \delta \text{ and } \|\mathbf{Y}_0 - \mathbf{Y}_{d0}\| < \delta \Longrightarrow$$
$$\|\mathbf{X}_{Pd}(t) - \mathbf{X}_P(t)\| < \varepsilon,\ \forall t \in \mathfrak{T}_R = [0, \tau_R], \text{ and}$$
$$\left\|\mathbf{Y}_d^k(t) - \mathbf{Y}^k(t)\right\| < \varepsilon,\ \forall t \in \mathfrak{T}_0,\ k \in \{0, 1, 2, ...\}. \tag{8.65}$$

The zero, $(k = 0)$, order stable state stabilizing tracking with the finite reachability time τ_R is simply called ***the stable state stabilizing tracking with the finite reachability time*** τ_R.

The k-th order stable state stabilizing tracking with the finite reachability time τ_R is ***global (in the whole) over*** $\mathfrak{D}^k \mathrm{x} \mathfrak{Y}^k$ *if, and only if, both the k-th order tracking with the finite reachability time τ_R is global and $\delta = \delta(\varepsilon, \mathbf{D}, \mathbf{U}, \mathbf{Y}_d) \to \infty$ as $\varepsilon \to \infty$ for every $[\mathbf{D}(.), \mathbf{Y}_d(.)] \in \mathfrak{D}^k \mathrm{x} \mathfrak{Y}^k$.*

If, and only if, the values of both δ and Δ depend at most on $\mathfrak{D}^k \mathrm{x} \mathfrak{Y}^k$ and $\mathbf{U}$, but not on a particular choice of $[\mathbf{D}(.), \mathbf{Y}_d(.)]$ from $\mathfrak{D}^k \mathrm{x} \mathfrak{Y}^k$, then the k-th order stable state stabilizing tracking with the finite reachability time τ_R is ***uniform over*** $\mathfrak{D}^k \mathrm{x} \mathfrak{Y}^k$.

Since the exponential convergence of the real state $\mathbf{X}_P(t)$ to the desired state $\mathbf{X}_{Pd}(t)$ does not imply the exponential convergence of the real output vector $\mathbf{Y}(t)$ to the desired output vector $\mathbf{Y}_d(t)$, we will not present the definition of the state exponential tracking with the finite scalar reachability time (for which see [106]).

ISO plant state elementwise tracking with the state finite vector reachability time

The notation

$$\mathbf{X}_P(\mathbf{t}^n) = \mathbf{X}_{Pd}(\mathbf{t}^n), \ \forall \mathbf{t}^n \in \mathfrak{T}^n_{R\infty} = [\boldsymbol{\tau}^n_R, \ \infty \mathbf{1}_n[\tag{8.66}$$

means that the real state vector $\mathbf{X}_P(\mathbf{t}^n)$ becomes equal element by element to $\mathbf{X}_{Pd}(\mathbf{t}^n)$ at latest at the state vector reachability time $\boldsymbol{\tau}^n_R$,

$$\boldsymbol{\tau}^n_R = [\tau_{R1} \ \tau_{R2} \ ... \ \tau_{Rn} \ ]^T \in (Int \ \mathfrak{T}_0)^n \ , \ \mathfrak{T}^n_R = \{\mathbf{t}^n : \ \mathbf{0}_n \leq \ \mathbf{t}^n \leq \boldsymbol{\tau}^n_R\}, \tag{8.67}$$

and thereafter they rest equal forever,

$$X_{Pi}(t) = X_{Pdi}(t), \ \forall t \in [\tau_{Ri}, \ \infty[, \ \forall i = 1, 2, ..., n,$$

i.e.,

$$\mathbf{X}_P(\mathbf{t}^n) = \mathbf{X}_{Pd}(\mathbf{t}^n), \ \forall \mathbf{t}^n \in \mathfrak{T}^n_{R\infty} = [\boldsymbol{\tau}^n_R, \ \infty \mathbf{1}_n[,$$

The state vector reachability time $\boldsymbol{\tau}^n_R$ (8.66) ensures that the real output vector $\mathbf{Y}(t)$ becomes elementwise equal to its desired vector value $\mathbf{Y}_d(t)$ at latest at the scalar output reachability time τ_{RM},

$$\mathbf{Y}(t) = \mathbf{Y}_d(t), \ \forall t \in [\tau_{RM}, \ \infty[,$$
$$\tau_{RM} = \max\{\tau_{R1} \ , \ \tau_{R2} \ , .. \ \tau_{Rn} \ \}. \tag{8.68}$$

$\boldsymbol{\tau}^n_R \in (Int\mathfrak{T}_0)^n$ is the state finite vector reachability time, but not the output vector reachability time $\boldsymbol{\tau}^N_R \in (Int\mathfrak{T}_0)^N$. They are different time vectors in general.

Definition 215 ***The state elementwise tracking with the state finite vector reachability time*** τ_R^n ***of the desired output of the*** *ISO plant* ***(2.29), (2.30)***

The ISO plant (2.29), (2.30) exhibits ***the state elementwise output tracking with the state finite vector reachability time*** $\boldsymbol{\tau}_R^n$, *for short* ***the state elementwise tracking with the state finite vector reachability time*** $\boldsymbol{\tau}_R^n$, ***over*** $\mathfrak{D}^k x \mathfrak{Y}^k$ *if, and only if, for every* $[\mathbf{D}(.), \mathbf{Y}_d(.)] \in \mathfrak{D}^k x \mathfrak{Y}^k$ *there exist positive real numbers* Δ_i, *or* $\Delta_i = \infty$, *which are the entries of the vector* $\boldsymbol{\Delta}_{ISO}^n$, $\boldsymbol{\Delta}_{ISO}^n = \boldsymbol{\Delta}_{ISO}^n(\mathbf{D}, \mathbf{U}, \mathbf{Y}_d) = [\Delta_1\ \Delta_2\ ...\ \Delta_n]^T$, *such that for* $|\mathbf{X}_{Pd0} - \mathbf{X}_{P0}| < \boldsymbol{\Delta}_{ISO}^n$, $\mathbf{X}_P(\mathbf{t}^n)$ *becomes equal to* $\mathbf{X}_{Pd}(\mathbf{t}^n)$ *at latest at the vector moment* $\boldsymbol{\tau}_R^n$, *after which they rest equal forever, i.e.,*

$$\begin{gathered}\forall [\mathbf{D}(.), \mathbf{Y}_d(.)] \in \mathfrak{D}^k x \mathfrak{Y}^k,\ \exists \boldsymbol{\Delta}_{ISO}^n \in \mathfrak{R}^{+^n} \cup \{\infty \mathbf{1}_n\}, \\ |\mathbf{X}_{Pd0} - \mathbf{X}_{P0}| < \boldsymbol{\Delta}_{ISO}^n \implies \\ \mathbf{X}_P(\mathbf{t}^n) = \mathbf{X}_{Pd}(\mathbf{t}^n),\ \forall \mathbf{t}^n \in \mathfrak{T}_{R\infty}^n = [\boldsymbol{\tau}_R^n,\ \infty \mathbf{1}_n[. \end{gathered} \tag{8.69}$$

If, and only if, additionally,

$$\left|\mathbf{Y}_d^k\left(\mathbf{t}^{(k+1)N}\right) - \mathbf{Y}^k\left(\mathbf{t}^{(k+1)N}\right)\right| = \mathbf{0}_{(k+1)N},\ \forall \mathbf{t}^{(k+1)N} \in [\boldsymbol{\tau}_{R[k]}^{(k+1)N},\ \infty \mathbf{1}_{(k+1)N}[,$$

then the ISO plant (2.29), (2.30) exhibits ***the k-th order state elementwise tracking with the state finite vector reachability time*** $\boldsymbol{\tau}_R^n$ ***and with the finite output vector reachability time*** $\boldsymbol{\tau}_{R[k]}^{(k+1)N}$ ***over*** $\mathfrak{D}^k x \mathfrak{Y}^k$

If, and only if, the value of $\boldsymbol{\Delta}_{ISO}^n$ *depends at most on* $\mathfrak{D}^k x \mathfrak{Y}^k$ *and* $\mathbf{U}$, *but not on a particular choice of* $[\mathbf{D}(.), \mathbf{Y}_d(.)]$ *from* $\mathfrak{D}^k x \mathfrak{Y}^k$, *then the state elementwise tracking with the state finite vector reachability time* $\boldsymbol{\tau}_R^n$ *is* ***uniform over*** $\mathfrak{D}^k x \mathfrak{Y}^k$.

The state elementwise tracking with the state finite vector reachability time $\boldsymbol{\tau}_R^n$ *is* ***global (in the whole) over*** $\mathfrak{D}^k x \mathfrak{Y}^k$ *if, and only if,* $\boldsymbol{\Delta}_{ISO}^n = \infty \mathbf{1}_n$. *It is also uniform.*

This tracking property implies attraction of the desired motion $\mathbf{X}_{Pd}(t)$ with the state finite vector reachability time $\boldsymbol{\tau}_R^n$, but attraction of $\mathbf{X}_{Pd}(t)$ does not guarantee the elementwise tracking with the finite output vector reachability time $\boldsymbol{\tau}_R^N$.

Definition 216 ***The k-th order elementwise state stable and stabilizing tracking with the state finite vector reachability time*** τ_R^n ***of the desired output of the*** *ISO* ***plant (2.29), (2.30)***

The ISO plant (2.29), (2.30) exhibits ***the k-th order elementwise state stable and stabilizing tracking with the state finite vector reachability time*** $\boldsymbol{\tau}_R^n$, $\boldsymbol{\tau}_R^n \in (Int\ \mathfrak{T}_0)^n$, ***over*** $\mathfrak{D}^k x \mathfrak{Y}^k$ *if, and only if, it exhibits the k-th order state elementwise tracking with the state finite vector reachability time* $\boldsymbol{\tau}_R^n$ *over* $\mathfrak{D}^k x \mathfrak{Y}^k$, *and for every positive real n vector* $\boldsymbol{\varepsilon}$ *and for every* $[\mathbf{D}(.), \mathbf{Y}_d(.)] \in \mathfrak{D}^k x \mathfrak{Y}^k$ *there exists a positive real n vector* $\boldsymbol{\delta}$, $\boldsymbol{\delta} = \boldsymbol{\delta}(\boldsymbol{\varepsilon}, \mathbf{D}, \mathbf{U}, \mathbf{Y}_d)$, *such that* $|\mathbf{X}_{P0} - \mathbf{X}_{Pd0}| < \boldsymbol{\delta}$ *guarantees* $|\mathbf{X}_P(\mathbf{t}^n) - \mathbf{X}_{Pd}(\mathbf{t}^n)| < \boldsymbol{\varepsilon}$, *for all* $\mathbf{t}^n \in \mathfrak{T}_R^n =$

$[\mathbf{0}_n, \boldsymbol{\tau}_R^n]$, *and that* $\left\|\mathbf{Y}^k(0) - \mathbf{Y}_d^k(0)\right\| < \|\boldsymbol{\delta}\|$ *implies* $\left\|\mathbf{Y}^k(t) - \mathbf{Y}_d^k(t)\right\| < \|\boldsymbol{\varepsilon}\|$, *for all* $t \in \mathfrak{T}_0$, *i.e.,*

$$\forall [\mathbf{D}(.), \mathbf{Y}_d(.)] \in \mathfrak{D}^k \mathrm{x} \mathfrak{Y}^k,\ \forall \boldsymbol{\varepsilon} \in \mathfrak{R}^{+^n},\ \exists \boldsymbol{\delta} \in \mathfrak{R}^{+^n},$$
$$\boldsymbol{\delta} = \boldsymbol{\delta}(\boldsymbol{\varepsilon}, \mathbf{D}, \mathbf{U}, \mathbf{Y}_d),\ |\mathbf{X}_{P0} - \mathbf{X}_{Pd0}| < \boldsymbol{\delta} \Longrightarrow$$
$$|\mathbf{X}_P(\mathbf{t}^n) - \mathbf{X}_{Pd}(\mathbf{t}^n)| < \boldsymbol{\varepsilon},\ \forall \mathbf{t}^n \in \mathfrak{T}_R^n = [\mathbf{0}_n, \boldsymbol{\tau}_R^n],\ and$$
$$\left\|\mathbf{Y}^k(0) - \mathbf{Y}_d^k(0)\right\| < \|\boldsymbol{\delta}\| \Longrightarrow \left\|\mathbf{Y}^k(t) - \mathbf{Y}_d^k(t)\right\| < \|\boldsymbol{\varepsilon}\|,\ \forall t \in \mathfrak{T}_0,$$
$$k \in \{0, 1, 2, ...\}. \tag{8.70}$$

The zero, $(k = 0)$, order elementwise state stable and stabilizing tracking with the state finite vector reachability time $\boldsymbol{\tau}_R^n$ is simply called ***the elementwise state stable and stabilizing tracking with the state finite vector reachability time*** $\boldsymbol{\tau}_R^n$.

The k-th order elementwise state stable and stabilizing tracking with the state finite vector reachability time $\boldsymbol{\tau}_R^n$ is ***global (in the whole) over*** $\mathfrak{D}^k \mathrm{x} \mathfrak{Y}^k$ *if, and only if, both the k-th order state elementwise tracking with the state finite vector reachability time $\boldsymbol{\tau}_R^n$ is global and $\boldsymbol{\delta}(\boldsymbol{\varepsilon}, \mathbf{D}, \mathbf{U}, \mathbf{Y}_d) \to \infty \mathbf{1}_n$ as $\boldsymbol{\varepsilon} \to \infty \mathbf{1}_n$, for every* $[\mathbf{D}(.), \mathbf{Y}_d(.)] \in \mathfrak{D}^k \mathrm{x} \mathfrak{Y}^k$,

The k-th order elementwise state stable and stabilizing tracking with the state finite vector reachability time $\boldsymbol{\tau}_R^n$ is ***uniform over*** $\mathfrak{D}^k \mathrm{x} \mathfrak{Y}^k$ *if, and only if, the state elementwise tracking with the state finite vector reachability time $\boldsymbol{\tau}_R^n$ is uniform and the value of $\boldsymbol{\delta}$ depends at most on $\mathfrak{D}^k \mathrm{x} \mathfrak{Y}^k$ and $\mathbf{U}$, but not on a particular choice of $[\mathbf{D}(.), \mathbf{Y}_d(.)]$ from $\mathfrak{D}^k \mathrm{x} \mathfrak{Y}^k$.*

This tracking property guarantees asymptotic stability of the desired motion $\mathbf{X}_d(t)$ with the state finite vector reachability time $\boldsymbol{\tau}_R^n$. Besides, it ensures asymptotic stability of the desired output behavior $\mathbf{Y}_{Pd}(t)$.

Comment 217 *An application of any of the preceding definitions requires the knowledge of the desired (nominal) motion $\mathbf{X}_d(t)$ related to every chosen desired output behavior $\mathbf{Y}_d(t)$.*

8.4.6 Tracking of the *ISO* plant and the target set

ISO plant tracking in Lyapunov sense and the target set

Equations (8.18) in Definition 172 (Subsection 8.3.3) determine **the** *k-th* **order target set** $\Upsilon_{ISO}^k(t; \mathbf{D}; \mathbf{U}; \mathbf{Y}_d^k)$ of the *ISO* plant (2.33), (2.34) (Subsection 2.2.2) by

$$\Upsilon_{ISO}^k(t; \mathbf{D}; \mathbf{U}; \mathbf{Y}_d^k) = \left\{\mathbf{X}_P:\ \mathbf{Y}^k(t) = \mathbf{Y}_d^k(t)\right\} =$$
$$= \left\{ \mathbf{X}_P: \left(\begin{array}{c} \mathbf{Y}(t) = C_P\mathbf{X}_P + H_P\mathbf{U}(t) + D_P\mathbf{D}(t) = \mathbf{Y}_d(t), \\ \mathbf{Y}^{(j)}(t) = C_P A_P^j \mathbf{X}_P + C_P\left[\sum_{i=1}^{i=j} A_P^{j-i} B_P \mathbf{U}^{(i-1)}(t)\right] + \\ +H_P\mathbf{U}^{(j)}(t) + C_P\left[\sum_{i=1}^{i=j} A_P^{j-i} L_P \mathbf{D}^{(i-1)}(t)\right] + \\ +D_P\mathbf{D}^{(j)}(t) = \mathbf{Y}_d^{(j)}(t),\ \forall j = 1, ..., k, \end{array} \right) \right\},$$
$$\Upsilon_{ISO}^k(t; \mathbf{D}; \mathbf{U}; \mathbf{Y}_d^k) \subset \mathfrak{R}^n,\ k \in \{0, 1, 2, ...\}. \tag{8.71}$$

We need the target set in order to define tracking properties (which concern the output behaviors) via the state space $\mathfrak{R}^n$ without determining the plant desired motion $\mathbf{X}_{Pd}\left(.;\mathbf{X}_{P0};\mathbf{D};\mathbf{U};\mathbf{Y}_d^k\right)$.

We define *the scalar distance* $d\left[\mathbf{Y}(t),\Upsilon_{ISO}^k\right]$ *of* $\mathbf{Y}(t)$ *from* the *k-th* order target set Υ_{ISO}^k at the moment $t \in \mathfrak{T}_0$,

$$d\left[\mathbf{Y}(t),\Upsilon_{ISO}^k(t;\mathbf{D};\mathbf{U};\mathbf{Y}_d^k)\right] = \inf[\|\mathbf{Y}(t) - \mathbf{w}\| : \mathbf{w} \in \Upsilon_{ISO}^k(t;\mathbf{D};\mathbf{U};\mathbf{Y}_d^k)]. \tag{8.72}$$

This is analogous to (8.48), Subsection 8.4.3.

$\mathbf{X}_P(t)$ denotes simply $\mathbf{X}_P(t;\mathbf{X}_{P0};\mathbf{D};\mathbf{U};\mathbf{Y}_d^k)$ in what follows. We use $\mathbf{Y}(t)$ for $\mathbf{Y}(t;\mathbf{Y}_0;\mathbf{D};\mathbf{U};\mathbf{Y}_d^k)$. $\mathbf{Y}^k(t)$ is the abbreviation of $\mathbf{Y}^k(t;\mathbf{Y}_0^k;\mathbf{D};\mathbf{U};\mathbf{Y}_d^k)$, and we utilize $\Upsilon_{ISO}^k(t)$ to replace $\Upsilon_{ISO}^k(t;\mathbf{D};\mathbf{U};\mathbf{Y}_d^k)$.

Note 218 *The following definitions do not need the knowledge of the desired (nominal) motion* $\mathbf{X}_{Pd}\left(.;\mathbf{X}_{P0};\mathbf{D};\mathbf{U};\mathbf{Y}_d^k\right)$ *related to every chosen desired output behavior* $\mathbf{Y}_d(t)$.

Definition 219 ***The k-th order tracking in Lyapunov sense of the desired output of the*** *ISO* ***plant (2.33), (2.34)***

The ISO plant (2.33), (2.34) exhibits ***the k-th order asymptotic output tracking,*** *for short* ***the k-th order tracking, over*** $\mathfrak{D}^k \mathrm{x} \mathfrak{Y}_d^k$ *if, and only if, for every* $[\mathbf{D}(.),\mathbf{Y}_d(.)] \in \mathfrak{D}^k \mathrm{x} \mathfrak{Y}_d^k$ *there exists a positive real number* Δ, $\Delta = \Delta(\mathbf{D},\mathbf{U},\mathbf{Y}_d)$, *or* $\Delta = \infty$, *such that* $d\left[\mathbf{X}_{P0},\Upsilon_{ISO}^k(0)\right] < \Delta$ *guarantees that* $\mathbf{X}_P(t)$ *approaches asymptotically* $\Upsilon_{ISO}^k(t)$ *as time t goes to infinity, i.e.,*

$$\begin{gathered}\forall\left[\mathbf{D}(.),\mathbf{Y}_d(.)\right] \in \mathfrak{D}^k \mathrm{x} \mathfrak{Y}_d^k,\ \exists \Delta \in \mathfrak{R}^+ \cup \{\infty\},\ \Delta = \Delta(\mathbf{D},\mathbf{U},\mathbf{Y}_d),\\ d\left[\mathbf{X}_{P0},\Upsilon_{ISO}^k(0)\right] < \Delta \Longrightarrow \\ \lim\left\langle d\left[\mathbf{X}_P(t),\Upsilon_{ISO}^k(t)\right] : t \to \infty\right\rangle = 0,\ k \in \{0,1,2,...\}.\end{gathered} \tag{8.73}$$

The zero order tracking is simply called ***tracking.***

If, and only if, the value of Δ *depends at most on* $\mathfrak{D}^k \mathrm{x} \mathfrak{Y}_d^k$ *and* $\mathbf{U}$, *but not on a particular choice of* $[\mathbf{D}(.),\mathbf{Y}_d(.)]$ *from* $\mathfrak{D}^k \mathrm{x} \mathfrak{Y}_d^k$, *then the k-th order tracking is* ***uniform over*** $\mathfrak{D}^k \mathrm{x} \mathfrak{Y}_d^k$.

The k-th order tracking is ***global (in the whole) over*** $\mathfrak{D}^k \mathrm{x} \mathfrak{Y}_d^k$ *if, and only if,* $\Delta = \infty$. *It is also uniform over* $\mathfrak{D}^k \mathrm{x} \mathfrak{Y}_d^k$.

In order to assure that the plant state $\mathbf{X}_P(t)$ stays in an accepted neighborhood of $\Upsilon_{ISO}^k(t)$ all time $t \in \mathfrak{T}_0$ we present the following.

Definition 220 ***The k-th order state stabilizing tracking in Lyapunov sense of the desired output of the*** *ISO* ***plant (2.33), (2.34)***

The ISO plant (2.33), (2.34) exhibits ***the k-th order state stabilizing asymptotic output tracking,*** *for short* ***the k-th order state stabilizing tracking, over*** $\mathfrak{D}^k \mathrm{x} \mathfrak{Y}_d^k$ *if, and only if, it exhibits* ***the k-th order tracking over*** $\mathfrak{D}^k \mathrm{x} \mathfrak{Y}_d^k$, *and for every positive real number* ε *and for every* $[\mathbf{D}(.),\mathbf{Y}_d(.)] \in$

$\mathfrak{D}^k \mathrm{x} \mathfrak{Y}_d^k$ *there exists a positive real number* δ, $\delta = \delta(\varepsilon, \mathbf{D}, \mathbf{U}, \mathbf{Y}_d)$, *such that* $d\left[\mathbf{X}_{P0}, \Upsilon_{ISO}^k(0)\right] < \delta$ *guarantees* $d\left[\mathbf{X}_P(t), \Upsilon_{ISO}^k(t)\right] < \varepsilon$ *for all* $t \in \mathfrak{T}_0$,

$$\forall \left[\mathbf{D}(.), \mathbf{Y}_d(.)\right] \in \mathfrak{D}^k \mathrm{x} \mathfrak{Y}_d^k,\ \forall \varepsilon \in \mathfrak{R}^+,\ \exists \delta \in \mathfrak{R}^+, \delta = \delta(\varepsilon, \mathbf{D}, \mathbf{U}, \mathbf{Y}_d),$$
$$d\left[\mathbf{X}_{P0}, \Upsilon_{ISO}^k(0)\right] < \delta \Longrightarrow d\left[\mathbf{X}_P(t), \Upsilon_{ISO}^k(t)\right] < \varepsilon,\ \forall t \in \mathfrak{T}_0,$$
$$k \in \{0, 1, 2, ...\}. \tag{8.74}$$

The zero, $(k = 0)$, *order state stabilizing tracking is simply called* ***the state stabilizing tracking.***

The k-th order state stabilizing tracking is ***global (in the whole) over*** $\mathfrak{D}^k \mathbf{x} \mathfrak{Y}_d^k$ *if, and only if, it is both the k-th order global tracking, and* $\delta = \delta(\varepsilon, \mathbf{D}, \mathbf{U}, \mathbf{Y}_d) \rightarrow \infty$ *as* $\varepsilon \rightarrow \infty$ *for every* $\left[\mathbf{D}(.), \mathbf{Y}_d(.)\right] \in \mathfrak{D}^k \mathrm{x} \mathfrak{Y}_d^k$.

If, and only if, the values of both δ *and* Δ *depend at most on* $\mathfrak{D}^k \mathrm{x} \mathfrak{Y}_d^k$ *and* $\mathbf{U}$, *but not on a particular choice of* $\left[\mathbf{D}(.), \mathbf{Y}_d(.)\right]$ *from* $\mathfrak{D}^k \mathrm{x} \mathfrak{Y}_d^k$, *then the k-th order state stabilizing tracking is* ***uniform over*** $\mathfrak{D}^k \mathrm{x} \mathfrak{Y}_d^k$.

In order to ensure state stabilizing tracking with an exponential rate of the convergence we present the following.

Definition 221 ***The k-th order state exponential tracking in Lyapunov sense of the desired output of the ISO plant (2.33), (2.34)***

The ISO plant (2.33), (2.34) exhibits ***the k-th order state exponential output tracking,*** *for short* ***the k-th order state exponential tracking, over*** $\mathfrak{D}^k \mathrm{x} \mathfrak{Y}_d^k$ *if, and only if, for every* $\left[\mathbf{D}(.), \mathbf{Y}_d(.)\right] \in \mathfrak{D}^k \mathrm{x} \mathfrak{Y}_d^k$ *there exist positive real numbers* $\alpha \geq 1$, β, *and* Δ, *or* $\Delta = \infty$, $\alpha = \alpha(\mathbf{D}, \mathbf{U}, \mathbf{Y}_d)$, $\beta = \beta(\mathbf{D}, \mathbf{U}, \mathbf{Y}_d)$, $\Delta = \Delta(\mathbf{D}, \mathbf{U}, \mathbf{Y}_d)$, *such that* $d\left[\mathbf{X}_{P0}, \Upsilon_{ISO}^k(0)\right] < \Delta$ *guarantees that* $\mathbf{X}_P(t)$ *approaches exponentially* $\Upsilon_{ISO}^k(t)$ *all the time, i.e.,*

$$\forall \left[\mathbf{D}(.), \mathbf{Y}_d(.)\right] \in \mathfrak{D}^k \mathrm{x} \mathfrak{Y}_d^k,\ \exists \alpha \in \mathfrak{R}^+,\ \alpha \geq 1,\ \exists \beta \in \mathfrak{R}^+,\ \exists \Delta \in \mathfrak{R}^+ \cup \{\infty\},$$
$$d\left[\mathbf{X}_{P0}, \Upsilon_{ISO}^k(0)\right] < \Delta \Longrightarrow$$
$$d\left[\mathbf{X}_P(t), \Upsilon_{ISO}^k(t)\right] \leq \alpha d\left[\mathbf{X}_{P0}, \Upsilon_{ISO}^k(0)\right] \exp\left(-\beta t\right),\ \forall t \in \mathfrak{T}_0,$$
$$k \in \{0, 1, 2, ...\}. \tag{8.75}$$

The zero, $(k = 0)$, *order state exponential tracking is simply called* ***the state exponential tracking.***

The k-th order state exponential tracking is ***global (in the whole) over*** $\mathfrak{D}^k \mathbf{x} \mathfrak{Y}_d^k$ *if, and only if,* $\Delta = \infty$.

If, and only if, the values of α, β, *and* Δ *depend at most on* $\mathfrak{D}^k \mathrm{x} \mathfrak{Y}_d^k$ *and* $\mathbf{U}$, *but not on a particular choice of* $\left[\mathbf{D}(.), \mathbf{Y}_d(.)\right]$ *from* $\mathfrak{D}^k \mathrm{x} \mathfrak{Y}_d^k$, *then the k-th order state exponential tracking is* ***uniform over*** $\mathfrak{D}^k \mathrm{x} \mathfrak{Y}_d^k$.

Since the target set $\Upsilon_{ISO}^k(t)$ represents a hyperplane in the state space at every instant $t \in \mathfrak{T}_0$, then the convergence of $\mathbf{X}_P(t)$ to it does not guarantee boundedness of $\mathbf{X}_P(t)$; i.e., $\|\mathbf{X}_P(t)\|$ can blow to infinity as time t goes to infinity. Better behavior is state stabilizing tracking. It guarantees that the state behaviors rest in bounded neighborhoods of a compact subset $\mathfrak{B}(t)$ of the target set $\Upsilon_{ISO}^{k}(t)$ for the corresponding initial states,

$$Cl\mathfrak{B}(t) = \mathfrak{B}(t), \ \mathfrak{B}(t) \subset \Upsilon_{ISO}^{k}(t), \forall t \in \mathfrak{T}_0. \tag{8.76}$$

This ensures a stability property of the desired output behavior $\mathbf{Y}_d(t)$.

A better behavior is *bounded state (stabilizing* or *exponential) tracking.*

Definition 222 ***The k-th order bounded state stabilizing tracking in Lyapunov sense of the desired output of the ISO plant (2.33), (2.34)***

The ISO plant (2.33), (2.34) exhibits ***the k-th order bounded state stabilizing asymptotic output tracking,*** *for short* ***the k-th order bounded state stabilizing tracking, over*** $\mathfrak{D}^k \mathrm{x} \mathfrak{Y}_d^k$ *if, and only if, both, it exhibits the k-th order state stabilizing tracking over* $\mathfrak{D}^k \mathrm{x} \mathfrak{Y}_d^k$, *and there is a compact subset* $\mathfrak{B}(t)$ *of* $\Upsilon_{ISO}^{k}(t)$, *(8.76), such that for every positive real number* ξ *and for every* $[\mathbf{D}(.), \mathbf{Y}_d(.)] \in \mathfrak{D}^k \mathrm{x} \mathfrak{Y}_d^k$ *there exists a positive real number* ψ, $\psi = \psi(\xi, \mathfrak{B}, \mathbf{D}, \mathbf{U}, \mathbf{Y}_d)$, *such that* $d[\mathbf{X}_{P0}, \mathfrak{B}(0)] < \psi$ *guarantees* $d[\mathbf{X}_P(t), \mathfrak{B}(t)] < \xi$ *for all* $t \in \mathfrak{T}_0$, *i.e.,*

$$\exists \mathfrak{B}(t) = Cl\mathfrak{B}(t) \subset \Upsilon_{ISO}^{k}(t), \ sup\left[\|\mathbf{x} - \mathbf{y}\| : \mathbf{x}, \mathbf{y} \in \mathfrak{B}(t)\right] < \infty, \ \forall t \in \mathfrak{T}_0,$$
$$\forall [\mathbf{D}(.), \mathbf{Y}_d(.)] \in \mathfrak{D}^k \mathrm{x} \mathfrak{Y}_d^k, \ \forall \xi \in \mathfrak{R}^+, \ \exists \psi \in \mathfrak{R}^+, \ \psi = \psi(\xi, \mathbf{D}, \mathbf{U}, \mathbf{Y}_d),$$
$$d[\mathbf{X}_{P0}, \mathfrak{B}(0)] < \psi \Longrightarrow d[\mathbf{X}_P(t), \mathfrak{B}(t)] < \xi, \ \forall t \in \mathfrak{T}_0, \ k \in \{0, 1, 2, ...\}. \tag{8.77}$$

The zero, (k = 0), order bounded state stabilizing tracking is simply called ***the bounded state stabilizing tracking.***

The k-th order bounded state stabilizing tracking is ***global (in the whole) over*** $\mathfrak{D}^k \mathrm{x} \mathfrak{Y}_d^k$ *if, and only if, the k-th order state stabilizing tracking over* $\mathfrak{D}^k \mathrm{x} \mathfrak{Y}_d^k$ *is global and* $\xi \longrightarrow \infty$ *implies* $\psi \longrightarrow \infty$ *for every* $[\mathbf{D}(.), \mathbf{Y}_d(.)] \in \mathfrak{D}^k \mathrm{x} \mathfrak{Y}_d^k$.

The k-th order bounded state stabilizing tracking is ***uniform over*** $\mathfrak{D}^k \mathrm{x} \mathfrak{Y}_d^k$ *if, and only if, the k-th order state stabilizing tracking over* $\mathfrak{D}^k \mathrm{x} \mathfrak{Y}_d^k$ *is uniform and the value of* ψ *depends at most on* $\mathfrak{D}^k \mathrm{x} \mathfrak{Y}_d^k$ *and* $\mathbf{U}$, *but not on a particular choice of* $[\mathbf{D}(.), \mathbf{Y}_d(.)]$ *from* $\mathfrak{D}^k \mathrm{x} \mathfrak{Y}_d^k$.

Definition 223 ***The k-th order bounded state exponential tracking in Lyapunov sense of the desired output of the ISO plant (2.33), (2.34)***

The ISO plant (2.33), (2.34) exhibits ***the k-th order bounded state exponential output tracking,*** *for short* ***the k-th order bounded state exponential tracking, over*** $\mathfrak{D}^k \mathrm{x} \mathfrak{Y}_d^k$ *if, and only if, it exhibits the k-th order state exponential tracking over* $\mathfrak{D}^k \mathrm{x} \mathfrak{Y}_d^k$, *there is a compact subset* $\mathfrak{B}(t)$ *of* $\Upsilon_{ISO}^{k}(t)$, *(8.76), such that for every* $[\mathbf{D}(.), \mathbf{Y}_d(.)] \in \mathfrak{D}^k \mathrm{x} \mathfrak{Y}_d^k$ *there exist positive real numbers* $\varsigma \geq 1$, γ *and* θ, *or* $\theta = \infty$, $\varsigma = \varsigma(\mathfrak{B}, \mathbf{D}, \mathbf{U}, \mathbf{Y}_d)$, $\gamma = \gamma(\mathfrak{B}, \mathbf{D}, \mathbf{U}, \mathbf{Y}_d)$, $\theta = \theta(\mathfrak{B}, \mathbf{D}, \mathbf{U}, \mathbf{Y}_d)$, *for which* $d[\mathbf{X}_{P0}, \mathfrak{B}(0)] < \theta$ *guarantees*

$$d[\mathbf{X}_P(t), \mathfrak{B}(t)] < \varsigma d[\mathbf{X}_{P0}, \mathfrak{B}(0)] \exp(-\gamma t) \ \ for \ all \ t \in \mathfrak{T}_0,$$

i.e.,

$$\exists\mathfrak{B}\,(t) = Cl\mathfrak{B}\,(t) \subset \Upsilon^k_{ISO}(t),\ sup\,[\|\mathbf{x}-\mathbf{y}\| : \mathbf{x},\mathbf{y} \in\mathfrak{B}(t)] < \infty,\ \forall t \in \mathfrak{T}_0,$$
$$\forall\,[\mathbf{D}(.),\mathbf{Y}_d(.)] \in \mathfrak{D}^k\mathrm{x}\mathfrak{Y}^k_d,\ \exists\,(\varsigma \geq 1) \in \mathfrak{R}^+,\ \exists\gamma \in \mathfrak{R}^+,\ \exists\theta \in \mathfrak{R}^+,\ or\ \theta = \infty,$$
$$\varsigma = \varsigma(\mathfrak{B},\mathbf{D},\mathbf{U},\mathbf{Y}_d),\ \gamma = \gamma(\mathfrak{B},\mathbf{D},\mathbf{U},\mathbf{Y}_d),\ \theta = \theta(\mathfrak{B},\mathbf{D},\mathbf{U},\mathbf{Y}_d),$$
$$d\,[\mathbf{X}_{P0},\mathfrak{B}(0)] < \theta \Longrightarrow d\,[\mathbf{X}_P(t),\mathfrak{B}\,(t)] < \varsigma d\,[\mathbf{X}_{P0},\mathfrak{B}(0)]\exp\left(-\gamma t\right),\ \forall t \in \mathfrak{T}_0,$$
$$k \in \{0,1,2,...\}\,. \tag{8.78}$$

The zero, $(k=0)$, order bounded state exponential tracking is simply called ***the bounded state exponential tracking.***

The k-th order bounded state exponential tracking is ***global (in the whole) over $\mathfrak{D}^k\mathrm{x}\mathfrak{Y}^k_d$*** *if, and only if, the k-th order state exponential tracking over $\mathfrak{D}^k\mathrm{x}\mathfrak{Y}^k_d$ is global and $\theta = \infty$.*

The k-th order bounded state exponential tracking is ***uniform*** *over $D^kxY^k_d$ if, and only if, the k-th order state exponential tracking over $D^kxY^k_d$ is uniform and the values of ς, γ, and θ depend at most on $D^kxY^k_d$ and U, but not on a particular choice of $[\mathbf{D}(.),\mathbf{Y}_d(.)]$ from $D^kxY^k_d$.*

ISO plant tracking with the finite scalar reachability time and the target set

The exponential convergence of the state vector $\mathbf{X}_P(t)$ to the *k-th* order target set $\Upsilon^k_{ISO}(t;\mathbf{D};\mathbf{U};\mathbf{Y}_d)$ does not imply the exponential convergence of the real output vector $\mathbf{Y}(t)$ to the desired output vector $\mathbf{Y}_d(t)$.

The state reachability time, i.e., the reachability time related to the state space, assures that the real state vector $\mathbf{X}(t)$ reaches the *k-th* order target set $\Upsilon^k_{ISO}(t;\mathbf{D};\mathbf{U};\mathbf{Y}_d)$ in the state space at the state reachability time. The output (scalar) reachability time τ_R, i.e., the reachability time τ_R related to the output space, is the same as the state (scalar) reachability time τ_{RISO}, $\tau_R = \tau_{RISO}$.

Definition 224 ***The tracking with the finite scalar reachability time of the desired output of the ISO plant (2.33), (2.34)***

The ISO plant (2.33), (2.34) exhibits ***the k-th order output tracking with the finite scalar reachability time*** *τ_R, $\tau_R \in Int\,\mathfrak{T}_0$, for short* ***the k-th order tracking with the finite reachability time*** *τ_R,* ***over $\mathfrak{D}^k\mathrm{x}\mathfrak{Y}^k_d$*** *if, and only if, for every $[\mathbf{D}(.),\mathbf{Y}_d(.)] \in \mathfrak{D}^k\mathrm{x}\mathfrak{Y}^k_d$ there exists a positive real number Δ, $\Delta = \Delta(\mathbf{D},\mathbf{U},\mathbf{Y}_d)$, or $\Delta = \infty$, such that $d\left[\mathbf{X}_{P0},\Upsilon^k_{ISO}(0;\mathbf{D};\mathbf{U};\mathbf{Y}_d)\right] < \Delta$ guarantees that $\mathbf{X}_P(t)$ reaches $\Upsilon^k_{ISO}(t;\mathbf{D};\mathbf{U};\mathbf{Y}_d)$ at latest at the moment τ_R, after which it rests in $\Upsilon^k_{ISO}(t;\mathbf{D};\mathbf{U};\mathbf{Y}_d)$ forever, i.e.,*

$$\forall\,[\mathbf{D}(.),\mathbf{Y}_d(.)] \in \mathfrak{D}^k\mathrm{x}\mathfrak{Y}^k_d,\ \exists\Delta \in \mathfrak{R}^+ \cup \{\infty\}\,,$$
$$d\left[\mathbf{X}_{P0},\Upsilon^k_{ISO}(0;\mathbf{D};\mathbf{U};\mathbf{Y}_d)\right] < \Delta \Longrightarrow \mathbf{X}_P(t)\in\Upsilon^k_{ISO}(t;\mathbf{D};\mathbf{U};\mathbf{Y}_d),\ \forall t \in [\tau_R,\ \infty[,$$
$$k \in \{0,1,2,...\}\,. \tag{8.79}$$

The zero, $(k=0)$, order tracking with the finite reachability time τ_R is simply called ***tracking with the finite reachability time*** *τ_R.*

*If, and only if, the value of Δ depends at most on $\mathfrak{D}^k \text{x} \mathfrak{Y}_d^k$ and $\mathbf{U}$, but not on a particular choice of $[\mathbf{D}(.), \mathbf{Y}_d(.)]$ from $\mathfrak{D}^k \text{x} \mathfrak{Y}_d^k$, then the k-th order tracking with the finite reachability time τ_R is **uniform over** $\mathfrak{D}^k \text{x} \mathfrak{Y}_d^k$.*

*The k-th order tracking with the finite reachability time τ_R is **global (in the whole) over** $\mathfrak{D}^k$ if, and only if, $\Delta = \infty$. It is also uniform.*

Definition 225 ***The k-th order state stabilizing tracking with the finite scalar reachability time τ_R of the desired output of the ISO plant (2.33), (2.34)***

*The ISO plant (2.33), (2.34) exhibits **the k-th order state stabilizing output tracking with the finite scalar reachability time** τ_R, $\tau_R \in \mathfrak{T}_0$, for short **the k-th order state stabilizing tracking with the finite reachability time** τ_R, **over** $\mathfrak{D}^k \text{x} \mathfrak{Y}_d^k$ if, and only if, it exhibits the k-th order tracking with the finite reachability time τ_R over $\mathfrak{D}^k \text{x} \mathfrak{Y}_d^k$, and for every positive real number ε and for every $[\mathbf{D}(.), \mathbf{Y}_d(.)] \in \mathfrak{D}^k \text{x} \mathfrak{Y}_d^k$ there exists a positive real number δ, $\delta = \delta(\varepsilon, \mathbf{D}, \mathbf{U}, \mathbf{Y}_d)$, such that $d\left[\mathbf{X}_{P0}, \Upsilon_{ISO}^k(0; \mathbf{D}; \mathbf{U}, \mathbf{Y}_d)\right] < \delta$ guarantees $d\left[\mathbf{X}_P(t), \Upsilon_{ISO}^k(t; \mathbf{D}; \mathbf{U}, \mathbf{Y}_d)\right] < \varepsilon$ for all $t \in \mathfrak{T}_0$, i.e.,*

$$\begin{gathered}\forall [\mathbf{D}(.), \mathbf{Y}_d(.)] \in \mathfrak{D}^k \text{x} \mathfrak{Y}_d^k,\ \forall \varepsilon \in \mathfrak{R}^+,\ \exists \delta \in \mathfrak{R}^+, \delta = \delta(\varepsilon, \mathbf{D}, \mathbf{U}, \mathbf{Y}_d), \\ d\left[\mathbf{X}_{P0}, \Upsilon_{ISO}^k(0; \mathbf{D}; \mathbf{U}, \mathbf{Y}_d)\right] < \delta \Longrightarrow d\left[\mathbf{X}_P(t), \Upsilon_{ISO}^k(t; \mathbf{D}; \mathbf{U}, \mathbf{Y}_d)\right] < \varepsilon, \\ \forall t \in \mathfrak{T}_R = [0, \tau_R],\ k \in \{0, 1, 2, ...\}. \end{gathered} \tag{8.80}$$

*The zero, ($k = 0$), order state stabilizing tracking with the finite reachability time τ_R is simply called **the state stabilizing tracking with the finite reachability time** τ_R.*

*The k-th order state stabilizing tracking with the finite reachability time τ_R is **global (in the whole) over** $\mathfrak{D}^k$ if, and only if, both the k-th order tracking with the finite reachability time τ_R is global and $\delta = \delta(\varepsilon, \mathbf{D}, \mathbf{U}, \mathbf{Y}_d) \to \infty$ as $\varepsilon \to \infty$ for every $[\mathbf{D}(.), \mathbf{Y}_d(.)] \in \mathfrak{D}^k \text{x} \mathfrak{Y}_d^k$.*

*If, and only if, the values of both δ and Δ depend at most on $\mathfrak{D}^k \text{x} \mathfrak{Y}_d^k$ and $\mathbf{U}$, but not on a particular choice of $[\mathbf{D}(.), \mathbf{Y}_d(.)]$ from $\mathfrak{D}^k \text{x} \mathfrak{Y}_d^k$, then the k-th order state stabilizing tracking with the finite reachability time τ_R is **uniform over** $\mathfrak{D}^k \text{x} \mathfrak{Y}_d^k$.*

Since the state exponential convergence does not imply the output exponential convergence, we will not present the definition of the state exponential tracking with the finite scalar reachability time (for which see [106]).

ISO plant state elementwise tracking with the state finite vector reachability time and the target set

Let, analogously to (8.54) (Subsection 8.4.3) $Dist\left[\mathbf{X}_P(\mathbf{t}^n), \Upsilon_{ISO}^k(t; \mathbf{D}, \mathbf{U}, \mathbf{Y}_d)\right]$ be **the vector** (i.e., **the elementwise**) **distance of the vector $\mathbf{X}_P(\mathbf{t}^n)$ from the target set** $\Upsilon_{ISO}^k(t; \mathbf{D}, \mathbf{U}, \mathbf{Y}_d)$,

$$Dist\left[\mathbf{X}_P(\mathbf{t}^n), \Upsilon_{ISO}^k(t; \mathbf{D}, \mathbf{U}, \mathbf{Y}_d)\right] = \inf[|\mathbf{X}_P(\mathbf{t}^n) - \mathbf{w}| : \mathbf{w} \in \Upsilon_{ISO}^k(t; \mathbf{D}, \mathbf{U}, \mathbf{Y}_d)],$$

where the infimum holds elementwise. The notation

$$\mathbf{X}_P(\mathbf{t}^n)\in\Upsilon^k_{ISO}(t;\mathbf{D},\mathbf{U},\mathbf{Y}_d),\ \forall\mathbf{t}^n\in[\boldsymbol{\tau}^n_R,\ \infty\mathbf{1}^n[$$

means that the real state vector $\mathbf{X}_P(\mathbf{t}^n)$ enters the target set $\Upsilon^k_{ISO}(t;\mathbf{D},\mathbf{U},\mathbf{Y}_d)$ at latest at the state vector reachability time $\boldsymbol{\tau}^n_R$ (8.67) (Subsection 8.4.5), and thereafter rests in $\Upsilon^k_{ISO}(t;\mathbf{D},\mathbf{U},\mathbf{Y}_d)$ forever, which ensures that the real output vector $\mathbf{Y}$ and only its first k derivatives become elementwise equal to their desired values at latest at the reachability time τ_{RM} (8.68) (Subsection 8.4.5),

$$\mathbf{X}_P(\mathbf{t}^n)\in\Upsilon^k_{ISO}(t;\mathbf{D},\mathbf{U},\mathbf{Y}_d),\ \forall\mathbf{t}^n\in\mathfrak{T}^n_{R\infty}=[\boldsymbol{\tau}^n_R,\ \infty\mathbf{1}_n[,$$
$$\Longleftrightarrow$$
$$\mathbf{Y}^k(t)=\mathbf{Y}^k_d(t),\ \forall t\in[\tau_{RM},\ \infty[,\ k=0,1,2,....$$

The state vector reachability time $\boldsymbol{\tau}^n_R\in(Int\ \mathfrak{T}_0)^n$, the output scalar reachability time τ_{RM} and the output vector reachability time $\boldsymbol{\tau}^{k+1)N}_{R[k]}\in(Int\ \mathfrak{T}_0)^{(k+1)N}$ are different reachability times in general. Their dimensions n, one and $(k+1)N$ are different. They are related to different spaces.

Definition 226 ***The k-th order state elementwise tracking with the state finite vector reachability time*** τ^n_R ***of the desired output of the*** *ISO plant (2.33), (2.34)*

The ISO plant (2.33), (2.34) exhibits ***the k-th order state elementwise output tracking with the state finite vector reachability time*** $\boldsymbol{\tau}^n_R$ *for short* ***the k-th order state elementwise tracking with the state finite vector reachability time*** $\boldsymbol{\tau}^n_R$, ***over*** $\mathfrak{D}^k\mathrm{x}\mathfrak{Y}^k_d$ *if, and only if, for every* $[\mathbf{D}(.),\mathbf{Y}_d(.)]\in\mathfrak{D}^k\mathrm{x}\mathfrak{Y}^k_d$ *there exist positive real numbers* Δ_i, *or* $\Delta_i=\infty$, *which are the entries of the vector* $\boldsymbol{\Delta}^n_{ISO}$, $\boldsymbol{\Delta}^n_{ISO}=\boldsymbol{\Delta}^n_{ISO}(\mathbf{D},\mathbf{U},\mathbf{Y}_d)$, *such that for*

$$Dist\left[\mathbf{X}_{P0},\Upsilon^k_{ISO}(0;\mathbf{D},\mathbf{U},\mathbf{Y}_d)\right]<\boldsymbol{\Delta}^n_{ISO},$$

$\mathbf{X}_P(\mathbf{t}^n)$ *reaches* $\Upsilon^k_{ISO}(\mathbf{t}^n;\mathbf{D},\mathbf{U},\mathbf{Y}_d)$ *at latest at the vector moment* τ^n_R, *after which it rests in* $\Upsilon^k_{ISO}(\mathbf{t}^n;\mathbf{D},\mathbf{U},\mathbf{Y}_d)$ *forever, i.e.,*

$$\begin{gathered}\forall\left[\mathbf{D}(.),\mathbf{Y}_d(.)\right]\in\mathfrak{D}^k\mathrm{x}\mathfrak{Y}^k_d,\ \exists\boldsymbol{\Delta}^n_{ISO}\in\mathfrak{R}^{+^n}\cup\{\infty\mathbf{1}_n\},\\ Dist\left[\mathbf{X}_{P0},\Upsilon^k_{ISO}(0;\mathbf{D},\mathbf{U},\mathbf{Y}_d)\right]<\boldsymbol{\Delta}^n_{ISO}\Longrightarrow\\ \mathbf{X}_P(\mathbf{t}^n)\in\Upsilon^k_{ISO}(\mathbf{t}^n;\mathbf{D},\mathbf{U},\mathbf{Y}_d),\ \forall\mathbf{t}^n\in\mathfrak{T}^n_{R\infty},\ k\in\{0,1,2,...\}.\end{gathered}\tag{8.81}$$

The zero, $(k=0)$, *order state elementwise tracking with the state finite vector reachability time* $\boldsymbol{\tau}^n_R$ *is simply called* ***the state elementwise tracking with the state finite vector reachability time*** $\boldsymbol{\tau}^n_R$.

If, and only if, the value of $\boldsymbol{\Delta}^n_{ISO}$ *depends at most on* $\mathfrak{D}^k\mathrm{x}\mathfrak{Y}^k_d$ *and* $\mathbf{U}$, *but not on a particular choice of* $[\mathbf{D}(.),\mathbf{Y}_d(.)]$ *from* $\mathfrak{D}^k\mathrm{x}\mathfrak{Y}^k_d$, *then the k-th order state elementwise tracking with the state finite vector reachability time* $\boldsymbol{\tau}^n_R$ *is* ***uniform over*** $\mathfrak{D}^k\mathrm{x}\mathfrak{Y}^k_d$.

The k-th order state elementwise tracking with the state finite vector reachability time $\boldsymbol{\tau}_R^n$ is ***global (in the whole) over*** $\mathfrak{D}^k$ *if, and only if,* $\boldsymbol{\Delta}_{ISO}^n = \infty\mathbf{1}^n$. *It is also uniform.*

This tracking property implies attraction of the target set $\Upsilon_{ISO}^k(t;\mathbf{D},\mathbf{U},\mathbf{Y}_d)$ with the state finite vector reachability time $\boldsymbol{\tau}_R^n$, but vice versa is not true.

Definition 227 ***The elementwise state stable tracking with the state finite vector reachability time*** $\boldsymbol{\tau}_R^n$ ***of the desired output of the ISO plant (2.33), (2.34)***

The ISO plant (2.33), (2.34) exhibits ***the elementwise state stable tracking with the state finite vector reachability time*** $\boldsymbol{\tau}_R^n$, ***over*** $\mathfrak{D}\text{x}\mathfrak{Y}_d$ *if, and only if, it exhibits the state elementwise tracking with the state finite reachability vector time* $\boldsymbol{\tau}_R^n$ *over* $\mathfrak{D}\text{x}\mathfrak{Y}_d$, *and for every positive real n vector* $\boldsymbol{\varepsilon}$ *and for every* $[\mathbf{D}(.),\mathbf{Y}_d(.)] \in \mathfrak{D}\text{x}\mathfrak{Y}_d$ *there exists a positive real n vector* $\boldsymbol{\delta}$, $\boldsymbol{\delta} = \boldsymbol{\delta}(\boldsymbol{\varepsilon},\mathbf{D},\mathbf{U},\mathbf{Y}_d)$, *such that* $Dist\,[\mathbf{X}_{P0},\Upsilon_{ISO}(0;\mathbf{D},\mathbf{U},\mathbf{Y}_d)] < \boldsymbol{\delta}$ *guarantees*

$$Dist\,[\mathbf{X}_P(\mathbf{t}^n),\Upsilon_{ISO}(\mathbf{t}^n;\mathbf{D},\mathbf{U},\mathbf{Y}_d)] < \boldsymbol{\varepsilon},$$

for every $\mathbf{t}^n \in \mathfrak{T}_R^n = [\mathbf{0}_n,\boldsymbol{\tau}_R^n]$, *i.e.,*

$$\begin{gathered}\forall\,[\mathbf{D}(.),\mathbf{Y}_d(.)] \in \mathfrak{D}\text{x}\mathfrak{Y}_d,\ \forall\boldsymbol{\varepsilon} \in \mathfrak{R}^{+^n},\ \exists\boldsymbol{\delta} \in \mathfrak{R}^{+^n},\\ \boldsymbol{\delta} = \boldsymbol{\delta}(\boldsymbol{\varepsilon},\mathbf{D},\mathbf{U},\mathbf{Y}_d),\ Dist\,[\mathbf{X}_{P0},\Upsilon_{ISO}(0;\mathbf{D},\mathbf{U},\mathbf{Y}_d)] < \boldsymbol{\delta} \Longrightarrow\\ Dist\,[\mathbf{X}_P(\mathbf{t}^n),\Upsilon_{ISO}(\mathbf{t}^n;\mathbf{D},\mathbf{U},\mathbf{Y}_d)] < \boldsymbol{\varepsilon},\ \forall\mathbf{t}^n \in \mathfrak{T}_R^n = [\mathbf{0}_n,\boldsymbol{\tau}_R^n].\end{gathered} \tag{8.82}$$

The elementwise state stable tracking with the state finite vector reachability time $\boldsymbol{\tau}_R^n$ *is* ***global (in the whole) over*** $\mathfrak{D}\text{x}\mathfrak{Y}_d$ *if, and only if, both the state elementwise tracking with the state finite vector reachability time* $\boldsymbol{\tau}_R^n$ *is global and* $\boldsymbol{\delta}(\boldsymbol{\varepsilon},\mathbf{D},\mathbf{U},\mathbf{Y}_d) \rightarrow \infty\mathbf{1}^n$ *as* $\boldsymbol{\varepsilon} \rightarrow \infty\mathbf{1}^n$ *for every* $[\mathbf{D}(.),\mathbf{Y}_d(.)] \in \mathfrak{D}\text{x}\mathfrak{Y}_d$.

The elementwise state stable tracking with the state finite vector reachability time $\boldsymbol{\tau}_R^n$ *is* ***uniform over*** $\mathfrak{D}\text{x}\mathfrak{Y}_d$ *if, and only if, the state elementwise tracking with the state finite vector reachability time* $\boldsymbol{\tau}_R^n$ *is uniform and the value of* $\boldsymbol{\delta}$ *depends at most on* $\mathfrak{D}\text{x}\mathfrak{Y}_d$ *and* $\mathbf{U}$, *but not on a particular choice of* $[\mathbf{D}(.),\mathbf{Y}_d(.)]$ *from* $\mathfrak{D}\text{x}\mathfrak{Y}_d$.

This tracking property guarantees asymptotic stability of the target set $\Upsilon_{ISO}(t;\mathbf{D},\mathbf{U},\mathbf{Y}_d)$ with the state finite vector reachability time $\boldsymbol{\tau}_R^n$. However, it does not ensure either stability or boundedness of the desired motion $\mathbf{X}_{Pd}(t)$.

Definition 228 ***The k-th order elementwise bounded state tracking with the state finite vector reachability time*** τ_R^k ***of the desired output of the*** *ISO plant (2.33), (2.34)*

The plant (2.33), (2.34) exhibits ***the k-th order elementwise, bounded state, tracking with the state finite vector reachability time*** $\boldsymbol{\tau}_R^n$, *for short* ***the k-th order elementwise bounded state tracking with the state finite vector reachability time*** $\boldsymbol{\tau}_R^n$, ***over*** $\mathfrak{D}^k\text{x}\mathfrak{Y}_d^k$ *if, and only if, it exhibits the k-th order state elementwise tracking with the state finite vector reachability* *time* $\boldsymbol{\tau}_R^n$ *over* $\mathfrak{D}^k \text{x} \mathfrak{Y}_d^k$ *and there is a compact subset* $\mathfrak{B}(t)$ *of* $\Upsilon_{ISO}^k(t; \mathbf{D}, \mathbf{U}, \mathbf{Y}_d)$, *(8.76), such that for every positive real vector* $\boldsymbol{\xi}^n$ *and for every* $[\mathbf{D}(.), \mathbf{Y}_d(.)] \in \mathfrak{D}^k \text{x} \mathfrak{Y}_d^k$ *there exists a positive real vector* $\boldsymbol{\psi}^n$, $\boldsymbol{\psi}^n = \boldsymbol{\psi}^n(\boldsymbol{\xi}^n, \mathfrak{B}, \mathbf{D}, \mathbf{U}, \mathbf{Y}_d)$, *such that* $Dist[\mathbf{X}_{P0}, \mathfrak{B}(0)] < \boldsymbol{\psi}^n$ *guarantees* $Dist[\mathbf{X}_P(\mathbf{t}^n), \mathfrak{B}(\mathbf{t}^n)] < \boldsymbol{\xi}^n$ *for all* $\mathbf{t}^n \in \mathfrak{T}_0^n$, *i.e.,*

$$
\begin{gathered}
\exists \mathfrak{B}(t) = Cl\ \mathfrak{B}(t) \subset \Upsilon_{ISO}^k(t; \mathbf{D}, \mathbf{U}, \mathbf{Y}_d),\ \forall t \in \mathfrak{T}_0, \\
sup\,[\|\mathbf{x} - \mathbf{y}\| : \mathbf{x}, \mathbf{y} \in \mathfrak{B}(t)] < \infty,\ \forall t \in \mathfrak{T}_0, \\
\forall [\mathbf{D}(.), \mathbf{Y}_d(.)] \in \mathfrak{D}^k \text{x} \mathfrak{Y}_d^k, \forall \boldsymbol{\xi}^n \in \mathfrak{R}^{+^n},\ \exists \boldsymbol{\psi}^n \in \mathfrak{R}^{+^n}, \\
\boldsymbol{\psi}^n = \boldsymbol{\psi}^n(\boldsymbol{\xi}^n, \mathfrak{B}, \mathbf{D}, \mathbf{U}, \mathbf{Y}_d),\ Dist[\mathbf{X}_{P0}, \mathfrak{B}(0)] < \boldsymbol{\psi}^n \implies \\
Dist[\mathbf{X}_P(\mathbf{t}^n), \mathfrak{B}(\mathbf{t}^n)] < \boldsymbol{\xi}^n,\ \forall \mathbf{t}^n \in [\boldsymbol{\tau}_R^n,\ \infty \mathbf{1}_n[, \\
k \in \{0, 1, 2, ...\}.
\end{gathered}
\tag{8.83}
$$

The zero, $(k = 0)$*, order elementwise bounded state tracking with the state finite vector reachability time* $\boldsymbol{\tau}_R^n$ *is simply called* **the elementwise bounded state tracking with the state finite vector reachability time** $\boldsymbol{\tau}_R^n$.

The k-th order elementwise bounded state tracking with the state finite vector reachability time $\boldsymbol{\tau}_R^n$ *is* **uniform over** $\mathfrak{D}^k \text{x} \mathfrak{Y}_d^k$ *if, and only if, the k-th order state elementwise tracking with the state finite vector reachability time* $\boldsymbol{\tau}_R^n$ *is uniform and the value of* $\boldsymbol{\psi}^n = \boldsymbol{\psi}^n(\boldsymbol{\xi}^n, \mathfrak{B}, \mathbf{D}, \mathbf{U}, \mathbf{Y}_d)$ *depends at most on* $\mathfrak{D}^k \text{x} \mathfrak{Y}_d^k$ *and* $\mathbf{U}$, *but not on a particular choice of* $[\mathbf{D}(.), \mathbf{Y}_d(.)]$ *from* $\mathfrak{D}^k \text{x} \mathfrak{Y}_d^k$.

The k-th order elementwise bounded state tracking with the state finite vector reachability time $\boldsymbol{\tau}_R^n$ *is* **global (in the whole) over** $\mathfrak{D}^k \text{x} \mathfrak{Y}_d^k$ *if, and only if, the k-th order state elementwise tracking with the state finite vector reachability time* $\boldsymbol{\tau}_R^n$ *over* $\mathfrak{D}^k \text{x} \mathfrak{Y}_d^k$ *is global and* $\boldsymbol{\xi}^n \longrightarrow \infty \mathbf{1}^n$ *implies* $\boldsymbol{\psi}^n \longrightarrow \infty \mathbf{1}^n$ *for every* $[\mathbf{D}(.), \mathbf{Y}_d(.)] \in \mathfrak{D}^k \text{x} \mathfrak{Y}_d^k$. *It is also uniform.*

This tracking property ensures asymptotic stability of the set $\mathfrak{B}(t)$ with the state finite vector reachability time $\boldsymbol{\tau}_R^n$, but vice versa is not valid.

Note 229 *Tracking is necessary for all other above tracking properties.*

Note 230 *State exponential tracking property signifies that the state vector converges exponentially to the target set* $\Upsilon_{ISO}^k(t; \mathbf{D}, \mathbf{U}, \mathbf{Y}_d)$ *in the state space.*

Note 231 *The scalar reachability time represents the latest moment when both the real state reaches the target set and the real output becomes equal to the desired output, respectively.*

Note 232 *The vector reachability time is the latest elementwise vector instant when the state vector is in the target set and stays in it forever, and then the real output vector* $\mathbf{Y}$ *becomes equal to the desired output vector* $\mathbf{Y}_d$ *forever.*

Note 233 *In the case when the real plant state* $\mathbf{X}_P$, *hence the desired plant state* $\mathbf{X}_{Pd}$, *is (or replaces formally) the real output* $\mathbf{Y}$, *hence the desired output* $\mathbf{Y}_d$, *respectively,* $\mathbf{Y} = \mathbf{X}_P$ *and* $\mathbf{Y}_d = \mathbf{X}_{Pd}$, *then the output tracking properties become*

the state tracking properties*. In such a case the target set* $\Upsilon(t;\mathbf{D};\mathbf{U};\mathbf{X}_{Pd})$ *becomes the singleton and takes the following form*

$$\Upsilon_{ISO}(t;\mathbf{D};\mathbf{U};\mathbf{X}_{Pd}) = \Upsilon_{ISO}(t;\mathbf{X}_{Pd}) = \{\mathbf{X}_P : \ \mathbf{X}_P = \mathbf{X}_{Pd}(t)\}, \ \forall t \in \mathfrak{T}_0; \tag{8.84}$$

the scalar distance $d\left[\mathbf{X}_P, \Upsilon_{ISO}(t;\mathbf{X}_{Pd})\right]$ *of the state* $\mathbf{X}_P$ *from* $\Upsilon_{ISO}(t;\mathbf{X}_{Pd})$ *becomes*

$$d\left[\mathbf{X}_P, \Upsilon_{ISO}(t;\mathbf{X}_{Pd})\right] = \|\mathbf{X}_P - \mathbf{X}_{Pd}(t)\| \equiv \|\mathbf{X}_{Pd}(t) - \mathbf{X}_P\|; \tag{8.85}$$

and the elementwise vector distance $Dist\left[\mathbf{X}_P(\mathbf{t}^n), \Upsilon_{ISO}(t;\mathbf{X}_{Pd})\right]$ *of the state* $\mathbf{X}_P$ *from the target set* $\Upsilon_{ISO}(t;\mathbf{X}_{Pd})$ *becomes*

$$Dist\left[\mathbf{X}_P(\mathbf{t}^n), \Upsilon_{ISO}(t;\mathbf{X}_{Pd})\right] = |\mathbf{X}_P(\mathbf{t}^n) - \mathbf{X}_{Pd}(\mathbf{t}^n)| \equiv |\mathbf{X}_{Pd}(\mathbf{t}^n) - \mathbf{X}_P(\mathbf{t}^n)|. \tag{8.86}$$

Then

- *tracking over* $\mathfrak{D}^k$ *of the desired motion* $\mathbf{X}_{Pd}(.)$ *guarantees its attraction,*
- *stable state tracking over* $\mathfrak{D}^k$ *of the desired motion* $\mathbf{X}_{Pd}(.)$ *guarantees its asymptotic stability,*
- *state exponential tracking over* $\mathfrak{D}^k$ *of the desired motion* $\mathbf{X}_{Pd}(.)$ *guarantees its exponential stability,*

but vice versa do not hold in general.

Chapter 9

Trackability theory

9.1 Trackability versus controllability

Kalman's concept of the state controllability has become a fundamental control concept, [199]-[203]. E. G. Gilbert [66] generalized it to the $MIMO$ systems. M. L. J. Hautus [186] established for them the simple form of the controllability criterion in the complex domain.

J. E. Bertram and P. E. Sarachik [14] broadened *the state controllability* concept to *the output controllability concept.*

Both the state controllability concept and the output controllability concept consider the system possibility of steering a state or an output from any initial state or from any initial output to another state or another output, in general, or to the zero state or to the zero output, in particular, respectively.

R. W. Brockett and M. D. Mesarović (Mesarovitch) [20] introduced *the concept of functional (output) reproducibility*, called also *the output function controllability* [8, page 313], [29, page 216], [319, pages 72 and 164], in which the target is not a particular output (e.g., the zero output) but a given function representing a reference (desired) output response. This concept concerns the systems free of any external disturbance action.

All these controllability concepts assume the nonexistence of any external perturbations acting on the system. The only external influences on the system are control actions.

Dynamic systems, in general, and plants, hence their control systems, in particular, are subject in reality to actions of unpredictable external perturbations (called usually *disturbances*).

Remark 234 ***Disturbance compensation***

*It is not surprising to find among students who begin to learn about control those who say that the control acting on the plant "**rejects**" the disturbance in the sense that it **eliminates** the disturbance, or **eliminates** the disturbance action (influence) on the plant. Since it is physically meaningless, it is a substantial physical and control mistake. The control action cannot reject (elimi-*

nate) the disturbance action (influence) on the plant. For example, there is not a controller that can reject (eliminate) the wind action (influence) on a flying plane.

*The control action can (at most, at best) **(fully) compensate**, i.e., **neutralize**, the disturbance action on the plant. The control action achieves it completely when the controller forces the plant to create an error in its behavior, which has the same magnitude as the error created by the disturbance and initial conditions and which is opposite in sign to the error caused by the disturbance action and by the influence of the initial conditions. The result is the zero error, i.e., the resulting real plant behavior coincides with its desired behavior. With this elementary, but crucial, fact, we will use the term **disturbance compensation** rather than the widely used popular expression "disturbance rejection".*

S. P. Bhattacharyya [16], [17], S. P. Bhattacharyya et al. [18], E. J. Davison [38]-[40], E. J. Davison et al. [41]-[45], [244], E. Fabian and W. M. Wonham [58], B. Porter et al. [279], R. Saeks and J. Murray [292], and S. Y. Zhang and C. T. Chen [327] studied largely the problem of the disturbance compensation.

The controllability problems and the disturbance compensation problems have been mainly studied separately. However, each of them does not satisfy the basic control goal that is to force the plant **subject to disturbance actions and to arbitrary initial conditions *to follow, i.e., to track***, its desired behavior. Since this is the very goal of the control in the real plant environment and under real operating conditions, it led to the introduction of a new control concept called ***trackability***.

The trackability concept explains whether the plant itself has a property to enable the existence of a control that can guarantee **tracking under arbitrary initial conditions (globally or from a domain) and under external perturbations** belonging to a set $\mathfrak{D}^i$ of permitted disturbances, as well as for every plant desired output response from a given functional family $\mathfrak{Y}_d^k$. It involves the controllability concept and the disturbance compensation concept.

Another plant property called ***natural trackability*** is a type of plant *trackability* that permits control synthesis and implementation without using information about the real values and forms of the disturbances and about the mathematical model of the plant internal dynamics. Such control that is continuous in *time* is ***Natural Tracking Control*** (***NTC***).

The concept of ***natural trackability*** was established and developed by discovering algorithms for synthesis of ***Natural Tracking Control*** in [85]-[87], [132]-[141], [149], [157], [159]-[161], [163], [164], [166], [167], [173], [176]-[180], [249]-[258].

PCUP (Principles 6 and 7, Chapter 1) summarizes the common general properties of physical variables. *TCUP* (*Time Continuity and Uniqueness Principle* 9, Chapter 1) jointly expresses these and the crucial properties of *time* [151], [164], [165], [170]-[172]. These principles enable effective *Natural Tracking Control synthesis* for linear and nonlinear dynamic plants.

9.2 Trackability definitions

9.2.1 Perfect trackability

What is the meaning of ***trackability*** and whether it can be perfect?

Definition 235 ***Definition of the k-th order perfect trackability of*** $\mathbf{Y}_d(.)$ ***for the given*** $\mathbf{D}(.)$

The desired output vector function $\mathbf{Y}_d(.)$ *is* ***the k-th order perfect trackable under the action of the given*** $\mathbf{D}(.)$ *if, and only if, there exists a control vector function* $\mathbf{U}(.)$ *such that the plant real output vector* $\mathbf{Y}(t)$ *and its first k derivatives are always equal to the desired plant output vector* $\mathbf{Y}_d(t)$ *and its first k derivatives, respectively, as soon as* $\mathbf{Y}^k(0) = \mathbf{Y}_d^k(0)$,

$$\text{given } \mathbf{D}(.),\ \exists \mathbf{U}(.) \text{ and } \mathbf{Y}^k(0) = \mathbf{Y}_d^k(0) \Longrightarrow \mathbf{Y}^k(t) = \mathbf{Y}_d^k(t),\ \forall t \in \mathfrak{T}_0. \quad (9.1)$$

We denote the *k-th* order right-hand side derivative of $\mathbf{Y}(t)$ at $t \in \mathfrak{T}_0$ with $D_r^k \mathbf{Y}(t)$.

Lemma 236 *If two functions* $\mathbf{Y}(.)$ *and* $\mathbf{Y}_d(.)$ *are defined, k-times continuously differentiable on* $]\sigma, \infty[$, $\sigma \in \mathfrak{T}_0$, $]\sigma, \infty[\subseteq Int\mathfrak{T}_0$, *as well as at* $t = \sigma$ *from the right-hand side, i.e., at* $t = \sigma^+$, *and identical on* $[\sigma, \infty[$, *then all their derivatives up to the order k included are also identical on* $]\sigma, \infty[$ *and at* $t = \sigma^+$.

Proof. Let $\sigma \in \mathfrak{T}_0$, $]\sigma, \infty[\subseteq Int\mathfrak{T}_0$, and let two functions $\mathbf{Y}(.)$ and $\mathbf{Y}_d(.)$ be defined, k-times continuously differentiable on $]\sigma, \infty[$, as well as at $t = \sigma^+$, and identical on $[\sigma, \infty[$,

$$\begin{gathered} \mathbf{Y}(.), \mathbf{Y}_d(.) \in \mathfrak{C}^{k-1}\left([\sigma, \infty[\right) \cup \mathfrak{C}^{k}\left(]\sigma, \infty[\right), \\ \mathbf{Y}(t) = \mathbf{Y}_d(t),\ \forall t \in [\sigma, \infty[. \end{gathered} \quad (9.2)$$

By the definition of the first derivative and in view of (9.2):

$$\frac{d\mathbf{Y}(t)}{dt} = \lim_{\theta \longrightarrow 0} \left[\frac{\mathbf{Y}(t+\theta) - \mathbf{Y}(t)}{\theta} \right] =$$

$$= \lim_{\theta \longrightarrow 0} \left[\frac{\mathbf{Y}_d(t+\theta) - \mathbf{Y}_d(t)}{\theta} \right] = \frac{d\mathbf{Y}_d(t)}{dt},\ \forall t \in]\sigma, \infty[,$$

and at $t = \sigma$

$$D_r^1 \mathbf{Y}(t)_{t=\sigma} = \lim_{\theta \longrightarrow 0^+} \left[\frac{\mathbf{Y}(\theta+\sigma) - \mathbf{Y}(\sigma)}{\theta} \right] =$$

$$= \lim_{\theta \longrightarrow 0^+} \left[\frac{\mathbf{Y}_d(\theta+\sigma) - \mathbf{Y}_d(\sigma)}{\theta} \right] = D_r^1 \mathbf{Y}_d(t)_{t=\sigma}.$$

The statement holds for $i = 0, 1$. Let it hold for any $i \in \{0, 1, ..., k-1\}$,

$$\begin{gathered} \mathbf{Y}^{(i)}(t) = \mathbf{Y}_d^{(i)}(t),\ \forall t \in]\sigma, \infty[, \\ D_r^i \mathbf{Y}(t)_{t=\sigma} = D_r^i \mathbf{Y}_d(t)_{t=\sigma}. \end{gathered}$$

These equations yield

$$\frac{d^{i+1}\mathbf{Y}(t)}{dt^{i+1}} = \lim_{\theta \longrightarrow 0}\left[\frac{\mathbf{Y}^{(i)}(t+\theta)-\mathbf{Y}^{(i)}(t)}{\theta}\right] =$$

$$= \lim_{\theta \longrightarrow 0}\left[\frac{\mathbf{Y}_d^{(i)}(t+\theta)-\mathbf{Y}_d^{(i)}(t)}{\theta}\right] = \frac{d^{i+1}\mathbf{Y}_d(t)}{dt^{i+1}},\ \forall t \in]\sigma,\infty[,$$

and

$$D_r^{i+1}\mathbf{Y}(t)_{t=\sigma} = \lim_{\theta \longrightarrow 0+}\left[\frac{D_r^i\mathbf{Y}(\theta+\sigma)-D_r^i\mathbf{Y}(\sigma)}{\theta}\right] =$$

$$= \lim_{\theta \longrightarrow 0+}\left[\frac{D_r^i\mathbf{Y}_d(\theta+\sigma)-D_r^i\mathbf{Y}_d(\sigma)}{\theta}_{t=0}\right] = D_r^{i+1}\mathbf{Y}_d(t)_{t=\sigma}.$$

Since the statement is true for $i = 0,1$ and for $i+1$ if it holds for $i \in \{0,1,...,k-1\}$, then by the mathematical induction it holds $\forall i \in \{0,1,...,k\}$ ■

Note 237 *Notice that this Lemma refines and generalizes Theorem 157 (Subsection 8.3.1) by allowing different vector values of* $\mathbf{Y}(t)$ *and of its derivatives from the left-hand side and the right-hand side of* $t=\sigma$, $\sigma \in \mathfrak{T}_0$.

Definition 52 (Subsection 3.3.1), Definition 235 and Lemma 236 imply directly the following claim.

Lemma 238 *If the desired output vector function* $\mathbf{Y}_d(.)$ *is differentiable at least up to the order k,* $\mathbf{Y}_d(t) \in \mathfrak{C}^k$, *then for it to be the k-th order perfect trackable under the action of the given* $\mathbf{D}(.)$ *it is necessary and sufficient to be realizable for the given* $\mathbf{D}(.)$, *equivalently, to be perfect trackable under the action of the given* $\mathbf{D}(.)$.

We are interested in perfect trackability of every plant desired output $\mathbf{Y}_d(.)$ from $\mathfrak{Y}_d^k$ rather than in perfect trackability of a single plant desired output.

Definition 239 ***The l-th order perfect trackability of the plant in*** $\mathfrak{D}^i$ ***on*** $\mathfrak{Y}_d^k$

The m-th order dynamic plant is ***the l-th order perfect trackable in*** $\mathfrak{D}^i$ ***on*** $\mathfrak{Y}_d^k$, $i,k \in \{0,1,...,m-1\}$, $l \in \{0,1,...,k\}$, *if, and only if, for every* $\mathbf{Y}_d(.) \in \mathfrak{Y}_d^k$ *there exist a disturbance vector function* $\mathbf{D}(.) \in \mathfrak{D}^i$ *and a control vector function* $\mathbf{U}(.)$ *such that the plant real output and its first l derivatives are always equal to the plant desired output and its first l derivatives, respectively, as soon as* $\mathbf{Y}^{m-1}(0) = \mathbf{Y}_d^{m-1}(0)$,

$$\mathbf{Y}^{m-1}(0) = \mathbf{Y}_d^{m-1}(0),\ \forall \mathbf{Y}_d(.) \in \mathfrak{Y}_d^k,\ \exists \mathbf{D}(.) \in \mathfrak{D}^i,\ \exists \mathbf{U}(.) \Longrightarrow$$
$$\mathbf{Y}^l(t) = \mathbf{Y}_d^l(t),\ \forall t \in \mathfrak{T}_0,\ i,k \in \{0,1,...,m-1\},\ l \in \{0,1,...,k\}. \quad (9.3)$$

From Definition 53, i.e., from Definition 54 (Subsection 3.3.1), Definition 239, Lemma 236 and Lemma 238 we deduce the following.

Lemma 240 ***The l-th order perfect trackability in $\mathfrak{D}^i$ on $\mathfrak{Y}_d^k$ and the perfect trackability in $\mathfrak{D}^i$ on $\mathfrak{Y}_d^k$***

For the m-th order plant to be the l-th order perfect trackable in $\mathfrak{D}^i$ on $\mathfrak{Y}_d^k$, $i,k \in \{0,1,...,m-1\}$, $l \in \{0,1,...,k\}$, it is necessary and sufficient that every $\mathbf{Y}_d(.) \in \mathfrak{Y}_d^l$ is realizable in $\mathfrak{D}^i$, equivalently, to be perfect trackable in $\mathfrak{D}^i$ on $\mathfrak{Y}_d^k$.

Definition 241 ***The l-th order perfect trackability of the plant on $\mathfrak{D}^i x\mathfrak{Y}_d^k$***

The m-th order dynamic plant is ***the l-th order perfect trackable on*** *$\mathfrak{D}^i x\mathfrak{Y}_d^k$, $i,k \in \{0,1,...,m-1\}$, $l \in \{0,1,...,k\}$, if, and only if, for every $[\mathbf{D}(.), \mathbf{Y}_d(.)] \in \mathfrak{D}^i x\mathfrak{Y}_d^k$ there exists a control vector function $\mathbf{U}(.)$ such that the plant real output and its first l derivatives are always equal to the plant desired output and its first l derivatives, respectively, as soon as $\mathbf{Y}^{m-1}(0) = \mathbf{Y}_d^{m-1}(0)$*

$$\mathbf{Y}^{m-1}(0) = \mathbf{Y}_d^{m-1}(0),\ \forall[\mathbf{D}(.), \mathbf{Y}_d(.)] \in \mathfrak{D}^i x\mathfrak{Y}_d^k,\ \exists \mathbf{U}(.) \Longrightarrow$$
$$\mathbf{Y}^l(t) = \mathbf{Y}_d^l(t),\ \forall t \in \mathfrak{T}_0,\ i,k \in \{0,1,...,m-1\}, l \in \{0,1,...,k\}. \tag{9.4}$$

From Definition 56 (Subsection 3.3.1), Lemma 236, Definition 241 and Lemma 238 the following results.

Lemma 242 ***The l-th order perfect trackability on $\mathfrak{D}^i x\mathfrak{Y}_d^k$ and the perfect trackability on $\mathfrak{D}^i x\mathfrak{Y}_d^k$***

For the m-th order dynamic plant to be ***the** l-th* ***order perfect trackable on*** *$\mathfrak{D}^k x\mathfrak{Y}_d^k$, $k \in \{0,1,...,m-1\}$, $l \in \{0,1,...,k\}$, it is necessary and sufficient that every $\mathbf{Y}_d(.) \in \mathfrak{Y}_d^k$ is realizable on $\mathfrak{D}^i$, equivalently, to be perfect trackable on $\mathfrak{D}^i x\mathfrak{Y}_d^k$.*

These lemmas express the equivalence between the realizability of the plant desired output and the plant perfect trackability. The form of the realizability conditions depends on the type of the plant, i.e., on the form of its mathematical model.

Except for the existence requirement, the preceding definitions do not impose any other condition on the control vector function $\mathbf{U}(.)$. Its existence means that its instantaneous vector value $\mathbf{U}(t)$ is defined at every moment $t \in \mathfrak{T}_0$. This permits piecewise continuity of $\mathbf{U}(t)$; i.e., it allows $\mathbf{U}(t) \in \mathfrak{C}^-(\mathfrak{T}_0)$. A piecewise continuous variable can be only a mathematical, but not a physical variable. It is not exactly physically realizable, which is explained by *PCUP* (Principles 6 and 7, Chapter 1). In order to be physically realizable, control variable $\mathbf{U}(.)$ should obey *PCUP*, equivalently *TCUP* (Principle 9, Chapter 1).

The preceding definitions determine the control vector function $\mathbf{U}(.)$ in terms of the disturbance vector function $\mathbf{D}(.)$.

The vector form and the instantaneous value of the disturbance variable $\mathbf{D}(.)$ are most often unknown, unpredictable, and their values can be also unmeasurable. These disturbance features cause the problem of the control realization if control is synthesized in terms of $\mathbf{D}(.)$.

9.2.2 Perfect natural trackability

Problem 243 ***Disturbance and the control synthesis problem***
Do the plant properties permit a control synthesis without using information about the real form and the value of the disturbance vector $\mathbf{D}(t)$ *at any* $t \in \mathfrak{T}_0$*?*

Mathematical models of plants, which are the starting point for the control synthesis, are approximative qualitatively (due to their nonlinear nature, their forms and dynamic complexity) and quantitatively (due to their order and parameter values).

Problem 244 ***Plant internal dynamics and the control synthesis problem***
Is it possible to synthesize control without using information about the mathematical model of the plant internal dynamics? Do the properties of the plant enable the existence of such control?

Comment 245 *The nature (e.g., the brain as a natural controller) creates very successfully time-continuous control (control of all organs) without using information about a mathematical model of the plant (of the organs). Besides, the nature (the brain) often does not have precise, or any, information about the forms and/or the values of disturbances. Since such control exists and since it is created by the nature, we call it* **Natural Control** *(NC) whatever the physical nature of the controller and regardless of the creator of the controller.*

The following definitions reply to the preceding questions.

Definition 246 ***The l-th order perfect natural trackability in*** $\mathfrak{D}^i$ ***on*** $\mathfrak{Y}_d^k$
The m-th order dynamic plant is ***the l-th order perfect natural trackable in*** $\mathfrak{D}^i$ ***on*** $\mathfrak{Y}_d^k$, $i,k \in \{0,1,...,m-1\}$, $l \in \{0,1,...,k\}$, *if, and only if, for every* $\mathbf{Y}_d(.) \in \mathfrak{Y}_d^k$ *there exist* $\mathbf{D}(.) \in \mathfrak{D}^i$ *and a control vector function* $\mathbf{U}(.)$ *obeying TCUP on* $\mathfrak{T}_0$*, which can be synthesized without using information about the form and the value of* $\mathbf{D}(.)$ *and about the mathematical model of the plant internal dynamics, such that the plant real output and its first l derivatives are always equal to the desired plant output and its first l derivatives, respectively, i.e., that (9.5) holds,*

$$\mathbf{Y}^{m-1}(0) = \mathbf{Y}_d^{m-1}(0),\ \forall \mathbf{Y}_d(.) \in \mathfrak{Y}_d^k,\ \exists \mathbf{D}(.) \in \mathfrak{D}^i,\ \exists \mathbf{U}(.) \in \mathfrak{C}(\mathfrak{T}_0) \Longrightarrow$$
$$\mathbf{Y}^l(t) = \mathbf{Y}_d^l(t),\ \forall t \in \mathfrak{T}_0,\ i,k \in \{0,1,...,m-1\},\ l \in \{0,1,...,k\}. \quad (9.5)$$

Such control is ***the l-th order perfect natural tracking control in*** $\mathfrak{D}^i$ ***on*** $\mathfrak{Y}_d^k$.

Comment 247 *The l-th order perfect trackability in* $\mathfrak{D}^i$ *on* $\mathfrak{Y}_d^k$ *is necessary for the l-th order perfect natural trackability in* $\mathfrak{D}^i$ *on* $\mathfrak{Y}_d^k$*, and the l-th order perfect natural trackability in* $\mathfrak{D}^i$ *on* $\mathfrak{Y}_d^k$ *is sufficient for the l-th order perfect trackability in* $\mathfrak{D}^i$ *on* $\mathfrak{Y}_d^k$.

Definition 248 ***The l-th order perfect natural trackability on $\mathfrak{D}^i x \mathfrak{Y}_d^k$***

The m-th order dynamic plant is ***the l-th order perfect natural trackable on*** $\mathfrak{D}^i x \mathfrak{Y}_d^k$ *if, and only if,* $\forall$ $[\mathbf{D}(.),\mathbf{Y}_d(.)] \in \mathfrak{D}^i x \mathfrak{Y}_d^k$, $i, k \in \{0, 1, ..., m-1\}$, $l \in \{0, 1, ..., k\}$, *there exists a control vector function* $\mathbf{U}(.)$ *obeying TCUP on* $\mathfrak{T}_0$, *which can be synthesized without using information about the form and the value of any* $\mathbf{D}(.) \in \mathfrak{D}^i$ *and about the mathematical model of the plant internal dynamics, such that the plant real output and its first l derivatives are always equal to the desired plant output and its first l derivatives, respectively, i.e., that (9.6) holds,*

$$\mathbf{Y}^{m-1}(0) = \mathbf{Y}_d^{m-1}(0),\ \forall[\mathbf{D}(.),\mathbf{Y}_d(.)] \in \mathfrak{D}^i x \mathfrak{Y}_d^k,\ \exists \mathbf{U}(.) \in \mathfrak{C}(\mathfrak{T}_0) \Longrightarrow$$
$$\mathbf{Y}^l(t) = \mathbf{Y}_d^l(t),\ \forall t \in \mathfrak{T}_0,\ i, k \in \{0, 1, ..., m-1\},\ l \in \{0, 1, ..., k\}. \tag{9.6}$$

Comment 249 *Definition 241 and Definition 248 imply that the l-th order perfect trackability on* $\mathfrak{D}^i x \mathfrak{Y}_d^k$ *is necessary for the l-th order perfect natural trackability on* $\mathfrak{D}^i x \mathfrak{Y}_d^k$, *and that the l-th order perfect natural trackability on* $\mathfrak{D}^i x \mathfrak{Y}_d^k$ *is sufficient for the l-th order perfect trackability on* $\mathfrak{D}^i x \mathfrak{Y}_d^k$.

We deduce the following directly from Lemma 236 and Definition 248.

Lemma 250 ***The l-th order perfect natural trackability on $\mathfrak{D}^i x \mathfrak{Y}_d^k$ and the perfect natural trackability on $\mathfrak{D}^i x \mathfrak{Y}_d^k$***

For the m-th order dynamic plant to be the l-th order perfect natural trackable on $\mathfrak{D}^i x \mathfrak{Y}_d^k$, $k \in \{0, 1, ..., m-1\}$, $l \in \{0, 1, ..., k\}$, *it is necessary and sufficient to be perfect trackable on* $\mathfrak{D}^i x \mathfrak{Y}_d^k$.

9.2.3 Imperfect trackability

We will present the conceptual definitions of imperfect trackability and of imperfect natural trackability. They determine that the real output deviates from the desired output until some moment, at and after which they become and stay equal.

Definition 251 ***The l-th order trackability in $\mathfrak{D}^i$ on $\mathfrak{Y}_d^k$***

The m-th order dynamic plant is ***the l-th order trackable in*** $\mathfrak{D}^i$ ***on*** $\mathfrak{Y}_d^k$, $i, k \in \{0, 1, ..., m-1\}$, $l \in \{0, 1, ..., k\}$, *if, and only if, there is* $\Delta \in \mathfrak{R}^+$, *or* $\Delta = \infty$, *such that for every plant output desired response* $\mathbf{Y}_d(.) \in \mathfrak{Y}_d^k$ *and for every instant* $\sigma \in Int\ \mathfrak{T}_0$, *there are a disturbance vector function* $\mathbf{D}(.) \in \mathfrak{D}^i$ *and a control vector function* $\mathbf{U}(.)$ *such that for every initial plant output vector* $\mathbf{Y}_0^{m-1}$ *in the* Δ *neighborhood of the plant initial desired output vector* $\mathbf{Y}_{d0}^{m-1}$, *the norm of the difference between* $\mathbf{Y}^k(t)$ *and* $\mathbf{Y}_d^k(t)$ *becomes equal to zero at latest at the moment* σ, *after which it rests equal to zero forever, i.e.,*

$$\exists \Delta \in]0,\ \infty],\ \forall \mathbf{Y}_d(.) \in \mathfrak{Y}_d^k,\ \forall \sigma \in Int\ \mathfrak{T}_0,$$
$$\exists \mathbf{D}(.) \in \mathfrak{D}^i,\ \mathbf{D}(t) = \mathbf{D}(t; \mathbf{Y}_d),\ \exists \mathbf{U}(.),\ \mathbf{U}(t) = \mathbf{U}(t; \sigma; \mathbf{D}, \mathbf{Y}_d) \Longrightarrow$$
$$\|\mathbf{Y}_0^m - \mathbf{Y}_{d0}^m\| < \Delta \Longrightarrow \left\|\mathbf{Y}^l(t) - \mathbf{Y}_d^l(t)\right\| = 0\ \forall (t \geq \sigma) \in \mathfrak{T}_0,$$
$$i, k \in \{0, 1, ..., m-1\},\ l \in \{0, 1, ..., k\}. \tag{9.7}$$

Such control is ***the l-th order tracking control in*** $\mathfrak{D}^i$ ***on*** $\mathfrak{Y}_d^k$.

The zero, ($l = 0$), order trackability in $\mathfrak{D}^i$ *on* $\mathfrak{Y}_d^k$ *is simply called* ***trackability in*** $\mathfrak{D}^i$ ***on*** $\mathfrak{Y}_d^k$.

The l-th order trackability in $\mathfrak{D}^i$ *on* $\mathfrak{Y}_d^k$ *is* ***global (in the whole)*** *if, and only if,* $\Delta = \infty$.

The l-th order trackability in $\mathfrak{D}^i$ *on* $\mathfrak{Y}_d^k$ *is* ***uniform*** *over* $\mathfrak{Y}_d^k$ *if, and only if,* $\mathbf{U}(.)$ *depends on* $\mathfrak{Y}_d^k$ *but not on an individual* $\mathbf{Y}_d(.)$ *from* $\mathfrak{Y}_d^k$, $\mathbf{U}(t) = \mathbf{U}(t; \mathbf{D}; \mathfrak{Y}_d^k)$.

What is the relationship between the *l-th* order (imperfect) trackability in $\mathfrak{D}^i$ on $\mathfrak{Y}_d^k$ and the *l-th* order perfect trackability in $\mathfrak{D}^i$ on $\mathfrak{Y}_d^k$?

Theorem 252 ***Perfect versus imperfect trackability in*** $\mathfrak{D}^i$ ***on*** $\mathfrak{Y}_d^k$

In order for the m-th order dynamic plant to be the l-th order perfect trackable in $\mathfrak{D}^i$ *on* $\mathfrak{Y}_d^k$, $i, k \in \{0, 1, ..., m-1\}$, $l \in \{0, 1, ..., k\}$, *it is necessary and sufficient that it is the l-th order trackable in* $\mathfrak{D}^i$ *on* $\mathfrak{Y}_d^k$.

Proof. *Necessity.* We prove the statement by the contradiction. Let the dynamic plant be the *l-th* order perfect trackable in $\mathfrak{D}^i$ on $\mathfrak{Y}_d^k$ but not the *l-th* order trackable in $\mathfrak{D}^i$ on $\mathfrak{Y}_d^k$. Definition 239 holds. Since the plant is not *l-th* order trackable in $\mathfrak{D}^i$ on $\mathfrak{Y}_d^k$, then, due to the violation of Definition 251, for $\mathbf{Y}_0^{m-1} = \mathbf{Y}_{d0}^{m-1}$,

$$\left\|\mathbf{Y}_0^{m-1} - \mathbf{Y}_{d0}^{m-1}\right\| = 0 < \Delta, \forall \Delta > 0,$$

and for every $\tau \in Int\ \mathfrak{T}_0$ there exist $j \in \{0, 1, ..., l\}$ and $(\gamma \geq \tau) \in \mathfrak{T}_0$ such that $\left\|\mathbf{Y}^j(\gamma) - \mathbf{Y}_d^j(\gamma)\right\| > 0$. This contradicts Definition 239, which is the consequence of the assumption that the plant is not the *l-th* order trackable in $\mathfrak{D}^i$ on $\mathfrak{Y}_d^k$. Hence, the dynamic plant is the *l-th* order trackable in $\mathfrak{D}^i$ on $\mathfrak{Y}_d^k$.

Sufficiency. Let the dynamic plant be the *l-th* order trackable in $\mathfrak{D}^i$ on $\mathfrak{Y}_d^k$. Definition 251 holds. Let $\sigma \in Int\ \mathfrak{T}_0$ and let it be arbitrarily small, i.e., $\sigma \longrightarrow 0^+$ in Definition 251 so that (9.7) becomes for such σ :

$$\begin{gathered}\exists \Delta \in]0,\ \infty],\ \forall \mathbf{Y}_d(.) \in \mathfrak{Y}_d^k,\ \forall \sigma \in Int\ \mathfrak{T}_0,\ \sigma \longrightarrow 0^+. \\ \exists \mathbf{D}(.) \in \mathfrak{D}^i,\ \mathbf{D}(t) = \mathbf{D}(t; \mathbf{Y}_d),\ \exists \mathbf{U}(.),\ \mathbf{U}(t) = \mathbf{U}(t; \sigma; \mathbf{D}, \mathbf{Y}_d) \Longrightarrow \\ \|\mathbf{Y}_0^m - \mathbf{Y}_{d0}^m\| < \Delta \Longrightarrow \left\|\mathbf{Y}^l(t) - \mathbf{Y}_d^l(t)\right\| = 0,\ \forall\, (t \geq \sigma) \in \mathfrak{T}_0, \\ i, k \in \{0, 1, ..., m-1\},\ l \in \{0, 1, ..., k\},\end{gathered} \tag{9.8}$$

Let $\mathbf{Y}_0^{m-1} = \mathbf{Y}_{d0}^{m-1}$, i.e., $\left\|\mathbf{Y}_0^{m-1} - \mathbf{Y}_{d0}^{m-1}\right\| = 0 < \Delta$. Such $\mathbf{Y}_0^{m-1}$ obeys (9.7) that permits $\sigma \longrightarrow 0^+$. They together imply $\left\|\mathbf{Y}^l(t) - \mathbf{Y}_d^l(t)\right\| = 0$, $\forall t \in \mathfrak{T}_0$, equivalently $\mathbf{Y}^l(t) = \mathbf{Y}_d^l(t)$, $\forall t \in \mathfrak{T}_0$. This satisfies Definition 239 and proves the sufficiency part of the theorem statement ■

In order to achieve fully the trackability examination we should explore whether the plant is trackable for every $[\mathbf{D}(.), \mathbf{Y}_d(.)] \in \mathfrak{D}^i \text{x} \mathfrak{Y}_d^k$.

Definition 253 ***The l-th order trackability on*** $\mathfrak{D}^i \text{x} \mathfrak{Y}_d^k$

The m-th order dynamic plant is ***the l-th order trackable on*** $\mathfrak{D}^i \text{x} \mathfrak{Y}_d^k$, $i, k \in \{0, 1, ..., m-1\}$, $l \in \{0, 1, ..., k\}$, *if, and only if, there is* $\Delta \in \mathfrak{R}^+$, *or*

$\Delta = \infty$, such that for every disturbance vector function $\mathbf{D}(.) \in \mathfrak{D}^i$, for every plant output desired response $\mathbf{Y}_d(.) \in \mathfrak{Y}_d^k$, and for every instant $\sigma \in Int\ \mathfrak{T}_0$, there is a control vector function $\mathbf{U}(.)$ such that for every plant initial output vector $\mathbf{Y}_0^{m-1}$ in the Δ neighborhood of the plant initial desired output vector $\mathbf{Y}_{d0}^{m-1}$, the norm of the difference between $\mathbf{Y}^l(t)$ and $\mathbf{Y}_d^l(t)$ becomes equal to zero at latest at the moment σ, after which it rests equal to zero forever, i.e.,

$$\begin{gathered}\exists\Delta \in]0,\ \infty],\ \forall\left[\mathbf{D}(.), \mathbf{Y}_d(.)\right] \in \mathfrak{D}^i \mathrm{x} \mathfrak{Y}_d^k,\ \forall\sigma \in Int\ \mathfrak{T}_0, \\ \exists\mathbf{U}(.),\ \mathbf{U}(t) = \mathbf{U}(t;\sigma;\mathbf{D};\mathbf{Y}_d) \implies \\ \left\|\mathbf{Y}_0^{m-1} - \mathbf{Y}_{d0}^{m-1}\right\| < \Delta \Longrightarrow \left\|\mathbf{Y}^l(t) - \mathbf{Y}_d^l(t)\right\| = 0,\ \forall\,(t \geq \sigma) \in \mathfrak{T}_0, \\ k \in \{0,1,...,m-1\},\ l \in \{0,1,...,k\}. \end{gathered} \tag{9.9}$$

Such control is ***the l-th order tracking control on $\mathfrak{D}^i \mathrm{x} \mathfrak{Y}_d^k$****, for short,* ***the l-th order tracking control.***

The zero, ($l = 0$), order trackability on $\mathfrak{D}^i \mathrm{x} \mathfrak{Y}_d^k$ is simply called ***trackability on $\mathfrak{D}^i \mathrm{x} \mathfrak{Y}_d^k$****. The zero, ($l = 0$), order tracking control on $\mathfrak{D}^i \mathrm{x} \mathfrak{Y}_d^k$ is simply called the tracking control on $\mathfrak{D}^i \mathrm{x} \mathfrak{Y}_d^k$, for short,* ***the tracking control.***

The l-th order trackability on $\mathfrak{D}^i \mathrm{x} \mathfrak{Y}_d^k$ is ***global (in the whole)*** *if, and only if, $\Delta = \infty$.*

The l-th order trackability on $\mathfrak{D}^i \mathrm{x} \mathfrak{Y}_d^k$ is ***uniform*** *over $\mathfrak{D}^i \mathrm{x} \mathfrak{Y}_d^k$ if, and only if, $\mathbf{U}(.)$ depends on $\mathfrak{D}^i \mathrm{x} \mathfrak{Y}_d^k$ but not on an individual pair $[\mathbf{D}(.), \mathbf{Y}_d(.)]$ from $\mathfrak{D}^i \mathrm{x} \mathfrak{Y}_d^k$, $\mathbf{U}(t) = \mathbf{U}(t;\sigma;\mathfrak{D}^k;\mathfrak{Y}_d^k)$.*

Theorem 254 ***Perfect versus imperfect trackability on $\mathfrak{D}^i \mathrm{x} \mathfrak{Y}_d^k$***

In order for the m-th order dynamic plant to be the l-th order perfect trackable on $\mathfrak{D}^i \mathrm{x} \mathfrak{Y}_d^k$, $k \in \{0,1,...,m-1\}$, $l \in \{0,1,...,k\}$, it is necessary and sufficient that it is the l-th order trackable on $\mathfrak{D}^i \mathrm{x} \mathfrak{Y}_d^k$.

Proof. *Necessity.* We prove the statement by the contradiction. Let the dynamic plant be the *l-th* order perfect trackable on $\mathfrak{D}^i \mathrm{x} \mathfrak{Y}_d^k$ but not the *l-th* order trackable on $\mathfrak{D}^i \mathrm{x} \mathfrak{Y}_d^k$. Definition 241 holds. Since the plant is not the *l-th* order trackable in $\mathfrak{D}^i \mathrm{x} \mathfrak{Y}_d^k$, then, due to the violation of Definition 253, for $\mathbf{Y}_0^{m-1} = \mathbf{Y}_{d0}^{m-1}$,

$$\left\|\mathbf{Y}_0^{m-1} - \mathbf{Y}_{d0}^{m-1}\right\| = 0 < \Delta, \forall\Delta > 0,$$

and for every $(\tau > 0) \in \mathfrak{T}_0$ there exist $j \in \{0,1,...,l\}$ and $(\gamma \geq \tau) \in \mathfrak{T}_0$ such that $\left\|\mathbf{Y}^j(\gamma) - \mathbf{Y}_d^j(\gamma)\right\| > 0$. This contradicts Definition 241. The contradiction disproves the assumption that plant is not the *l-th* order trackable on $\mathfrak{D}^i \mathrm{x} \mathfrak{Y}_d^k$. Hence, the dynamic plant is the *l-th* order trackable on $\mathfrak{D}^i \mathrm{x} \mathfrak{Y}_d^k$.

Sufficiency. Let the dynamic plant be the *l-th* order trackable on $\mathfrak{D}i\mathrm{x}\mathfrak{Y}_d^k$. Definition 253 holds. Let $(\sigma > 0) \in \mathfrak{T}_0$ and let it be arbitrarily small: $\sigma \longrightarrow 0^+$

in Definition 253 so that (9.9) becomes for such σ :

$$\begin{gathered}\exists\Delta\in]0,\ \infty],\ \forall\left[\mathbf{D}(.),\mathbf{Y}_d(.)\right]\in\mathfrak{D}^i\mathrm{x}\mathfrak{Y}_d^k,\ \forall\sigma\in Int\ \mathfrak{T}_0,\ \sigma\longrightarrow0^+\\ \exists\mathbf{U}(.),\ \mathbf{U}(t)=\mathbf{U}(t;\sigma;\mathbf{D};\mathbf{Y}_d)\implies\\ \left\|\mathbf{Y}_0^{m-1}-\mathbf{Y}_{d0}^{m-1}\right\|<\Delta\Longrightarrow\left\|\mathbf{Y}^l(t)-\mathbf{Y}_d^l(t)\right\|=0,\ \forall\left(t\geq\sigma\right)\in\mathfrak{T}_0,\\ k\in\left\{0,1,...,m-1\right\},\ l\in\left\{0,1,...,k\right\}.\end{gathered}\tag{9.10}$$

Let $\mathbf{Y}_0^{m-1}=\mathbf{Y}_{d0}^{m-1}$, i.e., $\left\|\mathbf{Y}_0^{m-1}-\mathbf{Y}_{d0}^{m-1}\right\|=0<\Delta$. Such $\mathbf{Y}_0^{m-1}$ obeys (9.9) so that they altogether imply $\left\|\mathbf{Y}^l(t)-\mathbf{Y}l(t)\right\|=0$, $\forall t\in\mathfrak{T}_0$, equivalently $\mathbf{Y}^l(t)=\mathbf{Y}_d^l(t)$, $\forall t\in\mathfrak{T}_0$. This satisfies Definition 241 and proves the sufficiency part of the theorem statement ■

Lemma 236 and Definition 253 directly imply the following.

Lemma 255 ***The l-th order trackability and the trackability***

For the m-th order dynamic plant to be the l-th order trackable on $\mathfrak{D}^i\mathrm{x}\mathfrak{Y}_d^k$, $k\in\{0,1,...,m-1\}$, $l\in\{0,1,...,k\}$, it is necessary and sufficient to be trackable on $\mathfrak{D}^i\mathrm{x}\mathfrak{Y}_d^k$.

9.2.4 Imperfect natural trackability

Perfect natural trackability properties assume that $\mathbf{Y}^{m-1}(0)=\mathbf{Y}_d^{m-1}(0)$, Definition 246 and Definition 248. We will consider the cases when this initial condition is not satisfied, which excludes the perfection of the natural trackability. What follows introduces imperfect natural trackability properties.

Definition 256 ***The l-th order natural trackability in $\mathfrak{D}^i$ on $\mathfrak{Y}_d^k$***

The m-th order dynamic plant is ***the l-th order natural trackable in $\mathfrak{D}^i$ on $\mathfrak{Y}_d^k$****, $i,k\in\{0,1,...,m-1\}$, $l\in\{0,1,...,k\}$, if, and only if, there is $\Delta\in\mathfrak{R}^+$, or $\Delta=\infty$, such that for every plant desired output response $\mathbf{Y}_d(.)\in\mathfrak{Y}_d^k$, and for every instant $\sigma\in Int\ \mathfrak{T}_0$ there are a disturbance vector function $\mathbf{D}(.)\in\mathfrak{D}^i$ and a control vector function $\mathbf{U}(.)$ obeying $TCUP$ on $\mathfrak{T}_0$, which can be synthesized without using information about the form and the value of $\mathbf{D}(.)$ and about the mathematical model of the plant internal dynamics, such that for every plant initial output vector $\mathbf{Y}_0^{m-1}$ in the Δ neighborhood of the plant initial desired output vector $\mathbf{Y}_{d0}^{m-1}$, the norm of the difference between $\mathbf{Y}^l(t)$ and $\mathbf{Y}_d^l(t)$ becomes equal to zero at latest at the moment σ, after which it rests equal to zero forever, i.e.,*

$$\begin{gathered}\exists\Delta\in]0,\ \infty],\ \forall\mathbf{Y}_d(.)\in\mathfrak{Y}_d^k,\ \forall\sigma\in Int\ \mathfrak{T}_0,\\ \exists\mathbf{D}(.)\in\mathfrak{D}^i,\ \exists\mathbf{U}(.),\ \mathbf{U}(t)=\mathbf{U}(t;\sigma;\mathbf{Y}_d)\in\mathfrak{C}\left(\mathfrak{T}_0\right)\Longrightarrow\\ \left\|\mathbf{Y}_0^{m-1}-\mathbf{Y}_{d0}^{m-1}\right\|<\Delta\Longrightarrow\left\|\mathbf{Y}^l(t)-\mathbf{Y}_d^l(t)\right\|=0,\ \forall\left(t\geq\sigma\right)\in\mathfrak{T}_0,\\ k\in\left\{0,1,...,m-1\right\},\ l\in\left\{0,1,...,k\right\}.\end{gathered}\tag{9.11}$$

Such control is ***the l-th order natural tracking control in $\mathfrak{D}^i$ on $\mathfrak{Y}_d^k$****.*

The zero, ($l=0$), order natural trackability in $\mathfrak{D}^i$ on $\mathfrak{Y}_d^k$ is simply called ***natural trackability in $\mathfrak{D}^i$ on $\mathfrak{Y}_d^k$****. The zero, ($l=0$), order natural tracking* *control in $\mathfrak{D}^i$ on $\mathfrak{Y}_d^k$ is simply called* ***the natural tracking control in $\mathfrak{D}^i$ on $\mathfrak{Y}_d^k$.***

The l-th order natural trackability in $\mathfrak{D}^i$ on $\mathfrak{Y}_d^k$ is ***global (in the whole)*** *if, and only if, $\Delta = \infty$.*

The l-th order natural trackability in $\mathfrak{D}^i$ on $\mathfrak{Y}_d^k$ is ***uniform*** *over $\mathfrak{Y}_d^k$ if, and only if, $\mathbf{U}(.)$ depends on $\mathfrak{Y}_d^k$ but not on an individual $\mathbf{Y}_d(.)$ from $\mathfrak{Y}_d^k$, $\mathbf{U}(t) = \mathbf{U}(t;\sigma;\mathfrak{Y}_d^k)$.*

Comment 257 *Definition 251 and Definition 256 show that the l-th order trackability in $\mathfrak{D}^i$ on $\mathfrak{Y}_d^k$ is necessary for natural trackability in $\mathfrak{D}^i$ on $\mathfrak{Y}_d^k$, and natural trackability in $\mathfrak{D}^i$ on $\mathfrak{Y}_d^k$ is sufficient for trackability in $\mathfrak{D}^i$ on $\mathfrak{Y}_d^k$.*

Theorem 258 ***Perfect natural trackability versus natural trackability in $\mathfrak{D}^i$ on $\mathfrak{Y}_d^k$***

In order for the m-th order dynamic plant to be the l-th order perfect natural trackable in $\mathfrak{D}^i$ on $\mathfrak{Y}_d^k$, $i,k \in \{0,1,...,m-1\}$, $l \in \{0,1,...,k\}$, it is necessary and sufficient that it is the l-th order natural trackable in $\mathfrak{D}^i$ on $\mathfrak{Y}_d^k$.

Proof. *Necessity.* We prove the statement by the contradiction. Let the dynamic plant be the *l-th* order perfect natural trackable in $\mathfrak{D}^i$ on $\mathfrak{Y}_d^k$ but not the *l-th* order natural trackable in $\mathfrak{D}^i$ on $\mathfrak{Y}_d^k$. Definition 246 holds. Since it is not the *l-th* order natural trackable in $\mathfrak{D}^i$ on $\mathfrak{Y}_d^k$, then, due to the violation of Definition 256, for $\mathbf{Y}_0^{m-1} = \mathbf{Y}_{d0}^{m-1}$,

$$\left\|\mathbf{Y}_0^{m-1} - \mathbf{Y}_{d0}^{m-1}\right\| = 0 < \Delta, \forall\Delta > 0,$$

and for every $(\tau > 0) \in \mathfrak{T}_0$ there exist $j \in \{0,1,...,l\}$ and $(\gamma \geq \tau) \in \mathfrak{T}_0$ such that $\left\|\mathbf{Y}^j(\gamma) - \mathbf{Y}_d^j(\gamma)\right\| > 0$. This contradicts Definition 246. The contradiction is the consequence of the assumption on the violation of Definition 256. Hence, the dynamic plant is the *l-th* order natural trackable in $\mathfrak{D}^i$ on $\mathfrak{Y}_d^k$.

Sufficiency. Let the dynamic plant be the *l-th* order natural trackable on $\mathfrak{D}^i$x$\mathfrak{Y}_d^k$. Definition 256 holds. Let $\sigma \in Int\ \mathfrak{T}_0$ and let it be arbitrarily small, i.e., $\sigma \longrightarrow 0^+$ in Definition 256 so that (9.11) becomes for such σ :

$$\begin{gathered}\exists\Delta \in]0,\ \infty],\ \forall\mathbf{Y}_d(.) \in \mathfrak{Y}_d^k,\ \forall\sigma \in Int\ \mathfrak{T}_0,\ \sigma \longrightarrow 0^+,\\ \exists\mathbf{D}(.) \in \mathfrak{D}^k,\ \exists\mathbf{U}(.),\ \mathbf{U}(t) = \mathbf{U}(t;\sigma;\mathbf{Y}_d) \in \mathfrak{C}(\mathfrak{T}_0) \Longrightarrow\\ \left\|\mathbf{Y}_0^{m-1} - \mathbf{Y}_{d0}^{m-1}\right\| < \Delta \Longrightarrow \left\|\mathbf{Y}^l(t) - \mathbf{Y}_d^l(t)\right\| = 0,\ \forall(t \geq \sigma) \in \mathfrak{T}_0,\\ k \in \{0,1,...,m-1\},\ l \in \{0,1,...,k\}.\end{gathered} \tag{9.12}$$

Let $\mathbf{Y}_0^{m-1} = \mathbf{Y}_{d0}^{m-1}$, i.e., $\left\|\mathbf{Y}_0^{m-1} - \mathbf{Y}_{d0}^{m-1}\right\| = 0 < \Delta$. Such $\mathbf{Y}_0^{m-1}$ obeys (9.11) so that they altogether imply $\left\|\mathbf{Y}^l(t) - \mathbf{Y}_d^l(t)\right\| = 0$, $\forall t \in \mathfrak{T}_0$, equivalently $\mathbf{Y}^l(t) = \mathbf{Y}_d^l(t)$, $\forall t \in \mathfrak{T}_0$. This satisfies Definition 246 and proves the sufficiency part of the theorem statement ■

We define now the (imperfect) natural trackability for all vector functions $[\mathbf{D}(.), \mathbf{Y}_d(.)] \in \mathfrak{D}^ix\mathfrak{Y}_d^k$.

Definition 259 ***The l-th order natural trackability on $\mathfrak{D}^i x\mathfrak{Y}_d^k$***

The m-th order dynamic plant is ***the l-th order natural trackable on*** $\mathfrak{D}^i x\mathfrak{Y}_d^k$, $i, k \in \{0, 1, ..., m-1\}$, $l \in \{0, 1, ..., k\}$, *if, and only if, there is* $\Delta \in \mathfrak{R}^+$, *or* $\Delta = \infty$, *such that for every disturbance vector function* $\mathbf{D}(.) \in \mathfrak{D}^i$, *for every plant output desired response* $\mathbf{Y}_d(.) \in \mathfrak{Y}_d^k$, *and for every instant* $\sigma \in Int\ \mathfrak{T}_0$, *there is a control vector function* $\mathbf{U}(.)$ *obeying* $TCUP$ *on* $\mathfrak{T}_0$, *which can be synthesized without using information about the form and the value of* $\mathbf{D}(.) \in \mathfrak{D}^i$ *and about the mathematical model of the plant internal dynamics, such that for every plant initial output vector* $\mathbf{Y}_0^{m-1}$ *in the* Δ *neighborhood of the plant initial desired output vector* $\mathbf{Y}_{d0}^{m-1}$, *the norm of the difference between* $\mathbf{Y}^l(t)$ *and* $\mathbf{Y}_d^l(t)$ *becomes equal to zero at latest at the moment* σ, *after which it rests equal to zero forever, i.e.,*

$$\begin{gathered}\exists\Delta \in]0,\ \infty],\ \forall\left[\mathbf{D}(.), \mathbf{Y}_d(.)\right] \in \mathfrak{D}^i x\mathfrak{Y}_d^k,\ \forall\sigma \in Int\ \mathfrak{T}_0,\ \sigma \longrightarrow 0^+, \\ \exists\mathbf{U}(.),\ \mathbf{U}(t) = \mathbf{U}(t;\sigma;\mathbf{Y}_d) \in \mathfrak{C}\left(\mathfrak{T}_0\right) \Longrightarrow \\ \left\|\mathbf{Y}_0^{m-1} - \mathbf{Y}_{d0}^{m-1}\right\| < \Delta \Longrightarrow \left\|\mathbf{Y}^l(t) - \mathbf{Y}_d^l(t)\right\| = 0,\ \forall\left(t \geq \sigma\right) \in \mathfrak{T}_0, \\ k \in \{0, 1, ..., m-1\},\ l \in \{0, 1, ..., k\}. \end{gathered} \tag{9.13}$$

Such control is ***the l-th order natural tracking control on*** $\mathfrak{D}^i x\mathfrak{Y}_d^k$, *for short* ***the l-th order natural tracking control.***

The zero, $(l = 0)$, *order natural trackability on* $\mathfrak{D}^i x\mathfrak{Y}_d^k$ *is called* ***natural trackability on*** $\mathfrak{D}^i x\mathfrak{Y}_d^k$. *The zero,* $(l = 0)$, *order natural tracking control on* $\mathfrak{D}^i x\mathfrak{Y}_d^k$ *is called for short* ***natural tracking control on*** $\mathfrak{D}^i x\mathfrak{Y}_d^k$, *or shorter* ***natural tracking control (NTC).***

The l-th order natural trackability on $\mathfrak{D}^i x\mathfrak{Y}_d^k$ *is* ***global (in the whole)*** *if, and only if,* $\Delta = \infty$.

The l-th order natural trackability on $\mathfrak{D}^i x\mathfrak{Y}_d^k$ *is* ***uniform*** *over* $\mathfrak{Y}_d^k$ *if, and only if, control* $\mathbf{U}(.)$ *depends on the* $\mathfrak{Y}_d^k$ *but not on an individual* $\mathbf{Y}_d(.)$ *from* $\mathfrak{Y}_d^k$, $\mathbf{U}(t) = \mathbf{U}(t;\sigma;\mathfrak{Y}_d^k)$.

Comment 260 *Definition 253 and Definition 259 show that the l-th order trackability on* $\mathfrak{D}^i x\mathfrak{Y}_d^k$ *is necessary for the l-th order natural trackability on* $\mathfrak{D}^i x\mathfrak{Y}_d^k$, *and natural trackability on* $\mathfrak{D}^i x\mathfrak{Y}_d^k$ *is sufficient for trackability on* $\mathfrak{D}^i x\mathfrak{Y}_d^k$.

Theorem 261 ***Perfect natural trackability versus the natural trackability on $\mathfrak{D}^i x\mathfrak{Y}_d^k$***

In order for the m-th order dynamic plant to be the l-th order perfect natural trackable on $\mathfrak{D}^i x\mathfrak{Y}_d^k$, $i, k \in \{0, 1, ..., m-1\}$, $l \in \{0, 1, ..., k\}$, *it is necessary and sufficient that it is the l-th order natural trackable on* $\mathfrak{D}^i x\mathfrak{Y}_d^k$.

Proof. *Necessity.* We prove the statement by the contradiction. Let the dynamic plant be the *l-th* order perfect natural trackable on $\mathfrak{D}^i x\mathfrak{Y}_d^k$ but not the *l-th* order natural trackable on $\mathfrak{D}^i x\mathfrak{Y}_d^k$. Definition 248 holds. Since the plant is not the *l-th* order natural trackable on $\mathfrak{D}^i x\mathfrak{Y}_d^k$, then, due to the violation of Definition 259, for $\mathbf{Y}_0^{m-1} = \mathbf{Y}_{d0}^{m-1}$,

$$\left\|\mathbf{Y}_0^{m-1} - \mathbf{Y}_{d0}^{m-1}\right\| = 0 < \Delta, \forall\Delta > 0,$$

and for every $(\tau > 0) \in \mathfrak{T}_0$ there exist $j \in \{0, 1, ..., k\}$ and $(\gamma \geq \tau) \in \mathfrak{T}_0$ such that $\left\| \mathbf{Y}^j(\gamma) - \mathbf{Y}_d^j(\gamma) \right\| > 0$. This contradicts Definition 248. The violation of Definition 248 is the consequence of the assumption on the violation of Definition 259. Hence, the dynamic plant is the *l-th* order natural trackable on $\mathfrak{D}^i \text{x} \mathfrak{Y}_d^k$.

Sufficiency. Let the dynamic plant be the *l-th* order natural trackable on $\mathfrak{D}^i \text{x} \mathfrak{Y}_d^k$. Definition 259 holds. Let $\sigma \in Int\ \mathfrak{T}_0$ and let it be arbitrarily small, i.e., $\sigma \longrightarrow 0^+$ in Definition 259 so that (9.13) becomes for such σ:

$$\begin{gathered}
\exists \Delta \in]0,\ \infty],\ \forall [\mathbf{D}(.), \mathbf{Y}_d(.)] \in \mathfrak{D}^i \text{x} \mathfrak{Y}_d^k,\ \forall \sigma \in Int\ \mathfrak{T}_0,\ \sigma \longrightarrow 0^+, \\
\exists \mathbf{U}(.),\ \mathbf{U}(t) = \mathbf{U}(t; \sigma; \mathbf{Y}_d) \in \mathfrak{C}(\mathfrak{T}_0) \Longrightarrow \\
\left\| \mathbf{Y}_0^{m-1} - \mathbf{Y}_{d0}^{m-1} \right\| < \Delta \Longrightarrow \left\| \mathbf{Y}^l(t) - \mathbf{Y}_d^l(t) \right\| = 0,\ \forall (t \geq \sigma) \in \mathfrak{T}_0, \\
k \in \{0, 1, ..., m-1\},\ l \in \{0, 1, ..., k\}.
\end{gathered} \tag{9.14}$$

Let $\mathbf{Y}_0^{m-1} = \mathbf{Y}_{d0}^{m-1}$, i.e., $\left\| \mathbf{Y}_0^{m-1} - \mathbf{Y}_{d0}^{m-1} \right\| = 0 < \Delta$. Such $\mathbf{Y}_0^{m-1}$ obeys (9.13) so that they altogether imply $\left\| \mathbf{Y}^l(t) - \mathbf{Y}_d^l(t) \right\| = 0$, $\forall t \in \mathfrak{T}_0$, equivalently $\mathbf{Y}^l(t) = \mathbf{Y}_d^l(t)$, $\forall t \in \mathfrak{T}_0$. This satisfies Definition 248 and proves the sufficiency part of the theorem statement ■

Lemma 236 and Definition 259 result in the following.

Lemma 262 ***The l-th order natural trackability and the natural trackability***

For the m-th order dynamic plant to be the l-th order natural trackable on $\mathfrak{D}^i \text{x} \mathfrak{Y}_d^k$, $i, k \in \{0, 1, ..., m-1\}$, $l \in \{0, 1, ..., k\}$, it is necessary and sufficient to be natural trackable on $\mathfrak{D}^i \text{x} \mathfrak{Y}_d^k$.

9.2.5 Imperfect elementwise trackability

The preceding definitions of the trackability properties take the following forms in the framework of elementwise tracking.

Definition 263 ***The l-th order elementwise trackability on $\mathfrak{D}^i \text{x} \mathfrak{Y}_d^k$***

The m-th order dynamic plant is ***the l-th order elementwise trackable on*** $\mathfrak{D}^i \text{x} \mathfrak{Y}_d^k$, *$i, k \in \{0, 1, ..., m-1\}$, $l \in \{0, 1, ..., k\}$, if, and only if, there is $\boldsymbol{\Delta}^{mN} \in \mathfrak{R}^{+mN}$, or $\boldsymbol{\Delta}^{mN} = \infty \mathbf{1}_{mN}$, such that for every disturbance vector function $\mathbf{D}(.) \in \mathfrak{D}^i$, for every plant output desired response $\mathbf{Y}_d(.) \in \mathfrak{Y}_d^k$, and for every vector instant $\boldsymbol{\sigma} \in (Int\ \mathfrak{T}_0)^{(l+1)N}$, there is a control vector function $\mathbf{U}(.)$ such that for every plant initial output vector $\mathbf{Y}_0^{m-1}$ in the $\boldsymbol{\Delta}$ elementwise neighborhood of the plant initial desired output vector $\mathbf{Y}_{d0}^{m-1}$, the real plant output response $\mathbf{Y}^l(t)$ becomes elementwise equal to $\mathbf{Y}_d^l(t)$ at latest at the vector* *moment $\boldsymbol{\sigma}$, after which they rest equal forever, i.e.,*

$$\exists \mathbf{\Delta}^{mN} \in]\mathbf{0}_{mN},\ \infty \mathbf{1}_{mN}],\ \forall [\mathbf{D}(.), \mathbf{Y}_d(.)] \in \mathfrak{D}^i \mathrm{x} \mathfrak{Y}_d^k,$$
$$\forall \boldsymbol{\sigma} \in (Int\ \mathfrak{T}_0)^{(l+1)N},\ \exists \mathbf{U}(.),$$
$$\mathbf{U}(t) = \mathbf{U}(t;\boldsymbol{\sigma};\mathbf{D};\mathbf{Y}) \in \mathfrak{C}(\mathfrak{T}_0) \Longrightarrow \left|\mathbf{Y}_0^{m-1} - \mathbf{Y}_{d0}^{m-1}\right| < \mathbf{\Delta}^{mN} \Longrightarrow$$
$$\mathbf{Y}^l(\mathbf{t}^{(l+1)N}) = \mathbf{Y}_d^l(\mathbf{t}^{(l+1)N}),\ \forall \left(\mathbf{t}^{(l+1)N} \geq \boldsymbol{\sigma}\right) \in \mathfrak{T}_0^{(l+1)N},$$
$$i,k \in \{0,1,...,m-1\},\ l \in \{0,1,...,k\}. \tag{9.15}$$

Such control is ***the l-th order elementwise tracking control on $\mathfrak{D}^i \mathrm{x} \mathfrak{Y}_d^k$.***

The zero, $(l = 0)$, order elementwise trackability on $\mathfrak{D}^i \mathrm{x} \mathfrak{Y}_d^k$ is called ***elementwise trackability on $\mathfrak{D}^i \mathrm{x} \mathfrak{Y}_d^k$.*** *The zero, $(l = 0)$, order elementwise tracking control on $\mathfrak{D}^i \mathrm{x} \mathfrak{Y}_d^k$ is called for short* ***elementwise tracking control on $\mathfrak{D}^i \mathrm{x} \mathfrak{Y}_d^k$****, or shorter,* ***elementwise tracking control.***

The l-th order elementwise trackability on $\mathfrak{D}^i \mathrm{x} \mathfrak{Y}_d^k$ is ***global (in the whole)*** *if, and only if, $\mathbf{\Delta}^{mN} = \infty \mathbf{1}_{mN}$.*

The l-th order elementwise trackability on $\mathfrak{D}^i \mathrm{x} \mathfrak{Y}_d^k$ is ***uniform*** *over $\mathfrak{D}^i \mathrm{x} \mathfrak{Y}_d^k$ if, and only if, $\mathbf{U}(.)$ depends on $\mathfrak{D}^i \mathrm{x} \mathfrak{Y}_d^k$ but not on an individual pair $[\mathbf{D}(.), \mathbf{Y}_d(.)]$ from $\mathfrak{D}^i \mathrm{x} \mathfrak{Y}_d^k$, $\mathbf{U}(t) = \mathbf{U}(t;\boldsymbol{\sigma};\mathfrak{D}^i;\mathfrak{Y}_d^k)$.*

Comment 264 *Although a norm of a vector is equal to zero if, and only if, all the vector entries are zero, the equivalence between the l-th order trackability and the l-th order elementwise trackability does not follow from Definition 251 and Definition 263. The former specifies the same scalar reachability time $\sigma \in Int\ \mathfrak{T}_0$ for all output variables and their derivatives. The latter associates different scalar reachability times with different output variables and their derivatives; i.e., the latter associates the vector reachability time $\boldsymbol{\sigma} \in (Int\ \mathfrak{T}_0)^{(l+1)N}$ with the output vector and its derivatives.*

Lemma 265 *If two functions $\mathbf{Y}(.)$ and $\mathbf{Y}_d(.)$ are defined, k-times continuously differentiable on $]\boldsymbol{\sigma}, \infty \mathbf{1}_N[$, $\boldsymbol{\sigma} \in \mathfrak{T}_0^N$, $]\boldsymbol{\sigma}, \infty \mathbf{1}_N[\subseteq (Int\mathfrak{T}_0)^N$, as well as at $\mathbf{t}^N = \boldsymbol{\sigma}$ from the right-hand side, i.e., at $\mathbf{t}^N = \boldsymbol{\sigma}^+$, and identical on $[\boldsymbol{\sigma}, \infty \mathbf{1}_N[$, $[\boldsymbol{\sigma}, \infty \mathbf{1}_N[\subseteq \mathfrak{T}_0^N$, then all their derivatives up to the order k included are also identical on $]\boldsymbol{\sigma}, \infty \mathbf{1}_N[$ and at $\mathbf{t}^N = \boldsymbol{\sigma}^+$.*

Proof. Let $\boldsymbol{\sigma} \in \mathfrak{T}_0^N$, $]\boldsymbol{\sigma}, \infty \mathbf{1}_N[\subseteq (Int\mathfrak{T}_0)^N$, $[\boldsymbol{\sigma}, \infty \mathbf{1}_N[\subseteq \mathfrak{T}_0^N$, and let two functions $\mathbf{Y}(.)$ and $\mathbf{Y}_d(.)$ be defined, k-times continuously differentiable on $]\boldsymbol{\sigma}, \infty \mathbf{1}_N[$, as well as at $\mathbf{t}^N = \boldsymbol{\sigma}^+$, and identical on $[\boldsymbol{\sigma}, \infty \mathbf{1}_N[$,

$$\mathbf{Y}(.), \mathbf{Y}_d(.) \in \mathfrak{C}^{k-1}([\boldsymbol{\sigma}, \infty \mathbf{1}_N[) \cup \mathfrak{C}^k(]\boldsymbol{\sigma}, \infty \mathbf{1}_N[),$$
$$\mathbf{Y}(\mathbf{t}^N) = \mathbf{Y}_d(\mathbf{t}^N),\ \forall \mathbf{t}^N \in [\boldsymbol{\sigma}, \infty \mathbf{1}_N[. \tag{9.16}$$

Let $T = diag\{t\ t\ ...t\} \in \mathfrak{T}_0^N \mathrm{x} \mathfrak{T}_0^N$, $\boldsymbol{\theta} = [\theta\ \theta\ ...\theta]^T \in \mathfrak{T}^N$ and $\Theta = diag\{\theta\ \theta\ ...\theta\} \in \mathfrak{T}^N \mathrm{x} \mathfrak{T}^N$.

By the definition of the first derivative and in view of (9.16):

$$dT^{-1}\left[d\mathbf{Y}(\mathbf{t}^N)\right] = \lim_{\Theta \longrightarrow O_N}\left[\Theta^{-1}\left(\mathbf{Y}(\mathbf{t}^N+\boldsymbol{\theta}) - \mathbf{Y}(\mathbf{t}^N)\right)\right] =$$
$$= \lim_{\Theta \longrightarrow O_N}\left[\Theta^{-1}\left(\mathbf{Y}_d(\mathbf{t}^N+\boldsymbol{\theta}) - \mathbf{Y}_d(\mathbf{t}^N)\right)\right] = dT^{-1}\left[d\mathbf{Y}_d(\mathbf{t}^N)\right],$$
$$\forall \mathbf{t}^N \in]\boldsymbol{\sigma}, \infty\mathbf{1}_N[,$$

and at $\mathbf{t}^N = \boldsymbol{\sigma}^+$:

$$D_r^1\mathbf{Y}(\mathbf{t}^N)_{\mathbf{t}^N=\boldsymbol{\sigma}} = \lim_{\Theta \longrightarrow O_N}\left[\Theta^{-1}\left(\mathbf{Y}(\boldsymbol{\theta}+\boldsymbol{\sigma}) - \mathbf{Y}(\boldsymbol{\sigma})\right)\right] =$$
$$= \lim_{\Theta \longrightarrow O_N}\left[\Theta^{-1}\left(\mathbf{Y}_d(\boldsymbol{\theta}+\boldsymbol{\sigma}) - \mathbf{Y}_d(\boldsymbol{\sigma})\right)\right] = D_r^1\mathbf{Y}_d(\mathbf{t}^N)_{\mathbf{t}^N=\boldsymbol{\sigma}}.$$

The statement holds for $i = 0, 1$. Let it hold for any $i \in \{0, 1, ..., k-1\}$,

$$\mathbf{Y}^{(i)}(\mathbf{t}^N) = \mathbf{Y}_d^{(i)}(\mathbf{t}^N),\ \forall \mathbf{t}^N \in]\boldsymbol{\sigma}, \infty\mathbf{1}_N[,$$
$$D_r^i\mathbf{Y}(\mathbf{t}^N)_{\mathbf{t}^N=\boldsymbol{\sigma}} = D_r^i\mathbf{Y}_d(\mathbf{t}^N)_{\mathbf{t}^N=\boldsymbol{\sigma}}.$$

These equations yield

$$dT^{-(i+1)}\left[d^{i+1}\mathbf{Y}(\mathbf{t}^N)\right] = \lim_{\Theta \longrightarrow O_N}\left[\Theta^{-1}\left(\mathbf{Y}^{(i)}(\mathbf{t}^N+\boldsymbol{\theta}) - \mathbf{Y}^{(i)}(\boldsymbol{\theta})\right)\right] =$$
$$= \lim_{\Theta \longrightarrow O_N}\left[\Theta^{-1}\left(\mathbf{Y}_d^{(i)}(\mathbf{t}^N+\boldsymbol{\theta}) - \mathbf{Y}_d^{(i)}(\boldsymbol{\theta})\right)\right] = dT^{-(i+1)}\left[d^{i+1}\mathbf{Y}_d(\mathbf{t}^N)\right],$$
$$\forall \mathbf{t}^N \in]\boldsymbol{\sigma}, \infty\mathbf{1}_N[,$$

and

$$D_r^{i+1}\mathbf{Y}(\mathbf{t}^N)_{\mathbf{t}^N=\boldsymbol{\sigma}} = \lim_{\Theta \longrightarrow O_N}\left[\Theta^{-1}\left(D_r^i\mathbf{Y}(\boldsymbol{\theta}+\boldsymbol{\sigma}) - D_r^i\mathbf{Y}(\boldsymbol{\sigma})\right)\right] =$$
$$= \lim_{\Theta \longrightarrow O_N}\left[\Theta^{-1}\left(D_r^i\mathbf{Y}_d(\boldsymbol{\theta}+\boldsymbol{\sigma}) - D_r^i\mathbf{Y}_d(\boldsymbol{\sigma})\right)\right] = D_r^{i+1}\mathbf{Y}_d(\mathbf{t}^N)_{\mathbf{t}^N=\boldsymbol{\sigma}}.$$

Since the statement is true for $i = 0, 1$, and for $i+1$ if it holds for $i \in \{0, 1, ..., k-1\}$, then by mathematical induction it holds $\forall i \in \{0, 1, ..., k\}$ ■

Note 266 *Notice that this Lemma is the vector generalization of Lemma 236.*

Lemma 265 and Definition 263 induce the following.

Lemma 267 ***The l-th order elementwise trackability and elementwise trackability***

For the m-th order dynamic plant to be the l-th order (global) elementwise trackable on $\mathfrak{D}^i x \mathfrak{Y}_d^k$, $i, k \in \{0, 1, ..., m-1\}$, $l \in \{0, 1, ..., k\}$, it is necessary and sufficient to be (global) elementwise trackable on $\mathfrak{D}^i x \mathfrak{Y}_d^k$.

Definition 268 ***The l-th order elementwise natural trackability on $\mathfrak{D}^i x \mathfrak{Y}_d^k$***

The m-th order dynamic plant is ***the l-th order elementwise natural trackable*** *on $\mathfrak{D}^i x \mathfrak{Y}_d^k$, $k \in \{0, 1, ..., m-1\}$, $l \in \{0, 1, ..., k\}$, if, and only if,* *there is* $\mathbf{\Delta}^{mN} \in \mathfrak{R}_+^{mN}$, *or* $\mathbf{\Delta}^{mN} = \infty \mathbf{1}_{mN}$, *such that for every disturbance vector function* $\mathbf{D}(.) \in \mathfrak{D}^i$, *for every plant output desired response* $\mathbf{Y}_d(.) \in \mathfrak{Y}_d^k$, *and for every vector instant* $\boldsymbol{\sigma} \in (Int\ \mathfrak{T}_0)^{(l+1)N}$, *there is control vector function* $\mathbf{U}(.)$ *obeying TCUP on* $\mathfrak{T}_0$, *which can be synthesized without using information about the form and value of* $\mathbf{D}(.) \in \mathfrak{D}^k$ *and about the mathematical model of the plant internal dynamics, such that for every plant initial output vector* $\mathbf{Y}_0^{m-1}$ *in the* $\mathbf{\Delta}$ *elementwise neighborhood of the plant initial desired output vector* $\mathbf{Y}_{d0}^{m-1}$, *the real plant output response* $\mathbf{Y}^l(t)$ *becomes elementwise equal to* $\mathbf{Y}_d^l(t)$ *at latest at the vector moment* $\boldsymbol{\sigma}$, *after which they rest equal forever, i.e.,*

$$\begin{gathered}
\exists \mathbf{\Delta}^{mN} \in]\mathbf{0}_{mN}, \infty \mathbf{1}_{mN}], \ \forall [\mathbf{D}(.), \mathbf{Y}_d(.)] \in \mathfrak{D}^i x \mathfrak{Y}_d^k, \\
\forall \boldsymbol{\sigma} \in (Int\ \mathfrak{T}_0)^{(l+1)N}, \ \exists \mathbf{U}(.), \mathbf{U}(t) \in \mathfrak{C}, \\
\mathbf{U}(t) = \mathbf{U}(t; \boldsymbol{\sigma}; \mathbf{Y}_d) \in \mathfrak{C}(\mathfrak{T}_0) \Longrightarrow \left| \mathbf{Y}_0^{m-1} - \mathbf{Y}_{d0}^{m-1} \right| < \mathbf{\Delta}^{mN} \Longrightarrow \\
\mathbf{Y}^l(\mathbf{t}^{(l+1)N}) = \mathbf{Y}_d^l(\mathbf{t}^{(l+1)N}) \ \forall \left(\mathbf{t}^{(l+1)N} \geq \boldsymbol{\sigma} \right) \in \mathfrak{T}_0^{(l+1)N}, \\
i, k \in \{0, 1, ..., m-1\}, \ l \in \{0, 1, ..., k\}. \qquad (9.17)
\end{gathered}$$

Such control is ***the l-th order elementwise natural tracking control on*** $\mathfrak{D}^i x \mathfrak{Y}_d^k$, *for short,* ***the l-th order elementwise natural tracking control.***

The zero, ($l = 0$*), order elementwise natural trackability on* $\mathfrak{D}^i x \mathfrak{Y}_d^k$ *is called* ***elementwise natural trackability on*** $\mathfrak{D}^i x \mathfrak{Y}_d^k$. *The zero, (*$l = 0$*), order elementwise natural tracking control on* $\mathfrak{D}^i x \mathfrak{Y}_d^k$ *is called* ***elementwise natural tracking control on*** $\mathfrak{D}^i x \mathfrak{Y}_d^k$, *for short,* ***elementwise natural tracking control.***

The l-th order elementwise natural trackability on $\mathfrak{D}^i x \mathfrak{Y}_d^k$ *is* ***global (in the whole)*** *if, and only if,* $\mathbf{\Delta}^{mN} = \infty \mathbf{1}_{mN}$.

The l-th order elementwise natural trackability on $\mathfrak{D}^i x \mathfrak{Y}_d^k$ *is* ***uniform*** *over* $\mathfrak{Y}_d^k$ *if, and only if,* $\mathbf{U}(.)$ *depends on* $\mathfrak{Y}_d^k$ *but not on an individual* $\mathbf{Y}_d(.)$ *from* $\mathfrak{Y}_d^k$, $\mathbf{U}(t) = \mathbf{U}(t; \boldsymbol{\sigma}; \mathfrak{Y}_d^k)$.

Comment 269 *Definition 263 and Definition 268 show the difference between the l-th order elementwise trackability and the l-th order elementwise natural trackability. The former is necessary for the latter, and the latter is sufficient for the former.*

Lemma 265 and Definition 268 imply the following.

Lemma 270 ***The l-th order elementwise natural trackability and elementwise natural trackability***

For the m-th order dynamic plant to be the l-th order (global) elementwise natural trackable on $\mathfrak{D}^i x \mathfrak{Y}_d^k$, $i, k \in \{0, 1, ..., m-1\}$, $l \in \{0, 1, ..., k\}$, *it is necessary and sufficient to be (global) elementwise natural trackable on* $\mathfrak{D}^i x \mathfrak{Y}_d^k$.

Note 271 *The trackability concept is equally important for the engineer who designs the plant and for the engineer who designs the controller, hence the control system, for the plant. The fact that there is not a technical plant that* *can by itself, without control, realize its goal under arbitrary, unpredictable initial conditions and inputs, obliges the designer of the plant to ensure its trackability. The fact that a kind of trackability is necessary for the corresponding tracking property obliges the designer of the controller, hence of the control system, for the plant to verify its trackability and to use it in its design. Trackability is the fundamental link between the producer of the plant and the producer of the controller, hence of the control system, for the plant. It is also the link between the course on the dynamics and mathematical modeling of the plant and the control courses.*

9.3 Perfect trackability conditions

9.3.1 *IO* plant perfect (natural) trackability

Conditions for perfect trackability

Which properties of the *IO* plant enable the existence of a control that can force the plant to exhibit perfect tracking as soon as the initial real output vector is equal to the initial desired output vector? In other words, which properties of the *IO* plant ensure its perfect trackability?

We present, at first, the time domain conditions for the perfect trackability properties of the *IO* object (2.15) (Subsection 2.1.2).

Theorem 272 ***Time-domain condition for the perfect trackability of the IO plant (2.15) in*** $\mathfrak{D}^{\mu_{Pd}}$ ***on*** $\mathfrak{Y}_d^{\nu}$

a) For the desired output response $\mathbf{Y}_d^{\nu}(.)$ *of the IO plant (2.15) to be (the* $(\nu$*-1) th order) perfect trackable in* $\mathfrak{D}^{\mu_{Pd}}$ *it is necessary and sufficient that there is an input vector functional pair* $[\mathbf{D}(.),\mathbf{U}(.)]$, $\mathbf{D}(.)\in\mathfrak{D}^{\mu_{Pd}}$*, which obeys the following differential equation:*

$$D_{Pd}^{(\mu_{Pd})}\mathbf{D}^{\mu_{Pd}}(t)+C_{Pu}^{(\mu_{Pu})}\mathbf{U}^{\mu_{Pu}}(t)=A_P^{(\nu)}\mathbf{Y}_d^{\nu}(t),\ \forall t\in\mathfrak{T}_0,$$
$$\text{under the condition } \mathbf{Y}_0^{\nu-1}=\mathbf{Y}_{d0}^{\nu-1}. \tag{9.18}$$

b) For the IO object (2.15) to be (the $(\nu$*-1) th order) perfect trackable in* $\mathfrak{D}^{\mu_{Pd}}$ ***on*** $\mathfrak{Y}_d^{\nu}$ *it is necessary and sufficient that a) holds for every* $\mathbf{Y}_d^{\nu}(.)\in\mathfrak{Y}_d^{\nu}$.

Proof. Theorem 65 (Subsection 3.3.2) and Lemma 238 (Section 9.2) prove a) of this theorem in view of Definition 235. The statement under b) results from Definition 239 and Lemma 240 ∎

In the case of the perfect trackability on $\mathfrak{D}^{\mu_{Pd}}\text{x}\mathfrak{Y}_d^{\nu}$ the condition becomes more severe.

Theorem 273 ***Time-domain condition for the perfect trackability on*** $\mathfrak{D}^{\mu_{Pd}}\text{ x}\mathfrak{Y}_d^{\nu}$ ***of the IO plant (2.15)***

For the IO object (2.15) to be (the $(\nu$*-1) th order) perfect trackable on* $\mathfrak{D}^{\mu_{Pd}}$ $\text{x}\mathfrak{Y}_d^{\nu}$ *it is necessary and sufficient that for every* $[\mathbf{D}(.),\mathbf{Y}_d^{\nu}(.)]\in\mathfrak{D}^{\mu_{Pd}}\text{x}\mathfrak{Y}_d^{\nu}$ *there* *is a control vector function* $\mathbf{U}(.)$ *that obeys the following differential equation:*

$$C_{Pu}^{(\mu_{Pu})}\mathbf{U}^{\mu_{Pu}}(t) = -D_{Pd}^{(\mu_{Pd})}\mathbf{D}^{\mu_{Pd}}(t) + A_P^{(\nu)}\mathbf{Y}_d^{\nu}(t),\ \forall t \in \mathfrak{T}_0,$$
$$\text{under the condition } \mathbf{Y}_0^{\nu-1} = \mathbf{Y}_{d0}^{\nu-1}. \tag{9.19}$$

Proof. This theorem follows from Theorem 66 (Subsection 3.3.2) and Lemma 242 ■

We can express the trackability conditions in terms of the *IO* object (2.15) transfer function matrix $G_{IOPU}(s)$ relative to control $\mathbf{U}$. For this we first refer to equations (7.65) and (7.66) (Subsection 7.3). They determine the full transfer function matrix $F_{IOP}(s)$ of the *IO* object (2.15),

$$F_{IOP}(s) = \left(A_P^{(\nu)}S_N^{(\nu)}(s)\right)^{-1} \bullet \begin{bmatrix} \left(D_{Pd}^{(\mu_{Pd})}S_d^{(\mu_{Pd})}(s)\right)^T \\ \left(C_{Pu}^{(\mu_{Pu})}S_r^{(\mu_{Pu})}(s)\right)^T \\ -\left(D_{Pd}^{(\mu_{Pd})}Z_d^{(\mu_{Pd}-1)}\right)^T \\ -\left(C_{Pu}^{(\mu_{Pu})}Z_r^{(\mu_{Pu}-1)}(s)\right)^T \\ \left(A^{(\nu)}Z_N^{(\nu-1)}(s)\right)^T \end{bmatrix}^T, \tag{9.20}$$

and its transfer function matrix $G_{IOPU}(s)$ relative to the control $\mathbf{U}$,

$$G_{IOPU}(s) = \left(A_P^{(\nu)}S_N^{(\nu)}(s)\right)^{-1} C_{Pu}^{(\mu_{Pu})}S_r^{(\mu_{Pu})}(s) \in \mathfrak{C}^{N\text{x}r}.$$

Since

$$\deg\left(adj A_P^{(\nu)}S_N^{(\nu)}(s)\right) = \nu(N-1),\ \deg\left[\det\left(A_P^{(\nu)}S_N^{(\nu)}(s)\right)\right] = \nu N,$$
$$\deg\left(C_{Pu}^{(\mu_{Pu})}S_r^{(\mu_{Pu})}(s)\right) = \mu_{Pu},$$

then

$$\det A_{P0} \neq 0 \Longrightarrow G_{IOPUo} = G_{IOPU}(0) = A_{P0}^{-1}C_{Pu0} \in \mathfrak{C}^{N\text{x}r},$$
$$\det A_{P\nu} \neq 0 \Longrightarrow G_{IOPU\infty} = \underbrace{\lim}_{s\longrightarrow\infty}\left[s^{\nu-\mu_{Pu}}G_{IOPU}(s)\right] = A_{P\nu}^{-1}C_{Pu\mu_{Pu}} \in \mathfrak{C}^{N\text{x}r}. \tag{9.21}$$

These definitions hold regardless of the stability property of the *IO* plant (2.15). For example, the *IO* plant (2.15) can be unstable but $\det A_{P0} \neq 0$.

Theorem 274 ***The IO plant (2.15) transfer function matrix*** $G_{IOPU}(s)$ ***and rank***

For the IO plant (2.15) the following statements hold:

$$rankC_{Puo} = N \leq r \Longrightarrow$$
$$rankG_{IOPU}(s) = rankG_{IOPUo} = rankC_{Pu}^{(\mu_{Pu})}S_r^{(\mu_{Pu})}(s) =$$
$$= rankC_{Pu}^{(\mu_{Pu})} = rankC_{Puo}, \tag{9.22}$$

and

$$rankC_{Pu\mu_{Pu}} = N \leq r \Longrightarrow$$

$$rankG_{IOPU}(s) = rankG_{IOPU\infty} = rankC_{Pu}^{(\mu_{Pu})} S_r^{(\mu_{Pu})}(s) =$$

$$= rankC_{Pu}^{(\mu_{Pu})} = rankC_{Pu\mu_{Pu}} \ . \tag{9.23}$$

Proof. For the IO plant (2.15) $detA_{P\nu} \neq 0$. It is stable due to Claim 155 (Section 8.2), which guarantees $detA_{P0} \neq 0$. We will consider two extreme cases determined by (9.21).

Case 1. At the origin $s = 0$ we use $G_{IOPUo} = G_{IOPU}(0)$ (9.21),

$$G_{IOPUo} = A_{P0}^{-1} C_{Pu0} \in \mathfrak{C}^{N\text{x}r},$$

which is well defined since A_{P0} is nonsingular due to the plant stability (Claim 155, Section 8.2), i.e., due to negative real parts of all roots of

$$\det\left(A_P^{(\nu)} S_N^{(\nu)}(s)\right) = 0.$$

They imply

$$\det A_{Po} \neq 0.$$

Hence, if

$$rankC_{Puo} = N \leq r,$$

then

$$rankG_{IOPUo} = rankC_{Puo} = N \leq r.$$

This guarantees

$$rankC_{Pu}^{(\mu_{Pu})} S_r^{(\mu_{Pu})}(s) = rankG_{OUo} = rankC_{Puo} = rankC_{Pu}^{(\mu_{Pu})} = N \leq r.$$

The equations (9.22) hold.

Case 2. For s at infinity, $s = \infty$, we define $G_{IOPU\infty}$ (9.21),

$$G_{IOPU\infty} = \underbrace{\lim}_{s \longrightarrow \infty} s^{\nu - \mu_{Pu}} \left(A_P^{(\nu)} S_N^{(\nu)}(s)\right)^{-1} C_{Pu}^{(\mu_{Pu})} S_r^{(\mu_{Pu})}(s) = A_{P\nu}^{-1} C_{Pu\mu_{Pu}}.$$

Hence,

$$rankG_{IOPU\infty} = rankA_{P\nu}^{-1} C_{Pu\mu_{Pu}} = rankC_{Pu\mu_{Pu}} = N \leq r,$$

and then

$$rankC_{Pu}^{(\mu_{Pu})} S_r^{(\mu_{Pu})}(s) = rankG_{IOPU\infty} = rankC_{Pu\mu_{Pu}} = rankC_{Pu}^{(\mu_{Pu})} = N \leq r \ .$$

This completes the proof ■

If we determine $G_{IOPU\infty}$, then its right numerator matrix is the matrix $C_{Pu\mu_{Pu}}$.

If we determine G_{IOPU0}, then its right numerator matrix is the matrix C_{Pu0}.

The complex domain conditions read as follows.

Theorem 275 ***Conditions for the perfect trackability on $\mathfrak{D}^{\mu_{Pd}}\text{x}\mathfrak{Y}_d^{\nu}$ of the IO plant (2.15)***

For the IO object (2.15) to be (the (ν-1)th order) perfect trackable on $\mathfrak{D}^{\mu_{Pd}}\text{x}\mathfrak{Y}_d^{\nu}$ it is necessary and sufficient that both

1) $N \leq r$,

and

2) $rankG_{IOPU}(s) = rank\left(C_{Pu}^{(\mu_{Pu})}S_r^{(\mu_{Pu})}(s)\right) = rankC_{Pu}^{(\mu_{Pu})} = N.$

Proof. Since the conditions are independent of $[\mathbf{D}(.),\mathbf{Y}_d^{\nu}(.)] \in \mathfrak{D}^{\mu_{Pd}}\text{x}\mathfrak{Y}_d^{\nu}$, they hold for every $[\mathbf{D}(.),\mathbf{Y}_d^{\nu}(.)] \in \mathfrak{D}^{\mu_{Pd}}\text{x}\mathfrak{Y}_d^{\nu}$ so that the theorem results from Theorem 68 (Subsection 3.3.2), Lemma 242 (Section 9.2) and Theorem 274 ■

Comment 276 *If the IO plant (2.15) is not stable, then the condition 2) of this Theorem is satisfied as soon as any of the following holds:*

a) $rankC_{Pu\mu_{Pu}} = N$,

b) $rankC_{Pu0} = N$.

This is due to the definitions of the matrix rank and of $C_{Pu}^{(\mu_{Pu})}$,

$$C_{Pu}^{(\mu_{Pu})} = \begin{bmatrix} C_{Pu0} & C_{Pu1} & C_{Pu2} & . & . & . & C_{Pu\mu_{Pu}} \end{bmatrix} \in \mathfrak{R}^{N\text{x}(\mu_{Pu}+1)r}.$$

This analysis shows how we can use the IO object (2.15) transfer function matrix $G_{IOPU}(s)$ relative to control $\mathbf{U}$ *to test the object perfect trackability on* $\mathfrak{D}^{\mu_{Pd}}\text{x}\mathfrak{Y}_d^{\nu}$.

Note 277 *The conditions for the perfect trackability on $\mathfrak{D}^{\mu_{Pd}}\text{x}\mathfrak{Y}_d^{\nu}$ of the IO object (2.15) do not impose any requirement on the internal dynamics of the object. The only requirement is that the IO object (2.15) transfer function matrix $G_{IOPU}(s)$ is well defined. Besides, the conditions do not impose any demand on the disturbance. They are independent of the disturbance.*

Comment 278 *The perfect trackability of the IO object (2.15) on $\mathfrak{D}^{\mu_{Pd}}\text{x}\mathfrak{Y}_d^{\nu}$ takes into account all disturbances $\mathbf{D}(.) \in \mathfrak{D}^{\mu_{Pd}}$, while the output function controllability is defined only for the ISO systems and under the condition that $\mathbf{D}(t) = \mathbf{0}_d$, $\forall t \in \mathfrak{T}_0$. To the author's knowledge, the output function controllability of the IO object (2.15) has not been studied.*

Conditions for the perfect natural trackability

Theorem 279 ***Conditions for the perfect natural trackability on $\mathfrak{D}^{\mu_{Pd}}\text{x}\mathfrak{Y}_d^{\nu}$ of the IO plant (2.15)***

For the IO object (2.15) to be perfect natural trackable on $\mathfrak{D}^{\mu_{Pd}}\text{x}\mathfrak{Y}_d^{\nu}$ it is necessary and sufficient that both

1) $N \leq r$,

and

2) $rankG_{IOPU}(s) = rank\left(C_{Pu}^{(\mu_{Pu})}S_r^{(\mu_{Pu})}(s)\right) = rankC_{Pu}^{(\mu_{Pu})} = N.$

Proof. *Necessity.* The necessity of the conditions of this theorem results directly from Lemma 242, Comment 247 and Theorem 274.

Sufficiency. Let the conditions hold. Let $\mathbf{Y}_d(.) \in \mathfrak{Y}_d^{\nu}$ be arbitrarily chosen. Let $\mathbf{Y}(0) = \mathbf{Y}_d(0)$. The *IO* object (2.15) is perfect trackable on $\mathfrak{D}^{\mu_{Pd}} \mathrm{x} \mathfrak{Y}_d^{\nu}$ (Theorem 275). We should show that the perfect tracking control can be synthesized without using information about the plant internal dynamics and about $\mathbf{D}(.) \in \mathfrak{D}^{\mu_{Pd}}$. Laplace transform of (2.15) (Subsection 2.1.2) reads

$$\left[A_P^{(\nu)} S_N^{(\nu)}(s)\right] \mathbf{Y}(s) = \left[C_{Pu}^{(\mu_{Pu})} S_r^{(\mu_{Pu})}(s)\right] \mathbf{U}(s) -$$
$$-C_{Pu}^{(\mu_{Pu})} Z_r^{(\mu_{Pu}-1)}(s)\mathbf{U}_0^{(\mu_{Pu}-1)} + \left[D_{Pd}^{(\mu_{Pd})} S_d^{(\mu_{Pd})}(s)\right] \mathbf{D}(s) -$$
$$-D_{Pd}^{(\mu_{Pd})} Z_d^{(\mu_{Pd}-1)}(s)\mathbf{D}_0^{(\mu_{Pd}-1)} + A_P^{(\nu)} Z_N^{(\nu-1)}(s)\mathbf{Y}_0^{(\nu-1)}. \tag{9.24}$$

Let $\sigma \in \mathfrak{R}^+$ be arbitrarily small, i.e., $\sigma \longrightarrow 0^+$. Let

$$\mathbf{U}(s) = (1 - e^{-\sigma s})^{-1}\Gamma(s)\left[\Phi(s) + \varepsilon(s)\right], \ \mathbf{\Phi}(s) = \mathcal{L}\{\phi(t)\} \Rightarrow$$
$$\mathbf{U}(s) = e^{-\sigma s}\mathbf{U}(s) + \Gamma(s)\left[\mathbf{\Phi}(s) + \varepsilon(s)\right], \tag{9.25}$$

$$\Gamma(s) = \left[C_{Pu}^{(\mu_{Pu})} S_r^{(\mu_{Pu})}(s)\right]^T \left\{\left[C_{Pu}^{(\mu_{Pu})} S_r^{(\mu_{Pu})}(s)\right]\left[C_{Pu}^{(\mu_{Pu})} S_r^{(\mu_{Pu})}(s)\right]^T\right\}^{-1}, \tag{9.26}$$

$$\phi(.) : \mathfrak{T}_0 \longrightarrow \mathfrak{R}^N, \ \phi(t) \in \mathfrak{C}, \ \phi(t) = \mathbf{0}_N, \ \forall t \in [\sigma, \infty[, \ \phi(0) = -\varepsilon(0). \tag{9.27}$$

Evidently, $\mathbf{U}(.)$ does not depend either on the plant internal dynamics or on $\mathbf{D}(.)$. The control $\mathbf{U}(.)$ is natural control (Comment 245, Section 9.2). We replace $\mathbf{U}(s)$ with the right-hand side of (9.25) into (9.24):

$$\left[A_P^{(\nu)} S_N^{(\nu)}(s)\right] \mathbf{Y}(s) = \left[C_{Pu}^{(\mu_{Pu})} S_r^{(\mu_{Pu})}(s)\right] \left\{e^{-\sigma s}\mathbf{U}(s) + \Gamma(s)\left[\mathbf{\Phi}(s) + \varepsilon(s)\right]\right\} -$$
$$-C_{Pu}^{(\mu_{Pu})} Z_r^{(\mu_{Pu}-1)}(s)\mathbf{U}_0^{(\mu_{Pu}-1)} + \left[D_{Pd}^{(\mu_{Pd})} S_d^{(\mu_{Pd})}(s)\right] \mathbf{D}(s) -$$
$$-D_{Pd}^{(\mu_{Pd})} Z_d^{(\mu_{Pd}-1)}(s)\mathbf{D}_0^{(\mu_{Pd}-1)} + A_P^{(\nu)} Z_N^{(\nu-1)}(s)\mathbf{Y}_0^{(\nu-1)}.$$

We subtract this equation from (9.24),

$$\left[A_P^{(\nu)} S_N^{(\nu)}(s)\right] \mathbf{Y}(s) - \left[A_P^{(\nu)} S_N^{(\nu)}(s)\right] \mathbf{Y}(s) =$$
$$= \left[C_{Pu}^{(\mu_{Pu})} S_r^{(\mu_{Pu})}(s)\right] \mathbf{U}(s) - \left[C_{Pu}^{(\mu_{Pu})} S_r^{(\mu_{Pu})}(s)\right] \left\{ \begin{array}{c} e^{-\sigma s}\mathbf{U}(s) + \\ +\Gamma(s)\left[\mathbf{\Phi}(s) + \varepsilon(s)\right] \end{array} \right\} -$$
$$-C_{Pu}^{(\mu_{Pu})} Z_r^{(\mu_{Pu}-1)}(s)\mathbf{U}_0^{(\mu_{Pu}-1)} + C_{Pu}^{(\mu_{Pu})} Z_r^{(\mu_{Pu}-1)}(s)\mathbf{U}_0^{(\mu_{Pu}-1)} +$$
$$+ \left[D_{Pd}^{(\mu_{Pd})} S_d^{(\mu_{Pd})}(s)\right] \mathbf{D}(s) - \left[D_{Pd}^{(\mu_{Pd})} S_d^{(\mu_{Pd})}(s)\right] \mathbf{D}(s) -$$
$$-D_{Pd}^{(\mu_{Pd})} Z_d^{(\mu_{Pd}-1)}(s)\mathbf{D}_0^{(\mu_{Pd}-1)} + D_{Pd}^{(\mu_{Pd})} Z_d^{(\mu_{Pd}-1)}(s)\mathbf{D}_0^{(\mu_{Pd}-1)} +$$
$$+A_P^{(\nu)} Z_N^{(\nu-1)}(s)\mathbf{Y}_0^{(\nu-1)} - A_P^{(\nu)} Z_N^{(\nu-1)}(s)\mathbf{Y}_0^{(\nu-1)}.$$

The result is

$$\mathbf{0}_N = \left[C_{Pu}^{(\mu_{Pu})} S_r^{(\mu_{Pu})}(s)\right] \left\{\left(1 - e^{-\sigma s}\right) \mathbf{U}(s) - \Gamma(s) \left[\mathbf{\Phi}(s) - \varepsilon(s)\right]\right\}.$$

For $\sigma \longrightarrow 0^+$ the preceding equation takes the following form:

$$\mathbf{0}_N = \left[C_{Pu}^{(\mu_{Pu})} S_r^{(\mu_{Pu})}(s)\right] \Gamma(s) \left[\mathbf{\Phi}(s) + \varepsilon(s)\right],$$

i.e.,

$$\mathbf{0}_N = \mathbf{\Phi}(s) + \varepsilon(s)$$

due to (9.26). This equation yields in the *time* domain:

$$\varepsilon(t) = -\phi(t), \ \forall t \in \mathfrak{T}_0.$$

Hence,

$$\varepsilon(t) = -\phi(t) = \mathbf{0}_N, \ \forall t \in \mathfrak{T}_0$$

due to (9.27), or equivalently:

$$\varepsilon(t) = \mathbf{Y}(t) = \mathbf{Y}_d(t) = \mathbf{0}_N, \ \forall t \in \mathfrak{T}_0.$$

The natural control $\mathbf{U}(.)$ (9.25) guarantees perfect tracking on $\mathfrak{D}^{\mu_{Pd}} \mathrm{x} \mathfrak{Y}_d^{\nu}$. The plant is perfect naturally trackable on $\mathfrak{D}^{\mu_{Pd}} \mathrm{x} \mathfrak{Y}_d^{\nu}$ ■

Comment 280 *Theorem 275 and Theorem 279 show that the IO plant (2.15) is perfect natural trackable on $\mathfrak{D}^{\mu_{Pd}} \mathrm{x} \mathfrak{Y}_d^{\nu}$ if, and only if, it is perfect trackable on $\mathfrak{D}^{\mu_{Pd}} \mathrm{x} \mathfrak{Y}_d^{\nu}$. This completes Comment 249, (Section 9.2).*

Note 281 *The equation (9.25) means that the controller uses*

- *information about the output error $\varepsilon(t) = \mathbf{Y}_d(t)$-$\mathbf{Y}(t)$, which is expressed through $\Gamma(s) \left[\mathbf{\Phi}(s) + \varepsilon(s)\right]$, for which there is the classical global negative feedback loop in the control system from the plant output to the controller input, and*

- *information about the control $\mathbf{U}(t^-) = \mathbf{U}(t\text{-}\sigma)_{\sigma \longrightarrow 0^+}$, with which the controller has just acted on the plant at the moment t and which is then an input to the controller itself, for which there is local, internal, unit positive feedback in the controller. The value of $\sigma \longrightarrow 0^+$ ($\sigma = 0$ in the ideal case meaning there is not any delay in the local feedback) expresses the infinitesimal duration of the time interval during which the controller output $\mathbf{U}(t)$ becomes its own input $\mathbf{U}(t^-) = \mathbf{U}(t\text{-}\sigma)$. Since the controller local feedback is positive and unit, the controller may not be ON when disconnected. It can work properly only when it is in the feedback connection with the plant through the global negative feedback.*

9.3.2 *ISO* plant perfect (natural) trackability

Conditions for perfect trackability

Which properties of the *ISO* plant permit the existence of a control that can force the plant to exhibit perfect tracking as soon as the initial output vector is equal to the initial desired output vector? Equivalently, which properties of the *ISO* plant guarantee its perfect trackability?

Definition 241 (Section 9.2) determines the *k-th* order **output** perfect trackability of the plant on $\mathfrak{D}^i \text{x} \mathfrak{Y}_d^k$. Lemma 242 (Section 9.2) reduces the study of the *l-th* order perfect trackability of the plant on $\mathfrak{D}^i \text{x} \mathfrak{Y}_d^k$ to the study of the perfect trackability of the plant on $\mathfrak{D}^i \text{x} \mathfrak{Y}_d^k$.

Note 282 *In case the state vector* $\mathbf{X}$ *is simultaneously the output vector* $\mathbf{Y}$*, i.e.,* $\mathbf{Y}(t) \equiv \mathbf{X}(t)$*, then Definition 241 becomes the definition of the l-th order perfect* **state** *trackability.*

The full transfer function matrix of the *ISO* plant (2.33), (2.34) (Subsection 2.2.2) is determined in (7.86), (7.87) (Subsection 7.4.1),

$$\begin{aligned} F_{ISOP}(s) &= \left[G_{ISOPD}(s) \vdots G_{ISOPU}(s) \vdots G_{ISOPXo}(s) \right] = \\ &= \begin{bmatrix} \left(C_P (sI_{n_P} - A_P)^{-1} L_P + D_P \right)^T \\ \left(C_P (sI_{n_P} - A_P)^{-1} B_P + H_P \right)^T \\ \left(C_P (sI_{n_P} - A_P)^{-1} \right)^T \end{bmatrix}^T . \end{aligned}$$

The plant transfer function matrix $G_{ISOPu}(s)$ relative to the control $\mathbf{U}$ follows as

$$G_{ISOPU}(s) = C_P (sI_{n_P} - A_P)^{-1} B_P + H_P. \tag{9.28}$$

It leads to the complex domain criterion for the perfect trackability of the plant on $\mathfrak{D}^k \text{x} \mathfrak{Y}_d^k$.

Two extreme cases deserve our attention.

a) If A_P is nonsingular, then at the origin $s = 0$ the plant transfer function matrix $G_{ISOPU}(0)$ denoted by G_{ISOPUo} is well defined,

$$\det A_P \neq 0 \Longrightarrow G_{ISOPU}(0) = G_{ISOPUo} = H_P - C_P A_P^{-1} B_P, \tag{9.29}$$

b) At infinity, $s = \infty$, the plant transfer function matrix $G_{ISOPU}(s)$ reduces to H_P,

$$G_{ISOPU}(\infty) = \lim_{s \longrightarrow \infty} G_{ISOPU}(s) = H_P. \tag{9.30}$$

We define also

$$G_{ISOPU\infty} = \lim_{s \longrightarrow \infty} s \left[G_{ISOPU}(s) - G_{ISOPU}(\infty) \right], \tag{9.31}$$

which is related to the system matrices by

$$G_{ISOPU\infty} = C_P B_P. \tag{9.32}$$

This is the first Markov parameter of $G_{ISOPU}(s)$.

Theorem 283 ***The ISO object transfer function matrix $G_{ISOPU}(s)$ and rank***

For the ISO plant (2.33), (2.34) the following statements are valid:

$$rankG_{ISOPUo} = N \leq r \Longrightarrow$$
$$rankG_{ISOPU}(s) = rankG_{ISOPUo} = rank\left(H_P - C_P A_P^{-1} B_P\right), \tag{9.33}$$

$$rankG_{ISOPU}(\infty) = N \leq r \Longrightarrow$$
$$rankG_{ISOPU}(s) = rankG_{ISOPU}(\infty) = rankH_P \tag{9.34}$$

and

$$rankG_{ISOPU\infty} = N \leq r \Longrightarrow rankG_{ISOPU\infty} = rankC_P B_P. \tag{9.35}$$

Proof. Stability of the *ISO* plant (2.33), (2.34) (Claim 155 in Section 8.2) ensures $\det A_P \neq 0$ so that $C_P A_P^{-1} B_P$ exists. From $rankG_{ISOPUo} = N \leq r$ it follows that $G_{ISOPU}(s)$ has the full rank at $s = 0$. It has the full rank for almost every $s \in \mathfrak{C}$, i.e., for every $s \in \mathfrak{C}$ except for a finite number of values of $s \in \mathfrak{C}$. Therefore,

$$rankG_{ISOPUo} = N \leq r$$

implies

$$rankG_{ISOPU}(s) = rankG_{ISOPUo},$$

which with (9.29) proves (9.33).

The definition (9.30) of $G_{ISOPU}(\infty)$ implies directly (9.34).

If

$$rankG_{ISOPU\infty} = N \leq r,$$

then (9.32) implies (9.35) ■

Note 284 *The equations (9.33) hold if, and only if, $\det A_P \neq 0$ regardless of stability properties of the ISO plant (2.33), (2.34). A nonsingular matrix A_P can have an eigenvalue with a nonnegative real part, i.e., an unstable matrix A_P can be nonsingular.*

The equations (9.30) and (9.32) are valid completely independently of the properties of the matrix A_P. They permit the rank condition on $G_{ISOPU}(s)$ in (9.33) to be replaced by (9.30) or by (9.32) as done in (9.34) and (9.35), respectively. This is important for the ISO plant (2.33), (2.34) if it is not stable.

Theorem 285 ***Condition for the perfect trackability on*** $\mathfrak{D}^k x\mathfrak{Y}_d^k$

In order for the ISO plant (2.33), (2.34) to be perfect trackable on $\mathfrak{D}^k x\mathfrak{Y}_d^k$ *it is necessary and sufficient that the plant transfer function matrix* $G_{ISOPU}(s)$ *relative to control has the full row rank* N*, so that the dimension* r *of the control vector* $\mathbf{U}$ *is not less than the dimension* N *of the output vector* $\mathbf{Y}$*, i.e.,*

$$rankG_{ISOPU}(s) = rank\left[C_P(sI_n - A_P)^{-1}B_P + H_P\right] = N \leq r. \tag{9.36}$$

Laplace transform $\mathbf{U}_N^{\mp}(s)$ *of the nominal control* $\mathbf{U}_N(.)$ *is then determined by*

$$\begin{gathered}\mathbf{U}_N^{\mp}(s) = G_{ISOPU}(s)^T\left\{G_{ISOPU}(s)G_{ISOPU}(s)^T\right\}^{-1} \bullet \\ \bullet\left\{\mathbf{Y}_d^{\mp}(s) - G_{ISOPD}(s)\mathbf{D}^{\mp}(s) - G_{ISOPXo}(s)\mathbf{X}_{0\mp}\right\}.\end{gathered} \tag{9.37}$$

Proof. The equations (7.84)-(7.87) (Subsection 7.4.1) determine the full transfer function matrix $F_{ISOP}(s)$ and Laplace transform of the output response of the plant (2.33), (2.34), respectively.

Necessity. Let the plant (2.33), (2.34) be perfect trackable on $\mathfrak{D}^k x\mathfrak{Y}_d^k$. Hence, the nominal control $\mathbf{U}_N(.)$ is well defined on $\mathfrak{D}^k x\mathfrak{Y}_d^k$, which means that (7.84), together with (7.86) and (7.87), is solvable in $\mathbf{U}^{\mp}(s)$. This implies that

$$\mathbf{Y}_d^{\mp}(s) = G_{ISOPU}(s)\mathbf{U}_N^{\mp}(s) + G_{ISOPD}(s)\mathbf{D}^{\mp}(s) + G_{ISOPXo}(s)\mathbf{X}_{0\mp}$$

is solvable in $\mathbf{U}_N^{\mp}(s)$ for every $[\mathbf{D}(.),\ \mathbf{Y}_d(.)] \in \mathfrak{D}^k x\mathfrak{Y}_d^k$, which yields (9.36) and (9.37).

Sufficiency. Let (9.36) and (9.37) hold. Let $\mathbf{U}(.)$ be the nominal control $\mathbf{U}_N(.)$. The equations (7.84)-(7.87) and (9.37) give

$$\begin{gathered}\mathbf{Y}^{\mp}(s) = G_{ISOPU}(s)\mathbf{U}_N^{\mp}(s) + G_{ISOPD}(s)\mathbf{D}^{\mp}(s) + G_{ISOPXo}(s)\mathbf{X}_{0\mp} = \\ +G_{ISOPD}(s)\mathbf{D}^{\mp}(s) + G_{ISOPXo}(s)\mathbf{X}_{0\mp} = \\ = G_{ISOPU}(s)G_{ISOPU}(s)^T\left\{G_{ISOPU}(s)G_{ISOPU}(s)^T\right\}^{-1} \bullet \\ \bullet\left\{\mathbf{Y}_d^{\mp}(s) - G_{ISOPD}(s)\mathbf{D}^{\mp}(s) - G_{ISOPXo}(s)\mathbf{X}_{0\mp}\right\} + \\ +G_{ISOPD}(s)\mathbf{D}^{\mp}(s) + G_{ISOPXo}(s)\mathbf{X}_{0\mp} = \mathbf{Y}_d^{\mp}(s),\end{gathered}$$

i.e.,

$$\mathbf{Y}(t) = \mathbf{Y}_d(t),\ \forall t \in \mathfrak{T}_0$$

■

The condition 9.36 of Theorem 285 is the necessary and sufficient for the output function controllability [8, p. 313]; [29, p. 216, Theorem 5-23]; [319, p. 164, Theorem 5.5.7].

Comment 286 *Equation (9.37) determines the nominal control in terms of the disturbance vector. This means that the disturbance vector should be measurable, which is rarely satisfied.*

Condition for the perfect natural trackability

Definition 248 (Section 9.2) determines the perfect natural trackability of the ISO plant (2.33), (2.34) on $\mathfrak{D}^k \text{x} \mathfrak{Y}_d^k$.

Theorem 287 ***Condition for the perfect natural trackability***
In order for the ISO plant (2.33), (2.34) to be perfect natural trackable on $\mathfrak{D}^k \text{x} \mathfrak{Y}_d^k$ it is necessary and sufficient that the plant transfer function matrix $G_{ISOPU}(s)$ relative to control has the full row rank N, so that the dimension r of the control vector $\mathbf{U}$ is not less than the dimension N of the output vector $\mathbf{Y}$, i.e., that (9.36) holds.

Proof. *Necessity.* Let the plant (2.33), (2.34) be perfect natural trackable on $\mathfrak{D}^k \text{x} \mathfrak{Y}_d^k$. Definition 241 and Definition 248 (Section 9.2) show that the plant is perfect trackable on $\mathfrak{D}^k \text{x} \mathfrak{Y}_d^k$. Theorem 285 implies necessity of the condition of the theorem statement, i.e., of (9.36).

Sufficiency. Let the condition of the theorem statement be valid, i.e., let (9.36) hold. The plant is perfect trackable on $\mathfrak{D}^k \text{x} \mathfrak{Y}_d^k$ due to Theorem 285. We should show that (9.36) guarantees the existence of time-continuous control that is independent of the plant internal dynamics and of the disturbance $\mathbf{D}(.)$ such that it ensures the perfect tracking on $\mathfrak{D}^k \text{x} \mathfrak{Y}_d^k$. Let $\sigma \in \mathfrak{R}^+$ be arbitrarily small, i.e., $\sigma \longrightarrow 0^+$; in the ideal case $\sigma = 0$. Let

$$\mathbf{Y}(0) = \mathbf{Y}_d(0), \tag{9.38}$$

and

$$\Upsilon(s) = G_{ISOPU}(s)^T \left\{ G_{ISOPU}(s) G_{ISOPU}(s)^T \right\}^{-1} \tag{9.39}$$

$$\boldsymbol{\phi}(.) : \mathfrak{T}_0 \longrightarrow \mathfrak{R}^N, \ \boldsymbol{\phi}(t) \in \mathfrak{C}, \ \boldsymbol{\phi}(t) = \mathbf{0}_N, \ \forall t \in [\sigma, \infty[, \ \boldsymbol{\phi}(0) = -\boldsymbol{\varepsilon}(0), \tag{9.40}$$

$$\mathbf{\Phi}(s) = \mathcal{L}\left\{\boldsymbol{\phi}(t)\right\}, \tag{9.41}$$

$$\mathbf{U}(s) = (1 - e^{-\sigma s})^{-1} \Upsilon(s) \left[\Phi(s) + \boldsymbol{\varepsilon}(s)\right], \tag{9.42}$$

so that

$$\mathbf{U}(s) = e^{-\sigma s}\mathbf{U}(s) + \Upsilon(s) \left[\mathbf{\Phi}(s) + \boldsymbol{\varepsilon}(s)\right]. \tag{9.43}$$

We replace $\mathbf{U}(s)$ by the right-hand side of (9.43) into (7.85):

$$\mathbf{Y}^{\mp}(s) = F_{ISOPl}(s) \left(e^{-\sigma s}\mathbf{U}(s) + \Upsilon(s) \left[\mathbf{\Phi}(s) + \boldsymbol{\varepsilon}(s)\right] \vdots \mathbf{D}^{\mp}(s) \vdots \mathbf{X}_{0\mp} \right)^T. \tag{9.44}$$

We present (7.85) as follows by applying the system full transfer function matrix (7.86):

$$\mathbf{Y}^{\mp}(s) = \left[G_{ISOPU}(s) \vdots G_{ISOPD}(s) \vdots G_{ISOPXo}(s) \right] \begin{bmatrix} \mathbf{U}^{\mp}(s) \\ \mathbf{D}^{\mp}(s) \\ \mathbf{X}_{0\mp} \end{bmatrix}. \tag{9.45}$$

The solution of this equation in $G_{ISOPU}(s)\mathbf{U}(s)$ reads

$$G_{ISOPU}(s)\mathbf{U}^{\mp}(s)=\mathbf{Y}^{\mp}(s)-G_{ISOPD}(s)\mathbf{D}^{\mp}(s)-G_{ISOPXo}(s)\mathbf{X}_{0\mp}=$$
$$=\left[I_N \vdots -G_{ISOPD}(s) \vdots -G_{ISOPXo}(s)\right]\begin{bmatrix}\mathbf{Y}^{\mp}(s)\\ \mathbf{D}^{\mp}(s)\\ \mathbf{X}_{0\mp}\end{bmatrix}. \qquad (9.46)$$

(7.86) and (9.44) furnish

$$\mathbf{Y}^{\mp}(s)=\left[G_{ISOPU}(s)\vdots G_{ISOPD}(s)\vdots G_{ISOPXo}(s)\right]\bullet$$
$$\bullet\begin{bmatrix}e^{-\sigma s}\mathbf{U}^{\mp}(s)+\Upsilon(s)\left[\mathbf{\Phi}(s)+\boldsymbol{\varepsilon}(s)\right](s)\\ \mathbf{D}^{\mp}(s)\\ \mathbf{X}_{0\mp}\end{bmatrix}\Longrightarrow$$

$$\mathbf{Y}^{\mp}(s)=e^{-\sigma s}G_{ISOPU}(s)\mathbf{U}^{\mp}(s)+$$
$$+\left[G_{ISOPU}(s)\vdots G_{ISOPD}(s)\vdots G_{ISOPXo}(s)\right]\bullet$$
$$\bullet\begin{bmatrix}\Upsilon(s)\left[\mathbf{\Phi}(s)+\boldsymbol{\varepsilon}(s)\right]\\ \mathbf{D}^{\mp}(s)\\ \mathbf{X}_{0\mp}\end{bmatrix}. \qquad (9.47)$$

We apply (9.46) to (9.47):

$$\mathbf{Y}^{\mp}(s)=e^{-\sigma s}\left[I_N \vdots -G_{ISOPD}(s) \vdots -G_{ISOPXo}(s)\right]\begin{bmatrix}\mathbf{Y}^{\mp}(s)\\ \mathbf{D}^{\mp}(s)\\ \mathbf{X}_{0\mp}\end{bmatrix}+$$
$$+\left[G_{ISOPU}(s)\vdots G_{ISOPD}(s)\vdots G_{ISOPXo}(s)\right]\begin{bmatrix}\Upsilon(s)\left[\mathbf{\Phi}(s)+\boldsymbol{\varepsilon}(s)\right]\\ \mathbf{D}^{\mp}(s)\\ \mathbf{X}_{0\mp}\end{bmatrix}.$$

Since $\boldsymbol{\varepsilon}(s)=\mathbf{Y}_d^{\mp}(s)$-$\mathbf{Y}^{\mp}(s)$, the right-hand side of the preceding equation simplifies so that, in view of (9.39),

$$\mathbf{Y}^{\mp}(s)=e^{-\sigma s}\mathbf{Y}^{\mp}(s)+\mathbf{\Phi}(s)+\mathbf{Y}_d^{\mp}(s)-\mathbf{Y}^{\mp}(s).$$

As $\sigma\longrightarrow 0^{+}$, i.e., in the ideal case $\sigma=0$,

$$\mathbf{Y}^{\mp}(s)=\mathbf{Y}^{\mp}(s)+\mathbf{\Phi}(s)+\mathbf{Y}_d^{\mp}(s)-\mathbf{Y}^{\mp}(s)\Longrightarrow$$

$$\mathbf{\Phi}(s)+\mathbf{Y}_d^{\mp}(s)-\mathbf{Y}^{\mp}(s)=\mathbf{0}_N,$$

equivalently in the time domain

$$\boldsymbol{\phi}(t)+\mathbf{Y}_d(t)-\mathbf{Y}(t)=\mathbf{0}_N\ ,\ \forall t\in\mathfrak{T}_0,$$

and, due to (9.40),

$$\mathbf{Y}_d(t) - \mathbf{Y}(t) = \mathbf{0}_N \ , \ \forall t \in [\sigma, \infty[.$$

This, (9.38) and $\sigma \longrightarrow 0^+$, i.e., in the ideal case $\sigma = 0$, prove

$$\mathbf{Y}_d(t) - \mathbf{Y}(t) = \mathbf{0}_N \ , \ \forall t \in \mathfrak{T}_0.$$

This holds for every $[\mathbf{D}(.),\ \mathbf{Y}_d(.)] \in \mathfrak{D}^k \mathrm{x} \mathfrak{Y}_d^k$. The conditions $rankG_{ISOPU}(s) = N \leq r$ and (9.36) guarantee the existence of control $\mathbf{U}(.)$ that is independent of the plant internal dynamics and of the disturbance $\mathbf{D}(.)$ such that it ensures the perfect tracking on $\mathfrak{D}^k \mathrm{x} \mathfrak{Y}_d^k$ ■

Comment 288 *Theorem 285 and Theorem 287 show that for the ISO plant (2.33), (2.34) to be perfect natural trackable on* $\mathfrak{D}^k \mathrm{x} \mathfrak{Y}_d^k$ *it is necessary and sufficient that the plant is perfect trackable on* $\mathfrak{D}^k \mathrm{x} \mathfrak{Y}_d^k$. *This completes Comment 249 (Section 9.2).*

Comment 289 *The preceding proof, (9.39) and (9.43), shows that the control implementation does not need any information about the disturbance vector. This means that the disturbance vector need not be measurable, which corresponds to the reality.*

Note 290 *The equation (9.43) means that the controller uses information about both*

- *the output error* $\varepsilon(t) = \mathbf{Y}_d(t)$-$\mathbf{Y}(t)$, *which is expressed by* $\Upsilon(s)[\mathbf{\Phi}(s) + \varepsilon(s)]$, *for which there is the global negative feedback loop from the plant output to the controller input, and*

- *the control* $\mathbf{U}(t^-) = \mathbf{U}(t\text{-}\sigma)_{\sigma \longrightarrow 0^+}$, *with which the controller has just acted on the plant at the moment t and which is then an input to the controller itself, for which there is local, internal, unit positive feedback in the controller. The value of* $\sigma \longrightarrow 0^+$ *expresses the infinitesimal duration of the time interval during which the controller output* $\mathbf{U}(t)$ *becomes its input* $\mathbf{U}(t^-) = \mathbf{U}(t\text{-}\sigma)$. *The positive unit controller local feedback does not allow the controller to be ON when it is disconnected. It should be connected with the plant in the global negative feedback.*

Note 291 *In the ideal case there is not any delay of the signal transmission from the controller output through the local unit positive feedback of the controller itself to its input. Hence, then* $\sigma = 0$.

Comment 292 *The results show how the transfer function matrix* $G_{ISOPU}(s)$ *and its extreme values* G_{ISOPUo}, $G_{ISOPU}(\infty)$ *and the induced* $G_{ISOPU\infty}$, *of the ISO plant (2.33), (2.34) can be applied effectively to the perfect (natural) trackability test.*

9.4 Imperfect trackability conditions

9.4.1 *IO* plant imperfect (natural) trackability

Conditions for imperfect trackability

The imperfect trackability expresses the plant ability to permit the existence of control that can steer the output vector from its arbitrary initial value to its desired value in a finite time, after which the real output vector stays always equal to the desired output vector. We will first present and prove the conditions for the imperfect trackability of the *IO* plant (2.15) (Subsection 2.1.2).

Theorem 293 ***Conditions for the*** $(\nu\text{-}1)\,th$ ***global (elementwise) trackability on*** $\mathfrak{D}^{\mu_{Pd}}\mathrm{x}\mathfrak{Y}_d^{\nu}$ ***of the IO plant (2.15)***

For the IO object (2.15) to be the $(\nu\text{-}1)\,th$ *order global (elementwise) trackable on* $\mathfrak{D}^{\mu_{Pd}}\mathrm{x}\mathfrak{Y}_d^{\nu}$ *it is necessary and sufficient that both*

1) $N \leq r,$

and

2) $rankG_{IOPU}(s) = rank\left(C_{Pu}^{(\mu_{Pu})}S_r^{(\mu_{Pu})}(s)\right) = rankC_{Pu}^{(\mu_{Pu})} = N.$

Proof. *Necessity.* Let the *IO* plant (2.15) be the $(\nu\text{-}1)\,th$ order global (elementwise) trackable on $\mathfrak{D}^{\mu_{Pd}}\mathrm{x}\mathfrak{Y}_d^{\nu}$. Then it is also perfect trackable on $\mathfrak{D}^{\mu_{Pd}}\mathrm{x}\mathfrak{Y}_d^{\nu}$ (Lemma 242, Definition 253, Theorem 254, Definition 263, Section 9.2). Theorem 275 (Subsection 9.3.1) proves the necessity of the conditions 1) and 2) for $(\nu\text{-}1)\,th$ order trackability on $\mathfrak{D}^{\mu_{Pd}}\mathrm{x}\mathfrak{Y}_d^{\nu}$, which is necessary for both $(\nu\text{-}1)\,th$ order global trackability on $\mathfrak{D}^{\mu_{Pd}}\mathrm{x}\mathfrak{Y}_d^{\nu}$ and $(\nu\text{-}1)\,th$ order global elementwise trackability on $\mathfrak{D}^{\mu_{Pd}}\mathrm{x}\mathfrak{Y}_d^{\nu}$.

Sufficiency. Let the conditions 1) and 2) hold. Let $[\mathbf{D}(.), \mathbf{Y}_d(.)] \in \mathfrak{D}^{\mu_{Pd}}\mathrm{x}\mathfrak{Y}_d^{\nu}$, $\boldsymbol{\sigma} \in (Int\ \mathfrak{T}_0)^{\nu}$, $\boldsymbol{\sigma} = [\sigma_1\ \sigma_2\ ...\ \sigma_{\nu}]^T$, and $\mathbf{Y}_0^{\nu-1} \in \mathfrak{R}^{\nu N}$ be arbitrarily chosen. Let the control be defined by

$$\mathbf{U}(t) = \Gamma\left[-D_{Pd}^{(\mu_{Pd})}\mathbf{D}^{\mu_{Pd}}(t) + A_P^{(\nu)}\mathbf{Y}^{\nu}(t) - \mathbf{Z}(t) + \sum_{j=0}^{j=\nu-1}\left|\boldsymbol{\varepsilon}^{(j)}(t)\right|\right] \in \mathfrak{C}(\mathfrak{T}_0),$$

$$\Gamma = C_{Pu}^{(\mu_{Pu})^T}\left[C_{Pu}^{(\mu_{Pu})}C_{Pu}^{(\mu_{Pu})^T}\right]^{-1},$$

together with

$$\mathbf{Z}(t) = \left\{\begin{array}{c} \sum_{j=0}^{j=\nu-1}\left|\boldsymbol{\varepsilon}^{(j)}(0)\right|\left(1-\frac{t}{\sigma_j}\right),\ t\mathbf{1}_{\nu} \in [0,\ \boldsymbol{\sigma}], \\ \mathbf{0}_N,\ \forall\,(t\mathbf{1}_{\nu} \geq \boldsymbol{\sigma}) \in \mathfrak{T}_0^{\nu} \end{array}\right\} \in \mathfrak{C}(\mathfrak{T}_0). \tag{9.48}$$

For such control the plant mathematical model (2.15),

$$A_P^{(\nu)}\mathbf{Y}^{\nu}(t) = C_{Pu}^{(\mu_{Pu})}\mathbf{U}(t) + D_{Pd}^{(\mu_{Pd})}\mathbf{D}^{\mu_{Pd}}(t),$$

becomes

$$A_P^{(\nu)}\mathbf{Y}^{\nu}(t) = C_{Pu}^{(\mu_{Pu})}C_{Pu}^{(\mu_{Pu})^T}\left[C_{Pu}^{(\mu_{Pu})}C_{Pu}^{(\mu_{Pu})^T}\right]^{-1}\bullet$$

$$\bullet\left[-D_{Pd}^{(\mu_{Pd})}\mathbf{D}^{\mu_{Pd}}(t) + A_P^{(\nu)}\mathbf{Y}^{\nu}(t) - \mathbf{Z}(t) + \sum_{j=0}^{j=\nu-1}\left|\varepsilon^{(j)}(t)\right|\right] + D_{Pd}^{(\mu_{Pd})}\mathbf{D}^{\mu_{Pd}}(t) =$$

$$= -D_{Pd}^{(\mu_{Pd})}\mathbf{D}^{\mu_{Pd}}(t) + A_P^{(\nu)}\mathbf{Y}^{\nu}(t) - \mathbf{Z}(t) + \sum_{j=0}^{j=\nu-1}\left|\varepsilon^{(j)}(t)\right| + D_{Pd}^{(\mu_{Pd})}\mathbf{D}^{\mu_{Pd}}(t),$$

i.e.,

$$\sum_{j=0}^{j=\nu-1}\left|\varepsilon^{(j)}(t)\right| = \mathbf{Z}(t),\ \forall t \in \mathfrak{T}_0.$$

This and (9.48) imply

$$\varepsilon^{(j)}(t) = \mathbf{0}_N,\ \forall(t\mathbf{1}_\nu \geq \sigma) \in \mathfrak{T}_0^{\nu},\ \forall j = 0, 1, ..., \nu - 1,$$

i.e.,

$$\mathbf{Y}^{(j)}(t) = \mathbf{Y}_d^{(j)}(t),\ \forall(t\mathbf{1}_\nu \geq \sigma) \in \mathfrak{T}_0^{\nu N},\ \forall j = 0, 1, ..., \nu - 1,$$

which proves the $(\nu\text{-}1)\,th$ order global elementwise trackability on $\mathfrak{D}^{\mu_{Pd}}\mathrm{x}\mathfrak{Y}_d^{\nu}$ of the *IO* plant (2.15). The plant is the $(\nu\text{-}1)\,th$ order global elementwise trackable, hence global trackable, on $\mathfrak{D}^{\mu_{Pd}}\mathrm{x}\mathfrak{Y}_d^{\nu}$ due to Definition 253 and Definition 263 ∎

Comment 294 *Theorem 293, Lemma 242, Definition 253, Theorem 254, Definition 263, Theorem 275 and Theorem 279 show that the conditions are the same for the* $(\nu\text{-}1)\,th$ *order:*

- *perfect trackability,*
- *perfect natural trackability,*
- *global trackability,*
- *global elementwise trackability*

on $\mathfrak{D}^{\mu_{Pd}}\mathrm{x}\mathfrak{Y}_d^{\nu}$ *of the IO plant (2.15).*

Conditions for natural trackability

Theorem 295 ***Conditions for the*** $(\nu\text{-}1)\,th$ ***global (elementwise) natural trackability on*** $\mathfrak{D}^{\mu_{Pd}}\mathrm{x}\mathfrak{Y}_d^{\nu}$ ***of the IO plant (2.15)***

For the IO object (2.15) to be the $(\nu\text{-}1)\,th$ *order global (elementwise) natural trackable on* $\mathfrak{D}^{\mu_{Pd}}\mathrm{x}\mathfrak{Y}_d^{\nu}$ *it is necessary and sufficient that both*

1) $N \leq r,$

and

2) $rankG_{IOPU}(s) = rank\left(C_{Pu}^{(\mu_{Pu})}S_r^{(\mu_{Pu})}(s)\right) = rankC_{Pu}^{(\mu_{Pu})} = N.$

Proof. *Necessity,* Let the *IO* plant (2.15) be the (ν-1) *th* order global (elementwise) natural trackable on $\mathfrak{D}^{\mu_{Pd}}\text{x}\mathfrak{Y}_d^{\nu}$. Then it is also perfect trackable on $\mathfrak{D}^{\mu_{Pd}}\text{x}\mathfrak{Y}_d^{\nu}$ (Lemma 242, Definition 253, Theorem 254, Definition 259, Theorem 261, Definition 263, Definition 268, Section 9.2). Theorem 275 proves the necessity of conditions 1) and 2) for (ν-1) *th* order trackability on $\mathfrak{D}^{\mu_{Pd}}\text{x}\mathfrak{Y}_d^{\nu}$, which is necessary for both (ν-1) *th* order global natural trackability on $\mathfrak{D}^{\mu_{Pd}}\text{x}\mathfrak{Y}_d^{\nu}$ and (ν-1) *th* order global elementwise natural trackability on $\mathfrak{D}^{\mu_{Pd}}\text{x}\mathfrak{Y}_d^{\nu}$.

Sufficiency. Let the conditions 1) and 2) hold. Let $[\mathbf{D}(.), \mathbf{Y}_d(.)] \in \mathfrak{D}^{\mu_{Pd}}\text{x}\mathfrak{Y}_d^{\nu}$, $\boldsymbol{\sigma} \in (Int\ \mathfrak{T}_0)^{\nu}$, $\boldsymbol{\sigma} = [\sigma_1\ \sigma_2\ ...\ \sigma_{\nu}]^T$, and $\mathbf{Y}_0^{\nu-1} \in \mathfrak{R}^{\nu N}$ be arbitrarily chosen. Let the control be defined by

$$\mathbf{U}(t) = \mathbf{U}(t^-) + \Gamma\left[-\mathbf{Z}(t) + \sum_{j=0}^{j=\nu-1}\left|\boldsymbol{\varepsilon}^{(j)}(t)\right|\right] \in \mathfrak{C}(\mathfrak{T}_0),$$

for $\mathbf{Z}(t)$ determined by (9.48). For such control the plant mathematical model (2.15),

$$A_P^{(\nu)}\mathbf{Y}^{\nu}(t) = C_{Pu}^{(\mu_{Pu})}\mathbf{U}(t) + D_{Pd}^{(\mu_{Pd})}\mathbf{D}^{\mu_{Pd}}(t),$$

becomes

$$A_P^{(\nu)}\mathbf{Y}^{\nu}(t) = C_{Pu}^{(\mu_{Pu})}\mathbf{U}(t^-) + C_{Pu}^{(\mu_{Pu})}C_{Pu}^{(\mu_{Pu})^T}\left[C_{Pu}^{(\mu_{Pu})}C_{Pu}^{(\mu_{Pu})^T}\right]^{-1}\bullet$$
$$\bullet\left[-\mathbf{Z}(t) + \sum_{j=0}^{j=\nu-1}\left|\boldsymbol{\varepsilon}^{(j)}(t)\right|\right] + D_{Pd}^{(\mu_{Pd})}\mathbf{D}^{\mu_{Pd}}(t),$$

i.e.,

$$A_P^{(\nu)}\mathbf{Y}^{\nu}(t) = C_{Pu}^{(\mu_{Pu})}\mathbf{U}(t^-) - \mathbf{Z}(t) + \sum_{j=0}^{j=\nu-1}\left|\boldsymbol{\varepsilon}^{(j)}(t)\right| + D_{Pd}^{(\mu_{Pd})}\mathbf{D}^{\mu_{Pd}}(t). \quad (9.49)$$

The plant mathematical model at $t^- \in \mathfrak{T}_0$ reads

$$A_P^{(\nu)}\mathbf{Y}^{\nu}(t^-) = C_{Pu}^{(\mu_{Pu})}\mathbf{U}(t^-) + D_{Pd}^{(\mu_{Pd})}\mathbf{D}^{\mu_{Pd}}(t^-).$$

We solve this for $C_{Pu}^{(\mu_{Pu})}\mathbf{U}(t^-)$,

$$C_{Pu}^{(\mu_{Pu})}\mathbf{U}(t^-) = A_P^{(\nu)}\mathbf{Y}^{\nu}(t^-) - D_{Pd}^{(\mu_{Pd})}\mathbf{D}^{\mu_{Pd}}(t^-)$$

so that (9.49) becomes

$$A_P^{(\nu)}\mathbf{Y}^{\nu}(t) = A_P^{(\nu)}\mathbf{Y}^{\nu}(t^-) - D_{Pd}^{(\mu_{Pd})}\mathbf{D}^{\mu_{Pd}}(t^-) -$$
$$-\mathbf{Z}(t) + \sum_{j=0}^{j=\nu-1}\left|\boldsymbol{\varepsilon}^{(j)}(t)\right| + D_{Pd}^{(\mu_{Pd})}\mathbf{D}^{\mu_{Pd}}(t).$$

Linearity of (2.15) and $[\mathbf{D}(.), \mathbf{Y}_d(.)] \in \mathfrak{D}^{\mu_{Pd}} \mathrm{x} \mathfrak{Y}_d^{\nu}$, imply, (also due to Principle 9, Section 1.1), continuity of all variables in (2.15) so that

$$\mathbf{Y}^{\nu}(t) = \mathbf{Y}^{\nu}(t^{-}),\ \mathbf{D}^{\mu_{Pd}}(t) = \mathbf{D}^{\mu_{Pd}}(t^{-}).$$

Hence,

$$\sum_{j=0}^{j=\nu-1} \left| \varepsilon^{(j)}(t) \right| = \mathbf{Z}(t) = \mathbf{0}_N,\ \forall (t\mathbf{1}_{\nu N} \geq \sigma) \in \mathfrak{T}_0^{\nu N},$$
$$\forall j = 0, 1, ..., \nu - 1,$$

i.e.,

$$\mathbf{Y}^{(j)}(t) = \mathbf{Y}_d^{(j)}(t),\ \forall (t\mathbf{1}_{\nu N} \geq \sigma) \in \mathfrak{T}_0^{\nu N},$$
$$\forall j = 0, 1, ..., \nu - 1,$$

which proves the (ν-1) *th* order global elementwise natural trackability on $\mathfrak{D}^{\mu_{Pd}} \mathrm{x} \mathfrak{Y}_d^{\nu}$ of the *IO* plant (2.15), hence the plant is the (ν-1) *th* order global natural trackable on $\mathfrak{D}^{\mu_{Pd}} \mathrm{x} \mathfrak{Y}_d^{\nu}$ ■

Comment 296 *The preceding theorems and comment show that the conditions are the same for:*

- *the (ν-1) th order perfect trackability,*
- *the (ν-1) th order perfect natural trackability,*
- *the (ν-1) th order global trackability,*
- *the (ν-1) th order global elementwise trackability,*
- *the (ν-1) th order global natural trackability,*
- *the (ν-1) th order global elementwise natural trackability*

on $\mathfrak{D}^{\mu_{Pd}} \mathrm{x} \mathfrak{Y}_d^{\nu}$ of IO plant (2.15).

We defined G_{IOPUo} and $G_{IOPU\infty}$ by (9.21) in Subsection 9.3.1:

$$\det A_{P0} \neq 0 \Longrightarrow G_{IOPUo} = G_{IOPU}(0) = A_{P0}^{-1} C_{Pu0} \in \mathfrak{C}^{N\mathrm{x}r},$$
$$\det A_{P\nu} \neq 0 \Longrightarrow G_{IOPU\infty} = \lim_{s \longrightarrow \infty} \left[s^{\nu - \mu_{Pu}} G_{IOPU}(s) \right] =$$
$$= A_{P\nu}^{-1} C_{Pu\mu_{Pu}} \in \mathfrak{C}^{N\mathrm{x}r}. \tag{9.50}$$

Conclusion 297 ***The trackability condition in terms of the IO object full transfer function matrix***

Theorem 274 (Subsection 9.3.1) shows how the full rank of the extreme matrix values G_{IOPUo} and $G_{IOPU\infty}$ of the plant transfer function matrix $G_{IOPU}(s)$ relative to control guarantee the plant perfect and imperfect (natural) trackability. Instead of using the plant matrices it is sufficient to know one of the extreme values of the plant transfer function matrix $G_{IOPU}(s)$ relative to control. The second condition in the above theorems can be expressed also in either of the following equivalent forms:

- *$rankG_{IOPUo} = N$ if $\det A_{Po} \neq 0$,*

- $rankG_{IOPU\infty} = N$ *if* $\det A_{P\nu} \neq 0$.

This conclusion holds also for an unstable IO plant (2.15) for which

$$\det A_{Po} \neq 0.$$

If $A_{P\nu}$ *is singular, then the condition*

$$rankG_{IOPU\infty} = N$$

loses the sense.

9.4.2 *ISO* plant imperfect (natural) trackability

Conditions for trackability

The necessary and sufficient conditions for various trackability properties of the *ISO* plant (2.33), (2.34) (Subsection 2.2.2) represent the topic of what follows. We accepted Claim 155 (Section 8.2) to be valid, i.e., that the plant is stable.

The definition of the matrix G_{ISOPUo} is in (9.29) (Subsection 9.3.2).

Theorem 298 ***Condition for global (elementwise) trackability on*** $\mathfrak{D}^1 x \mathfrak{Y}^1_d$
In order for the ISO plant (2.33), (2.34) to be global (elementwise) trackable on $\mathfrak{D}^1 x \mathfrak{Y}^1_d$ *it is necessary and sufficient that the plant transfer function matrix* $G_{ISOPU}(s)$ *relative to control has the full row rank N, so that the dimension r of the control vector* $\mathbf{U}$ *is not less than the dimension N of the output vector* $\mathbf{Y}$*, i.e.,*

$$\begin{aligned} rankG_{ISOUP}(s) &= rank\left[C_P(sI_n - A_P)^{-1}B_P + H_P\right] = \\ &= rankG_{ISOPUo} = N \leq r. \end{aligned} \tag{9.51}$$

Proof. *Necessity.* Let the *ISO* plant (2.33), (2.34) be (elementwise) trackable on $\mathfrak{D}^1 x \mathfrak{Y}^1_d$. It is also perfect trackable on $\mathfrak{D}^1 x \mathfrak{Y}^1_d$ due to Definition 253, Theorem 254 and Definition 263 (Section 9.2). The condition (9.51) is necessary in view of Theorem 285 (Subsection 9.3.2).

Sufficiency. Let the condition (9.51) be valid. Let $[\mathbf{D}(.), \mathbf{Y}_d(.)] \in \mathfrak{D}^1 x \mathfrak{Y}^1_d$, $\boldsymbol{\sigma} \in (Int\ \mathfrak{T}_0)^N$, $\boldsymbol{\sigma} = \sigma \mathbf{1}_N$, and $\mathbf{X}_0 \in \mathfrak{R}^n$ be arbitrarily chosen. Let the control be defined by

$$\mathbf{U}(s) = \Upsilon(s)\left[\mathbf{Y}(s) - G_{ISOPD}(s)\mathbf{D}(s) - G_{ISOPx_0}(s)\mathbf{X}_0 - \mathbf{Z}(s) + \mathbf{E}(s)\right],$$
$$\Upsilon(s) = G_{ISOPU}(s)^T\left\{G_{ISOPU}(s)G_{ISOPU}(s)^T\right\}^{-1},$$

for

$$E(0) = diag\{\varepsilon_1(0)\ \ \varepsilon_2(0)\ ...\ \varepsilon_N(0)\},$$
$$\Sigma = diag\{\sigma_1\ \ \sigma_2\ ...\ \sigma_N\},$$

$$\mathbf{Z}(t) = \left\{\begin{array}{c} E(0)\left(\mathbf{1}_N - \Sigma^{-1}t\mathbf{1}_N\right),\ t\mathbf{1}_N \in [0, \boldsymbol{\sigma}[, \\ \mathbf{0}_N,\ \forall (t\mathbf{1}_N \geq \sigma) \in \mathfrak{T}_0^N \end{array}\right\} \in \mathfrak{C}(\mathfrak{T}_0). \tag{9.52}$$

The equations (7.85)-(7.87) (Subsection 7.4.1) together with the above definition of $\mathbf{U}(s)$ yield

$$\mathbf{Y}(s) = G_{ISOPU}(s)\mathbf{U}(s) + G_{ISOPD}(s)\mathbf{D}(s) + G_{ISOPx_0}(s)\mathbf{X}_0 =$$
$$= G_{ISOPU}(s)G_{ISOPU}(s)^T\left\{G_{ISOPU}(s)G_{ISOPU}(s)^T\right\}^{-1}\bullet$$
$$\bullet\left\{\begin{array}{c}\mathbf{Y}(s) - G_{ISOPD}(s)\mathbf{D}(s) - G_{ISOPx_0}(s)\mathbf{X}_0 - \\ -\mathbf{Z}(s) + \mathbf{E}(s)\end{array}\right\}+$$
$$+G_{ISOPD}(s)\mathbf{D}(s) + G_{ISOPx_0}(s)\mathbf{X}_0 \Longrightarrow$$
$$\mathbf{Y}(s) = \mathbf{Y}(s) - \mathbf{Z}(s) + \mathbf{E}(s) \Longrightarrow \mathbf{E}(s) = \mathbf{Z}(s).$$

The last equation and (9.52) imply

$$\varepsilon(t) = \mathbf{Z}(t) = \mathbf{0}_N,\ \forall (t\mathbf{1}_N \geq \sigma) \in \mathfrak{T}_0^N.$$

This proves global elementwise trackability on $\mathfrak{D}^1\text{x}\mathfrak{Y}_d^1$ of the ISO plant (2.33), (2.34), which implies its global trackability on $\mathfrak{D}^1\text{x}\mathfrak{Y}_d^1$ due to Definition 253, Theorem 254 and Definition 263 ■

Note 299 *We can use Note 284 (Subsection 9.3.2) in this framework.*

More sophisticated property is the first order trackability (Definition 253).

We defined $G_{ISOPU}(\infty)$ by (9.30) and $G_{ISOPU\infty}$ by (9.31) in Subsection 9.3.2.

Theorem 300 ***Conditions for global first order (elementwise) trackability***

In order for the ISO plant (2.33), (2.34) to be global first order (elementwise) trackable on $\mathfrak{D}^1\text{x}\mathfrak{Y}_d^1$ it is necessary and sufficient that the plant transfer function matrix $G_{ISOPU}(s)$ relative to control has the full row rank N, so that the dimension r of the control vector $\mathbf{U}$ is not less than the dimension N of the output vector $\mathbf{Y}$, i.e., that (9.51) holds and that

$$rank\left[G_{ISOPU\infty} + sG_{ISOPU}(\infty)\right] = N \leq r. \tag{9.53}$$

Proof. *Necessity.* Let the ISO plant (2.33), (2.34) be global first order (elementwise) trackable on $\mathfrak{D}^1\text{x}\mathfrak{Y}_d^1$. It is global (elementwise) trackable on $\mathfrak{D}^1\text{x}\mathfrak{Y}_d^1$. The condition (9.51) holds due to Theorem 298. We differentiate (2.34),

$$\mathbf{Y}^{(1)}(t) = \mathbf{Y}_P^{(1)}(t) = C_P\mathbf{X}_P^{(1)}(t) + H_P\mathbf{U}^{(1)}(t) + D_P\mathbf{D}^{(1)}(t)$$

which, with (2.33), becomes

$$\mathbf{Y}^{(1)}(t) = C_P\left[A_P\mathbf{X}_P(t) + B_P\mathbf{U}(t) + L_P\mathbf{D}(t)\right] +$$
$$+H_P\mathbf{U}^{(1)}(t) + D_P\mathbf{D}^{(1)}(t),$$

i.e.,

$$\mathbf{Y}^{(1)}(t) = C_P A_P \mathbf{X}_P(t) + C_P B_P \mathbf{U}(t) + H_P \mathbf{U}^{(1)}(t) + \\ + C_P L_P \mathbf{D}(t) + D_P \mathbf{D}^{(1)}(t).$$

Laplace transform of this equation has the following form

$$\mathfrak{L}\left\{\mathbf{Y}^{(1)}(t)\right\} = C_P A_P \mathbf{X}_P(s) + (C_P B_P + sH_P)\mathbf{U}(s) - H_P \mathbf{U}_0 + \\ + (C_P L_P + D_P s)\mathbf{D}(s) - D_P \mathbf{D}_0.$$

Let $[\mathbf{D}(.), \mathbf{Y}_d(.)] \in \mathfrak{D}^1 \mathrm{x} \mathfrak{Y}_d^1$, and $\boldsymbol{\sigma} \in (Int\ \mathfrak{T}_0)^N$ be arbitrarily chosen. Definition 253 guarantees

$$\mathbf{Y}^{(1)}(t) = \mathbf{Y}_d^{(1)}(t),\ \forall (t\mathbf{1}_N \geq \boldsymbol{\sigma}) \in \mathfrak{T}_0^N. \tag{9.54}$$

Since $\boldsymbol{\sigma} \in (Int\ \mathfrak{T}_0)^N$ can be arbitrarily small elementwise, then for the choice of $\mathbf{X}_0 \in \mathfrak{R}^n$ such that $\mathbf{Y}_0^{(1)} = \mathbf{Y}_{d0}^{(1)}$ we can write due to (9.54):

$$\mathfrak{L}\left\{\mathbf{Y}_d^{(1)}(t)\right\} = C_P A_P \mathbf{X}_P(s) + (C_P B_P + sH_P)\mathbf{U}(s) - H_P \mathbf{U}_0 + \\ + (C_P L_P + D_P s)\mathbf{D}(s) - D_P \mathbf{D}_0.$$

This equation is solvable in $\mathbf{U}(s)$ due to global (elementwise) trackability on $\mathfrak{D}^1 \mathrm{x} \mathfrak{Y}_d^1$ *of* the *ISO* plant (2.33), (2.34). This implies

$$rank\,(C_P B_P + sH_P) = N \leq r,$$

i.e.,

$$rank\,[G_{ISOPU\infty} + sG_{ISOPU}(\infty)] = N \leq r.$$

This is (9.53).

Sufficiency. Let the conditions of the theorem statement hold. The *ISO* plant (2.33), (2.34) is global (elementwise) trackable on $\mathfrak{D}^1 \mathrm{x} \mathfrak{Y}_d^1$ (Theorem 298). Let $\mathbf{X}_0 \in \mathfrak{R}^n$ be arbitrarily chosen so that $\mathbf{Y}_0$ is also arbitrary. Let

$$\mathbf{U}(s) = \Upsilon(s) \left[\begin{array}{c} s\mathbf{Y}(s) - \mathbf{Y}_0 - C_P A_P \mathbf{X}_P(s) + H_P \mathbf{U}_0 - \\ -(C_P L_P + D_P s)\mathbf{D}(s) + D_P \mathbf{D}_0 - \mathbf{Z}(s) + \mathbf{E}(s) \end{array} \right]$$

for $\mathbf{Z}(t)$ given by (9.52) so that

$$\mathfrak{L}\left\{\mathbf{Y}^{(1)}(t)\right\} = C_P A_P \mathbf{X}_P(s) + (C_P B_P + sH_P)\mathbf{U}(s) - H_P \mathbf{U}_0 + \\ + (C_P L_P + D_P s)\mathbf{D}(s) - D_P \mathbf{D}_0$$

becomes

$$\mathfrak{L}\left\{\mathbf{Y}^{(1)}(t)\right\} = C_P A_P \mathbf{X}_P(s) + (C_P B_P + sH_P)\Upsilon \bullet \\ \bullet \left[\begin{array}{c} \mathfrak{L}\{\mathbf{Y}^{(1)}(t)\} - C_P A_P \mathbf{X}_P(s) + H_P \mathbf{U}_0 - \\ -(C_P L_P + D_P s)\mathbf{D}(s) + D_P \mathbf{D}_0 - \mathbf{Z}(s) + \mathbf{E}(s) \end{array} \right] - \\ -H_P \mathbf{U}_0 + (C_P L_P + D_P s)\mathbf{D}(s) - D_P \mathbf{D}_0.$$

This and (9.52) imply

$$\varepsilon(t) = \mathbf{Z}(t) = \mathbf{0}_N, \ \forall (t\mathbf{1}_N \geq \sigma) \in \mathfrak{T}_0^N.$$

The *ISO* plant (2.33), (2.34) is the first order global elementwise trackable on $\mathfrak{D}^1$x$\mathfrak{Y}_d^1$ (Definition 263). Hence, it is the first order elementwise global trackable on $\mathfrak{D}^1$x$\mathfrak{Y}_d^1$ in view of Definition 253 and due to Lemma 267 (Section 9.2) ■

Note 301 *Since*

$$s^{-1}G_{ISOPU}(s) = C_P B_P + sH_P = G_{ISOPU\infty} + sG_{ISOPU}(\infty),$$

then

$$rank\left[G_{ISOPU\infty} + sG_{ISOPU}(\infty)\right] = N \Longrightarrow$$
$$rankG_{ISOPU}(s) = ranks^{-1}G_{ISOPU}(s) = rank\left[C_P B_P + sH_P\right] =$$
$$= rank\left[G_{ISOPU\infty} + sG_{ISOPU}(\infty)\right] = N.$$

Conditions for natural trackability

Theorem 302 ***Condition for global (elementwise) natural trackability on*** $\mathfrak{D}^1$x$\mathfrak{Y}_d^1$

In order for the ISO plant (2.33), (2.34) to be global (elementwise) natural trackable on $\mathfrak{D}^1$x$\mathfrak{Y}_d^1$ *it is necessary and sufficient that the plant transfer function matrix* $G_{ISOPU}(s)$ *relative to control has the full row rank* N*, equations in (9.51), so that the dimension* r *of the control vector* $\mathbf{U}$ *is not less than the dimension* N *of the output vector* $\mathbf{Y}$*, i.e., that the inequality in (9.51) holds.*

Proof. *Necessity.* Let the *ISO* plant (2.33), (2.34) be (elementwise) natural trackable on $\mathfrak{D}^1$x$\mathfrak{Y}_d^1$. It is also perfect natural trackable on $\mathfrak{D}^1$x$\mathfrak{Y}_d^1$ due to Theorem 261 (Section 9.2). The condition (9.51) is necessary in view of Theorem 287, Subsection 9.3.2.

Sufficiency. Let the condition (9.51) be valid. Let $[\mathbf{D}(.), \mathbf{Y}_d(.)] \in \mathfrak{D}^1x\mathfrak{Y}_d^1$, $\varepsilon \in Int\, \mathfrak{T}_0$, $\varepsilon \longrightarrow 0^+$, and $\mathbf{X}_0 \in \mathfrak{R}^n$ be arbitrarily chosen. Let the control be defined by

$$\mathbf{U}(s) = e^{-\varepsilon s}\mathbf{U}(s)_{\varepsilon \longrightarrow 0^+} + \Upsilon(s)\left[-\mathbf{Z}(s) + \mathbf{E}(s)\right],$$

with $\mathbf{Z}(s)$ being Laplace transform of $\mathbf{Z}(t)$ defined by (9.52). This control is natural control. It is time-continuous and independent of the plant mathematical model and of the disturbance vector. The equations (7.85)-(7.87) together with the above definition of $\mathbf{U}(s)$ yield

$$\mathbf{Y}(s) = G_{ISOPU}(s)\mathbf{U}(s) +$$
$$+ G_{ISOPD}(s)\mathbf{D}(s) + G_{ISOPx_0}(s)\mathbf{X}_0 = e^{-\varepsilon s}G_{ISOPU}(s)\mathbf{U}(s)_{\varepsilon \longrightarrow 0^+} +$$
$$+ G_{ISOPU}(s)G_{ISOPU}(s)^T\left\{G_{ISOPU}(s)G_{ISOPU}(s)^T\right\}^{-1} \bullet$$
$$\bullet\left[-\mathbf{Z}(s) + \mathbf{E}(s)\right] +$$
$$+ G_{ISOPD}(s)\mathbf{D}^{\mp}(s) + G_{ISOPx_0}(s)\mathbf{X}_{0\mp} \Longrightarrow$$

$$\mathbf{Y}(s) = e^{-\varepsilon s} G_{ISOPU}(s)\mathbf{U}(s)_{\varepsilon \longrightarrow 0^+} - \mathbf{Z}(s) + \mathbf{E}(s) + \\ + G_{ISOPD}(s)\mathbf{D}(s) + G_{ISOPx_0}(s)\mathbf{X}_0. \tag{9.55}$$

The equations (7.85)-(7.87) imply

$$G_{ISOPU}(s)\mathbf{U}(s) = \mathbf{Y}(s) - G_{ISOPD}(s)\mathbf{D}(s) - G_{ISOPx_0}(s)\mathbf{X}_0$$

and transform (9.55) into

$$\mathbf{Y}(s) = e^{-\varepsilon s}\left[\mathbf{Y}(s) - G_{ISOPD}(s)\mathbf{D}(s) - G_{ISOPx_0}(s)\mathbf{X}_0\right]_{\varepsilon \longrightarrow 0^+} - \\ - \mathbf{Z}(s) + \mathbf{E}(s) + G_{ISOPD}(s)\mathbf{D}(s) + G_{ISOPx_0}(s)\mathbf{X}_0,$$

i.e.,

$$\left(1 - e^{-\varepsilon s}\right)_{\varepsilon \longrightarrow 0^+}\left[\mathbf{Y}(s) - G_{ISOPD}(s)\mathbf{D}(s) - G_{ISOPx_0}(s)\mathbf{X}_0\right] = \\ = -\mathbf{Z}(s) + \mathbf{E}(s).$$

In the limit as $\varepsilon \longrightarrow 0^+$

$$\mathbf{E}(s) = \mathbf{Z}(s),$$

i.e.,

$$\boldsymbol{\varepsilon}(t) = \mathbf{Z}(t) = \mathbf{0}_N, \ \forall (t\mathbf{1}_N \geq \sigma) \in \mathfrak{T}_0^N,$$

due to (9.52). This proves global elementwise natural trackability on $\mathfrak{D}^1 \text{x} \mathfrak{Y}_d^1$ of the *ISO* plant (2.33), (2.34) due to Definition 268 and Lemma 270 (Section 9.2) ■

Conclusion 303 *The full rank of the matrix values* G_{ISOPUo}*,* $G_{ISOPU}(\infty)$ *and* $G_{ISOPU\infty}$ *of the plant transfer function matrix* $G_{ISOPU}(s)$ *relative to control guarantees the plant perfect and imperfect (natural) trackability.*

Part IV

NOVEL TRACKING CONTROL SYNTHESIS

Chapter 10

Linear tracking control (LITC)

10.1 Generating theorem

The book [148] contains what follows.

A complex valued matrix function $F(.) : \mathfrak{C} \rightarrow \mathfrak{C}^{pxn}$ is real rational matrix function if, and only if, it becomes a real valued matrix for the real value of the complex variable s, i.e., for $s = \sigma \in \mathfrak{R}$, and every entry of $F(s)$ is a quotient of two polynomials in s.

Let $F(s)$ have μ different poles denoted by s_k^*, $k = 1, 2, ..., \mu$. The multiplicity of the pole s_k^* is designated by ν_k. We denote its real and imaginary part by $\mathrm{Re}\, s_k^*$ and $\mathrm{Im}\, s_k^*$, respectively.

Theorem 304 *Generating theorem*

Let $F(.) : \mathfrak{R} \rightarrow \mathfrak{R}^{pxn}$, $F(t) = [F_{ij}(t)]$, *have Laplace transform* $F(.) : \mathfrak{C} \rightarrow \mathfrak{C}^{pxn}$, $F(s) = [F_{ij}(s)]$, *that is real rational matrix function. In order for the norm* $\|F(t)\|$ *of the original* $F(t)$*:*

- *a)* *to be* ***bounded****, i.e.,:*

$$\exists \alpha \in \mathfrak{R}^+ \Longrightarrow \|F(t)\| < \alpha, \ \forall t \in \mathfrak{T}_0,$$

it is necessary and sufficient that:

1. the real parts of all poles of $F(s)$ *are nonnegative, Fig. 10.1,*

$$\mathrm{Re}\, s_i^* \leq 0, \ \forall i = 1, 2, ..., \ \mu;$$

2. all imaginary poles of $F(s)$ *are simple (i.e., with the multiplicity* ν_i *that is equal to one),*

$$\mathrm{Re}\, s_i^* = 0, \nu_i = 1;$$

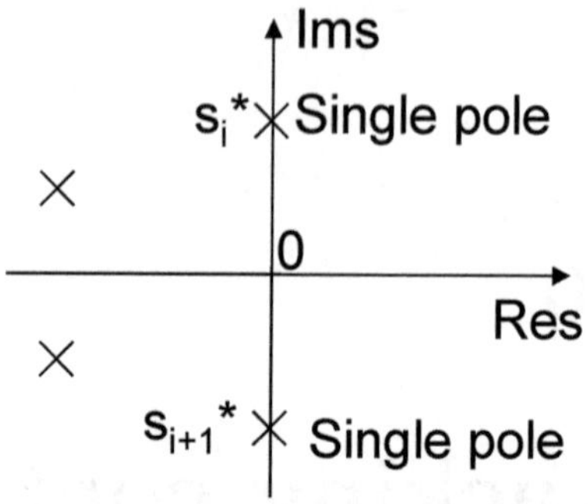

Figure 10.1: Poles with zero or negative real parts.

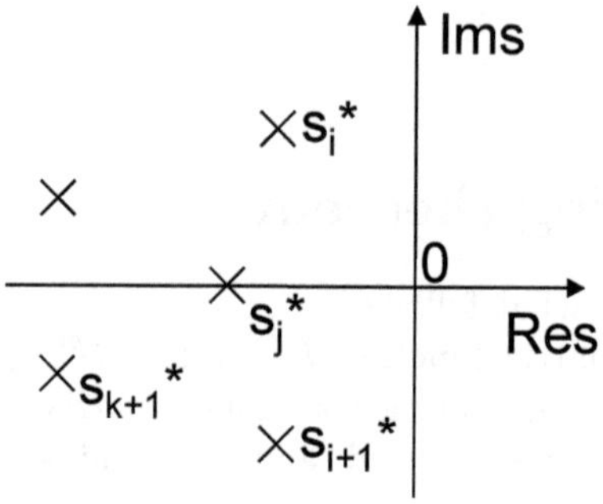

Figure 10.2: Poles with negative real parts

3. $F(s)$ is the zero matrix in the infinity, i.e., it is strictly proper,

$$F(\infty) = O_{pn} \in \mathfrak{R}^{pxn};$$

- *b) and in order for $\|F(t)\|$ to **vanish asymptotically**, i.e., in order for the following condition to hold:*

$$\lim[\|F(t)\| : t \longrightarrow \infty] = 0,$$

it is necessary and sufficient that:

1. the real parts of all poles of $F(s)$ are negative, Fig. 10.2,

$$\mathrm{Re}\, s_i^* < 0, \; \forall i = 1, 2, ..., \; \mu;$$

and

2. $F(s)$ is the zero matrix in the infinity, i.e., it is strictly proper,

$$F(\infty) = O_{pn} \in \mathfrak{R}^{pxn}.$$

Proof. Let Laplace transform $F(s)$ of $F(t)$ have μ different poles denoted by s_k^* with the multiplicity ν_k^*, $k = 1, 2, \ldots, \mu$. We know (from Heaviside expansion of $F(s)$) that the original $F(t)$ and its Laplace transform $F(s)$ are interrelated by the following formulae:

- in the matrix form

$$F(t) = \mathcal{L}^{-1}\{F(s)\} = \delta^-(t)R_0 + \sum_{k=1}^{k=\mu} e^{s_k^* t} \left[\sum_{r=1}^{r=\nu_k^*} \frac{1}{(r-1)!} t^{r-1} R_{kr} \right],$$

$$R_{kr} \in \mathfrak{R}^{p\times n}, \tag{10.1}$$

$$F(s) = L^-\{F(t)\} =$$

$$\mathcal{L}^- \left\{ \delta^-(t)R_0 + \sum_{k=1}^{k=\mu} e^{s_k^* t} \left[\sum_{r=1}^{r=\nu_k^*} \frac{1}{(r-1)!} t^{r-1} R_{kr} \right] \right\}, \tag{10.2}$$

- in the scalar form, where s_k^{im*} is one of poles of the μ_{im}-*th* entry $F_{im}(s)$ of $F(s)$, $k \in \{1, 2, \ldots, \mu\}$, the multiplicity of which is denoted by ν_k^{im*}:

$$F_{im}(t) = \mathcal{L}^{-1}\{F_{im}(s)\} = \delta^-(t)R_0^{im} +$$

$$+ \sum_{k=1}^{k=\mu_{im}} e^{s_k^{im*} t} \left[\sum_{r=1}^{r=\nu_k^{im*}} \frac{1}{(r-1)!} t^{r-1} R_{kr}^{im} \right], \quad R_{kr}^{im} \in \mathfrak{R}, \tag{10.3}$$

$$F_{im}(s) = \mathcal{L}^-\{F_{im}(t)\} =$$

$$= \mathcal{L}^- \left\{ \delta^-(t)R_0^{im} + \sum_{k=1}^{k=\mu_{im}} e^{s_k^{im*} t} \left[\sum_{r=1}^{r=\nu_k^{im*}} \frac{1}{(r-1)!} t^{r-1} R_{kr}^{im} \right] \right\}. \tag{10.4}$$

Necessity. a) Let $F(t)$ be bounded, i.e.,

$$\exists \alpha \in \mathfrak{R}^+ \Longrightarrow \|F(t)\| < \alpha, \ \forall t \in \mathfrak{T}_0. \tag{10.5}$$

We will apply the method of contradiction to complete the proof of the necessity. Let us assume that condition a-1) does not hold, i.e.,:

$$\exists s_k^{im*} = \sigma + j\omega \in \mathfrak{C} \Longrightarrow \operatorname{Re} s_k^{im*} = \sigma \in \mathfrak{R}^+. \tag{10.6}$$

This, $\delta(t)R_0^{im} = O$ for $t \neq 0$, and (10.3) imply

$$\left| e^{s_k^{im*} t} \left[\sum_{r=1}^{r=\nu_k^{im*}} \frac{1}{(r-1)!} t^{r-1} R_{kr}^{im} \right] \right| = e^{\sigma t} \left| e^{j\omega t} \right| \left| \left[\sum_{r=1}^{r=\nu_k^{im*}} \frac{1}{(r-1)!} t^{r-1} R_{kr}^{im} \right] \right| =$$

$$= e^{\sigma t} \left| \left[\sum_{r=1}^{r=\nu_k^{im*}} \frac{1}{(r-1)!} t^{r-1} R_{kr}^{im} \right] \right| \longrightarrow \infty \ as \ t \to \infty,$$

$$\Longrightarrow$$

$$\lim_{t\to\infty} |F_{im}(t)| = \lim_{t\to\infty} \left| \delta^-(t)R_0^{im} + \sum_{k=1}^{k=\mu_{im}} e^{s_k^{im*}t} \left[\sum_{r=1}^{r=\nu_k^{im*}} \frac{1}{(r-1)!} t^{r-1} R_{kr}^{im} \right] \right| = \infty ,$$

$$\Longrightarrow$$

$$\lim_{t\to\infty} \|F(t)\| = \infty .$$

It follows that $F(t)$ is not bounded, which contradicts (10.5). The contradiction is a consequence of (10.6) implying that (10.6) is incorrect. This proves necessity of a-1).

We continue with the method of contradiction. Let us suppose that the condition a-2) does not hold, i.e.,:

$$\exists s_k^{im*} = \sigma + j\omega \in \mathfrak{C} \Longrightarrow \operatorname{Re} s_k^{im*} = \sigma = 0 \textit{ and } \nu_k^{im*} \geq 2. \qquad (10.7)$$

Now, $\delta(t)R_0^{im} = O$ for $t \neq 0$, (10.3) and (10.7) imply

$$\lim_{t\to\infty} \left| e^{s_k^{im*}t} \left[\sum_{r=1}^{r=\nu_k^{im*}\geq 2} \frac{1}{(r-1)!} t^{r-1} R_{kr}^{im} \right] \right| =$$

$$= \lim_{t\to\infty} \left| e^{j\omega t} \right| \left\| \left[\sum_{r=1}^{r=\nu_k^{im*}\geq 2} \frac{1}{(r-1)!} t^{r-1} R_{kr}^{im} \right] \right\| =$$

$$= \lim_{t\to\infty} \left\| \left[\sum_{r=1}^{r=\nu_k^{im*}\geq 2} \frac{1}{(r-1)!} t^{r-1} R_{kr}^{im} \right] \right\| = \infty ,$$

$$\Longrightarrow$$

$$\lim_{t\to\infty} |F_{im}(t)| = \lim_{t\to\infty} \left| \delta^-(t)R_0^{im} + \sum_{k=1}^{k=\mu_{im}} e^{s_k^{im*}t} \left[\sum_{r=1}^{r=\nu_k^{im*}} \frac{1}{(r-1)!} t^{r-1} R_{kr}^{im} \right] \right| = \infty$$

$$\Longrightarrow$$

$$\lim_{t\to\infty} \|F(t)\| = \infty.$$

It follows that $F(t)$ is unbounded, which contradicts (10.5). The contradiction is a consequence of (10.7), which implies that (10.7) is not correct. This proves necessity of the condition a-2).

We continue further with the contradiction method. We assume that a-3) is not valid, i.e.,:

$$\lim_{s\to\infty} F(s) \neq O \Longrightarrow \exists i \in \{1,2,...,p\},\ \exists m \in \{1,2,...,n\} \Longrightarrow R_0^{im} \neq 0. \qquad (10.8)$$

Hence, (10.3) and (10.8) yield for $t = 0$:

$$|F_{im}(t)|_{t=0} = \left| \delta^-(t) R_0^{im} + \sum_{k=1}^{k=\mu_{ij}} e^{s_k^{im*}t} \left[\sum_{r=1}^{r=\nu_k^{im*}} \frac{1}{(r-1)!} t^{r-1} R_{kr}^{im} \right] \right|_{t=0} =$$

$$= \left| \delta^-(0) R_0^{im} + \sum_{k=1}^{k=\mu_{im}} R_{k1}^{im} \right|_{t=0} \in [0, \infty]$$

$$\Longrightarrow$$

$$\|F(t)\|_{t=0} \in [0, \infty].$$

This means that $F(t)$ is not bounded, which contradicts (10.5). The contradiction is a consequence of (10.8), which shows that (10.8) is incorrect. Hence,

$$R_0^{im} = 0, \ \forall i \in \{1, 2, ..., p\}, \ \forall m \in \{1, 2, ..., n\},$$

implying $R_0 = O_{pn}$ and $F(t)_{t=0} = O_{pn}$, or equivalently, $\lim_{s \to \infty} F(s) = O_{pn}$. This proves necessity of the condition a-3).

b) We keep on using the contradiction method. Let

$$\lim[\|F(t)\| : \ t \to \infty] = 0. \tag{10.9}$$

be true and let us suppose that the condition b-1) does not hold. If (10.6) were valid, then $F(t)$ would be unbounded as shown above in the proof of necessity of a-1), which would contradict (10.9). If (10.7) were valid, then $F(t)$ would be unbounded as shown above in the proof of necessity of a-2), which would again contradict (10.9). If

$$\exists s_k^{im*} = \sigma + j\omega \in \mathfrak{C} \Longrightarrow \operatorname{Re} s_k^{im*} = \sigma = 0 \ and \ \nu_k^{im*} = 1, \tag{10.10}$$

then

$$\lim_{t \to \infty} \left| e^{s_k^{im*}t} \left[\sum_{r=1}^{r=\nu_k^{im*}=1} \frac{1}{(r-1)!} t^{r-1} R_{kr}^{im} \right] \right| = \lim_{t \to \infty} \left| e^{j\omega t} R_{kr}^{im} \right| \in \mathfrak{R}^+ \Longrightarrow$$

$$\exists \xi \in \mathfrak{R}^+ \ such \ that \ \lim_{t \to \infty} |F_{im}(t)| =$$

$$= \lim_{t \to \infty} \left| \delta^-(t) R_0^{im} + \sum_{k=1}^{k=\mu_{im}} e^{s_k^{im*}t} \left[\sum_{r=1}^{r=\nu_k^{im*}=1} \frac{1}{(r-1)!} t^{r-1} R_{kr}^{im} \right] \right| = \xi \in \mathfrak{R}^+ ,$$

which would also contradict (10.9). Altogether, the validity of (10.9) proves the validity of the condition b-1). Necessity of the condition b-2) is proved under a-3).

Sufficiency. a) Let the conditions under a) hold. Then $F(s)$ is the zero matrix in the infinity,

$$F(\infty) = O_{pn} \in \mathfrak{R}^{pxn}.$$

Hence,

$$R_0^{im} = 0, \forall i \in \{1, 2, ..., p\}, \forall m \in \{1, 2, ..., n\}. \tag{10.11}$$

We recall the following facts:

1) If (10.10) holds, then

$$\lim_{t\to\infty} |F_{im}(t)| \in \mathfrak{R}^+. \tag{10.12}$$

2) If $\operatorname{Re} s_k^{im*} = \sigma_{im} < 0$, then

$$\lim_{t\to\infty} |F_{im}(t)| = \lim_{t\to\infty} \left\| \sum_{k=1}^{k=\mu_{im}} e^{s_k^{im*}t} \left[\sum_{r=1}^{r=\nu_k^{im*}=1} \frac{1}{(r-1)!} t^{r-1} R_{kr}^{im} \right] \right\| =$$

$$= \lim_{t\to\infty} \left| e^{-|\sigma_{im}|t} e^{j\omega t} \left[\sum_{r=1}^{r=\nu_k^{im*}} \frac{1}{(r-1)!} t^{r-1} R_{kr}^{im} \right] \right\| =$$

$$= \lim_{t\to\infty} \left| e^{-|\sigma_{im}|t} \left[\sum_{r=1}^{r=\nu_k^{im*}} \frac{1}{(r-1)!} t^{r-1} R_{kr}^{im} \right] \right\| = 0. \tag{10.13}$$

3) The results (10.11) through (10.13) prove boundedness of $\|F(t)\|$, i.e.,:

$$\exists \alpha \in \mathfrak{R}^+ \Longrightarrow \|F(t)\| < \alpha, \ \forall t \in \mathfrak{T}_0.$$

b) Let the conditions under b) hold. Now, $\operatorname{Re} s_k^{im*} = \sigma < 0$, $\forall i \in \{1, 2, ..., p\}$, $\forall m \in \{1, 2, ..., n\}$, so that (10.13) holds $\forall i \in \{1, 2, ..., p\}$, $\forall m \in \{1, 2, ..., n\}$, which proves that $\|F(t)\|$ vanishes asymptotically, i.e.,

$$\lim[\|F(t)\| : t \to \infty] = 0.$$

This completes the proof ■

Comment 305 ***Importance of the Generating theorem***

Qualitative stability properties (e.g., controllability, observability, optimality, stability, trackability) concern families of dynamic behaviors of a dynamic system, which are induced by sets of initial conditions and/or by sets of external actions. They take place in time. Their definitions are given in the time domain. It is impractical in the framework of linear systems (practically impossible in the framework of nonlinear systems) to use their definitions directly in order to test whether a given system possesses a requested qualitative dynamic property. It is preferable to establish conditions and criteria for them in the algebraic and/or in the complex domain, which enables us to test them without knowing individual system behaviors, i.e., without solving system mathematical model for every initial condition and for every external action. The Generating Theorem 304, which is in various forms well known, is the basis to establish such conditions and criteria in the complex domain for stability and tracking system properties.

10.2 LITC of the *IO* plants

We consider the *IO* plant P described by (2.15) (Subsection 2.1.2) controlled by the *IO* feedback controller CR (2.22) (Subsection 2.1.3). The full transfer function matrix of their control system relative to the output error vector $\varepsilon(s)$ (3.6) (Subsection 3.3.1) reads (7.71), (7.73)-(7.76), (7.78)-(7.80), i.e., (7.81) (Section 7.3):

$$F_{IOCS\varepsilon}(s) =$$

$$= \left[G_{IOCSd}(s) \vdots G_{IOCS\varepsilon yd}(s) \vdots G_{IOCSdo}(s) \vdots G_{IOCSuo}(s) \vdots G_{IOCS\varepsilon o}(s) \vdots G_{IOCSyo}(s)\right], \tag{10.14}$$

where

$$G_{IOCSd}(s) =$$

$$= \left(A_P^{(\nu)} S_N^{(\nu)}(s) + C_{Pu}^{(\mu_{Pu})} S_r^{(\mu_{Pu})}(s) \left(A_{CR}^{(\nu_C)} S_r^{(\nu_C)}(s)\right)^{-1} P_{CR}^{(\mu_{Cy})} S_N^{(\mu_{Cy})}(s)\right)^{-1} \bullet$$

$$\bullet D_{Pd}^{(\mu_{Pd})} S_d^{(\mu_{Pd})}(s) \tag{10.15}$$

$$G_{IOCS\varepsilon yd}(s) =$$

$$\left\langle A_P^{(\nu)} S_N^{(\nu)}(s) + C_{Pu}^{(\mu_{Pu})} S_r^{(\mu_{Pu})}(s) \left(A_{CR}^{(\nu_C)} S_r^{(\nu_C)}(s)\right)^{-1} P_{CR}^{(\mu_{Cy})} S_N^{(\mu_{Cy})}(s)\right\rangle^{-1} \bullet$$

$$\bullet A_P^{(\nu)} S_N^{(\nu)}(s), \tag{10.16}$$

$$G_{IOCSdo}(s) =$$

$$= -\left(A_P^{(\nu)} S_N^{(\nu)}(s) + C_{Pu}^{(\mu_{Pu})} S_r^{(\mu_{Pu})}(s) \left(A_{CR}^{(\nu_C)} S_r^{(\nu_C)}(s)\right)^{-1} P_{CR}^{(\mu_{Cy})} S_N^{(\mu_{Cy})}(s)\right)^{-1} \bullet$$

$$\bullet D_{Pd}^{(\mu_{Pd})} Z_d^{(\mu_{Pd}-1)}(s), \tag{10.17}$$

$$G_{IOCSuo}(s) =$$

$$= \left(A_P^{(\nu)} S_N^{(\nu)}(s) + C_{Pu}^{(\mu_{Pu})} S_r^{(\mu_{Pu})}(s) \left(A_{CR}^{(\nu_C)} S_r^{(\nu_C)}(s)\right)^{-1} P_{CR}^{(\mu_{Cy})} S_N^{(\mu_{Cy})}(s)\right)^{-1} \bullet$$

$$\bullet \left\{ \begin{array}{c} C_{Pu}^{(\mu_{Pu})} S_r^{(\mu_{Pu})}(s) \left(A_{CR}^{(\nu_C)} S_r^{(\nu_C)}(s)\right)^{-1} A_{CR}^{(\nu_C)} Z_r^{(\nu_C-1)}(s) - \\ - \left[C_{Pu}^{(\mu_{Pu})} Z_r^{(\mu_{Pu}-1)}(s) \vdots O_{N,\nu_C-\mu_{Pu}}\right] \end{array} \right\}, \tag{10.18}$$

$$G_{IOCS\varepsilon o}(s) =$$

$$= -\left(A_P^{(\nu)} S_N^{(\nu)}(s) + C_{Pu}^{(\mu_{Pu})} S_r^{(\mu_{Pu})}(s) \left(A_{CR}^{(\nu_C)} S_r^{(\nu_C)}(s)\right)^{-1} P_{CR}^{(\mu_{Cy})} S_N^{(\mu_{Cy})}(s)\right)^{-1} \bullet$$

$$\bullet C_{Pu}^{(\mu_{Pu})} S_r^{(\mu_{Pu})}(s) \left(A_{CR}^{(\nu_C)} S_r^{(\nu_C)}(s)\right)^{-1} P_{CR}^{(\mu_{Cy})} Z_N^{(\mu_{Cy}-1)}(s), \tag{10.19}$$

$$G_{IOCSyo}(s) =$$
$$= \left(A_P^{(\nu)} S_N^{(\nu)}(s) + C_{Pu}^{(\mu_{Pu})} S_r^{(\mu_{Pu})}(s) \left(A_{CR}^{(\nu_C)} S_r^{(\nu_C)}(s)\right)^{-1} P_{CR}^{(\mu_{Cy})} S_N^{(\mu_{Cy})}(s)\right)^{-1} \bullet$$
$$\bullet A_{CR}^{(\nu_C)} Z_N^{(\nu-1)}(s). \tag{10.20}$$

Altogether,

$$F_{IOCS\varepsilon}(s) =$$
$$= \left(A_P^{(\nu)} S_N^{(\nu)}(s) + C_{Pu}^{(\mu_{Pu})} S_r^{(\mu_{Pu})}(s) \left(A_{CR}^{(\nu_C)} S_r^{(\nu_C)}(s)\right)^{-1} P_{CR}^{(\mu_{Cy})} S_N^{(\mu_{Cy})}(s)\right)^{-1} \bullet$$
$$\bullet \left[\begin{array}{c} D_{Pd}^{(\mu_{Pd})} S_d^{(\mu_{Pd})}(s) \vdots A_P^{(\nu)} S_N^{(\nu)}(s) \vdots D_{Pd}^{(\mu_{Pd})} Z_d^{(\mu_{Pd}-1)}(s) \vdots \\ \vdots \left\{ \begin{array}{c} C_{Pu}^{(\mu_{Pu})} S_r^{(\mu_{Pu})}(s) \left(A_{CR}^{(\nu_C)} S_r^{(\nu_C)}(s)\right)^{-1} A_{CR}^{(\nu_C)} Z_r^{(\nu_C-1)}(s) - \\ - \left[C_{Pu}^{(\mu_{Pu})} Z_r^{(\mu_{Pu}-1)}(s) \vdots O_{N,\nu-\mu_{Pu}}\right] \end{array} \right\} \vdots \\ \vdots C_{Pu}^{(\mu_{Pu})} S_r^{(\mu_{Pu})}(s) \left(A_{CR}^{(\nu_C)} S_r^{(\nu_C)}(s)\right)^{-1} P_{CR}^{(\mu_{Cy})} Z_N^{(\mu_{Cy}-1)}(s) \vdots \\ \vdots A_{CR}^{(\nu_C)} Z_N^{(\nu-1)}(s), \end{array}\right],$$

or equivalently

$$F_{IOCS\varepsilon}(s) = \left(A_P^{(\nu)} S_N^{(\nu)}(s)\right)$$
$$= \left(\begin{array}{c} I_N + \left(A_P^{(\nu)} S_N^{(\nu)}(s)\right)^{-1} C_{Pu}^{(\mu_{Pu})} S_r^{(\mu_{Pu})}(s) \left(A_{CR}^{(\nu_C)} S_r^{(\nu_C)}(s)\right)^{-1} \bullet \\ \bullet P_{CR}^{(\mu_{Cy})} S_N^{(\mu_{Cy})}(s) \end{array} \right)^{-1} \bullet$$
$$\bullet \left[\begin{array}{c} \left[D_{Pd}^{(\mu_{Pd})} S_d^{(\mu_{Pd})}(s)\right]^T \\ \left[A_P^{(\nu)} S_N^{(\nu)}(s)\right]^T \\ \left[D_{Pd}^{(\mu_{Pd})} Z_d^{(\mu_{Pd}-1)}(s)\right]^T \\ \left[\begin{array}{c} A_P^{(\nu)} S_N^{(\nu)}(s) \bullet \\ \bullet \left\{ \begin{array}{c} C_{Pu}^{(\mu_{Pu})} S_r^{(\mu_{Pu})}(s) \left(A_{CR}^{(\nu_C)} S_r^{(\nu_C)}(s)\right)^{-1} A_{CR}^{(\nu_C)} Z_r^{(\nu_C-1)}(s) - \\ - \left[C_{Pu}^{(\mu_{Pu})} Z_r^{(\mu_{Pu}-1)}(s) \vdots O_{N,\nu-\mu_{Pu}}\right] \end{array} \right\} \end{array}\right]^T \\ \left[C_{Pu}^{(\mu_{Pu})} S_r^{(\mu_{Pu})}(s) \left(A_{CR}^{(\nu_C)} S_r^{(\nu_C)}(s)\right)^{-1} P_{CR}^{(\mu_{Cy})} Z_N^{(\mu_{Cy}-1)}(s)\right]^T \\ \left[A_{CR}^{(\nu_C)} Z_N^{(\nu-1)}(s)\right]^T \end{array}\right]^T.$$

Laplace transform $\varepsilon(s)$ of the output error vector $\varepsilon(t)$ has now a compact form

(7.77) (Section 7.3):

$$\varepsilon(s)=F_{IOCS\varepsilon}(s)\underbrace{\begin{bmatrix}\mathbf{D}(s)\\ \mathbf{Y}_d(s)\\ \mathbf{D}_0^{\mu_{Pd}-1}\\ \mathbf{U}_0^{\nu-1}\\ \varepsilon_0^{\nu-1}\\ \mathbf{Y}_0^{\nu-1}\end{bmatrix}}_{\mathbf{V}_{IOCS}(s)}=F_{IOCS\varepsilon}(s)\mathbf{V}_{IOCS}(s). \tag{10.21}$$

The following tracking criterion shows the difference between Lyapunov stability criteria and Lyapunov tracking criteria.

Theorem 306 ***Criterion for the IO linear control system tracking***

For the IO linear feedback control system (2.26) composed of the IO plant P (2.15) and of the IO feedback controller CR (2.22) to exhibit tracking over $[\mathfrak{D}\mathrm{x}\mathfrak{Y}_d]\cap\mathfrak{L}$ *it is necessary and sufficient that the real parts of all poles of Laplace transform* $\varepsilon(s)$ *of the output error vector* ε,

$$\varepsilon(s)=F_{IOCS\varepsilon}(s)\mathbf{V}_{IOCS}(s),$$

are negative for every $[\mathbf{D}(.),\mathbf{Y}_d(.)]\in[\mathfrak{D}\mathrm{x}\mathfrak{Y}_d]\cap\mathfrak{L}$.

Then, and only then, the IO linear feedback control system (2.26) composed of the IO plant P (2.15) and of the IO feedback controller CR (2.22) exhibits

- *global stablewise tracking over* $[\mathfrak{D}\mathrm{x}\mathfrak{Y}_d]\cap\mathfrak{L}$, *and*
- *global exponential tracking over* $[\mathfrak{D}\mathrm{x}\mathfrak{Y}_d]\cap\mathfrak{L}$.

Proof. Linearity of the *IO* linear feedback control system (2.26) composed of the *IO* plant P (2.15) and of the *IO* controller CR (2.22), Definitions 180, 182, 184, continuity and boundedness of every $[\mathbf{D}(.),\mathbf{Y}_d(.)]\in[\mathfrak{D}\mathrm{x}\mathfrak{Y}_d]\cap\mathfrak{L}$ by the definition, (2.12) (Subsection 2.1.1), (2.17), (2.19) (Subsection 2.1.2), and the Generating theorem 304 (Section 10.1) imply the statement of the theorem ■

Comment 307 *This theorem illustrates the necessity to use the full transfer function matrix* $F_{IOCS\varepsilon}(s)$ *of the control system with respect to the error* ε *rather than only its transfer function matrix* $G_{IOCSd}(s)$ *or* $G_{IOCS\varepsilon yd}(s)$. *The latter are insufficient for the analysis or synthesis of the control system from the tracking point of view, while the former is both necessary and sufficient.*

The poles of $F_{IOCS\varepsilon}(s)$, of $\mathbf{D}(s)$ and of $\mathbf{Y}_d(s)$ constitute the set of all poles of $\varepsilon(s)$ in view of (10.21). The only information about the disturbance vector function $\mathbf{D}(.)$ is that it belongs to $\mathfrak{D}$ and that the signs of the real parts of all poles of its Laplace transform $\mathbf{D}(s)$ are known. If some poles of $\mathbf{D}(s)$ have nonnegative real parts, then they should be known as well as their multiplicity.

Theorem 308 *For the IO linear feedback control system composed of the IO plant (2.15) and of the IO controller (2.22) to exhibit tracking over* $[\mathfrak{D}_{-}\mathrm{x}\mathfrak{Y}_{d-}]\cap\mathfrak{L}$ *it is necessary and sufficient that the real parts of all poles of the full transfer function matrix* $F_{IOCS\varepsilon}(s)$ *of the control system relative to the output error vector* $\varepsilon(t)$ *are negative. Then, and only then, the IO linear feedback control system exhibits*

- *global stablewise tracking over* $[\mathfrak{D}_{-}\mathrm{x}\mathfrak{Y}_{d-}]\cap\mathfrak{L}$,
- *global exponential tracking over* $[\mathfrak{D}_{-}\mathrm{x}\mathfrak{Y}_{d-}]\cap\mathfrak{L}$.

Proof. Equations (10.21) show that the poles of $F_{IOCS\varepsilon}(s)$, of $\mathbf{D}(s)$ and $\mathbf{Y}_d(.)$ compose the set of all poles of $\varepsilon(s)$. The definitions of $\mathfrak{D}_{-}$ and of $\mathfrak{Y}_{d-}$ determine that the real parts of all poles of $[\mathbf{D}(s),\mathbf{Y}_d(s)]\in\mathfrak{D}_{-}\mathrm{x}\mathfrak{Y}_{d-}$ are negative. This and $[\mathbf{D}(.),\mathbf{Y}_d(.)]\in[\mathfrak{D}_{-}\mathrm{x}\mathfrak{Y}_{d-}]\cap\mathfrak{L}$ reduce the condition of Theorem 306 to the demand that the real parts of all poles of the full transfer function matrix $F_{IOCS\varepsilon}(s)$ of the control system relative to the output error vector $\varepsilon(t)$ are negative ■

The exponential tracking is the best tracking quality that can be achieved with this approach. The controller action is smooth and robust relative to the disturbance action. Although the criterion demands the test of the real parts of all poles of the full transfer function matrix $F_{IOCS\varepsilon}(s)$ of the control system relative to the output error vector $\varepsilon(t)$ to be negative, it usually means that the characteristic polynomial of $F_{IOCS\varepsilon}(s)$ is to be known.

10.3 LITC of the *ISO* plants

We determined in (7.98)-(7.101) (Subsection 7.4.3), Laplace transform $\varepsilon(s)$ of the output error vector ε (3.6) (Subsection 3.3.1), of the *ISO* feedback control system (2.45), (2.46) (Subsection 2.2.4) composed of the *ISO* plant (2.33), (2.34) (Subsection 2.2.2), and *ISO* controller (2.37), (2.38) (Subsection 2.2.3),

$$\varepsilon(s)=F_{ISOCS\varepsilon}(s)\mathbf{V}_{ISOCS\varepsilon}(s)=$$

$$=\underbrace{\left[G_{ISOCS\varepsilon d}(s)\vdots G_{ISOCS\varepsilon yd}\vdots G_{ISOCS\varepsilon xcro}(s)\vdots G_{ISOCS\varepsilon xpo}(s)\right]}_{F_{ISOCS\varepsilon}(s)}\underbrace{\begin{bmatrix}\mathbf{D}(s)\\ \mathbf{Y}_d(s)\\ \mathbf{X}_{CR0}\\ \mathbf{X}_{P0}\end{bmatrix}}_{\mathbf{V}_{ISOCS}(s)} \tag{10.22}$$

$$G_{ISOCS\varepsilon d}(s)=-\left\langle\begin{array}{c} I_N+\left[C_P\left(sI_{n_P}-A_P\right)^{-1}B_P+H_P\right]\bullet\\ \bullet\left[C_{CR}\left(sI_{n_c}-A_{CR}\right)^{-1}B_{CR}+H_{CR}\right]\end{array}\right\rangle^{-1}\bullet$$

$$\bullet\left[C_P\left(sI_{n_P}-A_P\right)^{-1}L_P+D_P\right]=-G_{ISOCSyd}(s), \tag{10.23}$$

$$G_{ISOCS\varepsilon yd} = \left\langle \begin{array}{c} I_N + \left[C_P \left(sI_{n_P} - A_P \right)^{-1} B_P + H_P \right] \bullet \\ \bullet \left[C_{CR} \left(sI_{n_c} - A_{CR} \right)^{-1} B_{CR} + H_{CR} \right] \end{array} \right\rangle^{-1}, \tag{10.24}$$

$$G_{ISOCS\varepsilon xcro}(s) = -G_{ISOCSyxcro}(s) =$$

$$= - \left\langle \begin{array}{c} I_N + \left[C_P \left(sI_{n_P} - A_P \right)^{-1} B_P + H_P \right] \bullet \\ \bullet \left[C_{CR} \left(sI_{n_c} - A_{CR} \right)^{-1} B_{CR} + H_{CR} \right] \end{array} \right\rangle^{-1} \bullet$$

$$\bullet \left[C_P \left(sI_{n_P} - A_P \right)^{-1} B_P + H_P \right] C_{CR} \left(sI_{n_c} - A_{CR} \right)^{-1}, \tag{10.25}$$

$$G_{ISOCS\varepsilon xpo}(s) = -G_{ISOCSyxpo}(s) =$$

$$= - \left\langle \begin{array}{c} I_N + \left[C_P \left(sI_{n_P} - A_P \right)^{-1} B_P + H_P \right] \bullet \\ \bullet \left[C_{CR} \left(sI_{n_c} - A_{CR} \right)^{-1} B_{CR} + H_{CR} \right] \end{array} \right\rangle^{-1} C_P \left(sI_{n_P} - A_P \right)^{-1}. \tag{10.26}$$

The equations (10.22) through (10.26) determine both the full transfer function matrix $F_{ISOCS\varepsilon}(s)$ of the ISO feedback control system relative to the output error vector $\varepsilon(t)$ and its Laplace transform $\varepsilon(s)$, (7.102) (Subsection 7.4.3),

$$\varepsilon(s) = \left\langle \begin{array}{c} I_N + \left[C_P \left(sI_{n_P} - A_P \right)^{-1} B_P + H_P \right] \bullet \\ \bullet \left[C_{CR} \left(sI_{n_c} - A_{CR} \right)^{-1} B_{CR} + H_{CR} \right] \end{array} \right\rangle^{-1} \bullet$$

$$\bullet \left\langle \begin{array}{c} \mathbf{Y}_d(s) - \left[C_P \left(sI_{n_P} - A_P \right)^{-1} B_P + H_P \right] \left[C_{CR} \left(sI_{n_c} - A_{CR} \right)^{-1} \mathbf{X}_{CR0} \right] + \\ - \left[C_P \left(sI_{n_P} - A_P \right)^{-1} L_P + D_P \right] \mathbf{D}(s) - C_P \left(sI_{n_P} - A_P \right)^{-1} \mathbf{X}_{P0} \end{array} \right\rangle. \tag{10.27}$$

Note 309 *The full transfer function matrix $F_{ISOCS\varepsilon}(s)$ of the ISO feedback control system relative to the output error vector $\varepsilon(t)$ is different from the full transfer function matrix $F_{ISOCSy}(s)$ (7.92)-(7.97), (Subsection 7.4.3), of the system relative to the output vector $\mathbf{Y}(t)$. The former is adequate for tracking studies. The latter is adequate for Lyapunov and BI (Bounded-Input) stability studies [148].*

Theorem 310 ***Criterion for the ISO linear control system tracking***
For the ISO linear feedback control system (2.45), (2.46) composed of the ISO plant (2.33), (2.34) and of the ISO controller CR (2.37), (2.38) to exhibit tracking over $[\mathfrak{D}\mathrm{x}\mathfrak{Y}_d] \cap \mathcal{L}$ it is necessary and sufficient that the real parts of all poles of

$$\varepsilon(s) = F_{ISOCS\varepsilon}(s) \mathbf{V}_{ISOCS\varepsilon}(s) \tag{10.28}$$

are negative for every $[\mathbf{D}(.), \mathbf{Y}_d(.)] \in [\mathfrak{D}\mathrm{x}\mathfrak{Y}_d] \cap \mathcal{L}$.

Then, and only then, the ISO linear feedback control system (2.45), (2.46) composed of the ISO plant (2.33), (2.34) and of the ISO controller CR (2.37), (2.38) exhibits

- *global stablewise tracking over* $[\mathfrak{D}\mathrm{x}\mathfrak{Y}_d] \cap \mathcal{L}$,
- *global exponential tracking over* $[\mathfrak{D}\mathrm{x}\mathfrak{Y}_d] \cap \mathcal{L}$.

Proof. Linearity of the *ISO* linear feedback control system (2.45), (2.46) composed of the *IO* plant *P* (2.33), (2.34) and the *IO* controller *CR* (2.37), (2.38), Definitions 180, 182, 184 (Subsection 8.4.1), continuity and boundedness of every $[\mathbf{D}(.), \mathbf{Y}_d(.)] \in [\mathfrak{D}\mathrm{x}\mathfrak{Y}_d] \cap \mathcal{L}$ together with (2.12) (Section 2.1) and the Generating theorem 304 (Section 10.1) imply the statement of the theorem due to (10.28) that follows from (10.22)-(10.26) ■

Comment 311 *This theorem shows the necessity to use the full transfer function matrix* $F_{ISOCS\varepsilon}(s)$ *of the ISO system rather than only its transfer function matrix* $G_{ISOCSd}(s)$ *or* $G_{ISOCS\varepsilon yd}(s)$. *The latter are insufficient for the analysis or synthesis of the control system to assure the requested tracking, while the former is both necessary and sufficient.*

Equations (10.22) and (10.28) show that the poles of $F_{ISOCS\varepsilon}(s)$, of $\mathbf{D}(s)$ and of $\mathbf{Y}_d(.)$ compose the set of all poles of $\varepsilon(s)$. The only information about the disturbance vector function $\mathbf{D}(.)$ is that it belongs to $\mathfrak{D} \cap \mathcal{L}$ and that the signs of the real parts of all poles of its Laplace transform $\mathbf{D}(s)$ are known. If some poles of $\mathbf{D}(s)$ have nonnegative real parts then they should be known together with their multiplicities.

Theorem 312 *For the ISO linear feedback control system (2.45), (2.46) composed of the ISO plant (2.33), (2.34) and of the ISO controller (CR) (2.37), (2.38) to exhibit tracking over* $[\mathfrak{D}_{-}\mathrm{x}\mathfrak{Y}_{d-}] \cap \mathcal{L}$ *it is necessary and sufficient that the real parts of all poles of the full transfer function matrix* $F_{ISOCS\varepsilon}(s)$ *of the control system relative to the output error vector* $\varepsilon(t)$ *are negative.*

Then, and only then, the ISO linear feedback control system (2.45), (2.46) composed of the ISO plant (2.33), (2.34) and of the ISO controller (CR) (2.37), (2.38) exhibits

- *global stablewise tracking over* $[\mathfrak{D}_{-}\mathrm{x}\mathfrak{Y}_{d-}] \cap \mathcal{L}$,
- *global exponential tracking over* $[\mathfrak{D}_{-}\mathrm{x}\mathfrak{Y}_{d-}] \cap \mathcal{L}$.

Proof. Equations (10.22) and (10.28) show that the poles of $F_{ISOCS\varepsilon}(s)$, of $\mathbf{D}(s)$ and of $\mathbf{Y}_d(s)$ compose the set of all poles of $\varepsilon(s)$. The definitions of $\mathfrak{D}_{-}$, of $\mathcal{L}$ and of $\mathfrak{Y}_{d-}$ determine that the real parts of all poles of $[\mathbf{D}(s), \mathbf{Y}_d(s)]$ are negative due to $[\mathbf{D}(.), \mathbf{Y}_d(.)] \in [\mathfrak{D}_{-}\mathrm{x}\mathfrak{Y}_{d-}] \cap \mathcal{L}$. This reduces the condition of Theorem 310 to the demand that the real parts of all poles of the full transfer function matrix $F_{ISOCS\varepsilon}(s)$ of the control system relative to the output error vector $\varepsilon(t)$ are negative ■

The exponential tracking is the best tracking quality that can be achieved with this approach. The controller action is smooth and robust relative to the disturbance action. The criteria demand the test of the sign of the real parts of all poles of $F_{ISOCS\varepsilon}(s)$, which usually means that the characteristic polynomial of $F_{ISOCS\varepsilon}(s)$ should be known.

Note 313 *The classical LITC of the IO plants and of the ISO plants can guarantee tracking, but it cannot assure tracking with the finite reachability time. Therefore, we will examine other methods for tracking control synthesis.*

Chapter 11

Lyapunov Tracking Control (LTC)

11.1 Vector Lyapunov function (VLF)

11.1.1 Introduction to *VLF* concept

The concept of vector Lyapunov functions (VLF) was coincidentally introduced by R. Bellman [13] in the linear systems setting and by V. M. Matrosov [236] in the general nonlinear systems framework. Matrosov continued to develop the VLF concept to large-scale nonlinear systems [237], [238]. It became the basic mathematical tool for studying stability properties of complex (interconnected and large-scale) dynamic systems [131], [237], [243], [298]. The VLF was the mathematical mean to effectively construct a scalar Lyapunov function for the complex dynamic systems and to reduce their stability test to simple algebraic conditions imposed on constant matrices the dimension of which was reduced to the number of subsystems of a high dimensional overall system.

The analysis of the application of a scalar Lyapunov function for control synthesis meets the mathematical problem of how to separate the control from the Lyapunov function gradient and how to accommodate it to the tracking task. In order to overcome this drawback of the scalar Lyapunov function approach it was proposed in [149], [162], [173]-[175] to use the VLF in its real vector form without any need for the scalar Lyapunov function application to the whole system in order to ensure its tracking. We will present it in its simplified form adequate to the need of the tracking control synthesis in the framework of the linear systems.

11.1.2 Definitions of *VLF*'s

All vector and matrix equalities, inequalities and powers hold elementwise. We will generalize Lyapunov's concept of definite functions.

Definition 314 ***Definition of vector definite functions***

A vector function $\mathbf{v}(.) : R^N \to R^N$, $\mathbf{v}(\boldsymbol{\varepsilon}) = [v_1(\boldsymbol{\varepsilon}) \quad v_2(\boldsymbol{\varepsilon}) \quad ... \; v_N(\boldsymbol{\varepsilon})]^T$, $v_i(.) : \mathfrak{R}^N \longrightarrow \mathfrak{R}$, $\forall i = 1, 2, ..., N$, *is*

a) ***positive (negative) definite*** *if, and only if, there is a neighborhood* $\mathfrak{S}$ *of* $\boldsymbol{\varepsilon} = \mathbf{0}_N$, $\mathfrak{S} \subseteq \mathfrak{R}^N$, *such that (i) through (iii) hold:*

(i) $\mathbf{v}(.)$ *is defined and continuous on* $\mathfrak{S}$*:* $\mathbf{v}(\boldsymbol{\varepsilon}) \in \mathfrak{C}(\mathfrak{S})$,

(ii) $\mathbf{v}(\boldsymbol{\varepsilon}) \geq \mathbf{0}_N$, $(\mathbf{v}(\boldsymbol{\varepsilon}) \leq \mathbf{0}_N)$, $\forall \boldsymbol{\varepsilon} \in \mathfrak{S}$,

(iii) $v_i(\boldsymbol{\varepsilon}) = 0$ *for* $\boldsymbol{\varepsilon} \in \mathfrak{S}$ *if, and only if,* $\varepsilon_i = 0$, $\forall i = 1, 2, ..., N$.

b) ***global positive (negative) definite*** *if, and only if, (i) through (iii) hold for* $\mathfrak{S} = \mathfrak{R}^N$.

c) ***elementwise positive (negative) definite*** *if, and only if, it is positive (negative) definite and*

(iv) $v_i(.) : \mathfrak{R} \longrightarrow \mathfrak{R} \Longrightarrow v_i(\boldsymbol{\varepsilon}) \equiv v_i(\varepsilon_i)$, $\forall i = 1, 2, ..., N$.

d) ***global elementwise positive (negative) definite*** *if, and only if, (i) through (iv) hold for* $\mathfrak{S} = \mathfrak{R}^N$.

e) ***radially strictly increasing on*** $\mathfrak{S}$ *if, and only if,*

(v) $\mathbf{v}(\lambda_1 \boldsymbol{\varepsilon}) < \mathbf{v}(\lambda_2 \boldsymbol{\varepsilon})$, $0 < \lambda_1 < \lambda_2$, $\forall (\boldsymbol{\varepsilon} \neq \mathbf{0}_N) \in \mathfrak{S}$.

f) ***radially unbounded*** *if, and only if, the corresponding above property is global and*

(vi) $\mathbf{v}(\lambda \boldsymbol{\varepsilon}) \longrightarrow \infty \mathbf{1}_N$ *as* $\lambda \longrightarrow \infty$, $\forall \boldsymbol{\varepsilon} \in \mathfrak{R}^N$.

The conditions (i) through (iii) do not imply positive definiteness on $\mathfrak{S}$ of any entry $v_i(.) : \mathfrak{R}^N \longrightarrow \mathfrak{R}_+$ of $\mathbf{v}(.) : \mathfrak{R}^N \longrightarrow \mathfrak{R}^N_+$. However, they imply the positive semi-definiteness on $\mathfrak{S}$ of every entry $v_i(.)$ of $\mathbf{v}(.)$ because $v_i(.)$ is defined on $\mathfrak{R}^N$. The conditions (i) through (iii) imply $\mathbf{v}(\boldsymbol{\varepsilon}) = \mathbf{0}_N$ for $\boldsymbol{\varepsilon} \in \mathfrak{S}$ if, and only if, $\boldsymbol{\varepsilon} = \mathbf{0}_N$.

The conditions (i) through (iv) imply positive definiteness on $\mathfrak{S}_i$, $\mathfrak{S}_i \subseteq \mathfrak{R}^1$, of the entry $v_i(.)$ of $\mathbf{v}(.)$ because $v_i(.)$ is defined on $\mathfrak{R}^1$, $\forall i = 1, 2, ..., N$.

Definition 314 is compatible with Lyapunov's original definition of scalar definite functions [232], as well as with the concept of matrix definite functions introduced in [93].

The condition (iii) under a) can be relaxed if we accept the use of a scalar overall positive definite function $v : \mathfrak{R}^N \longrightarrow \mathfrak{R}^N$, $v(\boldsymbol{\varepsilon}) \in \mathfrak{C}(\mathfrak{R}^N)$,

$$\sum_{i=0}^{i=N} v_i(\boldsymbol{\varepsilon}) \geq 0, \ \forall \boldsymbol{\varepsilon} \in \mathfrak{R}^N, \ \sum_{i=0}^{i=N} v_i(\boldsymbol{\varepsilon}) = 0 \Longleftrightarrow \boldsymbol{\varepsilon} = \mathbf{0}_N.$$

In this case the functions $v_1(.)$, $v_2(.)$, ..., $v_N(.)$ can be each, but need not be each, (global) (radially unbounded) positive definite functions. However, their sum must be (global) (radially unbounded) positive definite function, which permits that some of them are only positive semidefinite functions permitting their dependence only on a subvector of the vector $\boldsymbol{\varepsilon}$.

Definition 315 ***Definition of vector Lyapunov functions***

A vector function $\mathbf{v}(.) : R^N \to R^N$ *is*

a) ***an error vector Lyapunov function of a given dynamic system*** *if, and only if, both (i) and (ii) hold:*

(i) $\mathbf{v}(.)$ *is positive definite,*

(ii) there is a neighborhood $\mathfrak{B}$ *of* $\varepsilon = \mathbf{0}_N$, $\mathfrak{B} \subseteq \mathfrak{R}^N$, *such that the following is valid,*

$$D^+\mathbf{v}(\varepsilon) \leq 0, \ \forall \left(\varepsilon, \varepsilon^{(1)}\right) \in \mathfrak{B}\mathrm{x}\mathfrak{B}. \tag{11.1}$$

If, and only if, additionally there is a positive definite vector function $\mathbf{\Psi}(.) : \mathfrak{R}^{2N} \rightarrow \mathfrak{R}^N$ *such that*

$$D^+\mathbf{v}(\varepsilon) \leq -\mathbf{\Psi}(\varepsilon^1), \ \forall \varepsilon^1 \in \mathfrak{B}\mathrm{x}\mathfrak{B}, \tag{11.2}$$

then the function v(.) is a ***strict error vector Lyapunov function of the system.***

b) ***an elementwise error vector Lyapunov function of the system*** *if, and only if, both (i) and (ii) hold:*

(i) $\mathbf{v}(.)$ *is elementwise positive definite,*

(ii) there is a neighborhood $\mathfrak{B}$ *of* $\varepsilon = \mathbf{0}_N$, $\mathfrak{B} \subseteq \mathfrak{R}^N$, *such that (11.1) is valid.*

If, and only if, additionally there is an elementwise positive definite vector function $\mathbf{\Psi}(.) : R^N \rightarrow R^N$ *such that (11.2) holds then the function* $\mathbf{v}(.)$ *is a* ***strict elementwise error vector Lyapunov function of the system.***

This definition is compatible with the concept of vector Lyapunov functions by R. Bellman [13] and V. M. Matrosov [236]-[238], as well as with the concept of matrix Lyapunov functions introduced in [93].

Note 316 *The vector function* $\mathbf{v}(.) : \mathfrak{R}^N \longrightarrow \mathfrak{R}^N$ *induces* $D^+\mathbf{v}(.) : \mathfrak{R}^{2N} \longrightarrow \mathfrak{R}^N$. *This means that* $\mathbf{v}(.)$ *depends on* ε, *while* $D^+\mathbf{v}(.)$ *is a function of* $\varepsilon^1 = \left[\varepsilon^T \ \ \varepsilon^{(1)^T}\right]^T$.

11.1.3 *VLF* generalization of the classical stability theorems

We denote the empty set by ϕ. Let $\mathbf{c} \in \mathfrak{R}^{+^N}$. The set $\mathfrak{V}_\mathbf{c}$, $\mathfrak{V}_\mathbf{c} \subseteq \mathfrak{R}^N$, is the largest open connected neighborhood of $\varepsilon = \mathbf{0}_N$ such that a vector function $\mathbf{v}(.)$ and the set $\mathfrak{V}_\mathbf{c}$ obey

$$\mathbf{v}(\varepsilon) < \mathbf{c}, \forall \varepsilon \in \mathfrak{V}_\mathbf{c}. \tag{11.3}$$

$Cl\mathfrak{V}_\mathbf{c}$ is the closure of the set $\mathfrak{V}_\mathbf{c}$, and $\partial\mathfrak{V}_\mathbf{c}$ is its boundary if the boundary exists. $\mathfrak{N}_{\mathbf{a}_i}$ is the $\mathbf{a}_i$-neighborhood of $\varepsilon = \mathbf{0}_N$ defined by

$$\mathfrak{N}_{\mathbf{a}_i} = \left\{\varepsilon : \varepsilon \in \mathfrak{R}^N, \ |\varepsilon| < \mathbf{a}_i\right\}, \ \mathbf{a}_i \in \mathfrak{R}^{+^N}. \tag{11.4}$$

Condition 317 *The sets* $\mathfrak{V}_{\mathbf{c}_i}$, $\mathbf{c}_i \in \mathfrak{R}^{+^N}$, $i = 1, 2$, *satisfy* ***a) through c):***

a) $Cl\mathfrak{V}_{\mathbf{c}_1} \subset Cl\mathfrak{V}_{\mathbf{c}_2}$, $\partial\mathfrak{V}_{\mathbf{c}_1} \cap \partial\mathfrak{V}_{\mathbf{c}_2} = \phi$, $\forall \mathbf{c}_i \in \mathfrak{R}^{+^N}$, $i = 1, 2$, $\mathbf{0}_N < \mathbf{c}_1 < \mathbf{c}_2$,

$$b) \quad \mathbf{c}_i \rightarrow \infty\mathbf{1},\ \mathbf{1} = (1\ \ 1 ...\ 1)^T \in \mathfrak{R}^{+^N} \Longrightarrow \mathfrak{V}_{\mathbf{c}_i} \rightarrow \mathfrak{R}^N,\ i = 1, 2,$$
$$c) \quad \forall \mathbf{c}_i \in \mathfrak{R}^{+^N},\ \exists \mathbf{a}_i \in \mathfrak{R}^{+^N} \Longrightarrow \mathfrak{V}_{\mathbf{c}_i} \subseteq \mathfrak{N}_{\mathbf{a}_i},\ i = 1, 2.$$

Note 318 *If the vector positive definite function* $\mathbf{v}(.)$ *is radially strictly increasing on* $\mathfrak{S}$*, then the sets* $\mathfrak{V}_{\mathbf{c}_i}$ *associated with* $\mathbf{v}(.)$ *satisfy Condition 317 on* $\mathfrak{S}$*.*

Theorem 319 *Let the condition 317 hold. In order for* $\boldsymbol{\varepsilon} = \mathbf{0}_N$ *of the system to be, respectively, {elementwise} asymptotically stable it is sufficient that there is a strict (elementwise) vector Lyapunov function* $\mathbf{v}(.)$ *of the system.*

If, additionally, $\mathfrak{B} = \mathfrak{S} = \mathfrak{R}^N$, $\mathbf{v}(.)$ *is also global strict {elementwise} vector Lyapunov function and radially unbounded, then* $\boldsymbol{\varepsilon} = \mathbf{0}_N$ *is globally (elementwise) asymptotically stable.*

Proof. Let the condition 317 be valid. Let $\boldsymbol{\epsilon} \in \mathfrak{R}^{+^N}$ be arbitrary elementwise positive vector. Let an elementwise positive vector $\mathbf{c} \in \mathfrak{R}^{+^N}$ be such that $\mathfrak{V}_{\mathbf{c}} \subset \mathfrak{N}_{\boldsymbol{\epsilon}} \cap \mathfrak{S} \cap \mathfrak{B}$. Let $\boldsymbol{\delta} \in \mathfrak{R}^{+^N}$ obey $\mathfrak{N}_{\boldsymbol{\delta}} \subset \mathfrak{V}_{\mathbf{c}}$. Hence, $\mathfrak{N}_{\boldsymbol{\delta}} \subset \mathfrak{N}_{\boldsymbol{\epsilon}}$ and $\boldsymbol{\delta} = \boldsymbol{\delta}(\boldsymbol{\varepsilon})$. Let $\boldsymbol{\varepsilon}_0$ be arbitrarily chosen under the condition that $\boldsymbol{\varepsilon}_0 \in \mathfrak{N}_{\boldsymbol{\delta}}$, i.e., $|\boldsymbol{\varepsilon}_0| < \boldsymbol{\delta}$ due to (11.4). Hence, $\boldsymbol{\varepsilon}_0 \in \mathfrak{V}_{\mathbf{c}}$, i.e., $\mathbf{v}(\boldsymbol{\varepsilon}_0) < \mathbf{c}$. Since $\mathbf{v}(.)$ is a strict (elementwise) vector Lyapunov function of the system, (11.1) implies

$$\mathbf{v}\left[\boldsymbol{\varepsilon}\left(t; \boldsymbol{\varepsilon}_0\right)\right] \leq \mathbf{v}(\boldsymbol{\varepsilon}_0) < \mathbf{c},\ \forall t \in \mathfrak{T}_0,$$

or equivalently

$$\boldsymbol{\varepsilon}\left(t; \boldsymbol{\varepsilon}_0\right) \in \mathfrak{V}_{\mathbf{c}} \subset \mathfrak{N}_{\boldsymbol{\epsilon}},\ \forall t \in \mathfrak{T}_0 \Longrightarrow \left|\boldsymbol{\varepsilon}\left(t; \boldsymbol{\varepsilon}_0\right)\right| < \boldsymbol{\epsilon},\ \forall t \in \mathfrak{T}_0.$$

We have proved

$$\forall \boldsymbol{\epsilon} \in \mathfrak{R}^{+^N},\ \exists \boldsymbol{\delta} \in \mathfrak{R}^{+^N},\ \boldsymbol{\delta} = \boldsymbol{\delta}\left(\boldsymbol{\varepsilon}\right) \Longrightarrow \left|\boldsymbol{\varepsilon}_0\right| < \boldsymbol{\delta} \Longrightarrow \left|\boldsymbol{\varepsilon}\left(t; \boldsymbol{\varepsilon}_0\right)\right| < \boldsymbol{\epsilon},\ \forall t \in \mathfrak{T}_0.$$

This proves (elementwise) stability of $\boldsymbol{\varepsilon} = \mathbf{0}_N$. Let $\boldsymbol{\Delta} \in \mathfrak{R}^{+^N}$ obey $\mathfrak{N}_{\boldsymbol{\Delta}} \subset \mathfrak{S} \cap \mathfrak{B}$. Since $\mathbf{v}(.)$ is a strict (elementwise) vector Lyapunov function of the system, (11.2) implies

$$\left|\boldsymbol{\varepsilon}_0\right| < \boldsymbol{\Delta} \Longrightarrow \underbrace{\lim}_{t \longrightarrow \infty} \left|\boldsymbol{\varepsilon}\left(t; \boldsymbol{\varepsilon}_0\right)\right| = \mathbf{0}_N.$$

The zero error vector $\boldsymbol{\varepsilon} = \mathbf{0}_N$ is attractive. Altogether, it is asymptotically stable. It is global if $\mathfrak{B} = \mathfrak{S} = \mathfrak{R}^N$ because then $\mathbf{v}(.)$ is also global strict (elementwise) vector Lyapunov function and radially unbounded. They permit both $\boldsymbol{\delta}\left(\boldsymbol{\varepsilon}\right) \longrightarrow \infty\mathbf{1}_N$ as $\boldsymbol{\varepsilon} \longrightarrow \infty\mathbf{1}_N$ and $\boldsymbol{\Delta} = \infty\mathbf{1}_N$ ■

11.1.4 *VLF* forms

Example 320 *The vector* $\boldsymbol{\varepsilon}^k$,

$$\boldsymbol{\varepsilon} = \left[\varepsilon_1\ \ \varepsilon_2\ \ ...\ \ \varepsilon_N\right]^T \in \mathfrak{R}^N \Longrightarrow \boldsymbol{\varepsilon}^{(i)} = \left[\varepsilon_1^{(i)}\ \ \varepsilon_2^{(i)}\ \ ...\ \ \varepsilon_N^{(i)}\right]^T \in \mathfrak{R}^N \Longrightarrow$$

$$\boldsymbol{\varepsilon}^k = \left[\boldsymbol{\varepsilon}^{(0)T}\ \ \boldsymbol{\varepsilon}^{(1)T}\ \ ...\ \ \boldsymbol{\varepsilon}^{(k)T}\right]^T \in \mathfrak{R}^{k+1},\ \boldsymbol{\varepsilon}^0 = \boldsymbol{\varepsilon},$$

determines the vector function $\mathbf{v}(.) : \mathfrak{R}^{(k+1)N} \longrightarrow \mathfrak{R}^{(k+1)N}$,

$$\mathbf{v}(\boldsymbol{\varepsilon}^k) = \frac{1}{2} E^k \boldsymbol{\varepsilon}^k,\ E^k = blocdiag\left\{E^{(0)} \quad E^{(1)} \ \dots \ E^{(k)}\right\},\ k \in \{0, 1, 2, .., \nu - 1\},$$
$$E^{(0)} = E,\ E^{(i)} = diag\left\{\varepsilon_1^{(i)} \quad \varepsilon_2^{(i)} \ \dots \ \varepsilon_N^{(i)}\right\} \in \mathfrak{R}^{N\mathrm{x}N},\ i \in \{0, 1, 2, .., k\},$$
$$E^k \in \mathfrak{R}^{(k+1)N\mathrm{x}(k+1)N}.$$

This $\mathbf{v}(\boldsymbol{\varepsilon}^k)$ *is an example of a global strict elementwise positive definite error vector function [149], [175] as a VLF candidate.*

Example 321 *Other possible forms of VLF follow:*

$$\mathbf{v}(\boldsymbol{\varepsilon}^k) = \frac{1}{2} E^k H \boldsymbol{\varepsilon}^k \in \mathfrak{R}^{(k+1)N},\ H = H^T > O_{(k+1)N},\ H \in \mathfrak{R}^{(k+1)N\mathrm{x}(k+1)N},$$

where $H > O_{(k+1)N}$ *denotes that* H, $H \in \mathfrak{R}^{(k+1)N\mathrm{x}(k+1)N}$, *is positive definite, or*

$$\mathbf{v}(\boldsymbol{\varepsilon}^k) = \left|\boldsymbol{\varepsilon}^k\right| \in \mathfrak{R}_+^{(k+1)N},$$

or

$$\mathbf{v}(\boldsymbol{\varepsilon}^k) = V(\boldsymbol{\varepsilon}^k) H \mathbf{V}(\boldsymbol{\varepsilon}^k) \in \mathfrak{R}_+^{(k+1)N},$$
$$V(\boldsymbol{\varepsilon}^k) = diag\left\{v_1\left(\varepsilon^k\right) \quad v_2\left(\varepsilon^k\right) \ \dots \ v_N\left(\varepsilon^k\right)\right\},$$
$$\mathbf{V}(\boldsymbol{\varepsilon}^k) = \left[v_1\left(\varepsilon^k\right) \quad v_2\left(\varepsilon^k\right) \ \dots \ v_N\left(\varepsilon^k\right)\right]^T \in \mathfrak{C}\left(\mathfrak{R}^{(k+1)N}\right),$$

or simply

$$\mathbf{v}(\boldsymbol{\varepsilon}^k) = \mathbf{V}(\boldsymbol{\varepsilon}^k) \in \mathfrak{R}_+^{(k+1)N}.$$

$\mathbf{V}(.) : \mathfrak{R}^{(k+1)N} \longrightarrow \mathfrak{R}^{(k+1)N}$ *is positive definite vector function on* $\mathfrak{R}^{(k+1)N}$.

Conclusion 322 *The vector definite functions and the vector Lyapunov functions introduced in the preceding definitions enable us to solve various stability problems of* $\boldsymbol{\varepsilon} = \mathbf{0}_N$ *of complex (interconnected and/or large-scale) systems without using a scalar Lyapunov function of the overall system. They permit us also to use the consistent Lyapunov methodology [79], [80], [82], [83], [88], [89], [91], [97], [113]-[117], [124], [152], [153] in order to find Lyapunov functions for the disconnected subsystems and then to use them as entries of a vector Lyapunov function.*

Moreover, the introduced vector Lyapunov functions enable us to ensure a high quality of stability properties such as elementwise asymptotic stability of a desired motion together with tracking, both with a finite reachability time [174].

11.2 LTC of the *IO* plant

11.2.1 Arbitrary scalar Lyapunov function

We will consider the classical application of Lyapunov method to tracking control synthesis for the *IO* plant (2.15) (Subsection 2.1.2),

$$A_P^{(\nu)}\mathbf{Y}^{\nu}(t) = C_{Pu}^{(\mu_{Pu})}\mathbf{U}^{\mu_{Pu}}(t) + D_{Pd}^{(\mu_{Pd})}\mathbf{D}^{\mu_{Pd}}(t),\ \det A_{P\nu} \neq 0,\ \forall t \in \mathfrak{T}_0,$$
$$\nu \geq \max\{\mu_{Pd},\ \mu_{Pu}\}. \tag{11.5}$$

The general consideration starts with the general form of scalar Lyapunov function $v(.) : \mathfrak{R}^{\nu N} \longrightarrow \mathfrak{R}_+$.

Assumption 323 *The dimension r of the control vector $\mathbf{U}$ is not less than the dimension N of the output vector $\mathbf{Y}$, $N \leq r$.*

This is a necessary condition for the *IO* plant trackability (Theorem 275 and Theorem 279 in Subsection 9.3.1, Theorem 293 and Theorem 295 in Subsection 9.4.1).

Let $v(\boldsymbol{\varepsilon}^{\nu-1}) \in \mathfrak{C}^1\left(\mathfrak{R}^{\nu N}\right)$. Its total time derivative $v^{(1)}(\boldsymbol{\varepsilon}^{\nu-1})$ along motions of the *IO* plant (11.5) is expressed via the gradient $gradv(\boldsymbol{\varepsilon}^{\nu-1})$ of $v(\boldsymbol{\varepsilon}^{\nu-1})$,

$$gradv(\boldsymbol{\varepsilon}^{\nu-1}) = \frac{\partial v(\boldsymbol{\varepsilon}^{\nu-1})}{\partial \boldsymbol{\varepsilon}^{\nu-1}} = \begin{bmatrix} \frac{\partial v(\boldsymbol{\varepsilon}^{\nu-1})}{\partial \boldsymbol{\varepsilon}} \\ \frac{\partial v(\boldsymbol{\varepsilon}^{\nu-1})}{\partial \boldsymbol{\varepsilon}^{(1)}} \\ \cdots \\ \frac{\partial v(\boldsymbol{\varepsilon}^{\nu-1})}{\partial \boldsymbol{\varepsilon}^{(v-2)}} \\ \frac{\partial v(\boldsymbol{\varepsilon}^{\nu-1})}{\partial \boldsymbol{\varepsilon}^{(\nu-1)}} \end{bmatrix},\ \frac{\partial v(\boldsymbol{\varepsilon}^{\nu-1})}{\partial \boldsymbol{\varepsilon}^{(k)}} = \begin{bmatrix} \frac{\partial v(\boldsymbol{\varepsilon}^{\nu-1})}{\partial \varepsilon_1^{(k)}} \\ \frac{\partial v(\boldsymbol{\varepsilon}^{\nu-1})}{\partial \varepsilon_2^{(k)}} \\ \cdots \\ \frac{\partial v(\boldsymbol{\varepsilon}^{\nu-1})}{\partial \varepsilon_{N-1}^{(k)}} \\ \frac{\partial v(\boldsymbol{\varepsilon}^{\nu-1})}{\partial \varepsilon_N^{(k)}} \end{bmatrix},$$

as follows:

$$v^{(1)}\left[\boldsymbol{\varepsilon}^{\nu-1}(t)\right] = \left[gradv(\boldsymbol{\varepsilon}^{\nu-1})\right]^T \frac{d\boldsymbol{\varepsilon}^{\nu-1}(t)}{dt} = \left[gradv(\boldsymbol{\varepsilon}^{\nu-1})\right]^T \begin{bmatrix} \boldsymbol{\varepsilon}^{(1)}(t) \\ \boldsymbol{\varepsilon}^{(2)}(t) \\ \cdots \\ \boldsymbol{\varepsilon}^{(\nu-1)}(t) \\ \boldsymbol{\varepsilon}^{(\nu)}(t) \end{bmatrix}.$$

We replace $\boldsymbol{\varepsilon}^{(\nu)}(t)$ by $\mathbf{Y}_d^{(\nu)}(t)$-$\mathbf{Y}^{(\nu)}(t)$,

$$v^{(1)}\left[\boldsymbol{\varepsilon}^{\nu-1}(t)\right] = \left[gradv(\boldsymbol{\varepsilon}^{\nu-1})\right]^T \begin{bmatrix} \boldsymbol{\varepsilon}^{(1)}(t) \\ \boldsymbol{\varepsilon}^{(2)}(t) \\ \cdots \\ \boldsymbol{\varepsilon}^{(\nu-1)}(t) \\ \mathbf{Y}_d^{(\nu)}(t) - \mathbf{Y}^{(\nu)}(t) \end{bmatrix}.$$

We solve (11.5) for $\mathbf{Y}^{(\nu)}(t)$, which is possible due to $detA_{P\nu} \neq 0$,

$$\mathbf{Y}^{(\nu)}(t) = A_{P\nu}^{-1}\left[C_{Pu}^{(\mu_{Pu})}\mathbf{U}^{\mu_{Pu}}(t) + D_{Pd}^{(\mu_{Pd})}\mathbf{D}^{\mu_{Pd}}(t) - A_P^{(\nu-1)}\mathbf{Y}^{\nu-1}(t)\right],$$

and eliminate $\mathbf{Y}^{(\nu)}(t)$ from the preceding equation of $v^{(1)}(\boldsymbol{\varepsilon}^{\nu-1})$,

$$v^{(1)}\left[\boldsymbol{\varepsilon}^{\nu-1}(t)\right] = \left[gradv(\boldsymbol{\varepsilon}^{\nu-1})\right]^T \begin{bmatrix} \boldsymbol{\varepsilon}^{(1)}(t) \\ \boldsymbol{\varepsilon}^{(2)}(t) \\ \dots \\ \boldsymbol{\varepsilon}^{(v-1)}(t) \\ \mathbf{Y}_d^{(\nu)}(t) - A_{P\nu}^{-1}\left\{ \begin{array}{c} C_{Pu}^{(\mu_{Pu})}\mathbf{U}^{\mu_{Pu}}(t)+ \\ +D_{Pd}^{(\mu_{Pd})}\mathbf{D}^{\mu_{Pd}}(t)- \\ -A_P^{(\nu-1)}\mathbf{Y}^{\nu-1}(t) \end{array} \right\} \end{bmatrix}, \tag{11.6}$$

which we can rearrange,

$$v^{(1)}\left[\boldsymbol{\varepsilon}^{\nu-1}(t)\right] = v^{(1)}(\boldsymbol{\varepsilon}^{\nu-1},\mathbf{D}) = \underbrace{\left[gradv(\boldsymbol{\varepsilon}^{\nu-1})\right]^T \begin{bmatrix} \mathbf{0}_N \\ \mathbf{0}_N \\ \dots \\ \mathbf{0}_N \\ -A_{P\nu}^{-1}D_{Pd}^{(\mu_{Pd})}\mathbf{D}^{\mu_{Pd}}(t) \end{bmatrix}}_{\psi(\mathbf{D}^{\mu_{Pd}},\boldsymbol{\varepsilon}^{\nu-1})} +$$

$$\underbrace{+\left[gradv(\boldsymbol{\varepsilon}^{\nu-1})\right]^T \begin{bmatrix} \boldsymbol{\varepsilon}^{(1)}(t) \\ \boldsymbol{\varepsilon}^{(2)}(t) \\ \dots \\ \boldsymbol{\varepsilon}^{(v-1)}(t) \\ \mathbf{0}_N \end{bmatrix}}_{\omega(\boldsymbol{\varepsilon}^{\nu-1})} +$$

$$+\left[gradv(\boldsymbol{\varepsilon}^{\nu-1})\right]^T \begin{bmatrix} \mathbf{0}_N \\ \mathbf{0}_N \\ \dots \\ \mathbf{0}_N \\ A_{P\nu}^{-1}\left\{A_{P\nu}\mathbf{Y}_d^{(\nu)}(t) - C_{Pu}^{(\mu_{Pu})}\mathbf{U}^{\mu_{Pu}}(t) + A_P^{(\nu-1)}\mathbf{Y}^{\nu-1}(t)\right\} \end{bmatrix}. \tag{11.7}$$

We should select $\mathbf{U}(t)$ to ensure that $v^{(1)}(\boldsymbol{\varepsilon}^{\nu-1},\mathbf{D})$ is negative definite for every $\mathbf{D}(.) \in \mathfrak{D}^{\mu_{Pd}}$.

Assumption 324 *The output vector $\mathbf{Y}(t)$ and its first $(\nu-1)$ derivatives are measurable, i.e., the vector $\mathbf{Y}^{\nu-1}(t)$ is measurable.*

This assumption can be too severe. The higher derivatives $\mathbf{Y}^{(k)}(t)$, $k \geq 2$, can be unmeasurable.

The following presentation is effective for the *IO* plants (11.5) that satisfy this assumption. Otherwise, other Lyapunov based methods (e.g., adaptive control methods, sliding motion approach or robust control synthesis methods) should be adapted to the tracking requirements.

We define $\mathbf{U}(.)$ by

$$C_{Pu}^{(\mu_{Pu})}\mathbf{U}^{\mu_{Pu}}(t) = A_{P\nu}\mathbf{Y}_d^{(\nu)}(t) + A_P^{(\nu-1)}\mathbf{Y}^{\nu-1}(t) + A_{P\nu}\mathbf{w}(t), \tag{11.8}$$

so that

$$v^{(1)}(\boldsymbol{\varepsilon}^{\nu-1}) = \psi(\mathbf{D}^{\mu_{Pd}}, \boldsymbol{\varepsilon}^{\nu-1}) + \omega\left(\boldsymbol{\varepsilon}^{\nu-1}\right) - \left[\frac{\partial v(\boldsymbol{\varepsilon}^{\nu-1})}{\partial \boldsymbol{\varepsilon}^{\nu-1}}\right]^T \mathbf{w}(t). \tag{11.9}$$

We have reduced the synthesis of control $\mathbf{U}(.)$ to the synthesis of the subsidiary control vector function $\mathbf{w}(.)$.

Let

$$\mathbf{D}_M^{\mu_{Pd}} > \left|\mathbf{D}^{\mu_{Pd}}(t)\right|, \ \forall [\mathbf{D}(.), t] \ \in \mathfrak{D}^{\mu_{Pd}} \mathrm{x} \mathfrak{T}_0,$$

$$\mathbf{w}(t) = \left\|\frac{\partial v(\boldsymbol{\varepsilon}^{\nu-1})}{\partial \boldsymbol{\varepsilon}^{\nu-1}}\right\|^{-2} \left[\frac{\partial v(\boldsymbol{\varepsilon}^{\nu-1})}{\partial \boldsymbol{\varepsilon}^{\nu-1}}\right] \left\{ \begin{array}{c} \omega\left(\boldsymbol{\varepsilon}^{\nu-1}\right) + \\ + \left|\frac{\partial v(\boldsymbol{\varepsilon}^{\nu-1})}{\partial \boldsymbol{\varepsilon}^{\nu-1}}\right|^T \left|A_{P\nu}^{-1} D_{Pd}^{(\mu_{Pd})}\right| \mathbf{D}_M^{\mu_{Pd}} + \\ +2\beta\zeta v(\boldsymbol{\varepsilon}^{\nu-1}) + \\ +\mu k \, sign v(\boldsymbol{\varepsilon}_0^{\nu-1}) + 2\eta k v^{1/2}(\boldsymbol{\varepsilon}^{\nu-1}) \end{array} \right\},$$

$$\beta, k \in \mathfrak{R}^+, \ \zeta, \mu, \eta \in \{0,1\}, \ \zeta + \mu + \eta = 1. \tag{11.10}$$

$\mathbf{D}_M^{\mu_{Pd}}$ is bounded, $\mathbf{D}_M^{\mu_{Pd}} \in \mathfrak{R}^{+(\mu_{Pd}+1)d}$, due to $\mathbf{D}(.) \in \mathfrak{D}^{\mu_{Pd}}$ and the fact that $\mathfrak{D}^{\mu_{Pd}}$ is, by the definition, the family of all μ_{Pd}- times continuously differentiable bounded vector functions.

Comment 325 *The use of* $\mathbf{D}_M^{\mu_{Pd}}$ *eliminates the need for the knowledge of the form of* $\mathbf{D}^{\mu_{Pd}}(t)$ *and for the measurement of the instantaneous value of the disturbance vector* $\mathbf{D}^{\mu_{Pd}}(t) \in \mathfrak{D}^{\mu_{Pd}}$*. Besides, it enables the full control robustness relative to the disturbance action on the plant.*

Comment 326 *Lyapunov method permits the use of* $\mathbf{D}_M^{\mu_{Pd}}$ *instead of* $\mathbf{D}^{\mu_{Pd}}(t)$ *as shown in the sequel. This is an important advantage of Lyapunov approach to tracking control synthesis.*

Comment 327 *The subsidiary control vector* $\mathbf{w}(t)$ *is unbounded because the term*

$$\left\|\frac{\partial v(\boldsymbol{\varepsilon}^{\nu-1})}{\partial \boldsymbol{\varepsilon}^{\nu-1}}\right\|^{-2} \left[\frac{\partial v(\boldsymbol{\varepsilon}^{\nu-1})}{\partial \boldsymbol{\varepsilon}^{\nu-1}}\right]$$

diverges to infinity as $\left\|\frac{\partial v(\boldsymbol{\varepsilon}^{\nu-1})}{\partial \boldsymbol{\varepsilon}^{(v-1)}}\right\|$ *tends to zero,*

$$\left\|\frac{\partial v(\boldsymbol{\varepsilon}^{\nu-1})}{\partial \boldsymbol{\varepsilon}^{\nu-1}}\right\|^{-2} \left[\frac{\partial v(\boldsymbol{\varepsilon}^{\nu-1})}{\partial \boldsymbol{\varepsilon}^{\nu-1}}\right] \longrightarrow \infty \mathbf{1}_{\nu N} \ as \ \left\|\frac{\partial v(\boldsymbol{\varepsilon}^{\nu-1})}{\partial \boldsymbol{\varepsilon}^{\nu-1}}\right\| \longrightarrow 0.$$

This is the principal and serious drawback of this approach. The essence of the drawback is unboundedness of the control. This obstacle is very difficult to overcome. It is the crucial reason that Lyapunov method has not achieved so *wide effective application to the control synthesis as was expected. We can be satisfied with a kind of practical tracking that demands only the output error vector to enter a final neighborhood $\mathfrak{P}_f \subset \mathfrak{R}^{\nu N}$ of the zero output error vector $\varepsilon^{\nu-1}$ in a finite, usually prespecified, time and to rest therein forever (really, until a given final moment that is usually finite) [98], [99], [111], [112]. Then we do not demand either that the real error vector approaches the zero error vector or that the gradient $gradv(\varepsilon^{\nu-1})$ of $v(\varepsilon^{\nu-1})$ becomes the zero vector. The neighborhood $\mathfrak{P}_f$ can be in any of the following forms:*

$$\mathfrak{P}_f = \left\{ \varepsilon^{(v-1)} : \left\| \frac{\partial v(\varepsilon^{\nu-1})}{\partial \varepsilon^{\nu-1}} \right\| < \mu, \ \mu \in \mathfrak{R}^+ \right\},$$

$$\mathfrak{P}_f = \left\{ \varepsilon^{(v-1)} : \sum_{i=1}^{i=N} \sum_{j=0}^{j=\nu-1} \left| \frac{\partial v(\varepsilon^{\nu-1})}{\partial \varepsilon_i^{(j)}} \right| < \mu, \ \mu \in \mathfrak{R}^+ \right\},$$

$$\mathfrak{P}_f = \left\{ \varepsilon^{(v-1)} : \left| \frac{\partial v(\varepsilon^{\nu-1})}{\partial \varepsilon^{\nu-1}} \right| < \boldsymbol{\mu}, \ \boldsymbol{\mu} \in \mathfrak{R}^{+^{\nu N}} \right\},$$

$$\boldsymbol{\mu} = [\mu_1 \ \ \mu_2 \ \cdots \ \mu_{\nu N} \ ]^T .$$

The numbers $\mu \in \mathfrak{R}^+$ and $\mu_i \in \mathfrak{R}^+$, $\forall i = 1, 2, ..., \nu N$, can be chosen small. The smaller their values, the bigger the maximum value of the needed control magnitude. Lyapunov approach is well effective for the control synthesis to guarantee that the control will steer the output error in the neighborhood $\mathfrak{P}_f$ and will keep it therein. With this in mind we continue to analyze Lyapunov method application to the tracking control synthesis.

Equations (11.8), (11.9), and (11.10) determine control $\mathbf{U}(t)$ and permit the simplification of $v^{(1)}\left[\varepsilon^{\nu-1}(t)\right]$,

$$v^{(1)}\left[\varepsilon^{\nu-1}(t)\right] = \psi(\mathbf{D}^{\mu_{Pd}}, \varepsilon^{\nu-1}) + \omega\left(\varepsilon^{\nu-1}\right) -$$
$$- \left[\frac{\partial v(\varepsilon^{\nu-1})}{\partial \varepsilon^{\nu-1}}\right]^T \left\| \frac{\partial v(\varepsilon^{\nu-1})}{\partial \varepsilon^{\nu-1}} \right\|^{-2} \left[\frac{\partial v(\varepsilon^{\nu-1})}{\partial \varepsilon^{\nu-1}}\right] \bullet$$
$$\bullet \left\{ \begin{array}{c} \omega\left(\varepsilon^{\nu-1}\right) + \left|\frac{\partial v(\varepsilon^{\nu-1})}{\partial \varepsilon^{\nu-1}}\right|^T \left| A_{P\nu}^{-1} D_{Pd}^{(\mu_{Pd})} \right| \mathbf{D}_M^{\mu_{Pd}} + \\ +2\beta\zeta v(\varepsilon^{\nu-1}) + \mu k sign v(\varepsilon_0^{\nu-1}) + 2\eta k v^{1/2}(\varepsilon^{\nu-1}) \end{array} \right\} \Longrightarrow$$

$$v^{(1)}\left[\varepsilon^{\nu-1}(t)\right] = \psi(\mathbf{D}^{\mu_{Pd}}) + \omega\left(\varepsilon^{\nu-1}\right) - \omega\left(\varepsilon^{\nu-1}\right) -$$
$$- \left| \frac{\partial v(\varepsilon^{\nu-1})}{\partial \varepsilon^{\nu-1}} \right|^T \left| A_{P\nu}^{-1} D_{Pd}^{(\mu_{Pd})} \right| \mathbf{D}_M^{\mu_{Pd}} -$$
$$-2\beta\zeta v(\varepsilon^{\nu-1}) - \mu k sign v(\varepsilon^{\nu-1}) - 2\eta k v^{1/2}(\varepsilon^{\nu-1}).$$

Let

$$\rho\left(\varepsilon^{\nu-1}, \mathbf{D}_M^{\mu_{Pd}}\right) = \left| \frac{\partial v(\varepsilon^{\nu-1})}{\partial \varepsilon^{\nu-1}} \right|^T \left| A_{P\nu}^{-1} D_{Pd}^{(\mu_{Pd})} \right| \mathbf{D}_M^{\mu_{Pd}}. \tag{11.11}$$

Since

$$\rho\left(\varepsilon^{\nu-1},\mathbf{D}_M^{\mu_{Pd}}\right)=\left|\frac{\partial v(\varepsilon^{\nu-1})}{\partial\varepsilon^{\nu-1}}\right|^T\left|A_{P\nu}^{-1}D_{Pd}^{(\mu_{Pd})}\right|\mathbf{D}_M^{\mu_{Pd}}\geq$$

$$\geq\left|\frac{\partial v(\varepsilon^{\nu-1})}{\partial\varepsilon^{\nu-1}}\right|^T\left|A_{P\nu}^{-1}D_{Pd}^{(\mu_{Pd})}\right|\left|\mathbf{D}^{\mu_{Pd}}(t)\right|\geq$$

$$\geq\left[gradv(\varepsilon^{\nu-1})\right]^T\begin{bmatrix}\mathbf{0}_N\\ \mathbf{0}_N\\ \cdots\\ \mathbf{0}_N\\ -A_{P\nu}^{-1}D_{Pd}^{(\mu_{Pd})}\mathbf{D}^{\mu_{Pd}}(t)\end{bmatrix}=\psi(\mathbf{D}^{\mu_{Pd}},\varepsilon^{\nu-1}),$$

$$\forall\left[\mathbf{D}(.),t\right]\ \in\mathfrak{D}^{\mu_{Pd}}\mathrm{x}\mathfrak{T}_0,$$

then

$$\psi(\mathbf{D}^{\mu_{Pd}})-\rho\left(\varepsilon^{\nu-1},\mathbf{D}_M^{\mu_{Pd}}\right)\leq 0,\forall\left[\mathbf{D}(.),t\right]\ \in\mathfrak{D}^{\mu_{Pd}}\mathrm{x}\mathfrak{T}_0,$$

and

$$v^{(1)}\left(\varepsilon^{\nu-1}\right)\leq-2\boldsymbol{\beta}\zeta v(\varepsilon^{\nu-1})-\mu k signv(\varepsilon_0^{\nu-1})-2\eta k v^{1/2}(\varepsilon^{\nu-1}).\qquad(11.12)$$

This is the crucial estimate of $v^{(1)}\left(\varepsilon^{\nu-1}\right)$. It shows that $v^{(1)}\left(\varepsilon^{\nu-1}\right)$ is globally negative definite function due to global positive definiteness of $v\left(\varepsilon^{\nu-1}\right)$ and due to

$$\boldsymbol{\beta},k\in\mathfrak{R}^+,\ \zeta,\mu,\eta\in\{0,1\},\ \zeta+\mu+\eta=1.$$

11.2.2 The first choice of a scalar Lyapunov function

We accept the quadratic form,

$$\varepsilon^{v-1^T}H\varepsilon^{\nu-1},\ H=H^T>O_{\nu N},$$

for $v(\varepsilon^{\nu-1})$

$$v(\varepsilon^{\nu-1})=\varepsilon^{v-1^T}H\varepsilon^{\nu-1}\Longrightarrow\frac{\partial v(\varepsilon^{\nu-1})}{\partial\varepsilon^{(v-1)}}=2H\varepsilon^{\nu-1}.$$

Equations (11.8), (11.10) and the choice of $\zeta,\mu,\eta\in\{0,1\}$ so that $\zeta+\mu+\eta=1$ determine the control $\mathbf{U}(t)$. We will analyze the impact of the selection of $\zeta,\mu,\eta\in\{0,1\}$ so that $\zeta+\mu+\eta=1$.

Note 328 *The subsidiary control vector $\mathbf{w}(t)$, (11.10), is unlimited. The term*

$$\left\|\frac{\partial v(\varepsilon^{\nu-1})}{\partial\varepsilon^{(\nu-1)}}\right\|^{-2}\left[\frac{\partial v(\varepsilon^{\nu-1})}{\partial\varepsilon^{(\nu-1)}}\right]=$$

$$=\frac{1}{2}\left\|\sum_{k=0}^{k=\nu-1}H_{k+1,\nu}^T\varepsilon^{(k)}(t)\right\|^{-2}\left(\sum_{k=0}^{k=\nu-1}H_{k+1,\nu}^T\varepsilon^{(k)}(t)\right)$$

diverges to infinity as $\left\|\varepsilon^{\nu-1}(t)\right\|$ tends to zero. The drawback of the approach rests for this choice of the function $v(.)$ (Comment 327).

Case 329 ***Control synthesis for global exponential tracking***

If

$$\zeta = 1 \Longrightarrow \mu = \eta = 0,$$

then

$$\frac{dV\left[\varepsilon^{\nu-1}(t)\right]}{dt} \leq -2\beta v\left[\varepsilon^{\nu-1}(t)\right], \forall\left[\mathbf{D}(.), t\right] \in \mathfrak{D}^{\mu_{Pd}} \mathrm{x} \mathfrak{T}_0$$

implies, together with

$$\lambda_m(H)\left\|\varepsilon^{\nu-1}\right\|^2 \leq v(\varepsilon^{\nu-1}) \leq \lambda_M(H)\left\|\varepsilon^{\nu-1}\right\|^2, \tag{11.13}$$

the following:

$$\lambda_m(H)\left\|\varepsilon^{\nu-1}(t)\right\|^2 \leq v\left[\varepsilon^{\nu-1}(t)\right] \leq e^{-2\beta t} v(\varepsilon_0^{\nu-1}) \leq$$
$$\leq e^{-\beta t}\lambda_M(H)\left\|\varepsilon_0^{\nu-1}\right\|^2,$$

i.e.,

$$\left\|\varepsilon^{\nu-1}(t)\right\| \leq \alpha e^{-\beta t}\left\|\varepsilon_0^{\nu-1}\right\|, \forall\left(t, \mathbf{D}(.), \varepsilon_0^{\nu-1}\right) \in \mathfrak{T}_0 \mathrm{x} \mathfrak{D}^{\mu_{Pd}} \mathrm{x} \mathfrak{R}^{\nu N},$$
$$\alpha = \sqrt{\lambda_M(H)\lambda_m^{-1}(H)}.$$

This proves the global exponential tracking on $\mathfrak{D}^{\mu_{Pd}}$.

Case 330 ***Control synthesis for global stablewise tracking with the finite reachability time***

If

$$\mu = 1 \Longrightarrow \zeta = \eta = 0,$$

then

$$\frac{dv\left[\varepsilon^{\nu-1}(t)\right]}{dt} \leq -k\,sign\,v(\varepsilon_0^{\nu-1}), \forall\left(t, \mathbf{D}(.), \varepsilon_0^{\nu-1}\right) \in \mathfrak{T}_0 \mathrm{x} \mathfrak{D}^{\mu_{Pd}} \mathrm{x} \mathfrak{R}^{\nu N}$$

yields

$$v(\left[\varepsilon^{\nu-1}(t)\right] \left\{\begin{array}{ll} \leq v(\varepsilon_0^{\nu-1}) - kt, & \left\langle\begin{array}{c} t \leq k^{-1} v(\varepsilon^{\nu-1}), \\ if\ \varepsilon_0^{\nu-1} \neq \mathbf{0}_{\nu N}, \end{array}\right\rangle \\ = 0, & \left(\begin{array}{c} \left\langle\begin{array}{c} t \geq k^{-1} v(\varepsilon_0^{\nu-1}), \\ if\ \varepsilon_0^{\nu-1} \neq \mathbf{0}_{\nu N}, \end{array}\right\rangle \\ \forall t \in \mathfrak{T}_0,\ if\ \varepsilon_0^{\nu-1} = \mathbf{0}_{\nu N} \end{array}\right) \end{array}\right\},$$
$$\forall\left(\mathbf{D}(.), \varepsilon_0^{\nu-1}\right) \in \mathfrak{D}^{\mu_{Pd}} \mathrm{x} \mathfrak{R}^{\nu N}$$

which, with (11.13), permits

$$\left\|\varepsilon^{\nu-1}(t)\right\| \leq \lambda_m^{-1/2}(H)\left\{\begin{array}{c}\sqrt{\varepsilon_0^{\nu-1^T} H\varepsilon_0^{\nu-1}} - kt, \\ t \leq k^{-1}\varepsilon_0^{\nu-1^T} H\varepsilon_0^{\nu-1^T}, \\ \forall t \in \mathfrak{T}_0,\ if\ \varepsilon_0^{\nu-1} \neq \mathbf{0}_{\nu N}\end{array}\right\},$$

$$\left\|\varepsilon^{\nu-1}(t)\right\| = 0, \quad \left(\begin{array}{c}\left\langle\begin{array}{c} t \geq k^{-1}\varepsilon_0^{\nu-1^T} H\varepsilon_0^{\nu-1}, \\ if\ \varepsilon_0^{\nu-1} \neq \mathbf{0}_{\nu N},\end{array}\right\rangle \\ \forall t \in \mathfrak{T}_0,\ if\ \varepsilon_0^{\nu-1} = \mathbf{0}_{\nu N},\end{array}\right),$$

$$\forall\left(\mathbf{D}(.), \varepsilon_0^{\nu-1}\right) \in \mathfrak{D}^{\mu_{Pd}} \mathrm{x} \mathfrak{R}^{\nu N}.$$

This proves the stablewise tracking on $\mathfrak{D}^{\mu_{Pd}}$ with the finite reachability time τ_R

$$\tau_R = k^{-1}\varepsilon_0^{\nu-1^T} H\varepsilon_0^{\nu-1} = \tau_R\left(\varepsilon_0^{\nu-1}; k\right),$$

which depends on $\varepsilon_0^{\nu-1}$. If τ_R is given, then we determine the gain k from

$$k = \tau_R^{-1}\varepsilon_0^{\nu-1^T} H\varepsilon_0^{\nu-1} = k\left(\varepsilon_0^{\nu-1}; \tau_R\right).$$

These equations express the relationship among the initial error vector $\varepsilon_0^{\nu-1}$, the reachability time τ_R and the gain k.

Case 331 ***Control synthesis for global stablewise tracking with the finite reachability time***

If

$$\eta = 1 \Longrightarrow \zeta = \mu = 0,$$

then

$$\frac{dv\left[\varepsilon^{\nu-1}(t)\right]}{dt} \leq -2kv^{1/2}\left[\varepsilon^{\nu-1}(t)\right],\ \forall\left[\mathbf{D}(.), t\right] \in \mathfrak{D}^{\mu_{Pd}} \mathrm{x} \mathfrak{T}_0,$$

i.e.,

$$dv^{1/2}\left[\varepsilon^{\nu-1}(t)\right] \leq -kdt,\ \forall\left[\mathbf{D}(.), t\right] \in \mathfrak{D}^{\mu_{Pd}} \mathrm{x} \mathfrak{T}_0.$$

The solution is

$$v^{1/2}\left[\varepsilon^{\nu-1}(t)\right] \leq \left\{\begin{array}{c} v^{1/2}(\varepsilon_0^{\nu-1}) - kt,\ t \in [0, \tau_R], \\ 0,\ t \geq \tau_R\end{array}\right\},$$

$$\tau_R = k^{-1}\varepsilon_0^{\nu-1^T} H\varepsilon_0^{\nu-1}.$$

Hence,

$$\left\|\varepsilon^{\nu-1}(t)\right\| \leq \left\{\begin{array}{c}\lambda_m^{-1/2}(H)\left[\sqrt{\varepsilon_0^{\nu-1^T} H\varepsilon_0^{\nu-1}} - kt\right]^2,\ t \in [0, \tau_R], \\ 0,\ t \geq \tau_R\end{array}\right\},$$

$$\forall\left(\mathbf{D}(.), \varepsilon^{\nu-1}\right) \in \mathfrak{D}^{\mu_{Pd}} \mathrm{x} \mathfrak{R}^{\nu N}.$$

This proves the stablewise tracking on $\mathfrak{D}^{\mu_{Pd}}$ *with the finite reachability time* τ_R

$$\tau_R = k^{-1}\sqrt{\varepsilon_0^{\nu-1^T} H \varepsilon_0^{\nu-1}} = \tau_R\left(\varepsilon_0^{\nu-1};k\right),$$

which depends on $\varepsilon_0^{\nu-1}$. *If* τ_R *is given, then we calculate the gain* k *from*

$$k = \tau_R^{-1}\sqrt{\varepsilon_0^{\nu-1^T} H \varepsilon_0^{\nu-1}} = k\left(\varepsilon_0^{\nu-1};\tau_R\right).$$

The smaller the reachability time τ_R*, the bigger the gain* k *for the given initial output error vector* $\varepsilon_0^{\nu-1}$*, and vice versa.*

Comment 332 *Equations (11.8)-(11.10) and (11.11) determine the feedback control for* $v(\varepsilon^{\nu-1}) = \varepsilon^{\nu-1^T} H \varepsilon^{\nu-1}$:

$$C_{Pu}^{(\mu_{Pu})}\mathbf{U}^{\mu_{Pu}}(t) = A_{P\nu}\mathbf{Y}_d^{(\nu)}(t) + A_P^{(\nu-1)}\mathbf{Y}^{\nu-1}(t) +$$
$$+A_{P\nu}\frac{1}{2}\left[\begin{array}{c} \omega\left(\varepsilon^{\nu-1}\right) + \rho\left(\varepsilon^{\nu-1},\mathbf{D}_M^{\mu_{Pd}}\right) + 2\beta\zeta(\varepsilon^{\nu-1^T} H\varepsilon^{\nu-1}) + \\ +\mu k sign(\varepsilon^{\upsilon-1^T} H\varepsilon^{\nu-1}) + 2\eta k(\varepsilon^{\upsilon-1^T} H\varepsilon^{\nu-1})^{1/2} \end{array}\right] \bullet$$
$$\bullet\left\| \sum_{k=0}^{k=\nu-1} H_{k+1,\nu}^T \varepsilon^{(k)}(t)\right\|^{-2} \left(\sum_{k=0}^{k=\nu-1} H_{k+1,\nu}^T \varepsilon^{(k)}(t)\right).$$

The right-hand side of this equation is well defined only for

$$\sum_{k=0}^{k=\nu-1} H_{k+1,\nu}^T \varepsilon^{(k)}(t) \neq \mathbf{0}_N,$$

which is the essential disadvantage of this approach. Control $\mathbf{U}(t)$ *is not defined for*

$$\sum_{k=0}^{k=\nu-1} H_{k+1,\nu}^T \varepsilon^{(k)}(t) = \mathbf{0}_N,$$

which is its drawback. Another problem of its realization emerges from the need to measure all derivatives of the output vector up to the order $(\nu-1)$.

11.2.3 The second choice of a scalar Lyapunov function

In order to try avoiding this problem we select another $v(\varepsilon^{\nu-1})$,

$$v(\varepsilon^{\nu-1}) = \mathbf{v}^T\left|\varepsilon^{\nu-1}\right|,\ \mathbf{v} \in \mathfrak{R}^{\nu N},\ \mathbf{v} > \mathbf{0}_{\nu N} \text{ elementwise},$$
$$\mathbf{v} = \left[\mathbf{v}_1^T\ \ \mathbf{v}_2^T\ \ldots\ \mathbf{v}_\nu^T\right]^T,\ \mathbf{v}_k \in \mathfrak{R}^{+N},\ \forall k = 1,2,\ldots,\nu, \tag{11.14}$$

and define

$$\sigma\left(\varepsilon_i^{(k)},\varepsilon_i^{(k+1)}\right) = \left\{\begin{array}{c} -1,\ \varepsilon_i^{(k)} < 0, \forall\varepsilon_i^{(k+1)} \in \mathfrak{R};\ \varepsilon_i^{(k)} = 0 \text{ and } \varepsilon_i^{(k+1)} < 0, \\ 0,\ \varepsilon_i^{(k)} = 0 \text{ and } \varepsilon_i^{(k+1)} = 0, \\ 1,\ \varepsilon_i^{(k)} > 0, \forall\varepsilon_i^{(k+1)} \in \mathfrak{R};\ \varepsilon_i^{(k)} = 0 \text{ and } \varepsilon_i^{(k+1)} > 0 \end{array}\right\}, \tag{11.15}$$

$$\Sigma\left(\boldsymbol{\varepsilon}^{(k)},\boldsymbol{\varepsilon}^{(k+1)}\right)=diag\left\{\sigma\left(\varepsilon_1^{(k)},\varepsilon_1^{(k+1)}\right)\ \sigma\left(\varepsilon_2^{(k)},\varepsilon_2^{(k+1)}\right)...\sigma\left(\varepsilon_N^{(k)},\varepsilon_N^{(k+1)}\right)\right\},$$
$$\forall k=1,2,...,\nu, \tag{11.16}$$

$$\Sigma\left(\boldsymbol{\varepsilon}^{\nu}\right)=blockdiag\left\{\Sigma\left(\boldsymbol{\varepsilon},\boldsymbol{\varepsilon}^{(1)}\right)\ \Sigma\left(\boldsymbol{\varepsilon}^{(1)},\boldsymbol{\varepsilon}^{(2)}\right)\ ...\ \Sigma\left(\boldsymbol{\varepsilon}^{(\nu-1)},\boldsymbol{\varepsilon}^{(\nu)}\right)\right\}, \tag{11.17}$$

so that

$$\begin{aligned}D^{+}v\left[\boldsymbol{\varepsilon}^{\nu-1}(t)\right] &= D^{+}\left(\mathbf{v}^T\left|\boldsymbol{\varepsilon}^{\nu-1}(t)\right|\right)=\mathbf{v}^T\Sigma\left(\boldsymbol{\varepsilon}^{\nu}\right)\begin{bmatrix}\boldsymbol{\varepsilon}^{(1)}(t)\\ \boldsymbol{\varepsilon}^{(2)}(t)\\ ...\\ \boldsymbol{\varepsilon}^{(\nu-1)}(t)\\ \boldsymbol{\varepsilon}^{(\nu)}(t)\end{bmatrix}=\\ &= \mathbf{v}^T\Sigma\left(\boldsymbol{\varepsilon}^{\nu}\right)\begin{bmatrix}\boldsymbol{\varepsilon}^{(1)}(t)\\ \boldsymbol{\varepsilon}^{(2)}(t)\\ ...\\ \boldsymbol{\varepsilon}^{(\nu-1)}(t)\\ \mathbf{Y}_d^{(\nu)}(t)-\mathbf{Y}^{(\nu)}(t)\end{bmatrix}\Longrightarrow\end{aligned}$$

$$D^{+}v\left[\boldsymbol{\varepsilon}^{\nu-1}(t)\right]=\mathbf{v}^T\Sigma\left(\boldsymbol{\varepsilon}^{v}\right)\bullet$$
$$\bullet\begin{bmatrix}\boldsymbol{\varepsilon}^{(1)}(t)\\ \boldsymbol{\varepsilon}^{(2)}(t)\\ ...\\ \boldsymbol{\varepsilon}^{(\nu-1)}(t)\\ \mathbf{Y}_d^{(\nu)}(t)-A_{P\nu}^{-1}\left\{\begin{array}{c}C_{Pu}^{(\mu_{Pu})}\mathbf{U}^{\mu_{Pu}}(t)+D_{Pd}^{(\mu_{Pd})}\mathbf{D}^{\mu_{Pd}}(t)-\\ -\sum_{k=0}^{k=\nu-1}A_k\mathbf{Y}^{(k)}(t)\end{array}\right\}\end{bmatrix}\Longrightarrow$$

$$D^{+}v(\boldsymbol{\varepsilon}^{\nu-1})=\underbrace{\mathbf{v}^T\Sigma\left(\boldsymbol{\varepsilon}^{v}\right)\begin{bmatrix}\mathbf{0}_N\\ \mathbf{0}_N\\ ...\\ \mathbf{0}_N\\ -A_{P\nu}^{-1}D_{Pd}^{(\mu_{Pd})}\mathbf{D}^{\mu_{Pd}}(t)\end{bmatrix}}_{\Psi(\mathbf{D}^{\mu_{Pd}},\boldsymbol{\varepsilon}^{\nu-1})}+\underbrace{\mathbf{v}^T\Sigma\left(\boldsymbol{\varepsilon}^{\nu}\right)\begin{bmatrix}\boldsymbol{\varepsilon}^{(1)}(t)\\ \boldsymbol{\varepsilon}^{(2)}(t)\\ ...\\ \boldsymbol{\varepsilon}^{(\nu-1)}(t)\\ \mathbf{0}_N\end{bmatrix}}_{\Xi(\boldsymbol{\varepsilon}^{\nu-1})}+$$
$$+\mathbf{v}^T\Sigma\left(\boldsymbol{\varepsilon}^{\nu}\right)\begin{bmatrix}\mathbf{0}_N\\ \mathbf{0}_N\\ ...\\ \mathbf{0}_N\\ \mathbf{Y}_d^{(\nu)}(t)+A_{P\nu}^{-1}\left[\sum_{k=0}^{k=\nu-1}A_{Pk}\mathbf{Y}^{(k)}(t)-C_{Pu}^{(\mu_{Pu})}\mathbf{U}^{\mu_{Pu}}(t)\right]\end{bmatrix}\Longrightarrow \tag{11.18}$$

$$D^{+}v(\varepsilon^{\nu-1})=\Psi(\mathbf{D}^{\mu_{Pd}},\varepsilon^{\nu-1})+\boldsymbol{\Xi}\left(\varepsilon^{\nu-1}\right)+\mathbf{v}_{\nu}^T\Sigma\left(\varepsilon^{(\nu-1)},\varepsilon^{(\nu)}\right)\bullet$$
$$\bullet A_{P\nu}^{-1}\left\{A_{P\nu}\mathbf{Y}_d^{(\nu)}(t)+\left[A_P^{(\nu-1)}\mathbf{Y}^{\nu-1}(t)-C_{Pu}^{(\mu_{Pu})}\mathbf{U}^{\mu_{Pu}}(t)\right]\right\}. \tag{11.19}$$

We define for $v(\varepsilon^{\nu-1}) = \mathbf{v}^T \left|\varepsilon^{\nu-1}\right|$:

$$C_{Pu}^{(\mu_{Pu})}\mathbf{U}^{\mu_{Pu}}(t) = A_{P\nu}\mathbf{Y}_d^{(\nu)}(t) + A_P^{(\nu-1)}\mathbf{Y}^{\nu-1}(t)+$$
$$+A_{P\nu}\left[\begin{array}{c}\Theta\left(\varepsilon^{\nu-1}, \mathbf{D}_M^{\mu_{Pd}}\right) + \Xi(\boldsymbol{\varepsilon}^{\nu-1}) + \beta\zeta(\mathbf{v}^T\left|\varepsilon^{\nu-1}\right|)+\\ +\mu k sign(\mathbf{v}^T\left|\varepsilon_0^{\nu-1}\right|) + 2\eta k(\mathbf{v}^T\left|\varepsilon^{\nu-1}\right|)^{1/2}\end{array}\right]\bullet$$
$$\bullet\left\|\Sigma\left(\varepsilon^{(\nu-1)}, \varepsilon^{(\nu)}\right)\mathbf{v}_\nu\right\|^{-2}\Sigma\left(\varepsilon^{(\nu-1)}, \varepsilon^{(\nu)}\right)\mathbf{v}_\nu,$$
$$\beta, k\in\mathfrak{R}^+, \zeta, \mu, \eta\in\{0,1\},\ \zeta+\mu+\eta = 1 \tag{11.20}$$

where $sign\ x = 0$ for $x = 0$, $sign\ x = |x|^{-1}x$ for $x \neq 0$, and

$$\Theta\left(\varepsilon^{\nu-1}, \mathbf{D}_M^{\mu_{Pd}}\right) = \left|\mathbf{v}^T\Sigma\left(\varepsilon^v\right)A_{P\nu}^{-1}\right|\mathbf{D}_M^{\mu_{Pd}}. \tag{11.21}$$

This control does not need information about the real value of the disturbance vector that can be unmeasurable. The tracking quality is good as illustrated in what follows. The main drawbacks are the need to measure all derivatives of the output vector up to the order ν and the unboundedness of control.

Note 333 *Since $\mathbf{v} > \mathbf{0}_{\nu N}$ elementwise, then*

$$\Sigma\left(\varepsilon^{(\nu-1)}, \varepsilon^{(\nu)}\right)\mathbf{v}_\nu = \mathbf{0}_N \iff \varepsilon^{(\nu-1)} = \varepsilon^{(\nu)} = \mathbf{0}_N.$$

The term

$$\left\|\Sigma\left(\varepsilon^{(\nu-1)}, \varepsilon^{(\nu)}\right)\mathbf{v}_\nu\right\|^{-2}\Sigma\left(\varepsilon^{(\nu-1)}, \varepsilon^{(\nu)}\right)\mathbf{v}_\nu$$

goes to infinity as $\left\|\varepsilon^{(\nu-1)}\right\| + \left\|\varepsilon^{(\nu)}\right\|$ approaches zero. The control vector becomes unlimited. The drawback rests. See Comment 327.

The control (11.20) transforms (11.19) into

$$D^+v(\varepsilon^{\nu-1}) = \Psi(\mathbf{D}^{\mu_{Pd}}, \varepsilon^{\nu-1}) + \boldsymbol{\Xi}\left(\varepsilon^{\nu-1}\right) - \Theta\left(\varepsilon^{\nu-1}, \mathbf{D}_M^{\mu_{Pd}}\right) -$$
$$-\Xi(\boldsymbol{\varepsilon}^{\nu-1}) - \beta\zeta v(\varepsilon^{\nu-1}) - \mu k sign v(\varepsilon_0^{\nu-1}) - 2\eta k v^{1/2}(\varepsilon^{\nu-1}) \Longrightarrow$$

$$D^+v(\varepsilon^{\nu-1}) \leq -\beta\zeta v(\varepsilon^{\nu-1}) - \mu k sign v(\varepsilon_0^{\nu-1}) - 2\eta k v^{1/2}(\varepsilon^{\nu-1}). \tag{11.22}$$

Equations (11.14), (11.18), (11.20), (11.21) and the choice of $\zeta, \mu, \eta\in\{0,1\}$ so that $\zeta+\mu+\eta = 1$ determine the control $\mathbf{U}(.)$. We will consider the influence of the choice of $\zeta, \mu, \eta\in\{0,1\}$ so that $\zeta+\mu+\eta = 1$.

Case 334 ***Control synthesis for global exponential tracking***
If

$$\zeta = 1 \Longrightarrow \mu = \eta = 0,$$

then (11.22) becomes

$$D^+v\left[\left(\varepsilon^{\nu-1}(t)\right] \leq -\beta v\left[\left(\varepsilon^{\nu-1}(t)\right].\right.\right.$$

The solution is

$$v\left[(\varepsilon^{\nu-1}(t)\right] \leq e^{-\beta t}v(\varepsilon_0^{\nu-1}),$$

i.e.,

$$\mathbf{v}^T\left|\varepsilon^{\nu-1}(t)\right| \leq e^{-\beta t}\mathbf{v}^T\left|\varepsilon_0^{\nu-1}\right|, \ \forall\left(t,\mathbf{D}(.),\varepsilon_0^{\nu-1}\right) \in \mathfrak{T}_0 \mathrm{x} \mathfrak{D}^{\mu_{Pd}} \mathrm{x} \mathfrak{R}^{\nu N}. \quad (11.23)$$

Notice that for $\mathbf{x} \in \mathfrak{R}^n$

$$\|\mathbf{x}\|_1 = \sum_{i=1}^{i=n}|x_i|$$

is the taxicab norm or Manhattan norm, and

$$\|\mathbf{x}\| = \|\mathbf{x}\|_2 = \left(\sum_{i=1}^{i=n}x_i^2\right)^{1/2}$$

is the Euclidean norm. The following relationships hold [8, p. 42]

$$\|\mathbf{x}\| \leq \|\mathbf{x}\|_1 \leq \sqrt{n}\,\|\mathbf{x}\|_2 .$$

Let ν_m *be the minimal entry of the vector* $\mathbf{v}$*, and* ν_M *be the maximal entry of the vector* $\mathbf{v}$*. Then,*

$$\nu_m\left\|\varepsilon^{\nu-1}\right\| \leq \nu_m\left\|\varepsilon^{\nu-1}\right\|_1 \leq \mathbf{v}^T\left|\varepsilon^{\nu-1}\right| \leq \nu_M\left\|\varepsilon^{\nu-1}\right\|_1 \leq \nu_M\sqrt{\nu N}\left\|\varepsilon^{\nu-1}\right\|,$$

In view of this we derive from (11.23)

$$\left\|\varepsilon^{\nu-1}(t)\right\| \leq \nu_m^{-1}\nu_M(\nu N)^{1/2}e^{-\beta t}\left\|\varepsilon_0^{\nu-1}\right\|, \ \forall\left(t,\mathbf{D}(.),\varepsilon_0^{\nu-1}\right) \in \mathfrak{T}_0 \mathrm{x} \mathfrak{D}^{\mu_{Pd}} \mathrm{x} \mathfrak{R}^{\nu N}.$$

This proves the global exponential tracking on $\mathfrak{D}^{\mu_{Pd}}$.

Case 335 ***Control synthesis for global stablewise tracking with the finite scalar reachability time***

If

$$\mu = 1 \Longrightarrow \zeta = \eta = 0,$$

then (11.22) reduces to

$$D^+v\left[\varepsilon^{\nu-1}(t)\right] \leq -ksignv(\varepsilon_0^{\nu-1}).$$

The solution reads

$$v\left[(\varepsilon^{\nu-1}(t)\right]\left\{\begin{array}{c} \leq v(\varepsilon^{\nu-1}) - ktsignv(\varepsilon_0^{\nu-1}), \\ t \leq k^{-1}v(\varepsilon_0^{\nu-1}), \\ = 0,\ t \geq k^{-1}v(\varepsilon_0^{\nu-1}) \end{array}\right\}, \left\|\varepsilon_0^{\nu-1}\right\| \neq 0,$$

$$v\left[(\varepsilon^{\nu-1}(t)\right] = 0, \qquad \forall t \in \mathfrak{T}_0, \qquad \left\|\varepsilon_0^{\nu-1}\right\| = 0,$$

i.e.,

$$\left\|\varepsilon^{\nu-1}(t)\right\| \left\{ \begin{array}{c} \leq \nu_m^{-1}\left(\mathbf{v}^T\left|\varepsilon^{\nu-1}\right| - ktsign\mathbf{v}^T\left|\varepsilon_0^{\nu-1}\right|\right), \\ t \leq ksign\mathbf{v}^T\left|\varepsilon_0^{\nu-1}\right|, \\ = 0,\ t \geq ksign\mathbf{v}^T\left|\varepsilon_0^{\nu-1}\right| \end{array} \right\}, \left\|\varepsilon_0^{\nu-1}\right\| \neq 0,$$

$$\left\|\varepsilon^{\nu-1}(t)\right\| = 0, \qquad \forall t \in \mathfrak{T}_0, \qquad \left\|\varepsilon_0^{\nu-1}\right\| = 0,$$

$$\forall\left(\mathbf{D}(.), \varepsilon_0^{\nu-1}\right) \in \mathfrak{D}^{\mu_{Pd}} x \mathfrak{R}^{\nu N}.$$

This proves the global stablewise tracking with the finite scalar reachability time $\tau_R = k^{-1}\mathbf{v}^T\left|\varepsilon_0^{\nu-1}\right|$,

$$\tau_R\left(\varepsilon_0^{\nu-1}, k\right) = k^{-1}sign\mathbf{v}^T\left|\varepsilon_0^{\nu-1}\right| = \left\{ \begin{array}{c} k^{-1}\mathbf{v}^T\left|\varepsilon_0^{\nu-1}\right|,\ \left\|\varepsilon_0^{\nu-1}\right\| \neq 0 \\ 0,\ \left\|\varepsilon_0^{\nu-1}\right\| = 0 \end{array} \right\}.$$

The smaller reachability time τ_R*, the bigger* k*, and vice versa, for fixed* $\varepsilon_0^{\nu-1}$.

The bigger $\left\|\varepsilon_0^{\nu-1}\right\|$, *the bigger reachability time* τ_R *for fixed* k.

Case 336 ***Control synthesis for global stablewise tracking with the finite scalar reachability time*** τ_R

If

$$\eta = 1 \Longrightarrow \zeta = \mu = 0,$$

then (11.22) reduces to

$$D^+v\left[\varepsilon^{\nu-1}(t)\right] \leq -2kV^{1/2}(\varepsilon^{\nu-1}).$$

The solution is

$$v\left[\varepsilon^{\nu-1}(t)\right] \leq \left\{ \begin{array}{c} \left[v^{1/2}(\varepsilon_0^{\nu-1}) - kt\right]^2,\ t \in [0, \tau_R], \\ 0,\ t \geq \tau_R \end{array} \right\},$$

i.e.,

$$\mathbf{v}^T\left|\varepsilon^{\nu-1}(t)\right| \leq \left\{ \begin{array}{c} \left[\left(\mathbf{v}^T\left|\varepsilon_0^{\nu-1}\right|\right)^{1/2} - kt\right]^2,\ t \in [0, \tau_R], \\ 0,\ t \geq \tau_R \end{array} \right\},$$

with

$$\tau_R = k^{-1}\sqrt{\mathbf{v}^T\left|\varepsilon_0^{\nu-1}\right|},\ \forall\left(\mathbf{D}(.), \varepsilon_0^{\nu-1}\right) \in \mathfrak{D}^{\mu_{Pd}} x \mathfrak{R}^{\nu N}.$$

The control system exhibits the global stablewise tracking with the finite scalar reachability time $\tau_R = k^{-1}\sqrt{\mathbf{v}^T\left|\varepsilon^{\nu-1}\right|}$.

11.2.4 Choice of a vector Lyapunov function

Various forms of a vector Lyapunov function are shown in Example 320 and in Example 321 (Subsection 11.1). The simplest one is

$$\mathbf{v}(\varepsilon) = \frac{1}{2}E\varepsilon.$$

Its total time derivative along $\varepsilon(t) = \varepsilon(t;\varepsilon_0) = \mathbf{Y}_d(t)$-$\mathbf{Y}(t)$ reads

$$\mathbf{v}^{(1)}(\varepsilon) = E\varepsilon^{(1)} = E\left[\mathbf{Y}_d^{(1)}(t) - \mathbf{Y}^{(1)}(t)\right].$$

We can solve the plant mathematical model (11.5) for $\mathbf{Y}^{(1)}(t)$ if, and only if,

$$\det A_{P1} \neq 0. \tag{11.24}$$

We continue the consideration for the subclass of the plants (11.5) for which (11.24) holds. Then

$$\mathbf{Y}^{(1)}(t) = A_{P1}^{-1}\left[C_{Pu}^{(\mu_{Pu})}\mathbf{U}^{\mu_{Pu}}(t) + D_{Pd}^{(\mu_{Pd})}\mathbf{D}^{\mu_{Pd}}(t) - A_P^{(\nu)}\mathbf{Y}^{\nu}(t) + A_{P1}\mathbf{Y}^{(1)}(t)\right],$$

so that

$$\mathbf{v}^{(1)}(\varepsilon) = E\left\{\mathbf{Y}_d^{(1)}(t) - A_{P1}^{-1}\left[\begin{array}{c} C_{Pu}^{(\mu_{Pu})}\mathbf{U}^{\mu_{Pu}}(t) + D_{Pd}^{(\mu_{Pd})}\mathbf{D}^{\mu_{Pd}}(t) - \\ -A_P^{(\nu)}\mathbf{Y}^{\nu}(t) + A_{P1}\mathbf{Y}^{(1)}(t). \end{array}\right]\right\}. \tag{11.25}$$

Notice that

$$A_P^{(\nu)}\mathbf{Y}^{\nu}(t) - A_{P1}\mathbf{Y}^{(1)}(t) = \sum_{i=0,\ i\neq 1}^{i=\nu} A_{Pi}\mathbf{Y}^{(i)}(t).$$

This shows that $\mathbf{v}^{(1)}(\varepsilon)$ does not depend on $\mathbf{Y}^{(1)}(t)$, but it depends on all other derivatives $\mathbf{Y}^{(k)}(t)$ of $\mathbf{Y}(t)$ up to the order ν. We accept (11.26),

$$C_{Pu}^{(\mu_{Pu})}\mathbf{U}^{\mu_{Pu}}(t) = A_{P1}\mathbf{Y}_d^{(\nu)}(t) + A_P^{(\nu)}\mathbf{Y}^{\nu}(t) - A_{P1}\mathbf{Y}^{(1)}(t) + A_{P1}\mathbf{w}(t). \tag{11.26}$$

It and (11.25) yield

$$\mathbf{v}^{(1)}(\varepsilon) = E\left\{\mathbf{Y}_d^{(1)}(t) - A_{P1}^{-1}\left[\begin{array}{c} A_{P1}\mathbf{Y}_d^{(\nu)}(t) + A_{P1}\mathbf{w}(t) + \\ +D_{Pd}^{(\mu_{Pd})}\mathbf{D}^{\mu_{Pd}}(t) \end{array}\right]\right\} \Longrightarrow$$

$$\mathbf{v}^{(1)}(\varepsilon) = -E\left[\mathbf{w}(t) + A_{P1}^{-1}D_{Pd}^{(\mu_{Pd})}\mathbf{D}^{\mu_{Pd}}(t)\right].$$

We define

$$\mathbf{w}(t) = \left[S(\varepsilon)\left|A_{P1}^{-1}D_{Pd}^{(\mu_{Pd})}\right|\mathbf{D}_M^{\mu_{Pd}} + 2\beta\zeta\varepsilon + 2\mu K sign\varepsilon\right],$$
$$K = diag\{k_1\ \ k_2\ ...\ k_N\} \in \mathfrak{R}^{+N\times N},$$
$$sign\varepsilon = [sign\varepsilon_1\ \ sign\varepsilon_2\ ...\ sign\varepsilon_N]^T,$$
$$S(\varepsilon) = diag\{sign\varepsilon_1\ \ sign\varepsilon_2\ ...\ sign\varepsilon_N\},$$

in order to get

$$\mathbf{v}^{(1)}(\varepsilon) \leq -E[2\beta\zeta\varepsilon + 2\mu K sign\varepsilon]. \tag{11.27}$$

The control is finally determined as the solution to the following differential equation:

$$C_{Pu}^{(\mu_{Pu})}\mathbf{U}^{\mu_{Pu}}(t) = A_{P1}\mathbf{Y}_d^{(\nu)}(t) + A_P^{(\nu)}\mathbf{Y}^{\nu}(t) - A_{P1}\mathbf{Y}^{(1)}(t) +$$
$$+A_{P1}\left[S(\varepsilon)\left|A_{P1}^{-1}D_{Pd}^{(\mu_{Pd})}\right|\mathbf{D}_M^{\mu_{Pd}} + 2\beta\zeta\varepsilon + 2\mu K\, sign\varepsilon\right],$$
$$\beta, k \in \mathfrak{R}^+,\ \zeta, \mu \in \{0,1\},\ \zeta + \mu = 1. \quad (11.28)$$

Comment 337 *Equation (11.28) determines the VLF dynamic controller. It generates the dynamic tracking control that is robust relative to the disturbance vector* $\mathbf{D}(t)$. *It ensures a good tracking quality as shown in the sequel. The control* $\mathbf{U}^{\mu_{Pu}}(t)$ *is bounded on* $\mathfrak{D}^{\mu_{Pd}}$. *The drawback is the need to measure all derivatives of the output vector up to the order* ν.

Case 338 ***Control synthesis for global elementwise exponential tracking***

If

$$\zeta = 1 \Longrightarrow \mu = 0,$$

then (11.27) becomes

$$\mathbf{v}^{(1)}(\varepsilon) \leq -2\beta E\varepsilon = -\, 2\beta\mathbf{v}(\varepsilon).$$

For

$$\mathbf{v}(\varepsilon) > \mathbf{0}_N,$$

i.e., for

$$V(\varepsilon) = diag\{v_1(\varepsilon) \ \ v_2(\varepsilon) \ \ ... \ v_N(\varepsilon)\} > O_N,$$

we can write

$$V^{-1}(\varepsilon)\mathbf{v}^{(1)}(\varepsilon) \leq -2\beta\mathbf{1}_N.$$

The solution

$$\ln \mathbf{v}(\varepsilon)_{\mathbf{v}(\varepsilon_0)}^{\mathbf{v}(\varepsilon)} = \ln\left[V^{-1}(\varepsilon_0)\mathbf{v}(\varepsilon(t;\varepsilon_0))\right] \leq -2\beta t\mathbf{1}_N$$

has another form:

$$V^{-1}(\varepsilon_0)\mathbf{v}\left[\varepsilon(t;\varepsilon_0)\right] \leq \exp(-2\beta t)\,\mathbf{1}_N,$$

i.e.,

$$\mathbf{v}\left[\varepsilon(t;\varepsilon_0)\right] \leq \exp(-2\beta t)\,\mathbf{v}(\varepsilon_0).$$

Equivalently,

$$E(t;\varepsilon_0)\,\varepsilon(t;\varepsilon_0) \leq \exp(-2\beta t)\,E_0\varepsilon_0,$$

or more simply,

$$|\varepsilon(t;\varepsilon_0)| \leq \exp(-\beta t)\,|\varepsilon_0|,\ \ \forall(\varepsilon_0, t) \in \mathfrak{R}^N \mathrm{x} \mathfrak{T}_0.$$

This expresses the global elementwise exponential tracking.

Case 339 ***Control synthesis for global elementwise stablewise tracking with the finite vector reachability time*** $\boldsymbol{\tau}_R^N$

If

$$\mu = 1 \Longrightarrow \zeta = 0,$$

then (11.27) takes the following form

$$\mathbf{v}^{(1)}(\boldsymbol{\varepsilon}) \leq -2KEsign\boldsymbol{\varepsilon} = -2K\left|\boldsymbol{\varepsilon}\right| = -2K\mathbf{v}^{1/2}(\boldsymbol{\varepsilon}),$$
$$\mathbf{v}^{1/2} = \left[v_1^{1/2}(\boldsymbol{\varepsilon}) \;\; v_2^{1/2}(\boldsymbol{\varepsilon}) \;\; ... \;\; v_N^{1/2}(\boldsymbol{\varepsilon})\right]^T,$$

or

$$v^{-1/2}(\boldsymbol{\varepsilon})\mathbf{v}^{(1)}(\boldsymbol{\varepsilon}) \leq -2K\mathbf{1}_N \Longrightarrow$$
$$\frac{1}{2}\frac{d\mathbf{v}^{1/2}(\boldsymbol{\varepsilon})}{dt} \leq -K\mathbf{1}_N.$$

We find the solution

$$\mathbf{v}^{1/2}\left[\boldsymbol{\varepsilon}\left(\mathbf{t}^N;\boldsymbol{\varepsilon}_0\right)\right] \leq \mathbf{v}^{1/2}(\boldsymbol{\varepsilon}_0) - K\mathbf{t}^N \Longrightarrow$$
$$E^{1/2}\left(\mathbf{t}^N;\boldsymbol{\varepsilon}_0\right)\boldsymbol{\varepsilon}^{1/2}\left(\mathbf{t}^N;\boldsymbol{\varepsilon}_0\right) \leq \left[E_0^{1/2}\boldsymbol{\varepsilon}_0^{1/2} - K\mathbf{t}^N\right] \Longrightarrow$$
$$\left|\boldsymbol{\varepsilon}\left(\mathbf{t}^N;\boldsymbol{\varepsilon}_0\right)\right|\left\{\begin{array}{ll} \leq \left(\left|\boldsymbol{\varepsilon}_0\right| - K\mathbf{t}^N\right), & \forall\mathbf{t}^N \in \left[\mathbf{0}_N,\, \boldsymbol{\tau}_R^N\right] \\ = \mathbf{0}_N, & \forall\mathbf{t}^N \in \left[\boldsymbol{\tau}_R^N,\, \infty\mathbf{1}_N\right[\end{array}\right\},\ \forall\boldsymbol{\varepsilon}_0 \in \mathfrak{R}^N,$$
$$\boldsymbol{\tau}_R^N = K^{-1}\left|\boldsymbol{\varepsilon}_0\right|.$$

The tracking is global elementwise stablewise with finite vector reachability time $\boldsymbol{\tau}_R^N$*. The output error convergence to the zero error vector is elementwise with the constant vector speed* $\mathbf{k}$*,* $\mathbf{k} = [k_1 \;\; k_2 \;\; ... \;\; k_N]^T \in \mathfrak{R}^{+N}$*. There are not oscillations, overshoot and undershoot.*

Comment 340 *LTC of the IO plant is nonlinear.*

11.3 LTC of the *ISO* plant

11.3.1 Arbitrary scalar Lyapunov function

The mathematical model of the *ISO* plant (2.33), (2.34) (Subsection 2.2.2),

$$\frac{d\mathbf{X}_P(t)}{dt} = A_P\mathbf{X}_P(t) + B_P\mathbf{U}(t) + L_P\mathbf{D}(t),\ \mathbf{X}_P \in R^{n_P}, \tag{11.29}$$
$$\mathbf{Y}(t) = \mathbf{Y}_P(t) = C_P\mathbf{X}_P(t) + H_P\mathbf{U}(t) + D_P\mathbf{D}(t),\ C_P \in R^{N\mathrm{x}n_P}, \tag{11.30}$$

does not contain a derivative of the output vector $\mathbf{Y}$. It appears reasonable to use a tentative Lyapunov function $v(.)$ dependent only on the output error vector $\boldsymbol{\varepsilon} = \mathbf{Y}_d - \mathbf{Y}$, $v(.) : \mathfrak{R}^N \longrightarrow \mathfrak{R}_+$. For the sake of the simplicity of the consideration let the function $v(.)$ be continuously differentiable, $v(\boldsymbol{\varepsilon}) \in \mathfrak{C}^1\left(\mathfrak{R}^N\right)$. Its total time derivative $v^{(1)}\left[\boldsymbol{\varepsilon}(t)\right]$ along motions of the ISO plant (11.29), (11.30) is expressed via its gradient $gradv(\boldsymbol{\varepsilon})$,

$$gradv(\boldsymbol{\varepsilon}) = \left[\begin{array}{ccccc} \frac{\partial v(\boldsymbol{\varepsilon})}{\partial \varepsilon_1} & \frac{\partial v(\boldsymbol{\varepsilon})}{\partial \varepsilon_2} & \cdots & \frac{\partial v(\boldsymbol{\varepsilon})}{\partial \varepsilon_{N-1}} & \frac{\partial v(\boldsymbol{\varepsilon})}{\partial \varepsilon_N} \end{array}\right]^T,$$

$$v^{(1)}\left[\boldsymbol{\varepsilon}(t)\right] = \frac{dv\left[\boldsymbol{\varepsilon}(t)\right]}{dt} = \left[gradv(\boldsymbol{\varepsilon})\right]^T \frac{d\boldsymbol{\varepsilon}(t)}{dt} = \left[gradv(\boldsymbol{\varepsilon})\right]^T \frac{d\left[\mathbf{Y}_d(t) - \mathbf{Y}(t)\right]}{dt},$$

i.e.,

$$v^{(1)}\left[\boldsymbol{\varepsilon}(t)\right] = \left[gradv(\boldsymbol{\varepsilon})\right]^T \frac{d\mathbf{Y}_d(t)}{dt} - \left[gradv(\boldsymbol{\varepsilon})\right]^T \frac{d\mathbf{Y}(t)}{dt}. \quad (11.31)$$

The derivative $\mathbf{Y}^{(1)}(t)$ follows from (11.30),

$$\mathbf{Y}^{(1)}(t) = C_P\mathbf{X}_P^{(1)}(t) + H_P\mathbf{U}^{(1)}(t) + D_P\mathbf{D}^{(1)}(t)$$

and from (11.29),

$$\mathbf{Y}^{(1)}(t) = C_P\left[A_P\mathbf{X}_P(t) + B_P\mathbf{U}(t) + L_P\mathbf{D}(t)\right] + H_P\mathbf{U}^{(1)}(t) + D_P\mathbf{D}^{(1)}(t)$$

i.e.,

$$\mathbf{Y}^{(1)}(t) = C_PA_P\mathbf{X}_P(t) + C_PB_P\mathbf{U}(t) + H_P\mathbf{U}^{(1)}(t) + C_PL_P\mathbf{D}(t) + D_P\mathbf{D}^{(1)}(t). \quad (11.32)$$

This transforms (11.31) into

$$v^{(1)}\left[\boldsymbol{\varepsilon}(t)\right] = \left[gradv(\boldsymbol{\varepsilon})\right]^T \left[\frac{d\mathbf{Y}_d}{dt} - C_PA_P\mathbf{X}_P(t) - C_PB_P\mathbf{U}(t) - H_P\mathbf{U}^{(1)}(t)\right] - \\ - \left[gradv(\boldsymbol{\varepsilon})\right]^T \left[C_PL_P\mathbf{D}(t) + D_P\mathbf{D}^{(1)}(t)\right]. \quad (11.33)$$

Assumption 341 *The state vector $\mathbf{X}_P$ of the ISO plant (11.29), (11.30) is measurable.*

This requirement is analogous to the demand (Assumption 324, Subsection 11.2.1) for the measurability of the output vector $\mathbf{Y}(t)$ and its derivatives in the framework of the IO plants (11.5) (Subsection 11.2.1).

Assumption 342 *i) The dimension r of the control vector $\mathbf{U}$ is not less than the dimension N of the output vector $\mathbf{Y}$,*

$$r \geq N.$$

ii) Either

$$rankH_P = N, \ if \ H_P \neq O_{N,r},$$

or

$$rankC_PB_P = N \ if \ H_P = O_{N,r}.$$

The condition i) is reasonable. In fact it is necessary in the case when every output variable should be controlled independently of other output variables. It is a necessary condition for the *ISO* plant trackability (Theorem 285 and Theorem 287 in Subsection 9.3.2, and Theorem 298, Theorem 300 and Theorem 302 in Subsection 9.4.2). The second condition restricts the further consideration to a special class of the *ISO* plants (11.29), (11.30). Under these conditions the control $\mathbf{U}$ is determined as follows.

We continue the consideration for the class of the *ISO* plants (11.29), (11.30) for which the preceding assumptions hold. Otherwise, other Lyapunov oriented methods (e.g., adaptive control schemes, sliding mode or robust control synthesis approaches) should be adapted to the tracking requirements.

Let

$$\mathbf{D}_M \geq \sup \left| C_P L_P \mathbf{D}(t) + D_P \mathbf{D}^{(1)}(t) \right|, \ \forall [\mathbf{D}(.), t] \in \mathfrak{D}^1 \mathrm{x} \mathfrak{T}_0. \tag{11.34}$$

The majorization of

$$\sup \left| C_P L_P \mathbf{D}(t) + D_P \mathbf{D}^{(1)}(t) \right|$$

over $[\mathbf{D}(.), t] \in \mathfrak{D}^1 \mathrm{x} \mathfrak{T}_0$ ensures the full control robustness relative to the disturbance vector $\mathbf{D}(t)$ and eliminates the need to measure the disturbance instantaneous value.

We define the control $\mathbf{U}(t)$ as the solution to

$$H_P \mathbf{U}^{(1)}(t) + C_P B_P \mathbf{U}(t) = \mathbf{w}(t), \tag{11.35}$$

so that (11.33) becomes

$$v^{(1)}\left[\varepsilon(t)\right] = \left[gradv(\varepsilon)\right]^T \left[\frac{d\mathbf{Y}_d}{dt} - C_P A_P \mathbf{X}_P(t) - \mathbf{w}(t)\right] -$$
$$- \left[gradv(\varepsilon)\right]^T \left[C_P L_P \mathbf{D}(t) + D_P \mathbf{D}^{(1)}(t)\right], \ \forall [\mathbf{D}(.), t] \in \mathfrak{D}^1 \mathrm{x} \mathfrak{T}_0.$$

If $H_P \neq O_{N,r}$, hence $rank\ H_P = N$ (Assumption 342), then the control $\mathbf{U}(t)$ is determined from (11.35) by

$$\mathbf{U}(t) = H_P^T \left(H_P H_P^T\right)^{-1} \mathbf{Z}(t), \ \mathbf{Z}^{(1)}(t) + C_P B_P H_P^T \left(H_P H_P^T\right)^{-1} \mathbf{Z}(t) = \mathbf{w}(t).$$

Such control is dynamic control.

If $H_P = O_{N,r}$ then $rank\ C_P B_P = N$ (Assumption 342), and the control $\mathbf{U}(t)$ is determined from (11.35) by

$$\mathbf{U}(t) = \left(C_P B_P\right)^T \left[C_P B_P \left(C_P B_P\right)^T\right]^{-1} \mathbf{w}(t).$$

This control is static control.

We define $\mathbf{w}(t)$ by

$$\mathbf{w}(t) = \frac{d\mathbf{Y}_d}{dt} - C_P A_P \mathbf{X}_P(t) +$$
$$+ \|gradv(\boldsymbol{\varepsilon})\|^{-2} [gradv(\boldsymbol{\varepsilon})] \left\{ \begin{array}{c} |gradv(\boldsymbol{\varepsilon})|^T \mathbf{D}_M + 2\boldsymbol{\beta}\zeta v(\boldsymbol{\varepsilon}) + \\ +\mu k sign v(\boldsymbol{\varepsilon}_0) + 2\eta k v^{1/2}(\boldsymbol{\varepsilon}) \end{array} \right\}$$
$$\boldsymbol{\beta}, k \in \mathfrak{R}^+, \zeta, \mu, \eta \in \{0,1\}, \ \zeta + \mu + \eta = 1. \tag{11.36}$$

This and (11.35) determine the control $\mathbf{U}(t)$. Equations (11.33)-(11.36) lead to

$$v^{(1)}[\boldsymbol{\varepsilon}(t)] \le -\left\{2\boldsymbol{\beta}\zeta v[\boldsymbol{\varepsilon}(t)] + \mu k sign v(\boldsymbol{\varepsilon}_0) + 2\eta k v^{1/2}[\boldsymbol{\varepsilon}(t)]\right\},$$
$$\forall [\mathbf{D}(.), t] \in \mathfrak{D}^1 x \mathfrak{T}_0. \tag{11.37}$$

Comments 325 through 327 (Subsection 11.2.1) are essentially applicable also herein.

11.3.2 Choice of a scalar Lyapunov function

The selection of Lyapunov function $v(.)$ is to be a quadratic form,

$$v(\boldsymbol{\varepsilon}) = \boldsymbol{\varepsilon}^T H \boldsymbol{\varepsilon}, \ H = H^T > O_N, \ gradv(\boldsymbol{\varepsilon}) = 2H\boldsymbol{\varepsilon}, \tag{11.38}$$

is simple. Equation (11.36) takes the following form

$$\mathbf{w}(t) = \frac{d\mathbf{Y}_d}{dt} - C_P A_P \mathbf{X}_P(t) +$$
$$+ \|H\boldsymbol{\varepsilon}\|^{-2} [H\boldsymbol{\varepsilon}] \left\{ \begin{array}{c} 2|H\boldsymbol{\varepsilon}|^T \mathbf{D}_M + 4\boldsymbol{\beta}\zeta \boldsymbol{\varepsilon}^T H \boldsymbol{\varepsilon} + \\ +\mu k sign(\boldsymbol{\varepsilon}_0^T H \boldsymbol{\varepsilon}_0) + 2\eta k (\boldsymbol{\varepsilon}^T H \boldsymbol{\varepsilon})^{1/2} \end{array} \right\}. \tag{11.39}$$

Inequality (11.37) becomes the following due to (11.38):

$$v^{(1)}[\boldsymbol{\varepsilon}(t)] \le -\left\{2\boldsymbol{\beta}\zeta \boldsymbol{\varepsilon}^T(t) H \boldsymbol{\varepsilon}(t) + \mu k sign(\boldsymbol{\varepsilon}_0^T H \boldsymbol{\varepsilon}_0) + 2\eta k \left[\boldsymbol{\varepsilon}^T(t) H \boldsymbol{\varepsilon}(t)\right]^{1/2}\right\},$$
$$\forall [\mathbf{D}(.), t] \in \mathfrak{D}^1 x \mathfrak{T}_0, \ \boldsymbol{\varepsilon}^T(t) H \boldsymbol{\varepsilon}(t) = v[\boldsymbol{\varepsilon}(t)]. \tag{11.40}$$

Case 343 ***Control synthesis for global exponential tracking***
If

$$\zeta = 1 \Longrightarrow \mu = \eta = 0,$$

then (11.40) reduces to

$$v^{(1)}[\boldsymbol{\varepsilon}(t)] \le -2\boldsymbol{\beta} v[\boldsymbol{\varepsilon}(t)], \ \forall [\mathbf{D}(.), t] \in \mathfrak{D}^1 x \mathfrak{T}_0, .$$

The solution reads

$$\|\boldsymbol{\varepsilon}(t)\| \le \alpha e^{-\beta t} \|\boldsymbol{\varepsilon}_0\|, \forall \left(t, \mathbf{D}(.), \boldsymbol{\varepsilon}_0^1\right) \in \mathfrak{T}_0 x \mathfrak{D}^1 x \mathfrak{R}^N,$$
$$\alpha = \sqrt{\lambda_M(H)\lambda_m^{-1}(H)}.$$

This proves the global exponential tracking on $\mathfrak{D}^1$.

Case 344 ***Control synthesis for global stablewise tracking with the finite reachability time***

If

$$\mu = 1 \Longrightarrow \zeta = \eta = 0,$$

then (11.40) becomes the following due to (11.38):

$$v^{(1)}\left[\varepsilon\left(t\right)\right] \leq -ksignv(\varepsilon_0),\ \forall\left(t, \mathbf{D}(.), \varepsilon_0\right) \in \mathfrak{T}_0 \text{x} \mathfrak{D}^1 \text{x} \mathfrak{R}^N,\ v(\varepsilon_0) = \varepsilon_0^T H \varepsilon_0.$$

The solution reads

$$\left\|\varepsilon\left(t\right)\right\| \leq \left\{\begin{array}{ll} \lambda_m^{-1/2}(H)\sqrt{\varepsilon_0^T H \varepsilon_0 - kt}, & t \leq k^{-1}\varepsilon_0 H \varepsilon_0, \\ 0, & t \geq k^{-1}\varepsilon_0 H \varepsilon_0 \end{array}\right\},$$
$$\forall\left(t, \mathbf{D}(.), \varepsilon_0\right) \in \mathfrak{T}_0 \text{x} \mathfrak{D}^1 \text{x} \mathfrak{R}^N.$$

This proves the stablewise tracking on $\mathfrak{D}^1$ with the finite scalar reachability time τ_R

$$\tau_R = k^{-1}\varepsilon_0 H \varepsilon_0 = \tau_R\left(\varepsilon_0; k\right),$$

which depends on ε_0. If the reachability time τ_R is given, then we determine the gain k from

$$k = \tau_R^{-1}\varepsilon_0 H \varepsilon_0 = k\left(\varepsilon_0; \tau_R\right).$$

These equations express the relationship among the initial error vector ε_0, the reachability time τ_R and the gain k.

Case 345 ***Control synthesis for global stablewise tracking with the finite reachability time***

If

$$\eta = 1 \Longrightarrow \zeta = \mu = 0,$$

then (11.40) takes the following form due to (11.38):

$$v^{(1)}\left[\varepsilon\left(t\right)\right] \leq -2kv^{1/2}\left[\varepsilon\left(t\right)\right],\ \forall\left(t, \mathbf{D}(.), \varepsilon_0\right) \in \mathfrak{T}_0 \text{x} \mathfrak{D}^1 \text{x} \mathfrak{R}^N.$$

We find the solution in the form

$$\left\|\varepsilon\left(t\right)\right\| \leq \left\{\begin{array}{ll} \lambda_m^{-1/2}(H)\left[\sqrt{\varepsilon_0 H \varepsilon_0} - kt\right]^2, & t \in [0, \tau_R], \\ 0, & t \geq \tau_R \end{array}\right\},$$
$$\tau_R = k^{-1}\sqrt{\varepsilon_0 H \varepsilon_0},\ \forall\left(t, \mathbf{D}(.), \varepsilon_0\right) \in \mathfrak{T}_0 \text{x} \mathfrak{D}^1 \text{x} \mathfrak{R}^N.$$

This proves the stablewise tracking on $\mathfrak{D}^1$ with the finite reachability time τ_R

$$\tau_R = k^{-1}\sqrt{\varepsilon_0 H \varepsilon_0} = \tau_R\left(\varepsilon_0; k\right),$$

which depends on ε_0. If τ_R is given, then we calculate the gain k from

$$k = \tau_R^{-1}\sqrt{\varepsilon_0 H \varepsilon_0} = k\left(\varepsilon_0; \tau_R\right).$$

The smaller the reachability time τ_R, the bigger the gain k for the given initial output error vector ε_0, and vice versa.

Comment 332 (Subsection 11.2) is valid also in this framework.

11.3.3 Choice of a vector Lyapunov function

The usage of the vector Lyapunov function

$$\mathbf{v}(\varepsilon) = \frac{1}{2}E\varepsilon$$

of Example 320 (Subsection 11.1.4) in this framework leads to

$$\mathbf{v}^{(1)}(\varepsilon) = E\varepsilon^{(1)} = E\left[\mathbf{Y}_d^{(1)}(t) - \mathbf{Y}^{(1)}(t)\right] =$$

$$= E\left[\begin{array}{c} \mathbf{Y}_d^{(1)}(t) - C_P A_P \mathbf{X}_P(t) - C_P B_P \mathbf{U}(t) - H_P \mathbf{U}^{(1)}(t) - \\ -C_P L_P \mathbf{D}(t) - D_P \mathbf{D}^{(1)}(t). \end{array}\right]$$

due to (11.32). Let

$$H_P \mathbf{U}^{(1)}(t) + C_P B_P \mathbf{U}(t) = \mathbf{Y}_d^{(1)}(t) - C_P A_P \mathbf{X}_P(t) + \mathbf{w}(t)$$

so that

$$\mathbf{v}^{(1)}(\varepsilon) = -E\left[\mathbf{w}(t) + C_P L_P \mathbf{D}(t) + D_P \mathbf{D}^{(1)}(t)\right].$$

Let $\mathbf{\Psi}(.) : \mathfrak{T}_0 \longrightarrow \mathfrak{R}^{+^N}$ obey

$$\mathbf{\Psi}(t) > \left|C_P L_P \mathbf{D}(t) + D_P \mathbf{D}^{(1)}(t)\right|, \ \forall (t, \mathbf{D}(.)) \in \mathfrak{T}_0 \mathrm{x} \mathfrak{D}^1,$$

and

$$\begin{array}{c} \mathbf{w}(t) = S(\varepsilon)\mathbf{\Psi}(t) + 2\boldsymbol{\beta}\zeta\varepsilon + \mu K sign\varepsilon, \\ \boldsymbol{\beta}, k \in \mathfrak{R}^+, \ \zeta, \mu \in \{0,1\}, \ \zeta + \mu = 1. \end{array}$$

The control is finally determined as the solution to

$$\begin{array}{c} H_P \mathbf{U}^{(1)}(t) + C_P B_P \mathbf{U}(t) = \mathbf{Y}_d^{(1)}(t) - C_P A_P \mathbf{X}_P(t) + \\ +S(\varepsilon)\mathbf{\Psi}(t) + +2\boldsymbol{\beta}\zeta\varepsilon + \mu K sign\varepsilon, \end{array}$$

and the derivative $\mathbf{v}^{(1)}(\varepsilon)$ satisfies

$$\mathbf{v}^{(1)}(\varepsilon) \leq -E\left(2\boldsymbol{\beta}\zeta\varepsilon + \mu K sign\varepsilon\right). \tag{11.41}$$

Case 346 ***Control synthesis for global elementwise exponential tracking***

This is the Case 338 in which

$$\zeta = 1 \Longrightarrow \mu = 0,$$

and the solution to (11.41) is found in the form

$$|\varepsilon(t;\varepsilon_0)| \leq \exp(-\boldsymbol{\beta}t)|\varepsilon_0|, \ \forall (\varepsilon_0, t) \in \mathfrak{R}^N \mathrm{x} \mathfrak{T}_0.$$

This proves the global elementwise exponential tracking.

Case 347 ***Control synthesis for global elementwise stablewise tracking with finite vector reachability time $\boldsymbol{\tau}_R^N$***

For

$$\mu = 1 \Longrightarrow \zeta = 0,$$

the solution to (11.41) is determined in the Case 339 as

$$|\boldsymbol{\varepsilon}(t;\boldsymbol{\varepsilon}_0)| \left\{ \begin{array}{c} \leq (|\boldsymbol{\varepsilon}_0| - tK\mathbf{1}_N), \ \forall t\mathbf{1}_N \in \left[\mathbf{0}_N, \ \boldsymbol{\tau}_R^N\right] \\ = \mathbf{0}_N, \ \forall t\mathbf{1}_N \in [\boldsymbol{\tau}_R^N, \ \infty\mathbf{1}_N[\end{array} \right\}, \ \forall \boldsymbol{\varepsilon}_0 \in \mathfrak{R}^N,$$
$$\boldsymbol{\tau}_R^N = K^{-1}|\boldsymbol{\varepsilon}_0|.$$

The tracking is global elementwise stablewise with the finite vector reachability time $\boldsymbol{\tau}_R^N$. The output error convergence is in the linear form with the constant speed K to the zero error vector. There are not oscillations, overshoot and undershoot in the output error vector.

The equation

$$\boldsymbol{\tau}_R^N = K^{-1}|\boldsymbol{\varepsilon}_0|$$

shows the elementwise trade off among $\boldsymbol{\tau}_R^N$, K and $|\boldsymbol{\varepsilon}_0|$.

Comment 348 *LTC of the linear ISO plant is nonlinear.*

Chapter 12

Natural Tracking Control (NTC)

12.1 Concept of NTC

What does the nature, i.e., the brain as the part of the nature, use to create control of any organ? It uses evidently information about the error ε of the real organ behavior $\mathbf{Y}(.)$ relative to its desired behavior $\mathbf{Y}_d(.)$. But, this is not the only information that the brain uses to create the control. For example, in order to control the position of a hand, of a finger, of a leg, the brain uses information about the difference between their desired and real positions, which is information about their position errors. However, the brain simultaneously uses information about the forces of the muscles acting on the organs. The muscle force is a control variable. The brain, as the central part of *the natural controller*, uses information about the (realized) control itself. This is one essential characteristic of the control created by the brain, i.e., by the nature.

The brain, in general the nature, does not have any information about a mathematical model of the controlled organ. This is another crucial characteristic of the control created by the brain, i.e., by the nature.

Definition 349 *Natural Control (NC)*

A control $\mathbf{U}$ *is* ***Natural Control (NC)*** *if, and only if:*

1. it obeys the Time Continuity and Uniqueness Principle (TCUP, Principle 9, Chapter 1),

2. its synthesis and effective implementation use information about both the output error vector $\boldsymbol{\varepsilon}$ *(and possibly its derivatives and/or its integral) and the control action* $\mathbf{U}$ *itself,*

3. its synthesis and effective implementation do not use information either about the plant mathematical model or about the mathematical description of the plant internal dynamics or about the real instantaneous values of disturbances,

$$\mathbf{U} = \mathbf{U}(\boldsymbol{\varepsilon}, \mathbf{U}),\ \mathbf{U}(t) \in \mathfrak{C}(\mathfrak{T}_0). \quad (12.1)$$

The controller should possess an internal local feedback from its output to its input in order to generate Natural Control. A mathematical rather than a physical consideration determines clearly and precisely the sign, the character and the strength of such local feedback. We refer to [85]-[87], [132]-[141], [149], [159]-[161], [176]-[180], [249]-[258] for the following definition.

Definition 350 ***Natural Tracking Control** (NTC)*

Natural Control is ***Natural Tracking Control*** *(NTC) if, and only if, it ensures a (demanded) type of tracking determined by a tracking algorithm described by an operator* $\boldsymbol{T}(.)$,

$$\mathbf{U} = \mathbf{U}(\boldsymbol{\varepsilon}, \mathbf{U}; \mathbf{T}), \mathbf{U}(t) \in \mathfrak{C}(\mathfrak{T}_0). \tag{12.2}$$

We will present and further develop the fundamentals of the *NTC* theory, the mathematical root of which is in the papers [105, Note 11, p. 19],[128, Note 11, p. S-38]. The papers showed the mathematical possibility to replace the internal object dynamics and the external disturbance action by the control used to compensate completely their influence on the object behavior. The mathematics showed that such control demands *the unit positive local feedback without delay in the controller.* The unit positive feedback without delay is forbidden in the control theory because such isolated feedback system is totally unstable and will blow immediately in reality. Z. B. Ribar and this author simulated effectively on an analog computer the *NTC* of a second order linear plant in the Laboratory of Automatic Control, Faculty of Mechanical Engineering, Belgrade University, Serbia (Spring 1988). The feedback *NT* controller is in the closed loop of the overall control system. Its local unit positive feedback operates in the full harmony with the global negative feedback of the control system. This is the control principle that is the basis of the life of every human cell and of the whole organism. Such control is *self adaptive control.* The further development of it showed that its more adequate name is the ***Natural Tracking Control*** (*NTC*) [85]-[87], [132]-[141], [149], [159]-[161], [176]-[180], [249]-[258]. In the papers [132]-[141], [176]-[180], [249]-[258]. *William Pratt Mounfield, Jr.* worked out all the examples by solving the difficult problem of digital simulations of the plant behavior controlled by time-continuous *NTC* that incorporates the local unit positive feedback. He was the first to do such simulations successfully and to show effective applications to technical plants. Other developments of the *NTC* and of its various applications to control of continuous-time technical plants can be found in the Ph. D./D. Sci. dissertations by A. Kökösy [206] and D. V. Lazitch [220], in the papers by N. Nedić (Neditch) and D. Pršić (Prshitch) [261]-[263], [285], Z. B. Ribar et al. [288], [289], and in the M. Sci. thesis by M. R. Jovanović (Yovanovitch) [325].

We can define a demanded type of tracking in the form of a solution to a differential equation in the output error $\boldsymbol{\varepsilon}$ (3.6) (Subsection 3.3.1) and its derivatives and/or its integral,

$$\mathbf{T}\left(t, \boldsymbol{\varepsilon}(t), \boldsymbol{\varepsilon}^{(1)}(t), ..., \boldsymbol{\varepsilon}^{(k)}(t), \int_{t_0=0}^{t} \boldsymbol{\varepsilon}(t)\, dt\right) =$$
$$= \mathbf{T}\left(t, \boldsymbol{\varepsilon}^{k}(t), \int_{t_0=0}^{t} \boldsymbol{\varepsilon}(t)\, dt\right) = \mathbf{0}_N, \forall t \in \mathfrak{T}_0,$$
$$k \in \{0, 1, 2, ...\}. \tag{12.3}$$

The use of the *time vector* $\mathbf{t}^{(k+1)N}$ in (8.38) (Subsection 8.4.2),

$$\mathbf{t}^{(k+1)N} = \left[\underbrace{t \;\; t \;\; ...t}_{(k+1)N-times}\right]^T \in \mathfrak{T}_0^{(k+1)N} \cup \{\infty\}^{(k+1)N}, \; k \in \{0, 1, 2, ...\},$$

simplifies formally mathematically the treatment of the vector relationships of the elementwise tracking with the finite vector reachability time $\boldsymbol{\tau}_R^{k+1)N}$. For the same reason let us introduce the error matrix $E(t)$,

$$E(t) = diag\{\varepsilon_1(t) \;\; \varepsilon_2(t) \;\; . \; . \; . \;\; \varepsilon_N(t)\} = E\left(\mathbf{t}^N\right). \tag{12.4}$$

This notation permits us to rewrite (12.3) as

$$\mathbf{T}\left(\mathbf{t}^N, \boldsymbol{\varepsilon}\left(\mathbf{t}^N\right), \boldsymbol{\varepsilon}^{(1)}\left(\mathbf{t}^N\right), ..., \boldsymbol{\varepsilon}^{(k)}\left(\mathbf{t}^N\right), \int_{\mathbf{t}_0^N=\mathbf{0}_N}^{\mathbf{t}^N} E\left(\mathbf{t}^N\right) d\mathbf{t}^N\right) =$$
$$= \mathbf{T}\left(\mathbf{t}^N, \boldsymbol{\varepsilon}^k\left(\mathbf{t}^{(k+1)N}\right), \int_{\mathbf{t}_0^N=\mathbf{0}_N}^{\mathbf{t}^N} E\left(\mathbf{t}^N\right) d\mathbf{t}^N\right) = \mathbf{0}_N,$$
$$\forall \mathbf{t}^{(k+1)N} \in \mathfrak{T}_0^{(k+1)N},$$
$$\mathbf{T}\left(\mathbf{t}^N, \boldsymbol{\varepsilon}^k\left(\mathbf{t}^{(k+1)N}\right), \int_{\mathbf{t}_0^N=\mathbf{0}_N}^{\mathbf{t}^N} E\left(\mathbf{t}^N\right) d\mathbf{t}^N\right) \in \mathfrak{C}\left(\mathfrak{R}^{(k+2)N}\right),$$
$$k \in \{0, 1, 2, ...\}. \tag{12.5}$$

The control should force the plant behavior to satisfy the demanded tracking quality specified by (12.5). The basic task of the control synthesis is to determine such control.

The following properties of *the vector tracking operator* $T(.)$ determine the class of *the tracking algorithms* that express demanded tracking qualities, and which will be herein the basis for the *NTC* synthesis.

Property 351 *If (12.5) holds, then the operator $T(.)$ guarantees that the solution $\boldsymbol{\varepsilon}\left(t; \boldsymbol{\varepsilon}_0^k\right)$ of (12.5) is continuous in time on $\mathfrak{T}_0$,*

$$\mathbf{T}\left(\mathbf{t}^N, \boldsymbol{\varepsilon}^k\left(\mathbf{t}^{(k+1)N}\right), \int_{\mathbf{t}_0^N=\mathbf{0}_N}^{\mathbf{t}^N} E\left(\mathbf{t}^N\right) d\mathbf{t}^N\right) = \mathbf{0}_N, \; \forall \mathbf{t}^{(k+1)N} \in \mathfrak{T}_0^{(k+1)N}$$
$$\Longrightarrow \boldsymbol{\varepsilon}\left(\mathbf{t}^N; \boldsymbol{\varepsilon}_0^k\right) \in \mathfrak{C}(\mathfrak{T}_0^N). \tag{12.6}$$

Property 352 *The operator $T(.)$ has the property to vanish at the origin at every moment,*

$$\varepsilon^k = \mathbf{0}_{(k+1)N} \Longrightarrow \mathbf{T}\left(\mathbf{t}^N, \mathbf{0}_{(k+1)N}, \int_{\mathbf{t}_0^N=\mathbf{0}_N}^{\mathbf{t}^N} O_N d\mathbf{t}^N\right) = \mathbf{0}_N,\ \forall \mathbf{t}^N \in \mathfrak{T}_0^N. \quad (12.7)$$

Property 353 *The solution of (12.5) for all zero initial conditions is identically equal to the zero vector,*

$$\varepsilon_0^k = \mathbf{0}_{(k+1)N} \text{ and}$$
$$\mathbf{T}\left(\mathbf{t}^N, \varepsilon^k\left(\mathbf{t}^{(k+1)N}\right), \int_{\mathbf{t}_0^N=\mathbf{0}_N}^{\mathbf{t}^N} E\left(\mathbf{t}^N\right) d\mathbf{t}^N\right) = \mathbf{0}_N,\ \forall \mathbf{t}^N \in \mathfrak{T}_0^N \Longrightarrow$$
$$\varepsilon^k\left(\mathbf{t}^{(k+1)N}; \mathbf{0}_{(k+1)N}\right) = \mathbf{0}_{(k+1)N},\ \forall \mathbf{t}^{(k+1)N} \in \mathfrak{T}_0^{(k+1)N}. \quad (12.8)$$

If (12.5) holds, then the condition $\forall \mathbf{t}^{(k+1)N} \in \mathfrak{T}_0^{(k+1)N}$ demands that the initial output error vector, its derivatives and its integral obey (12.5) at the initial moment $\mathbf{t}_0^N = \mathbf{0}_N$:

Property 354 *The initial output error vector, its initial derivatives and its integral obey (12.5) at the initial moment $t_0 = 0$, i.e., at $\mathbf{t}_0^{(k+1)N} = \mathbf{0}_{(k+1)N}$:*

$$\mathbf{T}\left(\mathbf{0}_N, \varepsilon_0^k, \int_{\mathbf{t}_0^N=\mathbf{0}_N}^{\mathbf{t}^N=\mathbf{0}_N} E\left(\mathbf{t}^N\right) d\mathbf{t}^N\right) = \mathbf{T}\left(\mathbf{0}_N, \varepsilon_0^k, \mathbf{0}_N\right) = \mathbf{0}_N,\ \forall \varepsilon_0^k \in \mathfrak{R}^{(k+1)N}. \quad (12.9)$$

Note 355 *The real, the actual, initial output error ε_0 and the initial output error derivatives $\varepsilon_0^{(1)}$, $\varepsilon_0^{(2)}$, ..., $\varepsilon_0^{(k)}$ are unpredictable, uncontrollable and arbitrary. They result from the past behavior of the plant, which is untouchable. They sometimes do not satisfy the condition (12.9), which then implies the violation of (12.5) and its nonrealizability at every $t \in \mathfrak{T}_0$ for such initial conditions.*

Problem 356 ***Matching the error vector, its derivatives and integral with the tracking algorithm on $\mathfrak{T}_0$***

How can the matching of the error vector, error vector derivatives and the error vector integral be ensured with the tracking algorithm on $\mathfrak{T}_0$?

In order to solve the problem we need *the scalar attainability time $\tau_{Ai(j)}$*, $\tau_{Ai(j)} \in \mathfrak{T}_0 \cup \{\infty\}$, of the *j-th* derivative of the *i-th* entry of the error vector ε:

$$0 \leq \tau_{Ai(j)} < \infty \Longrightarrow \left\{\begin{array}{c} \mathfrak{T}_{Ai(j)} = \left\{t : 0 \leq t \leq \tau_{Ai(j)}\right\} \subset \mathfrak{T}_0, \\ \mathfrak{T}_{Ai(j)\infty} = \left\{t :\ \infty > t \geq \tau_{Ai(j)}\right\} \subset \mathfrak{T}_0, \\ \mathfrak{T}_{Ai(j)} \cup \mathfrak{T}_{Ai(j)\infty} = \mathfrak{T}_0, \end{array}\right\},$$
$$\tau_{Ai(j)} = \infty \Longrightarrow \mathfrak{T}_{Ai(j)\infty} = \{\infty\},\ i = 1, 2, ..., N,\ j \in \{0, 1, .., k\}. \quad (12.10)$$

They induce *the vector attainability time* $\boldsymbol{\tau}_A^{(k+1)N} \in \mathfrak{T}_0^{(k+1)N} \cup \{\infty\}^{(k+1)N}$,

$$\boldsymbol{\tau}_A^{(k+1)N} = \left[\tau_{A1} \cdots \tau_{AN}\ \tau_{A1(1)} \cdots \tau_{AN(1)} \cdots\cdots \tau_{A1(k)} \cdots \tau_{AN(k)}\right]^T, \quad (12.11)$$

and the following *time* set products:

$$\mathfrak{T}_A^{k+1)N} = \left\{\mathbf{t}^{k+1)N} : \mathbf{0}_{k+1)N} \leq \mathbf{t}^{k+1)N} \leq \boldsymbol{\tau}_A^{k+1)N}\right\} \subset \mathfrak{T}_0^{k+1)N},$$

$$\mathfrak{T}_{A\infty}^{(k+1)N} = \left\{\mathbf{t}^{(k+1)N} : \mathbf{t}^{(k+1)N} \geq \boldsymbol{\tau}_A^{(k+1)N}\right\} \subseteq \mathfrak{T}_0^{(k+1)N} \cup \{\infty\}^{(k+1)N},$$

$$\mathfrak{T}_A^{(k+1)N} \cup \mathfrak{T}_{A\infty}^{(k+1)N} = \mathfrak{T}_0^{(k+1)N} \cup \{\infty\}^{(k+1)N}, \ k \in \{0,1,2,....\}, \quad (12.12)$$

so that

$$\mathbf{0}_{(k+1)N} \leq \boldsymbol{\tau}_A^{(k+1)N} < \infty \mathbf{1}_{(k+1)N} \Longrightarrow \left\{ \begin{array}{c} \mathfrak{T}_A^{(k+1)N} \subset \mathfrak{T}_0^{(k+1)N}, \\ \mathfrak{T}_{A\infty}^{(k+1)N} \subset \mathfrak{T}_0^{(k+1)N}, \\ \mathfrak{T}_A^{(k+1)N} \cup \mathfrak{T}_{A\infty}^{(k+1)N} = \mathfrak{T}_0^{(k+1)N}, \end{array} \right\},$$

$$\boldsymbol{\tau}_A^{(k+1)N} = \infty \mathbf{1}_{(k+1)N} \Longrightarrow \mathfrak{T}_{A\infty}^{(k+1)N} = \{\infty\}^{(k+1)N},$$
$$k \in \{0,1,2,....\}. \quad (12.13)$$

Notice that

$$\boldsymbol{\tau}_A^{k+1)N} = \infty \mathbf{1}_{k+1)N} \Longleftrightarrow \mathfrak{T}_{A\infty}^{k+1)N} = \{\infty\}^{k+1)N} = \left\{\infty \mathbf{1}_{k+1)N}\right\}. \quad (12.14)$$

Note 357 *The value* τ_A *is infinite,* $\tau_A = \infty$*; hence,* $\mathfrak{T}_{A\infty} = \{\infty\}$*, i.e.,* $\boldsymbol{\tau}_A^{(k+1)N} = \infty \mathbf{1}_{(k+1)N}$ *and* $\mathfrak{T}_{A\infty}^{(k+1)N} = \{\infty\}^{(k+1)N}$*, if, and only if, tracking should be asymptotic. Otherwise* $\tau_A \in \mathfrak{T}_0$*; hence,* $\mathfrak{T}_{A\infty} \subset \mathfrak{T}_0$*, and* $\mathbf{0}_{k+1)N} \leq \boldsymbol{\tau}_A^{k+1)N} < \infty \mathbf{1}_{k+1)N}$ *so that* $\mathfrak{T}_{A\infty}^{k+1)N} = \left\{\mathbf{t}^{k+1)N} : \infty \mathbf{1}_{k+1)N} > \mathbf{t}^{k+1)N} \geq \boldsymbol{\tau}_A^{k+1)N}\right\} \subset \mathfrak{T}_0^{(k+1)N}$.

Solution 358 *The root of the problem 356 is not in a particular initial error, but in the incompatibility of the initial errors with the tracking algorithm (12.5) at the initial moment* $t_0 = 0$ *due to the arbitrariness of the initial errors,*

$$\mathbf{T}\left(\mathbf{0}_N, \boldsymbol{\varepsilon}_0^k, \int_{\mathbf{t}_0^N = \mathbf{0}_N}^{\mathbf{t}^N = \mathbf{0}_N} E\left(\mathbf{t}^N\right) d\mathbf{t}^N\right) = \mathbf{T}\left(0, \boldsymbol{\varepsilon}_0^k, \mathbf{0}_N\right) \neq \mathbf{0}_N,$$
$$for\ some\ \boldsymbol{\varepsilon}_0^k \in \mathfrak{R}^{(k+1)N}. \quad (12.15)$$

It is a fact that control cannot influence such $\boldsymbol{\varepsilon}_0^k$*. Hence, let the demand be reduced so that (12.5) holds only on* $\mathfrak{T}_{A\infty}^{(k+1)N}$ *for such* $\boldsymbol{\varepsilon}_0^k$*,*

$$\mathbf{T}\left(\mathbf{t}^N, \boldsymbol{\varepsilon}^k\left(\mathbf{t}^{(k+1)N}\right), \int_{\mathbf{t}_0^N = \mathbf{0}_N}^{\mathbf{t}^N} E\left(\mathbf{t}^N\right) d\mathbf{t}^N\right) = \mathbf{0}_N, \ \forall \mathbf{t}^{(k+1)N} \in \mathfrak{T}_{A\infty}^{(k+1)N},$$
$$k \in \{0,1,2,...\}, \quad (12.16)$$

instead of on the whole $\mathfrak{T}_0^{(k+1)N}$*. This reduced demand should be satisfied for every output error vector* $\boldsymbol{\varepsilon}_0^k$ *that violates (12.9), (Property 354), i.e., for which (12.15) holds.*

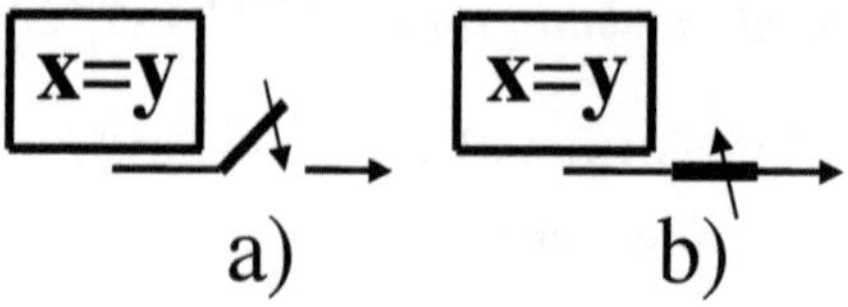

Figure 12.1: a) Switch *closes* if, and only if, $\mathbf{x} = \mathbf{y}$. b) Switch *opens* if, and only if, $\mathbf{x} = \mathbf{y}$.

The preceding analysis opens the need to modify the control goal.

Goal 359 ***Modified control goal***
The control should force the plant to behave so that the following tracking algorithm holds [instead of (12.5)]:

$$\mathbf{T}\left(\mathbf{t}^N, \boldsymbol{\varepsilon}^k, \int_{t_0^N=\mathbf{0}_N}^{\mathbf{t}^N} E d\mathbf{t}^N\right) = \mathbf{0}_N, \ \forall \mathbf{t}^N \left\{ \begin{array}{c} \in \mathfrak{T}_0^N \ if \ \mathbf{T}\left(\mathbf{0}_N; \boldsymbol{\varepsilon}_0^k, \mathbf{0}_N\right) = \mathbf{0}_N \\ \in \mathfrak{T}_{A\infty}^N \ if \ \mathbf{T}\left(\mathbf{0}_N \ ; \boldsymbol{\varepsilon}_0^k, \mathbf{0}_N\right) \neq \mathbf{0}_N \end{array} \right\}, \tag{12.17}$$

$$\boldsymbol{\varepsilon}^k\left(\mathbf{t}^{(k+1)N}; \boldsymbol{\varepsilon}_0^k\right) = -\mathbf{f}^k\left(\mathbf{t}^{(k+1)N}; \mathbf{f}_0^k\right), \ \left\{ \begin{array}{c} \forall \mathbf{t}^{(k+1)N} \in \mathfrak{T}_A^N, \ if \\ \mathbf{T}\left(\mathbf{0}_N, \boldsymbol{\varepsilon}_0^k, \mathbf{0}_N\right) \neq \mathbf{0}_N \end{array} \right\},$$
$$k \in \{0, 1, 2, ...\}, \tag{12.18}$$

for a given or to be determined both $\boldsymbol{\tau}_A^{(k+1)N} \in \mathfrak{T}_0^{(k+1)N} \cup \{\infty\}^{(k+1)N}$ *and for an appropriate subsidiary vector function* $\mathbf{f}(.) : \mathfrak{T}_0^N \longrightarrow \mathfrak{R}^N$.

Solution 360 *Let* $\mathbf{f}\left(t; \mathbf{f}_0^k\right)$ *obey*

$$\mathbf{f}(.) : \mathfrak{T}_0^N \longrightarrow \mathfrak{R}^N, \ \mathbf{f}\left(\mathbf{t}^N\right) \in \mathfrak{C}^i(\mathfrak{T}_0^N), \ i \in \{0, 1, 2, .., k, ...\}, \tag{12.19}$$

$$\mathbf{f}_0^k = \mathbf{f}^k\left(\mathbf{0}_{(k+1)N}; \mathbf{f}_0^k\right) = -\boldsymbol{\varepsilon}^k\left(\mathbf{0}_{(k+1)N}; \boldsymbol{\varepsilon}_0^k\right) = -\boldsymbol{\varepsilon}_0^k, \tag{12.20}$$

$$\mathbf{f}^k\left(\mathbf{t}^{(k+1)N}; \mathbf{f}_0^k\right) = \mathbf{0}_{(k+1)N}, \ \left\{ \begin{array}{c} \forall \mathbf{t}^{(k+1)N} = \mathfrak{T}_{A\infty}^{(k+1)N} \\ if \ \mathfrak{T}_{A\infty}^{(k+1)N} \subset \mathfrak{T}_0^{(k+1)N} \end{array} \right\},$$
$$\mathbf{f}^k\left(\mathbf{t}^{(k+1)N}; \mathbf{f}_0^k\right) \longrightarrow \mathbf{0}_{(k+1)N} \left\{ \begin{array}{c} as \ \mathbf{t}^{(k+1)N} \longrightarrow \infty \mathbf{1}_{(k+1)N} \\ if \ \mathfrak{T}_{A\infty}^{(k+1)N} = \{\infty\}^{(k+1)N}. \end{array} \right\}. \tag{12.21}$$

Fig. 12.1 explains the switch symbol used in Figures 12.2 and 12.3.
Fig. 12.2 shows the block diagram of the vector function $\mathbf{f}(.)$ generator.

Solution 361 *A particular form of the function* $\mathbf{f}(.)$ *in (12.19)-(12.21), is given in [170, pp. 141-146].*

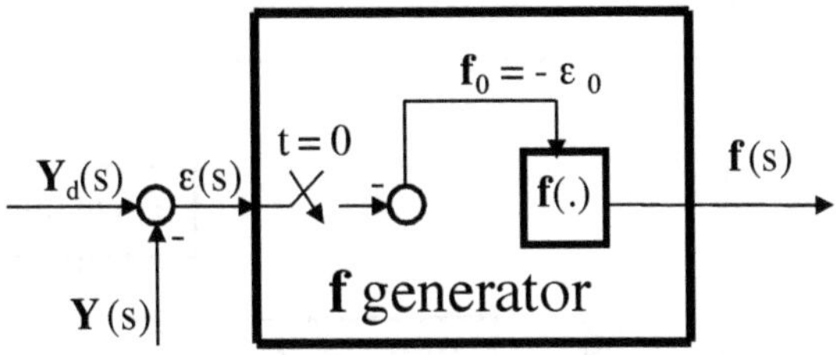

Figure 12.2: The block diagram of the vector function $\mathbf{f}(.)$ generator.

Solution 362 *We introduce* ***the reference output vector variable*** $\mathbf{Y}_R$*,*

$$\mathbf{Y}_R^k(\mathbf{t}^{(k+1)N}) = \mathbf{Y}_d^k\left(\mathbf{t}^{(k+1)N}\right) + \\ + \left\{ \begin{array}{ll} \mathbf{0}_{(k+1)N}\ , & \left(\begin{array}{c} \forall \mathbf{t}^{(k+1)N} \in \mathfrak{T}_0^{(k+1)N} \\ if\ \mathbf{T}\left(\mathbf{0}_N, \boldsymbol{\varepsilon}_0^k, \mathbf{0}_N\right) = \mathbf{0}_N, \end{array} \right) \\ \mathbf{f}^k\left(\mathbf{t}^{(k+1)N}\right), & \left(\begin{array}{c} \forall \mathbf{t}^{(k+1)N} \in \mathfrak{T}_0^{(k+1)N} \\ if\ \mathbf{T}\left(\mathbf{0}_N, \boldsymbol{\varepsilon}_0^k, \mathbf{0}_N\right) \neq \mathbf{0}_N, \end{array} \right) \end{array} \right\}, \tag{12.22}$$

$$\Longrightarrow$$

$$\mathbf{Y}_R^k(\mathbf{0}_{(k+1)N}) = \mathbf{Y}_d^k\left(\mathbf{0}_{(k+1)N}\right) + \\ + \left\{ \begin{array}{ll} \mathbf{0}_{(k+1)N} & if\ \ \mathbf{T}\left(\mathbf{0}_N, \boldsymbol{\varepsilon}_0^k, \mathbf{0}_N\right) = \mathbf{0}_N, \\ \mathbf{f}^k\left(\mathbf{0}_{(k+1)N}\right) & if\ \ \mathbf{T}\left(\mathbf{0}_N, \boldsymbol{\varepsilon}_0^k, \mathbf{0}_N\right) \neq \mathbf{0}_N, \end{array} \right\}, \tag{12.23}$$

and ***the induced subsidiary error vector*** $\boldsymbol{\epsilon}$*,*

$$\boldsymbol{\epsilon} = \left[\epsilon_1 \ \ \epsilon_2 \ \ ... \ \ \epsilon_N\right]^T \in \mathfrak{R}^N, \tag{12.24}$$

$$\boldsymbol{\epsilon}^k(\mathbf{t}^{(k+1)N}) = \mathbf{Y}_R^k(\mathbf{t}^{(k+1)N}) - \mathbf{Y}^k(\mathbf{t}^{(k+1)N}) \Longrightarrow \\ \boldsymbol{\epsilon}^k(\mathbf{t}^{(k+1)N}) = \boldsymbol{\varepsilon}^k(\mathbf{t}^{(k+1)N}) + \\ + \left\{ \begin{array}{ll} \mathbf{0}_{(k+1)N}, & \left(\begin{array}{c} \forall \mathbf{t}^{(k+1)N} \in \mathfrak{T}_0^{(k+1)N} \\ if\ \ \mathbf{T}\left(\mathbf{0}_{(k+1)N}, \boldsymbol{\varepsilon}_0^k, \mathbf{0}_N\right) = \mathbf{0}_N, \end{array} \right) \\ \mathbf{f}^k(\mathbf{t}^{(k+1)N}; \mathbf{f}_0^k), & \left(\begin{array}{c} \forall \mathbf{t}^{(k+1)N} \in \mathfrak{T}_0^{(k+1)N} \\ if\ \ \mathbf{T}\left(\mathbf{0}_{(k+1)N}, \boldsymbol{\varepsilon}_0^k, \mathbf{0}_N\right) \neq \mathbf{0}_N, \end{array} \right) \end{array} \right\}, \tag{12.25}$$

Fig. 12.3 represents the block diagram of the subsidiary error function $\epsilon\,(.)$ *generator: "*ϵ *generator".*

From (12.20), (12.22) and (12.23) follows:

$$\mathbf{Y}_R^k(\mathbf{0}_{(k+1)N}) = \mathbf{Y}_d^k\left(\mathbf{0}_{(k+1)N}\right) + \left\{ \begin{array}{ll} \mathbf{0}_{(k+1)N} & if\ \ \mathbf{T}\left(\mathbf{0}_N, \boldsymbol{\varepsilon}_0^k, \mathbf{0}_N\right) = \mathbf{0}_N, \\ -\boldsymbol{\varepsilon}^k\,(0) & if\ \ \mathbf{T}\left(\mathbf{0}_N, \boldsymbol{\varepsilon}_0^k, \mathbf{0}_N\right) \neq \mathbf{0}_N, \end{array} \right\}, \tag{12.26}$$

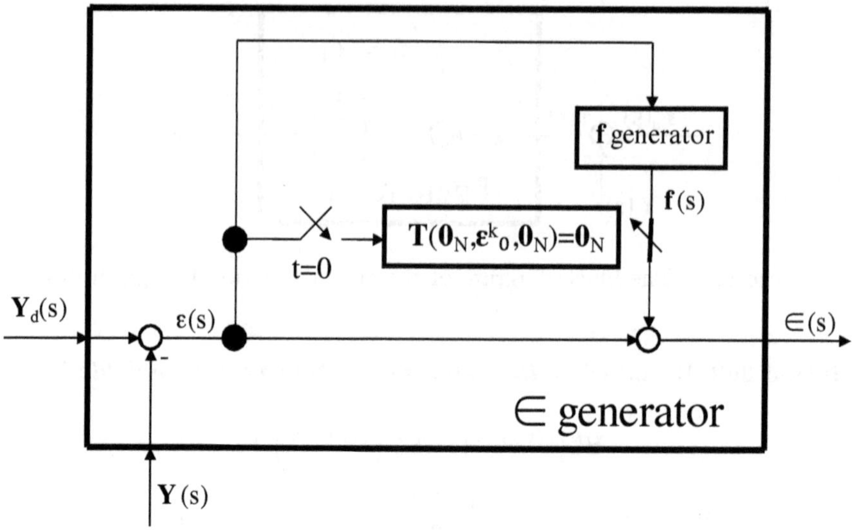

Figure 12.3: The block diagram of the subsidiary error function $\epsilon\,(.)$ generator.

i.e.,

$$\mathbf{Y}_R^k(\mathbf{0}_{(k+1)N}) = \mathbf{Y}_d^k\left(\mathbf{0}_{(k+1)N}\right) - \left\{ \begin{array}{c} \mathbf{0}_{(k+1)N} \; if \; \mathbf{T}\left(\mathbf{0}_N, \varepsilon_0^k, \mathbf{0}_N\right) = \mathbf{0}_N, \\ \left\{ \begin{array}{c} \left[\mathbf{Y}_d^k(0) - \mathbf{Y}^k(0)\right] \; if \\ \mathbf{T}\left(\mathbf{0}_N, \varepsilon_0^k, \mathbf{0}_N\right) \neq \mathbf{0}_N, \end{array} \right\} \end{array} \right\} = $$
$$= \mathbf{Y}^k\left(\mathbf{0}_{(k+1)N}\right), \tag{12.27}$$

and

$$\epsilon_0^k = \varepsilon_0^k + \left\{ \begin{array}{ll} \mathbf{0}_{(k+1)N} & if \;\; \mathbf{T}\left(\mathbf{0}_N, \varepsilon_0^k, \mathbf{0}_N\right) = \mathbf{0}_N, \\ \mathbf{f}^k(0; \mathbf{f}_0^k) & if \;\; \mathbf{T}\left(\mathbf{0}_N, \varepsilon_0^k, \mathbf{0}_N\right) \neq \mathbf{0}_N, \end{array} \right\} =$$
$$= \varepsilon_0^k + \left\{ \begin{array}{ll} \mathbf{0}_{(k+1)N} & if \;\; \mathbf{T}\left(\mathbf{0}_N, \varepsilon_0^k, \mathbf{0}_N\right) = \mathbf{0}_N, \\ -\varepsilon_0^k & if \;\; \mathbf{T}\left(\mathbf{0}_N, \varepsilon_0^k, \mathbf{0}_N\right) \neq \mathbf{0}_N, \end{array} \right\}.$$

It follows that

$$\epsilon_0^k\left(\varepsilon_0^k\right) = \left\{ \begin{array}{ll} \varepsilon_0^k & if \;\; \mathbf{T}\left(\mathbf{0}_N, \varepsilon_0^k, \mathbf{0}_N\right) = \mathbf{0}_N, \\ \mathbf{0}_{(k+1)N} & if \;\; \mathbf{T}\left(\mathbf{0}_N, \varepsilon_0^k, \mathbf{0}_N\right) \neq \mathbf{0}_N, \end{array} \right\},$$
$$\forall \varepsilon_0^k \in \mathfrak{R}^{(k+1)N}, \; k \in \{0, 1, 2, ...\}. \tag{12.28}$$

This shows that the initial value $\epsilon(\mathbf{0}_N; \varepsilon_0^k) = \epsilon_0(\varepsilon_0^k)$ *of the subsidiary error vector* $\epsilon\left(\mathbf{t}^N; \varepsilon_0^k\right)$ *and the initial values* $\epsilon_0^{(1)}\left(\varepsilon_0^k\right), \epsilon_0^{(2)}\left(\varepsilon_0^k\right), ..., \epsilon_0^{(k)}\left(\varepsilon_0^k\right)$ *of the derivatives* $\epsilon^{(1)}(\mathbf{t}^N), \epsilon^{(2)}(\mathbf{t}^N), ..., \epsilon^{(k)}(\mathbf{t}^N)$ *are all equal to the zero vector* $\mathbf{0}_N$ *for every real initial error vector* ε_0 *and for every initial value* $\varepsilon_0^{(1)}, \varepsilon_0^{(2)}, ..., \varepsilon_0^{(k)}$ *of the derivatives* $\varepsilon^{(1)}(t), \varepsilon^{(2)}(t), ..., \varepsilon^{(k)}(t)$.

Comment 363 *The reference output variable $\mathbf{Y}_R$ will replace the desired output variable $\mathbf{Y}_d$. The subsidiary error vector $\boldsymbol{\epsilon}$ will then replace the real error vector $\boldsymbol{\varepsilon}$. Equations (12.22)-(12.28) establish relations among them.*

We associate the matrix Ξ with the subsidiary vector $\boldsymbol{\epsilon}$ by

$$\Xi = diag\{\epsilon_1 \;\; \epsilon_2 \;\; ... \;\; \epsilon_N\}. \tag{12.29}$$

Theorem 364 ***The main theorem on the tracking algorithm and initial conditions***

Let (12.19) through (12.25) be valid. In order for the tracking algorithm determined by (12.17), (12.18) in terms of the real error vector $\boldsymbol{\varepsilon}$, its derivatives and integral, to hold, it is necessary and sufficient that the tracking algorithm $\boldsymbol{T}(.)$ (12.5) expressed in terms of the subsidiary error vector $\boldsymbol{\epsilon}$ (12.24), (12.25), its derivatives and integral, holds:

$$\mathbf{T}\left(\mathbf{t}^N, \boldsymbol{\epsilon}\left(\mathbf{t}^N\right), \boldsymbol{\epsilon}^{(1)}\left(\mathbf{t}^N\right), ..., \boldsymbol{\epsilon}^{(k)}\left(\mathbf{t}^N\right), \int_{\mathbf{t}_0^N=\mathbf{0}_N}^{\mathbf{t}^N} \Xi\left(\mathbf{t}^N\right) d\mathbf{t}^N\right) =$$
$$= \mathbf{T}\left(\mathbf{t}^N, \boldsymbol{\epsilon}^k(\mathbf{t}^{(k+1)N}), \int_{\mathbf{t}_0^N=\mathbf{0}_{(k+1)N}}^{\mathbf{t}^N} \Xi\left(\mathbf{t}^N\right) d\mathbf{t}^N\right) = \mathbf{0}_N, \ \forall \mathbf{t}^{(k+1)N} \in \mathfrak{T}_0^{(k+1)N}. \tag{12.30}$$

Proof. Notice that the tracking algorithm (12.30) possesses Properties 351 through 353. Let (12.19) through (12.25) be valid. Hence, (12.28) is also valid, which proves that the tracking algorithm (12.30) possesses also Property 354.

Necessity. Let (12.17), (12.18) hold. We separate the case $\mathbf{T}\left(\mathbf{0}_N, \boldsymbol{\varepsilon}_0^k, \mathbf{0}_N\right) = \mathbf{0}_N$ from the case $\mathbf{T}\left(\mathbf{0}_N, \boldsymbol{\varepsilon}_0^k, \mathbf{0}_N\right) \neq \mathbf{0}_N$. If $\boldsymbol{\varepsilon}_0^k$ is such that $\mathbf{T}\left(\mathbf{0}_N, \boldsymbol{\varepsilon}_0^k, \mathbf{0}_N\right) = \mathbf{0}_N$, then $\boldsymbol{\varepsilon}^k(\mathbf{t}^{(k+1)N}) = \boldsymbol{\epsilon}^k(\mathbf{t}^{(k+1)N})$ for every $\mathbf{t}^{(k+1)N} \in \mathfrak{T}_0^{(k+1)N}$ due to (12.24), (12.25) so that

$$\mathbf{T}\left(\mathbf{t}^N, \boldsymbol{\varepsilon}^k(\mathbf{t}^{(k+1)N}), \int_{\mathbf{t}_0^N=\mathbf{0}_N}^{\mathbf{t}^N} E\left(\mathbf{t}^N\right) d\mathbf{t}^N\right) =$$
$$= \mathbf{T}\left(\mathbf{t}^N, \boldsymbol{\epsilon}^k(\mathbf{t}^{(k+1)N}), \int_{\mathbf{t}_0^N=\mathbf{0}_N}^{\mathbf{t}^N} \boldsymbol{\Xi}\left(\mathbf{t}^N\right) d\mathbf{t}^N\right),$$
$$\forall \mathbf{t}^{(k+1)N} \in \mathfrak{T}_0^{(k+1)N}.$$

This proves necessity of (12.30). If $\boldsymbol{\varepsilon}_0^k$ violates (12.9), i.e., $\mathbf{T}\left(\mathbf{0}_N, \boldsymbol{\varepsilon}_0^k, \mathbf{0}_N\right) \neq \mathbf{0}_N$ holds, then the condition (12.18) guarantees $\boldsymbol{\epsilon}(\mathbf{t}^{(k+1)N}) = \mathbf{0}_{(k+1)N}$, $\forall \mathbf{t}^{(k+1)N} \in [\mathbf{0}_{(k+1)N}, \boldsymbol{\tau}_A^{(k+1)N}]$. Furthermore, (12.17) and (12.19) through (12.25) imply $\boldsymbol{\epsilon}^k(\mathbf{t}^{(k+1)N}) = \mathbf{0}_{(k+1)N}$, $\forall \mathbf{t}^{(k+1)N} \in \mathfrak{T}_{A\infty}^{(k+1)N}$. Altogether, $\boldsymbol{\epsilon}^k(\mathbf{t}^{(k+1)N}) = \mathbf{0}_{(k+1)N}$, $\forall \mathbf{t}^{(k+1)N} \in \mathfrak{T}_0^{(k+1)N}$, which proves that $\boldsymbol{\epsilon}^k(\mathbf{t}^{(k+1)N})$ satisfies (12.30), i.e., that (12.30) holds.

Sufficiency. Let (12.30) be valid. (12.19) through (12.25) imply (12.28). If ε_0^k is such that $\mathbf{T}\left(\mathbf{0}_N, \varepsilon_0^k, \mathbf{0}_N\right) = \mathbf{0}_N$, then $\boldsymbol{\epsilon}^k(\mathbf{t}^{(k+1)N}) = \boldsymbol{\varepsilon}^k(\mathbf{t}^{(k+1)N})$ for every $\mathbf{t}^{(k+1)N} \in \mathfrak{T}_0^{(k+1)N}$ due to (12.25) so that, in view of (12.30),

$$\mathbf{T}\left(\mathbf{t}^N, \boldsymbol{\epsilon}^k(\mathbf{t}^{(k+1)N}), \int_{\mathbf{t}_0^N=\mathbf{0}_N}^{\mathbf{t}^N} \Xi\left(\mathbf{t}^N\right) d\mathbf{t}^N\right) =$$
$$= \mathbf{T}\left(\mathbf{t}^N, \boldsymbol{\varepsilon}^k(\mathbf{t}^{(k+1)N}), \int_{\mathbf{t}_0^N=\mathbf{0}_N}^{\mathbf{t}^N} E\left(\mathbf{t}^N\right) d\mathbf{t}^N\right) = \mathbf{0}_N,$$
$$\forall \mathbf{t}^{(k+1)N} \in \mathfrak{T}_0^{(k+1)N}. \tag{12.31}$$

This proves validity of (12.17) for the case $\mathbf{T}\left(\mathbf{0}_N, \varepsilon_0^k, \mathbf{0}_N\right) = \mathbf{0}_N$. If ε_0^k violates (12.9), i.e., $\mathbf{T}\left(\mathbf{0}_N, \varepsilon_0^k, \mathbf{0}_N\right) \neq \mathbf{0}_N$ holds, then (12.30), (12.28) and (12.8) imply $\boldsymbol{\epsilon}^k(\mathbf{t}^{(k+1)N}; \boldsymbol{\epsilon}_0^k) = \mathbf{0}_{(k+1)N}$ for every $\mathbf{t}^{(k+1)N} \in \mathfrak{T}_0^{(k+1)N}$; hence, $\boldsymbol{\varepsilon}^k(\mathbf{t}^{(k+1)N}; \boldsymbol{\varepsilon}_0^k) + \mathbf{f}^k(\mathbf{t}^{(k+1)N}; \mathbf{f}_0^k) = \mathbf{0}_{(k+1)N}$ for every $\mathbf{t}^{(k+1)N} \in \mathfrak{T}_0^{(k+1)N}$ due to (12.25). This proves (12.18) and, together with (12.21), proves $\boldsymbol{\varepsilon}^k(\mathbf{t}^{(k+1)N}) = \mathbf{0}_{(k+1)N}$ for every $\mathbf{t}^{(k+1)N} \in \mathfrak{T}_{A\infty}^{(k+1)N}$. Therefore, (12.17) holds due to Property 352 ∎

We will present several characteristic simple forms of the tracking algorithm $T(.)$. They satisfy (12.6) through (12.9), i.e., they obey Properties 351-354. They satisfy also Solution 360 and Solution 362.

Comment 365 *If we allow the parameters of the tracking algorithm to depend on the initial error vector ε_0^k, then we can define them in terms of ε_0^k so that the tracking algorithm $T(.)$ (12.5) obeys (12.9). It is possible only if there are $i \in \{0, 1, ..., k\}$ and $j \in \{0, 1, ..., k\}$, $i \neq j$, such that $\varepsilon^{(i)}(0)\varepsilon^{(j)}(0) \neq 0$.*

Example 366 ***The first order linear elementwise exponential tracking algorithm***

In this case $\tau_A = \infty$, i.e., $\mathfrak{T}_{A\infty} = \{\infty\}$, (12.14). The following tracking algorithm

$$\mathbf{T}\left(t, \boldsymbol{\varepsilon}, \boldsymbol{\varepsilon}^{(1)}\right) = T_1\boldsymbol{\varepsilon}^{(1)}(t) + K_0\boldsymbol{\varepsilon}(t) = \mathbf{0}_N,\ \forall t \in \mathfrak{T}_0,\ if\ \mathbf{T}\left(0, \varepsilon_0^k, \mathbf{0}_N\right) = \mathbf{0}_N,$$
$$\boldsymbol{\varepsilon}^1\left(t; \varepsilon_0^1\right) = -\mathbf{f}^1(t),\ \forall t \in \mathfrak{T}_0,\ if\ \mathbf{T}\left(0, \varepsilon_0^1, \mathbf{0}_N\right) \neq \mathbf{0}_N,$$
$$T_1 = diag\left\{\tau_1\ \tau_2\ ...\tau_N\ \right\} > O_N,\ K_0 = diag\left\{k_{01}\ k_{02}\ ...k_{0N}\ \right\} > O_N,$$

determines the global exponential tracking, which is illustrated by the solution

$$\boldsymbol{\varepsilon}(t) = e^{-tK_0T_1^{-1}}\boldsymbol{\varepsilon}_0,\ \forall\left(\boldsymbol{\varepsilon}_0, t\right) \in \mathfrak{R}^N x\mathfrak{T}_0,$$
$$e^{-tK_0T_1^{-1}} = \exp\left(-tK_0T_1^{-1}\right) = diag\left\{e^{-tk_{01}\tau_1^{-1}}\ e^{-tk_{02}\tau_2^{-1}}\ ...\ e^{-tk_{0N}\tau_N^{-1}}\right\},$$

to

$$T_1\boldsymbol{\varepsilon}^{(1)}(t) + K_0\boldsymbol{\varepsilon}(t) = \mathbf{0}_N,\ \forall t \in \mathfrak{T}_0.$$

The reachability time is infinite. The convergence to the zero error vector is elementwise and exponentially asymptotic. Such tracking is stablewise.

We can apply Comment 365 by choosing $T_1 \in \mathfrak{R}^+$ and/or $K_0 \in \mathfrak{R}^+$ to obey

$$T_1(\varepsilon_0^1)\varepsilon_0^{(1)} + K_0(\varepsilon_0^1)\varepsilon_0 = \mathbf{0}_N \ if, \ and \ only \ if, \ \varepsilon_0^{(1)}\varepsilon_0 < 0,$$

e.g.,

$$T_1(\varepsilon_0^1) = -K_0\varepsilon_0\left(\varepsilon_0^{(1)}\right)^{-1} \in \mathfrak{R}^+$$

or

$$K_0(\varepsilon_0^1) = -T_1\varepsilon_0^{-1}\varepsilon_0^{(1)} \in \mathfrak{R}^+.$$

Such a choice of $T_1 \in \mathfrak{R}^+$ and/or $K_0 \in \mathfrak{R}^+$ assures that the tracking algorithm $T(.)$ obeys (12.9) for the given ε_0^1.

Let

$$E^{(\eta)} = [E_0 \quad E_1 \ . \ . \ . \ E_\eta] \in \mathfrak{R}^{N\mathsf{x}(\eta+1)N}.$$

Example 367 ***The higher order linear elementwise exponential tracking algorithm***

The attainability time $\tau_A = \infty$, i.e., $\mathfrak{T}_{A\infty} = \{\infty\}$, (12.14). We define the higher order linear elementwise exponential tracking algorithm by

$$\mathbf{T}\left(t, \varepsilon, \varepsilon^{(1)}, .., \varepsilon^{(\eta)}, \int_{t_0=0}^{t} \varepsilon dt\right) = \sum_{k=0}^{k=\eta\leq\nu-1} E_k\boldsymbol{\varepsilon}^{(k)}(t) = E^{(\eta)}\boldsymbol{\varepsilon}^{\eta}(t) = \mathbf{0}_N,$$
$$\forall t \in \mathfrak{T}_0 \ if \ \mathbf{T}\left(0, \boldsymbol{\varepsilon}_0^k, \mathbf{0}_N\right) = \mathbf{0}_N,$$
$$\boldsymbol{\varepsilon}^{\eta}\left(t; \boldsymbol{\varepsilon}_0^{\eta}\right) = -\mathbf{f}^{\eta}\left(t\right), \ \forall t \in \mathfrak{T}_0 \ if \ \mathbf{T}\left(0, \boldsymbol{\varepsilon}_0^{\eta}, \mathbf{0}_N\right) \neq \mathbf{0}_N,$$

with the matrices $E_k \in \mathfrak{R}^{N\mathsf{x}N}$ such that the real parts of the roots of its characteristic polynomial $f(s)$,

$$f(s) = \det\left(\sum_{k=0}^{k=\eta\leq\nu} E_k s^k\right),$$

are negative.

Being the linear differential equation with the constant coefficients the above differential equation has the unique solution for every initial condition $\boldsymbol{\varepsilon}^{\eta-1}(0) \in \mathfrak{R}^{\eta N}$.

The reachability time is infinite. The convergence to the zero error vector is elementwise, exponential and asymptotic. Such tracking is stablewise.

The application of Comment 365 is possible in the case that there are $i \in \{0, 1, ..., \eta\}$ and $j \in \{0, 1, ..., \eta\}$, $i \neq j$, such that $\varepsilon^{(i)}(0)\varepsilon^{(j)}(0) < 0$. Then we select $E_i \in \mathfrak{R}^+$ and $E_j \in \mathfrak{R}^+$ to assure that $E^{(\eta)}\boldsymbol{\varepsilon}^{\eta}(0) = \mathbf{0}_N$.

Example 368 ***The sharp elementwise stablewise tracking with the finite vector reachability time τ_R^{2N}***

In this case we accept the vector attainability time $\boldsymbol{\tau}_A^N$ and that it is equal to the vector reachability time $\boldsymbol{\tau}_R^N$, $\boldsymbol{\tau}_R^N = [\tau_{R1} \ \ \tau_{R2} \ \dots \ \tau_{RN}]^T$, (8.32) (Subsection 8.4.2),

$$\boldsymbol{\tau}_A^N = [\tau_{A1} \ \ \tau_{A2} \ \dots \ \tau_{AN}]^T = \boldsymbol{\tau}_R^N = \boldsymbol{\tau}_{R[0]}^N, \tag{12.32}$$

so that

$$\mathfrak{T}_A^N = [\mathbf{0}_N,\ \boldsymbol{\tau}_A^N] = \mathfrak{T}_R^N,\ \mathfrak{T}_{A\infty}^N = [\boldsymbol{\tau}_A^N,\ \infty\mathbf{1}_N[= \mathfrak{T}_{R\infty}^N, \tag{12.33}$$

and, in general,

$$\begin{aligned}\mathfrak{T}_A^{(k+1)N} &= [\mathbf{0}_{(k+1)N},\ \boldsymbol{\tau}_A^{(k+1)N}] = \mathfrak{T}_R^{(k+1)N},\\ \mathfrak{T}_{A\infty}^{(k+1)N} &= [\boldsymbol{\tau}_A^{(k+1)N},\ \infty\mathbf{1}_{(k+1)N}[= \mathfrak{T}_{R\infty}^{(k+1)N}.\end{aligned} \tag{12.34}$$

The algorithm for the elementwise stablewise tracking with the finite vector reachability time $\boldsymbol{\tau}_R^{2N}$,

$$\boldsymbol{\tau}_R^N = \boldsymbol{\tau}_{R[0]}^N = T_1K_0^{-1}\left|\boldsymbol{\varepsilon}_0\right|,\ \boldsymbol{\tau}_R^{2N} = \begin{bmatrix}\boldsymbol{\tau}_R^N\\ \boldsymbol{\tau}_R^N\end{bmatrix},$$

$$\mathfrak{T}_R^{2N} = \left\{\mathbf{t}^{2N}:\ \mathbf{0}_{2N}\le \mathbf{t}^{2N}\le \boldsymbol{\tau}_R^{2N}\right\},\ \mathfrak{T}_{R\infty}^{2N} = \left\{\mathbf{t}^{2N}:\ \boldsymbol{\tau}_R^{2N}\le \mathbf{t}^{2N}\le \infty\mathbf{1}_{2N}\right\},$$

$$\left|\boldsymbol{\varepsilon}_0\right| = \left[\left|\varepsilon_{10}\right|\ \ \left|\varepsilon_{20}\right|\ \ldots\ \left|\varepsilon_{N0}\right|\right]^T \in \mathfrak{R}^N,$$

reads

$$\begin{gathered}\mathbf{T}\left(\mathbf{t}^N,\boldsymbol{\varepsilon},\boldsymbol{\varepsilon}^{(1)}\right) = T_1\boldsymbol{\varepsilon}^{(1)}(\mathbf{t}^N) + K_0 sign\boldsymbol{\varepsilon}(\mathbf{0}_N) = \mathbf{0}_N,\\ \forall\mathbf{t}^N\in\mathfrak{T}_0^N\ if\ \mathbf{T}\left(\mathbf{0}_N,\boldsymbol{\varepsilon}_0^1,\mathbf{0}_N\right) = \mathbf{0}_N,\\ \boldsymbol{\varepsilon}^1\left(\mathbf{t}^{2N};\boldsymbol{\varepsilon}_0^1\right) = -\mathbf{f}^1\left(\mathbf{t}^{2N}\right),\ \forall\mathbf{t}^{2N}\in\mathfrak{T}_0^{2N},\ if\ \mathbf{T}\left(\mathbf{0}_N,\boldsymbol{\varepsilon}_0^1,\mathbf{0}_N\right)\neq\mathbf{0}_N,\end{gathered} \tag{12.35}$$

where

$$sign\varepsilon_k = \left\{\begin{array}{ll}\left|\varepsilon_k\right|^{-1}\varepsilon_k, & \varepsilon_k\neq 0,\\ 0, & \varepsilon_k = 0\end{array}\right\},\ sign\boldsymbol{\varepsilon} = \left[sign\varepsilon_1\ \ sign\varepsilon_2\ \ldots\ sign\varepsilon_N\right]^T.$$

The solution $\boldsymbol{\varepsilon}(\mathbf{t}^N;\boldsymbol{\varepsilon}_0)$,

$$\boldsymbol{\varepsilon}(\mathbf{t}^N;\boldsymbol{\varepsilon}_0) = \left\{\begin{array}{l}\left\{\begin{array}{c}\left\{\begin{array}{c}\boldsymbol{\varepsilon}_0 - T_1^{-1}K_0S\left(\boldsymbol{\varepsilon}_0\right)\mathbf{t}^N\ if\\ \mathbf{T}\left(\mathbf{0}_N,\boldsymbol{\varepsilon}_0^1,\mathbf{0}_N\right) = \mathbf{0}_N\end{array}\right\},\\ -\mathbf{f}\left(\mathbf{t}^N\right)\ if\ \mathbf{T}\left(\mathbf{0}_N,\boldsymbol{\varepsilon}_0^1,\mathbf{0}_N\right)\neq\mathbf{0}_N\end{array}\right\},\ \mathbf{t}^N\in\mathfrak{T}_R^N,\\ \mathbf{0}_N,\qquad\qquad \mathbf{t}^N\in\mathfrak{T}_{R\infty}^N\end{array}\right\}\Longrightarrow$$

$$\boldsymbol{\tau}_R^N = T_1K_0^{-1}\left|\boldsymbol{\varepsilon}_0\right|,\ \boldsymbol{\tau}_R^{2N} = \begin{bmatrix}\boldsymbol{\tau}_R^N\\ \boldsymbol{\tau}_R^N\end{bmatrix},$$

$$S\left(\boldsymbol{\varepsilon}_0\right) = diag\left\{sign\varepsilon_{10}\ \ sign\varepsilon_{20}\ \ldots\ sign\varepsilon_{N0}\right\},$$

to (12.35) determines the output error behavior that approaches sharply the zero error vector in the linear form (along a straight line) with the nonzero constant velocity $T_1^{-1}K_0S\left(\boldsymbol{\varepsilon}_0\right)\mathbf{1}_N$ *if* $\mathbf{T}\left(\mathbf{0}_N,\boldsymbol{\varepsilon}_0^1,\mathbf{0}_N\right) = \mathbf{0}_N$. *Then the convergence to the zero error vector is elementwise, strictly monotonous, continuous and*

$$\begin{gathered}\left|\boldsymbol{\varepsilon}(t;\boldsymbol{\varepsilon}_0)\right|\le\left|\boldsymbol{\varepsilon}_0\right|,\ \forall t\in\mathfrak{T}_0\Longrightarrow\\ \left\|\boldsymbol{\varepsilon}(t;\boldsymbol{\varepsilon}_0)\right\|\le\left\|\boldsymbol{\varepsilon}_0\right\|,\ \forall t\in\mathfrak{T}_0\Longrightarrow\delta\left(\varepsilon\right)=\varepsilon\Longrightarrow\\ \forall\varepsilon\in\mathfrak{R}^+,\ \exists\delta\in\mathfrak{R}^+,\ \delta=\delta\left(\varepsilon\right)=\varepsilon\Longrightarrow\\ \left\|\boldsymbol{\varepsilon}_0\right\|<\varepsilon\Longrightarrow\left\|\boldsymbol{\varepsilon}(t;\boldsymbol{\varepsilon}_0)\right\|\le\varepsilon,\ \forall t\in\mathfrak{T}_0.\end{gathered}$$

The tracking is stablewise if $\mathbf{T}\left(\mathbf{0}_N, \boldsymbol{\varepsilon}_0^1, \mathbf{0}_N\right) = \mathbf{0}_N$.

The bigger K_0, *the smaller* $\boldsymbol{\tau}_R^N$ *for fixed* T_1 *and* $\boldsymbol{\varepsilon}_0$, *and vice versa. The smaller* T_1, *the smaller* $\boldsymbol{\tau}_R^N$ *for fixed* K_0 *and* $\boldsymbol{\varepsilon}_0$, *and vice versa. The bigger* $|\boldsymbol{\varepsilon}_0|$, *the bigger* $\boldsymbol{\tau}_R^N$ *for fixed* T_1 *and* K_0, *and vice versa. These relationships hold elementwise.*

In this case we can adjust (Comment 365) the matrix parameters T_1 *and* K_0 *of the tracking algorithm so that elementwise*

$$T_1 E^{(1)}(0) + K_0 S\left(E_0\right) = O_N, \ if, \ and \ only \ if, \ E_0^{(1)} E_0 < 0,$$

where

$$E = diag\left\{\varepsilon_1 \ \ \varepsilon_2 \ \ . \ . \ . \ \ \varepsilon_N\right\}, \ S\left(E_0\right) = S\left(\boldsymbol{\varepsilon}_0\right).$$

In the scalar form,

$$\tau_{1i}\varepsilon_{i0}^{(1)} + k_{0i} sign\varepsilon_{i0} = 0, \ if, \ and \ only \ if, \ \varepsilon_{i0}^{(1)}\varepsilon_{i0} < 0,$$
$$\forall i = 1, 2, ..., N.$$

This means either

$$\tau_{1i}\left(\boldsymbol{\varepsilon}_{i0}^1\right) = -k_{0i}\left(\varepsilon_{i0}^{(1)}\right)^{-1} sign\varepsilon_{i0} = k_{0i}\left|\varepsilon_{i0}^{(1)}\right|^{-1} \in \mathfrak{R}^+$$

or

$$k_{0i}(\boldsymbol{\varepsilon}_{i0}^1) = -\tau_{1i}\varepsilon_{i0}^{(1)} sign\varepsilon_{i0} = \tau_{1i}\left|\varepsilon_{i0}^{(1)}\right| \in \mathfrak{R}^+$$

for every $i = 1, 2, ..., N$.

Example 369 ***The first power smooth elementwise stablewise tracking with the finite vector reachability time*** $\boldsymbol{\tau}_R^{2N}$

We accept (12.32). If the control acting on the plant ensures

$$\mathbf{T}\left(\mathbf{t}^N, \boldsymbol{\varepsilon}, \boldsymbol{\varepsilon}^{(1)}\right) = T_1\boldsymbol{\varepsilon}^{(1)}(\mathbf{t}^N) + 2K_0\left|E\left(\mathbf{t}^N\right)\right|^{1/2} sign\boldsymbol{\varepsilon}_0 = \mathbf{0}_N,$$
$$\forall \mathbf{t}^N \in \mathfrak{T}_0^N \ if \ \mathbf{T}\left(\mathbf{0}_N, \boldsymbol{\varepsilon}_0^1, \mathbf{0}_N\right) = \mathbf{0}_N,$$
$$\boldsymbol{\varepsilon}^1\left(\mathbf{t}^{2N}; \boldsymbol{\varepsilon}_0^1\right) = -\mathbf{f}^1\left(\mathbf{t}^{2N}\right), \ \forall \mathbf{t}^{2N} \in \mathfrak{T}_0^{2N}, \ if \ \mathbf{T}\left(\mathbf{0}_N, \boldsymbol{\varepsilon}_0^1, \mathbf{0}_N\right) \neq \mathbf{0}_N,$$

where

$$\left|E\left(\mathbf{t}^N\right)\right|^{1/2} = diag\left\{\left|\varepsilon_1(t)\right|^{1/2} \ \ \left|\varepsilon_2(t)\right|^{1/2} \ \ . \ . \ . \ \ \left|\varepsilon_N(t)\right|^{1/2}\right\},$$

then the plant exhibits the elementwise stablewise tracking with the finite vector reachability time $\boldsymbol{\tau}_R^{2N}$,

$$\boldsymbol{\tau}_R^N = T_1 K_0^{-1}\left|E_0\right|^{1/2}\mathbf{1}_N, \ \boldsymbol{\tau}_R^{2N} = \begin{bmatrix} \boldsymbol{\tau}_R^N \\ \boldsymbol{\tau}_R^N \end{bmatrix},$$

which is determined by the output error behavior

$$\varepsilon(\mathbf{t}^N;\varepsilon_0) = \left\{ \begin{array}{c} \left\{ \begin{array}{ll} \left[|E_0|^{1/2} - T_1^{-1}K_0T\right]^2 sign\varepsilon_0, \forall \mathbf{t}^N \in \mathfrak{T}_R^N, \\ \mathbf{0}_N, & \forall \mathbf{t}^N \in \mathfrak{T}_{R\infty}^N \\ \quad if\ \mathbf{T}\left(\mathbf{0}_N, \varepsilon_0^1, \mathbf{0}_N\right) = \mathbf{0}_N, \end{array} \right\}, \\ -\mathbf{f}\left(\mathbf{t}^N\right),\ \forall \mathbf{t}^N \in \mathfrak{T}_0^N,\ if\ \mathbf{T}\left(\mathbf{0}_N, \varepsilon_0^1, \mathbf{0}_N\right) \neq \mathbf{0}_N \end{array} \right\},$$

$$T = diag\left\{t\ \ t\ \ ...\ \ t\right\} \in \mathfrak{T}_0^{N\times N},\ \boldsymbol{\tau}_R^N = T_1K_0^{-1}\left|\varepsilon_0\right|^{1/2},$$

$$\left|\varepsilon_0\right|^{1/2} = \left[\left|\varepsilon_1(0)\right|^{1/2}\ \ \left|\varepsilon_2(0)\right|^{1/2}\ \ ...\ \ \left|\varepsilon_N(0)\right|^{1/2}\right]^T, \tag{12.36}$$

which implies

$$sign\varepsilon(\mathbf{t}^N) = sign\varepsilon_0,\ \forall \mathbf{t}^N \in [\mathbf{0}_N, \boldsymbol{\tau}_R^N[.$$

The output error vector approaches smoothly elementwise the zero output vector in the finite vector reachability time $\boldsymbol{\tau}_R^N$. *The convergence is strictly monotonous and continuous. It is also without any oscillation, overshoot or undershoot if* $\mathbf{T}\left(\mathbf{0}_N, \varepsilon_0^1, \mathbf{0}_N\right) = \mathbf{0}_N$. *Then the solution (12.36) obeys the following:*

$$\begin{gathered} \|\varepsilon(t;\varepsilon_0)\| \leq \|\varepsilon_0\|,\ \forall t \in \mathfrak{T}_0 \Longrightarrow \delta(\varepsilon) = \varepsilon \Longrightarrow \\ \forall \varepsilon \in \mathfrak{R}^+,\ \exists \delta \in \mathfrak{R}^+,\ \delta = \delta(\varepsilon) = \varepsilon \Longrightarrow \\ \|\varepsilon_0\| < \varepsilon \Longrightarrow \|\varepsilon(t;\varepsilon_0)\| \leq \varepsilon,\ \forall t \in \mathfrak{T}_0. \end{gathered}$$

Therefore, such tracking is stablewise if $\mathbf{T}\left(\mathbf{0}_N, \varepsilon_0^1, \mathbf{0}_N\right) = \mathbf{0}_N$.

The bigger K_0, *the smaller* $\boldsymbol{\tau}_R^N$ *for fixed* T_1 *and* ε_0, *and vice versa. The smaller* T_1, *the smaller* $\boldsymbol{\tau}_R^N$ *for fixed* K_0 *and* ε_0, *and vice versa. The bigger* $|\varepsilon_0|$, *the bigger* $\boldsymbol{\tau}_R^N$ *for fixed* T_1 *and* K_0, *and vice versa. These claims are in the elementwise sense.*

In this case we can adjust (Comment 365) the matrix parameters T_1 *and* K_0 *of the tracking algorithm so that elementwise*

$$T_1E^{(1)}(0) + 2K_0\left|E(0)\right|^{1/2}S(E_0) = O_N,\ if,\ and\ only\ if,\ E_0^{(1)}E_0 < 0.$$

In the scalar form,

$$\tau_{1i}\varepsilon_{i0}^{(1)} + 2k_{0i}\left|\varepsilon_{i0}\right|^{1/2} sign\varepsilon_{i0} = 0,\ if,\ and\ only\ if,\ \varepsilon_{i0}^{(1)}\varepsilon_{i0} < 0,$$
$$\forall i = 1, 2, ..., N.$$

This means either

$$\tau_{1i}(\varepsilon_{i0}^1) = -2k_{0i}\left|\varepsilon_{i0}\right|^{1/2}\left(\varepsilon_{i0}^{(1)}\right)^{-1} sign\varepsilon_{i0} = 2k_{0i}\left|\varepsilon_{i0}\right|^{1/2}\left|\varepsilon_{i0}^{(1)}\right|^{-1} \in \mathfrak{R}^+$$

or

$$k_{0i}(\varepsilon_{i0}^1) = -\frac{1}{2}\tau_{1i}\left|\varepsilon_{i0}\right|^{-1/2}\varepsilon_{i0}^{(1)} sign\varepsilon_{i0} = \frac{1}{2}\tau_{1i}\left|\varepsilon_{i0}\right|^{-1/2}\left|\varepsilon_{i0}^{(1)}\right| \in \mathfrak{R}^+$$

for every $i = 1, 2, ..., N$.

Example 370 ***The higher power smooth elementwise stablewise tracking with the finite vector reachability time*** $\boldsymbol{\tau}_R^{2N}$

Let (12.32) hold. Let the tracking algorithm be

$$\mathbf{T}\left(\mathbf{t}^N, \boldsymbol{\varepsilon}, \boldsymbol{\varepsilon}^{(1)}\right) = T_1 \boldsymbol{\varepsilon}^{(1)}(\mathbf{t}^N) + K_0 \left|E\left(\mathbf{t}^N\right)\right|^{I-K^{-1}} sign\boldsymbol{\varepsilon}_0 = \mathbf{0}_N,$$
$$\forall \mathbf{t}^N \in \mathfrak{T}_0^N \ if \ \mathbf{T}\left(\mathbf{0}_N, \boldsymbol{\varepsilon}_0^1, \mathbf{0}_N\right) = \mathbf{0}_N,$$
$$\boldsymbol{\varepsilon}^1\left(\mathbf{t}^{2N}; \boldsymbol{\varepsilon}_0^1\right) = -\mathbf{f}^1\left(\mathbf{t}^{2N}\right), \ \forall \mathbf{t}^{2N} \in \mathfrak{T}_0^{2N}, \ if \ \mathbf{T}\left(\mathbf{0}_N, \boldsymbol{\varepsilon}_0^1, \mathbf{0}_N\right) \neq \mathbf{0}_N,$$
$$K = diag\left\{k_1 \ k_2 \ ...k_N\right\}, \ k_i \in \{2, \ 3, \ ...\}, \ \forall i = 1, 2, \ ... \ , N,$$
$$\left|E\left(\mathbf{t}^N\right)\right|^{I-K^{-1}} = diag\left\{\left|\varepsilon_1(t)\right|^{1-k_1^{-1}} \ \left|\varepsilon_2(t)\right|^{1-k_2^{-1}} \ ... \ \left|\varepsilon_N(t)\right|^{1-k_N^{-1}}\right\}. \tag{12.37}$$

If $\mathbf{T}\left(\mathbf{0}_N, \boldsymbol{\varepsilon}_0^1, \mathbf{0}_N\right) = \mathbf{0}_N$, *then the solution* $\boldsymbol{\varepsilon}(\mathbf{t}^N; \boldsymbol{\varepsilon}_0)$ *to* $T\left(\mathbf{t}^N, \boldsymbol{\varepsilon}, \boldsymbol{\varepsilon}^{(1)}\right) = \mathbf{0}_N$, $\forall \mathbf{t}^N \in \mathfrak{T}_0^N$, *reads*

$$\boldsymbol{\varepsilon}(\mathbf{t}^N; \boldsymbol{\varepsilon}_0) =$$
$$= \frac{1}{2} S\left(\boldsymbol{\varepsilon}_0\right) \left\{I_N + S\left[\left|\boldsymbol{\varepsilon}_0\right|^{K^{-1}} - T_1^{-1} K_0 \mathbf{t}^N\right]\right\} \left[\left|\boldsymbol{\varepsilon}_0\right|^{K^{-1}} - T_1^{-1} K_0 \mathbf{t}^N\right]^K =$$
$$= \left\{ \begin{array}{ll} \left[\left|E_0\right|^{K^{-1}} - T_1^{-1} K_0 T\right]^K sign\boldsymbol{\varepsilon}_0, & \mathbf{t}^N \in \mathfrak{T}_R^N, \\ \mathbf{0}_N, & \mathbf{t}^N \in \mathfrak{T}_{R\infty}^N \end{array} \right\} \Longrightarrow$$
$$\boldsymbol{\tau}_R^N = T_1 K_0^{-1} \left|\boldsymbol{\varepsilon}_0\right|^{K^{-1}}, \ \boldsymbol{\tau}_R^{2N} = \begin{bmatrix} \boldsymbol{\tau}_R^N \\ \boldsymbol{\tau}_R^N \end{bmatrix}, \tag{12.38}$$

where

$$\left[\left|\boldsymbol{\varepsilon}_0\right|^{K^{-1}} - T_1^{-1} K_0 \mathbf{t}^N\right]^K = \begin{bmatrix} \left[\left|\varepsilon_{10}\right|^{k_1^{-1}} - t\tau_1^{-1} k_{01}\right]^{k_1} \\ \left[\left|\varepsilon_{20}\right|^{k_2^{-1}} - t\tau_2^{-1} k_{02}\right]^{k_2} \\ \cdots\cdots\cdots \\ \left[\left|\varepsilon_{N0}\right|^{k_N^{-1}} - t\tau_N^{-1} k_{0N}\right]^{k_N} \end{bmatrix} \in \mathfrak{R}^N. \tag{12.39}$$

This expresses the elementwise nonlinear convergence to the zero error vector if $\mathbf{T}\left(\mathbf{0}_N, \boldsymbol{\varepsilon}_0^1, \mathbf{0}_N\right) = \mathbf{0}_N$. *Then the convergence is strictly monotonous and continuous, without any oscillation, overshoot or undershoot. Therefore, such tracking is stablewise. The errors enter the zero values smoothly. Besides, (12.38) and (12.39) imply*

$$\left\|\boldsymbol{\varepsilon}(t; \boldsymbol{\varepsilon}_0)\right\| \leq \left\|\boldsymbol{\varepsilon}_0\right\|, \ \forall t \in \mathfrak{T}_0 \Longrightarrow \delta\left(\varepsilon\right) = \varepsilon \Longrightarrow$$
$$\forall \varepsilon \in \mathfrak{R}^+, \ \exists \delta \in \mathfrak{R}^+, \ \delta = \delta\left(\varepsilon\right) = \varepsilon \Longrightarrow$$
$$\left\|\boldsymbol{\varepsilon}_0\right\| < \varepsilon \Longrightarrow \left\|\boldsymbol{\varepsilon}(t; \boldsymbol{\varepsilon}_0)\right\| \leq \varepsilon, \ \forall t \in \mathfrak{T}_0.$$

The tracking is stablewise if $\mathbf{T}\left(\mathbf{0}_N, \boldsymbol{\varepsilon}_0^1, \mathbf{0}_N\right) = \mathbf{0}_N$.

The bigger K_0, the smaller $\boldsymbol{\tau}_R^N$ for fixed T_1 and ε_0, and vice versa. The smaller T_1, the smaller $\boldsymbol{\tau}_R^N$ for fixed K_0 and ε_0, and vice versa. The bigger $|\varepsilon_0|$, the bigger $\boldsymbol{\tau}_R^N$ for fixed T_1 and K_0, and vice versa.

If we wish to adjust (Comment 365) the matrix parameters T_1 and K_0 of the tracking algorithm, then they should satisfy

$$T_1 E^{(1)}(t) + K_0 |E_0|^{I-K^{-1}} S(E_0) = O_N,\ if,\ and\ only\ if,\ E_0^{(1)} E_0 < 0,$$

equivalently in the scalar form

$$\tau_{1i}\varepsilon_{i0}^{(1)} + k_{0i} |\varepsilon_{i0}|^{1-k_i^{-1}} sign\varepsilon_{i0} = 0,\ if,\ and\ only\ if,\ \varepsilon_{i0}^{(1)}\varepsilon_{i0} < 0,$$
$$\forall i = 1, 2, ..., N.$$

This means either

$$\tau_{1i}(\boldsymbol{\varepsilon}_{i0}^1) = -k_{0i} |\varepsilon_{i0}|^{1-k_i^{-1}} \left(\varepsilon_{i0}^{(1)}\right)^{-1} sign\varepsilon_{i0} = k_{0i} |\varepsilon_{i0}|^{1-k_i^{-1}} \left|\varepsilon_{i0}^{(1)}\right|^{-1} \in \mathfrak{R}^+$$

or

$$k_{0i}(\boldsymbol{\varepsilon}_{i0}^1) = -\tau_{1i} |\varepsilon_{i0}|^{-(1-k_i^{-1})} \varepsilon_{i0}^{(1)} sign\varepsilon_{i0} = \tau_{1i} |\varepsilon_{i0}|^{-(1-k_i^{-1})} \left|\varepsilon_{i0}^{(1)}\right| \in \mathfrak{R}^+$$

for every $i = 1, 2, ..., N$.

Example 371 ***Sharp absolute error vector value tracking elementwise and stablewise with the finite vector reachability time $\boldsymbol{\tau}_R^{2N}$***

We accept (12.32) to hold. The solution to the following tracking algorithm (in which we use (11.15), (11.16), Subsection 12.2.2):

$$\mathbf{T}\left(\mathbf{t}^N, \boldsymbol{\varepsilon}, \boldsymbol{\varepsilon}^{(1)}\right) = T_1 \Sigma\left(\boldsymbol{\varepsilon}, \boldsymbol{\varepsilon}^{(1)}\right) \boldsymbol{\varepsilon}^{(1)} + K_0 sign |\boldsymbol{\varepsilon}_0| = \mathbf{0}_N,$$
$$\forall \mathbf{t}^N \in \mathfrak{T}_0^N\ if\ \mathbf{T}\left(\mathbf{0}_N, \boldsymbol{\varepsilon}_0^1, \mathbf{0}_N\right) = \mathbf{0}_N,$$
$$\boldsymbol{\varepsilon}^1\left(\mathbf{t}^{2N}; \boldsymbol{\varepsilon}_0^1\right) = -\mathbf{f}^1\left(\mathbf{t}^{2N}\right),\ \forall \mathbf{t}^{2N} \in \mathfrak{T}_0^{2N},\ if\ \mathbf{T}\left(\mathbf{0}_N, \boldsymbol{\varepsilon}_0^1, \mathbf{0}_N\right) \neq \mathbf{0}_N,$$

reads for $\mathbf{T}\left(\mathbf{0}_N, \boldsymbol{\varepsilon}_0^1, \mathbf{0}_N\right) = \mathbf{0}_N$:

$$\left|\boldsymbol{\varepsilon}(\mathbf{t}^N; \boldsymbol{\varepsilon}_0)\right| = \left\{ \begin{array}{ll} |\boldsymbol{\varepsilon}_0| - T_1^{-1} K_0 T sign |\boldsymbol{\varepsilon}_0|, & \mathbf{t}^N \in [\mathbf{0}_N, T_1 K_0^{-1} |\boldsymbol{\varepsilon}_0|], \\ \mathbf{0}_N, & \mathbf{t}^N \in [T_1 K_0^{-1} |\boldsymbol{\varepsilon}_0|,\ \infty \mathbf{1}_N[. \end{array} \right\},$$
$$\boldsymbol{\tau}_R^N = T_1 K_0^{-1} |\boldsymbol{\varepsilon}_0|,\ \boldsymbol{\tau}_R^{2N} = \left[\begin{array}{c} \boldsymbol{\tau}_R^N \\ \boldsymbol{\tau}_R^N \end{array}\right], \tag{12.40}$$

which permits

$$|\boldsymbol{\varepsilon}(t; \boldsymbol{\varepsilon}_0)| \leq |\boldsymbol{\varepsilon}_0|,\ \forall t \in \mathfrak{T}_0, \Longrightarrow$$
$$\|\boldsymbol{\varepsilon}(t; \boldsymbol{\varepsilon}_0)\| \leq \|\boldsymbol{\varepsilon}_0\|,\ \forall t \in \mathfrak{T}_0 \Longrightarrow$$
$$\forall \varepsilon \in \mathfrak{R}^+,\ \exists \delta \in \mathfrak{R}^+,\ \delta = \delta(\varepsilon) = \varepsilon \Longrightarrow$$
$$\|\boldsymbol{\varepsilon}_0\| < \varepsilon \Longrightarrow \|\boldsymbol{\varepsilon}(t; \boldsymbol{\varepsilon}_0)\| \leq \varepsilon,\ \forall t \in \mathfrak{T}_0.$$

Then the tracking is stablewise and elementwise with the finite vector reachability time $\boldsymbol{\tau}_R^N = T_1 K_0^{-1} |\varepsilon_0|$. It is strictly monotonous and continuous without oscillation, overshoot and undershoot.

In order to follow Comment 365 we adjust the matrix parameters T_1 and K_0 of the tracking algorithm to satisfy

$$T_1 \Sigma \left(\varepsilon_0, \varepsilon_0^{(1)}\right) E_0^{(1)} + K_0 S\left(E_0\right) = O_N, \ if, \ and \ only \ if, \ E_0^{(1)} E_0 < 0,$$

equivalently in the scalar form

$$\tau_{1i}\sigma\left(\varepsilon_{i0}, \varepsilon_{i0}^{(1)}\right) \varepsilon_{i0}^{(1)} + k_{0i} sign \varepsilon_{i0} = 0, \ if, \ and \ only \ if, \ \varepsilon_{i0}^{(1)} \varepsilon_{i0} < 0,$$
$$\forall i = 1, 2, ..., N.$$

This means either

$$\tau_{1i}(\varepsilon_{i0}^1) = -k_{0i}\left[\sigma\left(\varepsilon_{i0}, \varepsilon_{i0}^{(1)}\right) \varepsilon_{i0}^{(1)}\right]^{-1} sign \varepsilon_{i0} = k_{0i}\left|\varepsilon_{i0}^{(1)}\right|^{-1} \in \mathfrak{R}^+$$

or

$$k_{0i}(\varepsilon_{i0}^1) = -\tau_{1i}\sigma\left(\varepsilon_{i0}, \varepsilon_{i0}^{(1)}\right) \varepsilon_{i0}^{(1)} sign \varepsilon_{i0} = \tau_{1i}\left|\varepsilon_{i0}^{(1)}\right| \in \mathfrak{R}^+$$

for every $i = 1, 2, ..., N$.

Example 372 ***The exponential absolute error vector value tracking elementwise and stablewise with the finite vector reachability time $\boldsymbol{\tau}_R^{2N}$***

We accept the validity of (12.32). The tracking algorithm is in terms of the elementwise absolute value of the error vector,

$$\mathbf{T}\left(\varepsilon, \varepsilon^{(1)}\right) = T_1 D^+ |\varepsilon| + K\left(|\varepsilon| + K_0 sign\,|\varepsilon_0|\right) =$$
$$= T_1 \Sigma\left(\varepsilon, \varepsilon^{(1)}\right) \varepsilon^{(1)} + K\left(|\varepsilon| + K_0 sign\,|\varepsilon_0|\right) = \mathbf{0}_N,$$
$$\forall \mathbf{t}^N \in \mathfrak{T}_0^N \ if \ \mathbf{T}\left(\mathbf{0}_N, \varepsilon_0^1, \mathbf{0}_N\right) = \mathbf{0}_N,$$
$$\varepsilon^1\left(\mathbf{t}^{2N}; \varepsilon_0^1\right) = -\mathbf{f}^1\left(\mathbf{t}^{2N}\right), \ \forall \mathbf{t}^{2N} \in \mathfrak{T}_0^{2N}, \ if \ \mathbf{T}\left(\mathbf{0}_N, \varepsilon_0^1, \mathbf{0}_N\right) \neq \mathbf{0}_N.$$

The solution of the differential equation written in the matrix diagonal form

$$D^+\left[|E| + K_0 S\left(|\varepsilon_0|\right)\right] = -T_1^{-1} K\left[|E| + K_0 S\left(|\varepsilon_0|\right)\right] \Longrightarrow$$
$$\left[|E| + K_0 S\left(|\varepsilon_0|\right)\right]^{-1} D^+\left[|E| + K_0 S\left(|\varepsilon_0|\right)\right] = -T_1^{-1} K \Longrightarrow$$
$$D^+\left\{\ln\left[|E| + K_0 S\left(|\varepsilon_0|\right)\right]\right\} = -T_1^{-1} K,$$

reads in the matrix form

$$\ln\left\{\left[|E| + K_0 S\left(|\varepsilon_0|\right)\right]\left[|E_0| + K_0 S\left(|\varepsilon_0|\right)\right]^{-1}\right\} = -T_1^{-1} K T, \ \forall T \in \mathfrak{T}_0^{N\text{x}N}.$$

The final form of the solution is

$$|E(t;E_0)| =$$
$$= \left\{ \begin{array}{ll} e^{-T_1^{-1}KT}\left[|E_0| + K_0 S\left(|\varepsilon_0|\right)\right] - K_0 S\left(|\varepsilon_0|\right), & \forall T \in [O_N, T_R], \\ O_N, & \forall T \in [T_R,\ \infty I_N[, \\ \multicolumn{2}{c}{where\ 0\infty = 0,} \end{array} \right\}, \quad (12.41)$$

$$T_R = \left\{ \begin{array}{ll} T_1 K^{-1} \ln\left\{K_0^{-1} S^{-1}\left(|\varepsilon_0|\right)\left[|E_0| + K_0 S\left(|\varepsilon_0|\right)\right]\right\}, & \varepsilon_0 \neq \mathbf{0}_N, \\ O_N, & \varepsilon_0 = \mathbf{0}_N \end{array} \right\}, \quad (12.42)$$

$$\boldsymbol{\tau}_R^N = \left[\tau_{R1}\ \ \tau_{R2}\ \ .\ .\ .\ \tau_{RN}\right]^T \Longleftrightarrow T_R = diag\left\{\tau_{R1}\ \ \tau_{R2}\ \ .\ .\ .\ \tau_{RN}\right\},$$
$$\tau_R^{2N} = \begin{bmatrix} \boldsymbol{\tau}_R^N \\ \boldsymbol{\tau}_R^N \end{bmatrix}. \quad (12.43)$$

We can set the solution (12.41) in the equivalent vector form

$$\left|\boldsymbol{\varepsilon}(\mathbf{t}^N;\boldsymbol{\varepsilon}_0)\right| = \left\{ \begin{array}{c} e^{-T_1^{-1}KT}\left[|\boldsymbol{\varepsilon}_0| + K_0 sign\left(|\boldsymbol{\varepsilon}_0|\right)\right] - K_0 sign\left(|\boldsymbol{\varepsilon}_0|\right), \\ \forall \mathbf{t}^N \in \left[\mathbf{0}_N, \boldsymbol{\tau}_R^N\right],\ i.e.,\ \forall T \in [O_N,\ T_R], \\ \mathbf{0}_N,\ \ \forall \mathbf{t}^N \in [\boldsymbol{\tau}_R^N,\ \infty \mathbf{1}_N[,\ i.e.,\ \forall T \in [T_R,\ \infty I_N]. \end{array} \right\}.$$

The solution is continuous and monotonous without oscillation, overshoot and undershoot, and obeys

$$|\boldsymbol{\varepsilon}(t;\boldsymbol{\varepsilon}_0)| \leq |\boldsymbol{\varepsilon}_0|,\ \forall t \in \mathfrak{T}_0 \Longrightarrow$$
$$\forall \varepsilon \in \mathfrak{R}^+,\ \exists \delta \in \mathfrak{R}^+,\ \delta = \delta(\varepsilon) = \varepsilon \Longrightarrow$$
$$\|\boldsymbol{\varepsilon}_0\| < \varepsilon \Longrightarrow \|\boldsymbol{\varepsilon}(t;\boldsymbol{\varepsilon}_0)\| \leq \varepsilon,\ \forall t \in \mathfrak{T}_0,$$

if $\mathbf{T}\left(\mathbf{0}_N, \varepsilon_0^1, \mathbf{0}_N\right) = \mathbf{0}_N$. *Then the tracking is stablewise. It converges with the exponential rate to the zero error vector and reaches elementwise the origin in finite vector reachability time* $\boldsymbol{\tau}_R^N$ *(12.42), (12.43).*

Comment 365 leads to the following adjustment of the matrix parameters T_1 *and* K_0 *of the tracking algorithm:*

$$T_1 \Sigma\left(\varepsilon_0, \varepsilon_0^{(1)}\right) E_0^{(1)} + K\left[|E_0| + K_0 S\left(|E_0|\right)\right] = O_N,\ \ if,\ and\ only\ if,\ E_0^{(1)} E_0 < 0,$$

equivalently in the scalar form

$$\tau_{1i}\sigma\left(\varepsilon_{i0}, \varepsilon_{i0}^{(1)}\right) \varepsilon_{i0}^{(1)} + k_i\left(|\varepsilon_{i0}| + k_{0i} sign\left|\varepsilon_{i0}\right|\right) = 0,\ \ if,\ and\ only\ if,\ \varepsilon_{i0}^{(1)} \varepsilon_{i0} < 0,$$
$$\forall i = 1, 2, ..., N.$$

This means either

$$\tau_{1i}(\varepsilon_{i0}^1) = -k_i\left(|\varepsilon_{i0}| + k_{0i} sign\left|\varepsilon_{i0}\right|\right)\left[\sigma\left(\varepsilon_{i0}, \varepsilon_{i0}^{(1)}\right) \varepsilon_{i0}^{(1)}\right]^{-1} =$$
$$= k_i\left(|\varepsilon_{i0}| + k_{0i} sign\left|\varepsilon_{i0}\right|\right)\left|\varepsilon_{i0}^{(1)}\right|^{-1} \in \mathfrak{R}^+$$

or

$$k_{0i}(\varepsilon_{i0}^{1}) = -k_i^{-1}\left[k_i\left|\varepsilon_{i0}\right| + \tau_{1i}\sigma\left(\varepsilon_{i0},\varepsilon_{i0}^{(1)}\right)\varepsilon_{i0}^{(1)}\right] sign\varepsilon_{i0} =$$
$$= k_i^{-1}\left[k_i\left|\varepsilon_{i0}\right| + \tau_{1i}\left|\varepsilon_{i0}^{(1)}\right|\right] \in \mathfrak{R}^{+}$$

for every $i = 1, 2, ..., N$.

Note 373 *We can effectively use the tracking algorithms proposed in the preceding examples, Example 366 through Example 372, also when we synthesize tracking control by applying other design (e.g., Lyapunov like, adaptive control, sliding mode) methods.*

The system behavior results from the actions of control, of disturbances and of initial conditions. The control cannot influence disturbances and initial conditions, but can take into account their consequences that are the plant internal dynamic behavior and the output error.

The principle of the natural control is to use information about the realized control action on the plant instead of information about the plant internal dynamic behavior and instead of information about the disturbance value. We will show mathematically that this requires the solvability of the plant mathematical model in control. We established the necessary and sufficient simple, algebraic, conditions for the mathematical model solvability in control separately for the *IO* plants in Subsections 9.3.1 and 9.4.1, as well as for the *ISO* plants in Subsections 9.3.2 and 9.4.2. They will be a part of the necessary and sufficient simple, algebraic, conditions for *NTC* to guarantee various requested tracking properties, hence, for various tracking qualities.

12.2 NTC of the *IO* plant

12.2.1 General consideration

We recall Note 204 (Subsection 8.4.3). Notice that Claim 155 (Section 8.2) is valid if it is not said otherwise. This means that we adopt that the plant is stable.

Our aim is to synthesize only tracking control in what follows.

We will consider the *IO* plant (2.15) (Subsection 2.1.2) at an arbitrary moment $t \in \mathfrak{T}_0$,

$$A_P^{(\nu)}\mathbf{Y}^{\nu}(t) = C_{Pu}^{(\mu_{Pu})}\mathbf{U}^{\mu_{Pu}}(t) + D_{Pd}^{(\mu_{Pd})}\mathbf{D}^{\mu_{Pd}}(t),\ \det A_{P\nu} \neq 0,\ \forall t \in \mathfrak{T}_0,$$
$$\nu \geq \max\left\{\mu_{Pd},\ \mu_{Pu}\right\}, \tag{12.44}$$

and at the moment $t^{-} \in \mathfrak{T}_0$ just preceding the moment $t \in \mathfrak{T}_0$. It is in the ideal case the beginning of the duration of the moment t itself,

$$A_P^{(\nu)}\mathbf{Y}^{\nu}(t^{-}) = C_{Pu}^{(\mu_{Pu})}\mathbf{U}^{\mu_{Pu}}(t^{-}) + D_{Pd}^{(\mu_{Pd})}\mathbf{D}^{\mu_{Pd}}(t^{-}),\ \det A_{P\nu} \neq 0,\ t^{-} \in \mathfrak{T}_0,$$
$$t^{-} = t - \sigma,\ 0 < \sigma <<< 1,\ i.e.,\ \sigma \rightarrow 0^{+},\ in\ the\ ideal\ case\ \sigma = 0. \tag{12.45}$$

Property 351 through Property 354 (Section 12.1) form the basis for the following theorems, Theorem 374, Theorem 377 through Theorem 383.

Theorem 374 ***General NTC synthesis***
Let (12.19) through (12.25) (Section 12.1) be valid.
In order for the IO plant (12.44) to be controlled by the natural tracking control $\mathbf{U}$ *to exhibit tracking on* $\mathfrak{D}^{\mu_{Pd}}\text{x}\mathfrak{Y}_d^{\nu}$ *determined by the tracking algorithm* $\mathbf{T}(.)$, *(12.17), (12.18), (Section 12.1), it is necessary and sufficient that*
1) $N \leq r$,
2) $rankG_{IOPU}(s) = rank\left(C_{Pu}^{(\mu_{Pu})} S_r^{(\mu_{Pu})}(s)\right) = rankC_{Pu}^{(\mu_{Pu})} = N,$

3) $$\mathbf{U}^{\mu_{Pu}}(t) = \mathbf{U}^{\mu_{Pu}}(t^-) + \Gamma T\left(t, \boldsymbol{\epsilon}(t), \boldsymbol{\epsilon}^{(1)}(t), ..., \int_{t_0=0}^{t} \boldsymbol{\epsilon}(t)dt\right), \tag{12.46}$$

$$\Gamma = C_{Pu}^{(\mu_{Pu})T}\left(C_{Pu}^{(\mu_{Pu})} C_{Pu}^{(\mu_{Pu})T}\right)^{-1}, \tag{12.47}$$

where (12.25) defines $\boldsymbol{\epsilon}(t;\boldsymbol{\epsilon}_0)$ *(Section 12.1), or equivalently*

$$\mathbf{U}(s) = e^{-\sigma s}\mathbf{U}(s) + \Gamma(s)\mathbf{T}(s, \boldsymbol{\epsilon}(\mathbf{s})),\ 0 < \sigma <<< 1,\ i.e.,\ \sigma \longrightarrow 0^+, \tag{12.48}$$

$$\mathbf{T}(s, \boldsymbol{\epsilon}(\mathbf{s})) = \mathcal{L}\left\{\mathbf{T}\left(t, \boldsymbol{\epsilon}(t), \boldsymbol{\epsilon}^{(1)}(t), ..., \int_{t_0=0}^{t} \boldsymbol{\epsilon}(t)dt\right)\right\}, \tag{12.49}$$

$$\Gamma(s) = \left[C_{Pu}^{(\mu_{Pu})} S_r^{(\mu_{Pu})}(s)\right]^T \left\{\left[C_{Pu}^{(\mu_{Pu})} S_r^{(\mu_{Pu})}(s)\right]\left[C_{Pu}^{(\mu_{Pu})} S_r^{(\mu_{Pu})}(s)\right]^T\right\}^{-1}. \tag{12.50}$$

Proof. *Necessity.* Let the *IO* object (12.44) controlled by the natural tracking control $\mathbf{U}$ exhibit tracking on $\mathfrak{D}^{\mu_{Pd}}\text{x}\mathfrak{Y}_d^{\nu}$ determined by (12.17), (12.18). Then, due to Theorem 364 (Section 12.1), the *NTC* of the *IO* plant (12.44) ensures that the plant exhibits tracking on $\mathfrak{D}^{\mu_{Pd}}\text{x}\mathfrak{Y}_d^{\nu}$ determined by (12.30) (Section 12.1). Equations (12.5) and (12.9) ensure continuity of the tracking algorithm in terms of $\boldsymbol{\epsilon}$ at the initial moment $t_0 = 0$ due to (12.28). The necessity of the conditions 1) and 2) of this theorem results directly from Theorem 68 (Subsection 3.3.2), Lemma 242, Theorem 254 (Subsection 9.2.3), and Theorem 275 (Subsection 9.3.1). Since the plant controlled by the natural tracking control $\mathbf{U}$ exhibits tracking on $\mathfrak{D}^{\mu_{Pd}}\text{x}\mathfrak{Y}_d^{\nu}$ determined by (12.30), then we can rewrite (12.44) as

$$A_P^{(\nu)}\mathbf{Y}^{\nu}(t) = C_{Pu}^{(\mu_{Pu})}\mathbf{U}^{\mu_{Pu}}(t) + D_{Pd}^{(\mu_{Pd})}\mathbf{D}^{\mu_{Pd}}(t) - \mathbf{T}\left(t, \boldsymbol{\epsilon}, \boldsymbol{\epsilon}^{(1)}, ..., \int_{t_0=0}^{t} \boldsymbol{\epsilon}dt\right).$$

This and (12.45) yield

$$A_P^{(\nu)}\left[\mathbf{Y}^{\nu}(t) - \mathbf{Y}^{\nu}(t^-)\right] = C_{Pu}^{(\mu_{Pu})}\left[\mathbf{U}^{\mu_{Pu}}(t) - \mathbf{U}^{\mu_{Pu}}(t^-)\right] +$$
$$+ D_{Pd}^{(\mu_{Pd})}\left[\mathbf{D}^{\mu_{Pd}}(t) - \mathbf{D}^{\mu_{Pd}}(t^-)\right] - \mathbf{T}\left(t, \boldsymbol{\epsilon}, \boldsymbol{\epsilon}^{(1)}, ..., \int_{t_0=0}^{t} \boldsymbol{\epsilon}dt\right).$$

Linearity of the system and continuity of all variables yield, [equivalently, Principle 9 (Section 1.1) yields],

$$\mathbf{Y}^{\nu}(t)=\mathbf{Y}^{\nu}(t^{-}),\ \mathbf{D}^{\mu_{Pd}}(t)=\mathbf{D}^{\mu_{Pd}}(t^{-}),\ \forall t\in\mathfrak{T}_{0}, \tag{12.51}$$

so that

$$C_{Pu}^{(\mu_{Pu})}\left[\mathbf{U}^{\mu_{Pu}}(t)-\mathbf{U}^{\mu_{Pu}}(t^{-})\right]=\mathbf{T}\left(t,\boldsymbol{\epsilon},\boldsymbol{\epsilon}^{(1)},...,\int_{t_0=0}^{t}\boldsymbol{\epsilon}dt\right). \tag{12.52}$$

The conditions 1) and 2) permit the transformation of (12.52) into (12.46) with Γ defined in (12.47) for the following reason. Let

$$\mathbf{U}^{\mu_{Pu}}=C_{Pu}^{(\mu_{Pu})T}\mathbf{W}. \tag{12.53}$$

Hence, (12.52) becomes

$$\left(C_{Pu}^{(\mu_{Pu})}C_{Pu}^{(\mu_{Pu})T}\right)\left[\mathbf{W}(t)-\mathbf{W}(t^{-})\right]=\mathbf{T}\left(t,\boldsymbol{\epsilon},\boldsymbol{\epsilon}^{(1)},...,\int_{t_0=0}^{t}\boldsymbol{\epsilon}dt\right). \tag{12.54}$$

This is solvable in $[\mathbf{W}(t)-\mathbf{W}(t^{-})]$ due to the conditions 1) and 2),

$$\left[\mathbf{W}(t)-\mathbf{W}(t^{-})\right]=\left(C_{Pu}^{(\mu_{Pu})}C_{Pu}^{(\mu_{Pu})T}\right)^{-1}\mathbf{T}\left(t,\boldsymbol{\epsilon},\boldsymbol{\epsilon}^{(1)},...,\int_{t_0=0}^{t}\boldsymbol{\epsilon}dt\right). \tag{12.55}$$

Multiplying this on the left by $C_{Pu}^{(\mu_{Pu})T}$,

$$\left[C_{Pu}^{(\mu_{Pu})T}\mathbf{W}(t)-C_{Pu}^{(\mu_{Pu})T}\mathbf{W}(t^{-})\right]=\mathbf{U}^{\mu_{Pu}}(t)-\mathbf{U}^{\mu_{Pu}}(t^{-})=$$

$$=C_{Pu}^{(\mu_{Pu})T}\left(C_{Pu}^{(\mu_{Pu})}C_{Pu}^{(\mu_{Pu})T}\right)^{-1}\mathbf{T}\left(t,\boldsymbol{\epsilon},\boldsymbol{\epsilon}^{(1)},...,\int_{t_0=0}^{t}\boldsymbol{\epsilon}dt\right) \tag{12.56}$$

we prove (12.46) due to (12.47).

In the complex domain (12.52) has the following form

$$\left[C_{Pu}^{(\mu_{Pu})}S_{r}^{(\mu_{Pu})}(s)\right]\left[\mathbf{U}(s)-e^{-\sigma s}\mathbf{U}(s)\right]=\mathbf{T}\left(s,\mathbf{E}(\mathbf{s})\right).$$

Since $\mathbf{U}(s)$ satisfies this equation, then it obeys (12.48)-(12.50) in view of the conditions 1) and 2). To show this let

$$\mathbf{U}(s)=\left[C_{Pu}^{(\mu_{Pu})}S_{r}^{(\mu_{Pu})}(s)\right]^{T}W(s)$$

so that

$$\left[C_{Pu}^{(\mu_{Pu})}S_{r}^{(\mu_{Pu})}(s)\right]\left[\mathbf{U}(s)-e^{-\sigma s}\mathbf{U}(s)\right]=$$

$$=\left[C_{Pu}^{(\mu_{Pu})}S_{r}^{(\mu_{Pu})}(s)\right]\left[C_{Pu}^{(\mu_{Pu})}S_{r}^{(\mu_{Pu})}(s)\right]^{T}\left[W(s)-e^{-\sigma s}\mathbf{U}(s)\right]=$$

$$=\mathbf{T}\left(s,\mathbf{E}(\mathbf{s})\right)\Longrightarrow$$

$$\left[W(s) - e^{-\sigma s}\mathbf{W}(s)\right] =$$
$$\left(\left[C_{Pu}^{(\mu_{Pu})}S_r^{(\mu_{Pu})}(s)\right]\left[C_{Pu}^{(\mu_{Pu})}S_r^{(\mu_{Pu})}(s)\right]^T\right)^{-1}\mathbf{T}\left(s,\mathbf{E(s)}\right) \Longrightarrow$$

$$\left[C_{Pu}^{(\mu_{Pu})}S_r^{(\mu_{Pu})}(s)\right]^T\left[W(s) - e^{-\sigma s}\mathbf{U}(s)\right] =$$
$$\left[C_{Pu}^{(\mu_{Pu})}S_r^{(\mu_{Pu})}(s)\right]^T\left(\left[C_{Pu}^{(\mu_{Pu})}S_r^{(\mu_{Pu})}(s)\right]\left[C_{Pu}^{(\mu_{Pu})}S_r^{(\mu_{Pu})}(s)\right]^T\right)^{-1}\mathbf{T}\left(s,\mathbf{E(s)}\right) \Longrightarrow$$
$$\left[\mathbf{U}(s) - e^{-\sigma s}\mathbf{U}(s)\right] = \mathbf{T}\left(s,\mathbf{E(s)}\right) \Longrightarrow$$
$$\mathbf{U}(s) = e^{-\sigma s}\mathbf{U}(s) + \Gamma(s)\mathbf{T}\left(s,\mathbf{E(s)}\right).$$

Sufficiency. Let the conditions 1) through 3) hold. The *IO* object (12.44) is perfect trackable on $\mathfrak{D}^{\mu_{Pd}}\mathrm{x}\mathfrak{Y}_d^{\nu}$ (Theorem 275) (Subsection 9.3.1). We should show that the natural control (12.46), (12.47), i.e., (12.48), (12.49), ensures tracking on $\mathfrak{D}^{\mu_{Pd}}\mathrm{x}\mathfrak{Y}_d^{\nu}$ determined by (12.17), (12.18) for every $\varepsilon_0^k \in \mathfrak{R}^{(k+1)N}$. We multiply (12.48) by

$$C_{Pu}^{(\mu_{Pu})}S_r^{(\mu_{Pu})}(s)$$

on the left. The result is

$$(1 - e^{-\sigma s})C_{Pu}^{(\mu_{Pu})}S_r^{(\mu_{Pu})}(s)\mathbf{U}(s) = \mathbf{T}\left(s,\boldsymbol{\epsilon}(\mathbf{s})\right), \textit{ i.e.},$$
$$C_{Pu}^{(\mu_{Pu})}\left[\mathbf{U}^{\mu_{Pu}}(t) - \mathbf{U}^{\mu_{Pu}}(t^-)\right] = \mathbf{T}\left(t,\boldsymbol{\epsilon},\boldsymbol{\epsilon}^{(1)},...,\int_{t_0=0}^{t}\boldsymbol{\epsilon}dt\right),$$

which for $t^- = t\text{-}\sigma$, $0 < \sigma <<< 1$, i.e., for $\sigma \longrightarrow 0^+$, or in the ideal case for $\sigma = 0$, becomes

$$\mathbf{T}\left(s,\boldsymbol{\epsilon}(\mathbf{s})\right) = \mathbf{0}_N,$$

or equivalently in the time domain

$$\mathbf{T}\left(t,\boldsymbol{\epsilon},\boldsymbol{\epsilon}^{(1)},...,\int_{t_0=0}^{t}\boldsymbol{\epsilon}dt\right) = \mathbf{0}_N,\ \forall t \in \mathfrak{T}_0. \tag{12.57}$$

This implies the validity of (12.17), (12.18) for every $\varepsilon_0^k \in \mathfrak{R}^{(k+1)N}$due to Theorem 364, which completes the proof ■

Notice that we can prove (12.57) by starting with (12.44) from which we subtract (12.45) and we apply (12.46), (12.47),

$$A_P^{(\nu)}\left[\mathbf{Y}^{\nu}(t) - \mathbf{Y}^{\nu}(t^-)\right] = C_{Pu}^{(\mu_{Pu})}\left[\mathbf{U}^{\mu_{Pu}}(t) - \mathbf{U}^{\mu_{Pu}}(t^-)\right] +$$
$$+D_{Pd}^{(\mu_{Pd})}\left[\mathbf{D}^{\mu_{Pd}}(t) - \mathbf{D}^{\mu_{Pd}}(t^-)\right] - \mathbf{T}\left(t,\boldsymbol{\epsilon},\boldsymbol{\epsilon}^{(1)},...,\int_{t_0=0}^{t}\boldsymbol{\epsilon}dt\right),$$
$$\forall t \in \mathfrak{T}_0.$$

(12.51), equivalently Principle 9, reduces the preceding equation to (12.57)

Fig. 12.4 shows the full block diagram of the *Natural Tracking* (*NT*) *Controller* of the *IO* plant (12.44).

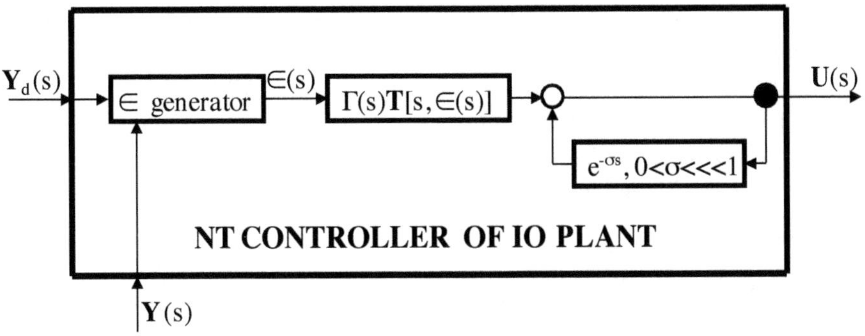

Figure 12.4: The full block diagram of the Natural Tracking (NT) Controller of the IO plant (12.44).

Comment 375 *The time domain form (12.46), (12.47) of the control algorithm (12.48),*

$$\mathbf{U}(t) = \mathbf{U}(t^{-}) + C_{Pu}^{(\mu_{Pu})T}\left(C_{Pu}^{(\mu_{Pu})}C_{Pu}^{(\mu_{Pu})T}\right)^{-1-1}\mathbf{T}\left(t, \boldsymbol{\epsilon}, \boldsymbol{\epsilon}^{(1)}, ..., \int_{t_0=0}^{t} \boldsymbol{\epsilon} dt\right), \tag{12.58}$$

is the general form of (12.1) (Section 12.1) for the IO plant (12.44). It determines the unit local positive feedback in the controller, which is determined by

$$\mathbf{U}(t) = \mathbf{U}(t^{-}) + ...$$

and the global negative feedback in the overall control system specified by

$$\mathbf{U}(t) = ... + C_{Pu}^{(\mu_{Pu})T}\left(C_{Pu}^{(\mu_{Pu})}C_{Pu}^{(\mu_{Pu})T}\right)^{-1-1}\mathbf{T}\left(t, \boldsymbol{\epsilon}, \boldsymbol{\epsilon}^{(1)}, ..., \int_{t_0=0}^{t} \boldsymbol{\epsilon} dt\right),$$

$$\boldsymbol{\epsilon} = \mathbf{Y}_R - \mathbf{Y}.$$

Since the unit positive feedback makes the disconnected controller unstable, it can be ON only when it is connected with the plant in the global negative feedback. Then, the local unit positive feedback and the global negative feedback ensure the requested tracking quality.

Comment 376 *The implementation of the NTC (12.58) does not need any information about the internal dynamics of the plant, about the plant mathematical model or about the disturbance vector value. This expresses the superiority of the NTC over other tracking control approaches, e.g., over the LITC synthesized by using the full transfer function matrices of the IO plant and of the IO controller, and over the LTC synthesized by Lyapunov method.*

12.2.2 Control synthesis for specific tracking qualities

All following theorems result directly from Theorem 374.

Theorem 377 ***NTC for the first order linear elementwise exponential tracking, Example 366 (Section 12.1)***

Let (12.19) through (12.25) be valid. Let the attainability time $\tau_A = \infty$, i.e., $\mathfrak{T}_{A\infty} = \{\infty\}$, (12.14) (Section 12.1).

For the IO object (12.44) to be controlled by the natural tracking control $\mathbf{U}$ *in order to exhibit tracking on $\mathfrak{D}^{\mu_{Pd}} x \mathfrak{Y}_d^{\nu}$ determined by the following tracking algorithm*

$$\mathbf{T}\left(t, \varepsilon, \varepsilon^{(1)}\right) = T_1 \varepsilon^{(1)}(t) + K_0 \varepsilon(t) = \mathbf{0}_N,\ \forall t \in \mathfrak{T}_0,\ if\ \mathbf{T}\left(0, \varepsilon_0^k, \mathbf{0}_N\right) = \mathbf{0}_N,$$
$$\varepsilon^1\left(t; \varepsilon_0^1\right) = -\mathbf{f}^1(t),\ \forall t \in \mathfrak{T}_0,\ if\ \mathbf{T}\left(0, \varepsilon_0^1, \mathbf{0}_N\right) \neq \mathbf{0}_N,$$
$$T_1 = diag\{\tau_1\ \tau_2\ ...\tau_N\} > O_N,\ K_0 = diag\{k_{01}\ k_{02}\ ...k_{0N}\} > O_N, \quad (12.59)$$

it is necessary and sufficient that

1) $N \leq r$,

2) $rankG_{IOPU}(s) = rank\left(C_{Pu}^{(\mu_{Pu})} S_r^{(\mu_{Pu})}(s)\right) = rankC_{Pu}^{(\mu_{Pu})} = N$, *and*

3) *the control obeys in the temporal domain:*

$$\mathbf{U}^{\mu_{Pu}}(t) = \mathbf{U}^{\mu_{Pu}}(t^-) + \Gamma\left[T_1 \epsilon^{(1)}(t) + K_0 \epsilon(t)\right], \quad (12.60)$$
$$\Gamma = C_{Pu}^{(\mu_{Pu})T}\left(C_{Pu}^{(\mu_{Pu})} C_{Pu}^{(\mu_{Pu})T}\right)^{-1}, \quad (12.61)$$

where (12.25) defines $\epsilon(t; \epsilon_0)$, or equivalently in the complex domain:

$$\mathbf{U}(s) = e^{-\sigma s}\mathbf{U}(s) + \Gamma(s)\mathbf{T}(s, \epsilon(\mathbf{s})),\ 0 < \sigma <<< 1,\ \sigma \longrightarrow 0^+, \quad (12.62)$$
$$\mathbf{T}(s, \epsilon(\mathbf{s})) = \mathcal{L}\{T_1 \epsilon(t) + K_0 \epsilon(t)\}, \quad (12.63)$$
$$\Gamma(s) = G_{IOPU}^T(s)\left[G_{IOPU}(s) G_{IOPU}^T(s)\right]^{-1}. \quad (12.64)$$

Proof. The tracking algorithm (12.59) is a special case of the tracking algorithm $\mathbf{T}(.)$, (12.17), (12.18). All conditions of Theorem 374 hold. Theorem 374 implies Theorem 377 ∎

We will use

$$S_N^{(\eta)}(s) = \left[s^0 I_N \vdots s^1 I_i \vdots s^2 I_N \vdots ... \vdots s^\eta I_N\right]^{\mathbf{T}} \in \mathfrak{C}^{(\eta+1)N x N}.$$

Theorem 378 ***NTC for the higher order linear elementwise exponential tracking, Example 367 (Section 12.1)***

Let (12.19) through (12.25) be valid. Let the attainability time $\tau_A = \infty$, i.e., $\mathfrak{T}_{A\infty} = \{\infty\}$, (12.14).

For the IO object (12.44) to be controlled by the natural tracking control $\mathbf{U}$ *in order to exhibit tracking on $\mathfrak{D}^{\mu_{Pd}} x \mathfrak{Y}_d^{\nu}$ determined by the tracking algorithm*

(12.65),

$$\mathbf{T}\left(t,\varepsilon,\varepsilon^{(1)},..,\varepsilon^{(\eta)},\int_{t_0=0}^{t}\varepsilon dt\right)=\sum_{k=0}^{k=\eta\leq\nu-1}E_k\varepsilon^{(k)}(t)=E^{(\eta)}\varepsilon^{\eta}(t)=\mathbf{0}_N,$$
$$\forall t\in\mathfrak{T}_0\ if\ \mathbf{T}\left(0,\varepsilon_0^k,\mathbf{0}_N\right)=\mathbf{0}_N,$$
$$\varepsilon^{\eta}\left(t;\varepsilon_0^{\eta}\right)=-\mathbf{f}^{\eta}\left(t\right),\ \forall t\in\mathfrak{T}_0\ if\ \mathbf{T}\left(0,\varepsilon_0^{\eta},\mathbf{0}_N\right)\neq\mathbf{0}_N, \tag{12.65}$$

with the matrices $E_k\in\mathfrak{R}^{N\times N}$ *such that the real parts* $\mathrm{Re}\,s_i$ *of the roots* s_i *of its characteristic polynomial* $f\left(s\right)$,

$$f\left(s\right)=\det\left(E^{(\eta)}S_N^{(\eta)}\left(s\right)\right), \tag{12.66}$$

are negative,

$$f\left(s_i\right)=\det\left(E^{(\eta)}S_N^{(\eta)}\left(s_i\right)\right)=0\Longrightarrow\mathrm{Re}\,s_i<0,\ \forall i=1,2,...,\eta N, \tag{12.67}$$

it is necessary and sufficient that

1) $N\leq r$,
2) $rankG_{IOPU}(s)=rank\left(C_{Pu}^{(\mu_{Pu})}S_r^{(\mu_{Pu})}(s)\ \right)=rankC_{Pu}^{(\mu_{Pu})}=N$, *and*
3) *the control obeys in the temporal domain:*

$$\mathbf{U}^{\mu_{Pu}}(t)=\mathbf{U}^{\mu_{Pu}}(t^-)+\Gamma E^{(\eta)}\epsilon^{\eta}(t), \tag{12.68}$$
$$\Gamma=C_{Pu}^{(\mu_{Pu})T}\left(C_{Pu}^{(\mu_{Pu})}C_{Pu}^{(\mu_{Pu})T}\right)^{-1}, \tag{12.69}$$

where (12.25) defines $\epsilon\left(t;\epsilon_0\right)$, *or equivalently in the complex domain:*

$$\mathbf{U}(s)=e^{-\sigma s}\mathbf{U}(s)+\Gamma(s)E^{(\eta)}S_N^{(\eta)}(s),\ 0<\sigma<<<1,\ \sigma\longrightarrow0^+, \tag{12.70}$$
$$\Gamma(s)=G_{IOPU}^T(s)\left[G_{IOPU}(s)G_{IOPU}^T(s)\right]^{-1}. \tag{12.71}$$

Proof. The tracking algorithm (12.65) is a special form of the tracking algorithm $\mathbf{T}\left(.\right)$, (12.17), (12.18). The conditions of Theorem 374 hold. Theorem 374 implies Theorem 378 ∎

Theorem 379 ***NTC for the sharp elementwise stablewise tracking with the finite vector reachability time*** τ_R^{2N}, ***Example 368 (Section 12.1)***

Let (12.19) through (12.25) be valid.

For the IO object (12.44) to be controlled by the natural tracking control $\mathbf{U}$ *in order to exhibit tracking on* $\mathfrak{D}^{\mu_{Pd}}x\mathfrak{Y}_d^{\nu}$ *determined by the algorithm*

$$\mathbf{T}\left(\mathbf{t}^N,\varepsilon,\varepsilon^{(1)}\right)=T_1\varepsilon^{(1)}(\mathbf{t}^N)+K_0sign\varepsilon(\mathbf{0}_N)=\mathbf{0}_N,$$
$$\forall\mathbf{t}^N\in\mathfrak{T}_0^N\ if\ \mathbf{T}\left(\mathbf{0}_N,\varepsilon_0^1,\mathbf{0}_N\right)=\mathbf{0}_N,$$
$$\varepsilon^1\left(\mathbf{t}^{2N};\varepsilon_0^1\right)=-\mathbf{f}^1\left(\mathbf{t}^{2N}\right),\ \forall\mathbf{t}^{2N}\in\mathfrak{T}_0^{2N},\ if\ \mathbf{T}\left(\mathbf{0}_N,\varepsilon_0^1,\mathbf{0}_N\right)\neq\mathbf{0}_N, \tag{12.72}$$

for the sharp elementwise stablewise tracking with the finite vector reachability time τ_R^{2N},

$$\boldsymbol{\tau}_R^N = \boldsymbol{\tau}_{R[0]}^N = T_1 K_0^{-1} \left|\boldsymbol{\varepsilon}_0\right|, \ \boldsymbol{\tau}_R^{2N} = \begin{bmatrix} \boldsymbol{\tau}_R^N \\ \boldsymbol{\tau}_R^N \end{bmatrix}, \tag{12.73}$$

it is necessary and sufficient that

1) $N \leq r$,

2) $rankG_{IOPU}(s) = rank\left(C_{Pu}^{(\mu_{Pu})} S_r^{(\mu_{Pu})}(s)\right) = rankC_{Pu}^{(\mu_{Pu})} = N$, *and*

3) *the control obeys in the temporal domain:*

$$\mathbf{U}^{\mu_{Pu}}(t) = \mathbf{U}^{\mu_{Pu}}(t^-) + \Gamma\left[T_1 \boldsymbol{\epsilon}^{(1)}(t) + K_0 sign\boldsymbol{\epsilon}(0)\right], \tag{12.74}$$

$$\Gamma = C_{Pu}^{(\mu_{Pu})T}\left(C_{Pu}^{(\mu_{Pu})} C_{Pu}^{(\mu_{Pu})T}\right)^{-1}, \tag{12.75}$$

where (12.25) defines $\boldsymbol{\epsilon}(t;\boldsymbol{\epsilon}_0)$, *or equivalently in the complex domain:*

$$\mathbf{U}(s) = e^{-\sigma s}\mathbf{U}(s) + \Gamma(s)\mathfrak{L}\left\{T_1 \boldsymbol{\epsilon}^{(1)}(t) + K_0 sign\boldsymbol{\epsilon}(0)\right\}, \ 0 < \sigma <<< 1, \ \sigma \longrightarrow 0^+, \tag{12.76}$$

$$\Gamma(s) = G_{IOPU}^T(s)\left[G_{IOPU}(s) G_{IOPU}^T(s)\right]^{-1}. \tag{12.77}$$

Proof. The tracking algorithm $\mathbf{T}(.)$, (12.17), (12.18) takes the form of the tracking algorithm (12.65) in this case. The conditions of Theorem 374 hold. Theorem 374 implies Theorem 379 ■

We defined

$$\left|E(t)\right|^{1/2} = diag\left\{\left|\varepsilon_1(t)\right|^{1/2} \ \ \left|\varepsilon_2(t)\right|^{1/2} \ \ \ldots \ \ \left|\varepsilon_N(t)\right|^{1/2}\right\},$$

$$\boldsymbol{\epsilon} = \left[\epsilon_1 \ \ \epsilon_2 \ \ \ldots \ \ \epsilon_N\right]^{\mathbf{T}} \Longrightarrow \left|\Xi\right|^{1/2} = diag\left\{\left|\epsilon_1\right|^{1/2} \ \ \left|\epsilon_2\right|^{1/2} \ \ \ldots \ \ \left|\epsilon_N\right|^{1/2}\right\},$$

in Example 369 (Section 12.1).

Theorem 380 ***The NTC synthesis for the first power smooth elementwise stablewise tracking with the finite vector reachability time*** $\boldsymbol{\tau}_R^{2N}$, ***Example 369 (Section 12.1)***

Let (12.19) through (12.25) be valid.

For the IO object (12.44) to be controlled by the natural tracking control $\mathbf{U}$ *in order to exhibit tracking on* $\mathfrak{D}^{\mu_{Pd}} x \mathfrak{Y}_d^{\nu}$ *determined by*

$$\mathbf{T}\left(\mathbf{t}^N, \varepsilon, \varepsilon^{(1)}\right) = T_1 \varepsilon^{(1)}(\mathbf{t}^N) + 2K_0 \left|E\left(\mathbf{t}^N\right)\right|^{1/2} sign\varepsilon_0 = \mathbf{0}_N,$$

$$\forall \mathbf{t}^N \in \mathfrak{T}_0^N \ if \ \mathbf{T}\left(\mathbf{0}_N, \varepsilon_0^1, \mathbf{0}_N\right) = \mathbf{0}_N,$$

$$\varepsilon^1\left(\mathbf{t}^{2N}; \varepsilon_0^1\right) = -\mathbf{f}^1\left(\mathbf{t}^{2N}\right), \ \forall \mathbf{t}^{2N} \in \mathfrak{T}_0^{2N}, \ if \ \mathbf{T}\left(\mathbf{0}_N, \varepsilon_0^1, \mathbf{0}_N\right) \neq \mathbf{0}_N,$$

$$\left|E\left(\mathbf{t}^N\right)\right|^{1/2} = diag\left\{\left|\varepsilon_1(t)\right|^{1/2} \ \ \left|\varepsilon_2(t)\right|^{1/2} \ \ \ldots \ \ \left|\varepsilon_N(t)\right|^{1/2}\right\}, \tag{12.78}$$

for the first power smooth elementwise stablewise tracking with the finite vector reachability time $\boldsymbol{\tau}_R^{2N}$,

$$\boldsymbol{\tau}_R^N = T_1 K_0^{-1} \left|E_0\right|^{1/2} \mathbf{1}_N, \ \boldsymbol{\tau}_R^{2N} = \begin{bmatrix} \boldsymbol{\tau}_R^N \\ \boldsymbol{\tau}_R^N \end{bmatrix},$$

it is necessary and sufficient that

1) $N \leq r$,
2) $rankG_{IOPU}(s) = rank\left(C_{Pu}^{(\mu_{Pu})} S_r^{(\mu_{Pu})}(s)\right) = rankC_{Pu}^{(\mu_{Pu})} = N$, *and*
3) *the control obeys in the temporal domain:*

$$\mathbf{U}^{\mu_{Pu}}(t) = \mathbf{U}^{\mu_{Pu}}(t^-) + \Gamma\left[T_1 \boldsymbol{\epsilon}(t) + 2K_0 \left|\Xi(t)\right|^{1/2} sign\boldsymbol{\epsilon}_0\right], \tag{12.79}$$

$$\Gamma = C_{Pu}^{(\mu_{Pu})T}\left(C_{Pu}^{(\mu_{Pu})} C_{Pu}^{(\mu_{Pu})T}\right)^{-1}, \tag{12.80}$$

where (12.25) defines $\boldsymbol{\epsilon}(t;\boldsymbol{\epsilon}_0)$, *or equivalently in the complex domain:*

$$\mathbf{U}(s) = e^{-\sigma s}\mathbf{U}(s) + \Gamma(s)\mathfrak{L}\left\{T_1 \boldsymbol{\epsilon}(t) + 2K_0 \left|\Xi(t)\right|^{1/2} sign\boldsymbol{\epsilon}_0\right\}, \tag{12.81}$$

$$0 < \sigma <<< 1, \ \sigma \longrightarrow 0^+,$$

$$\Gamma(s) = G_{IOPU}^T(s)\left[G_{IOPU}(s) G_{IOPU}^T(s)\right]^{-1}. \tag{12.82}$$

Proof. Theorem 374 implies Theorem 380 since the tracking algorithm $\mathbf{T}(.)$, (12.17), (12.18) has become the tracking algorithm (12.78) so that all conditions of Theorem 374 are satisfied ■

We defined in (12.37),

$$K = diag\{k_1\ k_2\ ...k_N\},\ k_i \in \{2,\ 3,\ ...\},\ \forall i = 1, 2,\ ...\ , N,$$
$$\left|E\left(\mathbf{t}^N\right)\right|^{I-K^{-1}} = diag\left\{\left|\varepsilon_1(t)\right|^{1-k_1^{-1}}\ \left|\varepsilon_2(t)\right|^{1-k_2^{-1}}\ ...\ \left|\varepsilon_N(t)\right|^{1-k_N^{-1}}\right\},$$

Example 370 (Section 12.1).

Theorem 381 ***The NTC synthesis for the higher power smooth elementwise stablewise tracking with the finite vector reachability time*** $\boldsymbol{\tau}_R^{2N}$, ***Example 370 (Section 12.1)***

Let (12.19) through (12.25) be valid.

For the IO object (12.44) to be controlled by the natural tracking control $\mathbf{U}$ *in order to exhibit tracking on* $\mathfrak{D}^{\mu_{Pd}}x\mathfrak{Y}_d^{\nu}$ *determined by the tracking algorithm*

$$\mathbf{T}\left(\mathbf{t}^N, \varepsilon, \varepsilon^{(1)}\right) = T_1 \varepsilon^{(1)}(\mathbf{t}^N) + K_0 \left|E\left(\mathbf{t}^N\right)\right|^{I-K^{-1}} sign\varepsilon_0 = \mathbf{0}_N,$$
$$\forall \mathbf{t}^N \in \mathfrak{T}_0^N \ if\ \mathbf{T}\left(\mathbf{0}_N, \varepsilon_0^1, \mathbf{0}_N\right) = \mathbf{0}_N,$$
$$\varepsilon^1\left(\mathbf{t}^{2N};\varepsilon_0^1\right) = -\mathbf{f}^1\left(\mathbf{t}^{2N}\right),\ \forall \mathbf{t}^{2N} \in \mathfrak{T}_0^{2N},\ if\ \mathbf{T}\left(\mathbf{0}_N, \varepsilon_0^1, \mathbf{0}_N\right) \neq \mathbf{0}_N, \tag{12.83}$$

for the higher power smooth elementwise stablewise tracking with the finite vector reachability time $\boldsymbol{\tau}_R^{2N}$,

$$\boldsymbol{\tau}_R^N = T_1 K_0^{-1} |\boldsymbol{\varepsilon}_0|^{K^{-1}}, \ \boldsymbol{\tau}_R^{2N} = \begin{bmatrix} \boldsymbol{\tau}_R^N \\ \boldsymbol{\tau}_R^N \end{bmatrix},$$

it is necessary and sufficient that

1) $N \le r$,

2) $rankG_{IOPU}(s) = rank\left(C_{Pu}^{(\mu_{Pu})} S_r^{(\mu_{Pu})}(s)\right) = rankC_{Pu}^{(\mu_{Pu})} = N$, *and*

3) the control obeys in the temporal domain:

$$\mathbf{U}^{\mu_{Pu}}(t) = \mathbf{U}^{\mu_{Pu}}(t^-) + \Gamma\left[T_1 \boldsymbol{\epsilon}(t) + K_0 |\Xi(t)|^{I-K^{-1}} sign\boldsymbol{\epsilon}_0\right], \tag{12.84}$$

$$\Gamma = C_{Pu}^{(\mu_{Pu})T}\left(C_{Pu}^{(\mu_{Pu})} C_{Pu}^{(\mu_{Pu})T}\right)^{-1}, \tag{12.85}$$

where (12.25) defines $\boldsymbol{\epsilon}(t;\boldsymbol{\epsilon}_0)$, *i.e., in the complex domain:*

$$\mathbf{U}(s) = e^{-\sigma s}\mathbf{U}(s) + \Gamma(s)\mathfrak{L}\left\{T_1 \boldsymbol{\epsilon}(t) + K_0 |\Xi(t)|^{I-K^{-1}} sign\boldsymbol{\epsilon}_0\right\}, \tag{12.86}$$

$$0 < \sigma <<< 1, \ \sigma \longrightarrow 0^+,$$

$$\Gamma(s) = G_{IOPU}^T(s)\left[G_{IOPU}(s) G_{IOPU}^T(s)\right]^{-1}. \tag{12.87}$$

Proof. The tracking algorithm $\mathbf{T}(.)$, (12.17), (12.18) reduces to the tracking algorithm (12.83). Since all conditions of Theorem 374 are valid, it implies Theorem 381 ■

Theorem 382 ***The NTC synthesis for the sharp absolute error vector value tracking elementwise and stablewise with the finite vector reachability time*** $\boldsymbol{\tau}_R^{2N}$, ***Example 371 (Section 12.1)***

Let (12.19) through (12.25) be valid.

For the IO object (12.44) to be controlled by the natural tracking control $\mathbf{U}$ *in order to exhibit tracking on* $\mathfrak{D}^{\mu_{Pd}} x \mathfrak{Y}_d^{\nu}$ *determined by the following tracking algorithm*

$$\mathbf{T}\left(\mathbf{t}^N, \boldsymbol{\varepsilon}, \boldsymbol{\varepsilon}^{(1)}\right) = T_1 \Sigma\left(\boldsymbol{\varepsilon}, \boldsymbol{\varepsilon}^{(1)}\right) \boldsymbol{\varepsilon}^{(1)} + K_0 sign\,|\boldsymbol{\varepsilon}_0| = \mathbf{0}_N,$$

$$\forall \mathbf{t}^N \in \mathfrak{T}_0^N \ if \ \mathbf{T}\left(0, \boldsymbol{\varepsilon}_0^1, \mathbf{0}_N\right) = \mathbf{0}_N,$$

$$\boldsymbol{\varepsilon}^1\left(\mathbf{t}^{2N}; \boldsymbol{\varepsilon}_0^1\right) = -\mathbf{f}^1\left(\mathbf{t}^{2N}\right), \ \forall \mathbf{t}^{2N} \in \mathfrak{T}_0^{2N}, \ if \ \mathbf{T}\left(\mathbf{0}_N, \boldsymbol{\varepsilon}_0^1, \mathbf{0}_N\right) \ne \mathbf{0}_N, \tag{12.88}$$

for the sharp absolute error vector value tracking elementwise stablewise with the finite vector reachability time $\boldsymbol{\tau}_R^{2N}$,

$$\boldsymbol{\tau}_R^N = T_1 K_0^{-1} |\boldsymbol{\varepsilon}_0|, \ \boldsymbol{\tau}_R^{2N} = \begin{bmatrix} \boldsymbol{\tau}_R^N \\ \boldsymbol{\tau}_R^N \end{bmatrix},$$

it is necessary and sufficient that

1) $N \le r$,

2) $rankG_{IOPU}(s) = rank\left(C_{Pu}^{(\mu_{Pu})}S_r^{(\mu_{Pu})}(s)\right) = rankC_{Pu}^{(\mu_{Pu})} = N$, *and*

3) *the control obeys in the temporal domain:*

$$\mathbf{U}^{\mu_{Pu}}(t) = \mathbf{U}^{\mu_{Pu}}(t^-) + \Gamma\left[T_1\Sigma\left(\boldsymbol{\epsilon},\boldsymbol{\epsilon}^{(1)}\right)\boldsymbol{\epsilon}^{(1)} + K_0 sign\left|\boldsymbol{\epsilon}\right|\right],$$
$$\Gamma = C_{Pu}^{(\mu_{Pu})T}\left(C_{Pu}^{(\mu_{Pu})}C_{Pu}^{(\mu_{Pu})T}\right)^{-1},$$

where (12.25) defines $\boldsymbol{\epsilon}\left(t;\boldsymbol{\epsilon}_0\right)$, *or equivalently in the complex domain:*

$$\mathbf{U}(s) = e^{-\sigma s}\mathbf{U}(s) + \Gamma(s)\mathfrak{L}\left\{T_1\Sigma\left(\boldsymbol{\epsilon},\boldsymbol{\epsilon}^{(1)}\right)\boldsymbol{\epsilon}^{(1)} + K_0 sign\left|\boldsymbol{\epsilon}\right|\right\},$$
$$0 < \sigma <<< 1,\ \sigma \longrightarrow 0^+,$$
$$\Gamma(s) = G_{IOPU}^T(s)\left[G_{IOPU}(s)G_{IOPU}^T(s)\right]^{-1}.$$

Proof. The tracking algorithm $\mathbf{T}(.)$, (12.17), (12.18) takes the form of the tracking algorithm (12.83). Theorem 374 implies Theorem 382 since all its conditions hold ∎

Theorem 383 ***NTC synthesis for the exponential absolute error vector value tracking elementwise and stablewise with the finite vector reachability time*** τ_R^{2N}***, Example 372 (Section 12.1)***

Let (12.19) through (12.25) be valid.

For the IO object (12.44) to be controlled by the natural tracking control $\mathbf{U}$ *in order to exhibit tracking on* $\mathfrak{D}^{\mu_{Pd}}x\mathfrak{Y}_d^{\nu}$ *determined by*

$$\mathbf{T}\left(\varepsilon,\varepsilon^{(1)}\right) = T_1 D^+\left|\varepsilon\right| + K\left(\left|\varepsilon\right| + K_0 sign\left|\varepsilon_0\right|\right) =$$
$$= T_1\Sigma\left(\varepsilon,\varepsilon^{(1)}\right)\varepsilon^{(1)} + K\left(\left|\varepsilon\right| + K_0 sign\left|\varepsilon_0\right|\right) = \mathbf{0}_N,$$
$$\forall \mathbf{t}^N \in \mathfrak{T}_0^N\ if\ \mathbf{T}\left(\mathbf{0}_N,\varepsilon_0^1,\mathbf{0}_N\right) = \mathbf{0}_N,$$
$$\varepsilon^1\left(\mathbf{t}^{2N};\varepsilon_0^1\right) = -\mathbf{f}^1\left(\mathbf{t}^{2N}\right),\ \forall \mathbf{t}^{2N} \in \mathfrak{T}_0^{2N},\ if\ \mathbf{T}\left(\mathbf{0}_N,\varepsilon_0^1,\mathbf{0}_N\right) \neq \mathbf{0}_N, \quad (12.89)$$

for the exponential absolute error vector value tracking elementwise and stablewise with the finite vector reachability time τ_R^{2N} *(12.42), (12.43) (Section 12.1), it is necessary and sufficient that*

1) $N \leq r$,

2) $rankG_{IOPU}(s) = rank\left(C_{Pu}^{(\mu_{Pu})}S_r^{(\mu_{Pu})}(s)\right) = rankC_{Pu}^{(\mu_{Pu})} = N$, *and*

3) *the control obeys in the temporal domain:*

$$\mathbf{U}^{\mu_{Pu}}(t) = \mathbf{U}^{\mu_{Pu}}(t^-) + \Gamma\left[T_1\Sigma\left(\boldsymbol{\epsilon},\boldsymbol{\epsilon}^{(1)}\right)\boldsymbol{\epsilon}^{(1)} + K\left(\left|\boldsymbol{\epsilon}\right| + K_0 sign\left|\boldsymbol{\epsilon}_0\right|\right)\right],$$
$$\Gamma = C_{Pu}^{(\mu_{Pu})T}\left(C_{Pu}^{(\mu_{Pu})}C_{Pu}^{(\mu_{Pu})T}\right)^{-1},$$

where (12.25) defines $\boldsymbol{\epsilon}(t;\boldsymbol{\epsilon}_0)$*, i.e., in the complex domain:*

$$\mathbf{U}(s) = e^{-\sigma s}\mathbf{U}(s) + \Gamma(s)\mathfrak{L}\left\{T_1\Sigma\left(\boldsymbol{\epsilon},\boldsymbol{\epsilon}^{(1)}\right)\boldsymbol{\epsilon}^{(1)} + K\left(|\boldsymbol{\epsilon}| + K_0 sign\,|\boldsymbol{\epsilon}_0|\right)\right\},$$
$$0 < \sigma <<< 1, \;,\; \sigma \longrightarrow 0^+,$$
$$\Gamma(s) = G_{IOPU}^T(s)\left[G_{IOPU}(s)G_{IOPU}^T(s)\right]^{-1}.$$

Proof. The tracking algorithm $\mathbf{T}(.)$, (12.17), (12.18) is in the form of the tracking algorithm (12.83). All conditions of Theorem 374 are fulfilled so that it implies Theorem 383 ■

Comment 384 *The above condition 2) is expressed in terms of the rank of the IO plant (12.44) transfer function matrix* $G_{IOPU}(s)$ *relative to the control* $\mathbf{U}$*. It does not need the knowledge of the internal dynamics of the IO plant.*

Comment 385 ***Implementation of the NTC algorithm***

The controller should comprise the corresponding component that generates the subsidiary vector function $\mathbf{f}(.)$ *defined by (12.19)-(12.21), Fig. 12.2 (Section 12.1).*

The controller uses information about the subsidiary vector function $\mathbf{f}(.)$ *and about the real output error vector* $\boldsymbol{\varepsilon}$*, determines the subsidiary reference output vector* $\mathbf{Y}_R$*, (12.22), and the subsidiary error vector* $\boldsymbol{\epsilon}$*, (12.25) (Section 12.1), and uses them to generate the NTC according to the accepted algorithm.*

Conclusion 297 (Subsection 9.4.1), shows that the controller does not use information about the plant mathematical model. It does not use any information about the disturbance vector or about high derivatives of the real output vector $\mathbf{Y}(t)$ *at the moment* t*. The NTC is fully robust. The preceding control algorithms verify these claims.*

The usage of $\mathbf{U}(t^-)$ *in order to generate* $\mathbf{U}(t)$ *at the moment* t *expresses a kind of memory and expresses the adaptability of the NTC.*

The transmission of the signal from the controller output to its input without delay, or with negligible delay, is the crucial demand on the controller.

We defined G_{IOPUo} and $G_{IOPU\infty}$ by (9.21) in Subsection 9.3.1:

$$\det A_{P0} \neq 0 \Longrightarrow G_{IOPUo} = G_{IOPU}(0) = A_{P0}^{-1}C_{Pu0} \in \mathfrak{C}^{Nxr},$$
$$\det A_{P\nu} \neq 0 \Longrightarrow G_{IOPU\infty} = \lim_{s \longrightarrow \infty}\left[s^{\nu-\mu_{Pu}}G_{IOPU}(s)\right] = A_{P\nu}^{-1}C_{Pu\mu_{Pu}} \in \mathfrak{C}^{Nxr}. \tag{12.90}$$

Comment 386 ***The simplification of the NTC condition***

Conclusion 297 (Subsection 9.4.1) permits us to replace the above condition 2) by

- $rankG_{OUo} = N \leq r$ *if* $\det A_{Po} \neq 0$,
- $rankG_{IOPU\infty} = N \leq r$ *if* $\det A_{P\nu} \neq 0$.

Conclusion 387 *This holds regardless of the stability property of the IO plant (2.15) under the condition*

$$\det A_{Po} \neq 0.$$

If $A_{P\nu}$ is singular, then the condition

$$rankG_{IOPU\infty} = N$$

is meaningless and inapplicable.

Comment 388 *The control gain $\Gamma(s)$, equivalently Γ, is the same for all considered tracking algorithms. They are expressed in terms of the plant transfer function matrix $G_{IOPU}(s)$ relative to control, equivalently G_{IOPUo}, respectively. The control gain Γ, equivalently $\Gamma(s)$, is the same as for trackability of the IO plant (12.44).*

The control gain Γ, hence $\Gamma(s)$, resolves the complex problem of the channeling the control action on the MIMO IO plant (12.44).

The NTC algorithms do not need any information about the plant whole internal dynamics vector, about the plant internal dynamics description or about the form and value of the disturbance vector function $\mathbf{D}(.)$.

Comment 389 *The higher order linear or nonlinear NTC algorithms need the measurement of the higher output error vector derivatives that are not often accessible for the measurement. For this reason we presented herein the nonlinear NTC algorithms of only the first order. They ensure high tracking qualities including the finite reachability time with the full robustness relative to the fluctuations of the plant internal dynamics and relative to the external disturbances.*

12.3 NTC of the *ISO* plant

12.3.1 General consideration

We recall the mathematical model of the *ISO* plant (2.33), (2.34) (Subsection 2.2.2),

$$\frac{d\mathbf{X}_P(t)}{dt} = A_P\mathbf{X}_P(t) + B_P\mathbf{U}(t) + L_P\mathbf{D}(t),\ \mathbf{X}_P \in \mathfrak{R}^{n_P}, \tag{12.91}$$

$$\mathbf{Y}(t) = \mathbf{Y}_P(t) = C_P\mathbf{X}_P(t) + H_P\mathbf{U}(t) + D_P\mathbf{D}(t),\ C_P \in \mathfrak{R}^{Nxn_P}. \tag{12.92}$$

Claim 155 (Section 8.2) holds if it is not said otherwise.

Since the internal dynamics of the *ISO* plant (12.91), (12.92) occurs in the state space $\mathfrak{R}^{n_P}$ and the tracking takes place in the output space $\mathfrak{R}^N$ in the course of *time,* we introduce an additional (formal, artificial) output variable Y_L defined on the state space $\mathfrak{R}^{n_P}$ as follows.

Definition 390 ***Definition of Lyapunov output variable*** Y_L

An output variable Y is the output variable Y_L called ***Lyapunov output variable*** *if, and only if, it satisfies:*

1. $Y(.) : \mathfrak{R}^{n_P} \longrightarrow \mathfrak{R}_+$ *depends on the difference* $\mathbf{X}_P - \mathbf{X}_{Pd}$ *between the plant real state vector* $\mathbf{X}_P \in \mathfrak{R}^{n_P}$ *and the plant desired state vector* $\mathbf{X}_{Pd} \in \mathfrak{R}^{n_P}$*, where* $\mathbf{X}_{Pd}(.)$ *is the desired (nominal) system motion induced by the desired output behavior* $\mathbf{Y}_d(.)$*.*

2. $Y(\mathbf{X}_P - \mathbf{X}_{Pd})$ *is globally positive definite and radially unbounded on* $\mathfrak{R}^{n_P}$*, that is:*

a) $Y(\mathbf{X}_P - \mathbf{X}_{Pd}) \in \mathfrak{C}(\mathfrak{R}^{n_P})$,
b) $Y(\mathbf{X}_P - \mathbf{X}_{Pd}) \geq 0, \forall \mathbf{X}_P \in \mathfrak{R}^{n_P}$,
c) $Y(\mathbf{X}_P - \mathbf{X}_{Pd}) = 0 \Longleftrightarrow \mathbf{X}_P = \mathbf{X}_{Pd}$,
d) $\|\mathbf{X}_P - \mathbf{X}_{Pd}\| \longrightarrow \infty \Longrightarrow Y(\mathbf{X}_P - \mathbf{X}_{Pd}) \longrightarrow \infty$.

3. $Y(t)$ *is defined by*

$$Y(t) = Y\left[\mathbf{X}_P(t) - \mathbf{X}_{Pd}(t)\right].$$

4. *The desired value* $Y_d(t)$ *of the output variable* $Y(t)$ *is its zero value,*

$$Y_d(t) = 0, \ \forall t \in \mathfrak{T}_0,$$

i.e.,

$$Y(t) \equiv Y_L(t) \Longleftrightarrow 1. - 4. \ \textit{hold}.$$

The error ε_L of Lyapunov variable Y_L is evidently

$$\varepsilon_L = Y_{Ld} - Y_L = 0 - Y_L = -Y_L.$$

$Y_{Ld} = 0$ does not necessarily imply $\mathbf{X}_{Pd} = \mathbf{0}_n$, but implies $\mathbf{X}_P$-$\mathbf{X}_{Pd} = \mathbf{0}_n$.

Condition 391 ***Lyapunov variable for state stability and tracking***
When we wish to ensure a stability property to the desired motion $\mathbf{X}_{Pd}(t)$*, which is induced by the desired output behavior* $\mathbf{Y}_d(t)$*, simultaneously with a tracking property of the ISO plant (12.91), (12.92), then we accept Lyapunov variable* Y_L *to be the* N*-th output variable,* $Y_N = Y_L$*, so that*

$$\mathbf{Y} = [Y_1 \ Y_2 \ . \ . \ . \ Y_{N-1} \ Y_L]^{\mathbf{T}} \in \mathfrak{R}^N.$$

The usage of Lyapunov variable will ensure a stability property to the desired motion $\mathbf{X}_{Pd}(t)$. This approach demands the knowledge of the desired motion $\mathbf{X}_{Pd}(t)$ and that the control vector dimension r is not less than the output vector dimension N, i.e., $r \geq N$.

However, by dividing the control $\mathbf{U}$ into stabilizing control $\mathbf{U}_S$ and tracking control $\mathbf{U_T}$, we can use the former to steer $Y_L[\mathbf{X}_P(t) - \mathbf{X}_{Pd}(t)]$ to zero, i.e., $\mathbf{X}_P(t)$ to $\mathbf{X}_{Pd}(t)$, Note 154 (Section 8.2). Then we do not use Lyapunov variable Y_L for the tracking control synthesis.

The mathematical description of the *ISO* plant (12.91), (12.92) at the moment t^- does not change its form, structure, order, parameters or character,

$$\frac{d\mathbf{X}_P(t^-)}{dt} = A_P\mathbf{X}_P(t^-) + B_P\mathbf{U}(t^-) + L_P\mathbf{D}(t^-), \ \mathbf{X}_P \in R^{n_P}, \tag{12.93}$$

$$\mathbf{Y}(t^-) = \mathbf{Y}_P(t^-) = C_P\mathbf{X}_P(t^-) + H_P\mathbf{U}(t^-) + D_P\mathbf{D}(t^-), \ C_P \in R^{N\text{x}n_P}. \tag{12.94}$$

We determined in (7.85)-(7.87) (Section 7.4), the transfer function matrix $G_{ISOPU}(s)$ of the ISO plant (12.91), (12.92) relative to the control $\mathbf{U}$,

$$G_{ISOPU}(s) = C_P\left(sI_{n_P} - A_P\right)^{-1} B_P + H_P. \tag{12.95}$$

It implies the transfer function matrix G_{ISOPUo} (9.29) (Subsection 9.3.2) of the ISO plant (12.91), (12.92) relative to the control $\mathbf{U}$ at the origin $s = 0$ if, and only if, $\det A_P \neq 0$,

$$\det A_P \neq 0 \Longrightarrow G_{ISOPU}(0) = G_{ISOPUo} = H_P - C_P A_P^{-1} B_P, \tag{12.96}$$

as well as $G_{ISOPU}(\infty)$ at $s = \infty$, (9.30) (Subsection 9.3.2),

$$G_{ISOPU}(\infty) = H. \tag{12.97}$$

which induces $G_{ISOPU\infty}$ at $s = \infty$, (9.31) (Subsection 9.3.2),

$$G_{ISOPU\infty} = \lim_{s \longrightarrow \infty} s\left[G_{ISOPU}(s) - G_{ISOPU}(\infty)\right], \tag{12.98}$$

i.e., as shown in (9.32) (Subsection 9.3.2),

$$G_{ISOPU\infty} = C_P B_P. \tag{12.99}$$

We accept in what follows the validity of Property 351 through Property 353 (Section 12.1).

Theorem 392 ***General NTC synthesis***

Let (12.19) through (12.25) (Section 12.1) be valid.

In order for the ISO plant (12.91), (12.92) to be controlled by the natural tracking control $\mathbf{U}$ *and to exhibit tracking on on* $\mathfrak{D}^1 \mathrm{x} \mathfrak{Y}_d^1$ *determined by the tracking algorithm* $\mathbf{T}(.)$*, (12.17), (12.18) (Section 12.1), it is necessary and sufficient that*

1) $N \leq r$,

2) $rankG_{ISOUP}(s) = rankG_{ISOPUo} = N$,

3)
$$\mathbf{U}(t) = \mathbf{U}(t^-) + \Upsilon T\left(t, \boldsymbol{\epsilon}, \boldsymbol{\epsilon}^{(1)}, ..., \int_{t_0=0}^{t} \boldsymbol{\epsilon} dt\right), \tag{12.100}$$

$$\Upsilon = G_{ISOPUo}^T\left\{G_{ISOPUo} G_{ISOPUo}^T\right\}^{-1}, \tag{12.101}$$

where (12.25) defines $\boldsymbol{\epsilon}(t; \boldsymbol{\epsilon}_0)$ *(Section 12.1), or equivalently*

$$\mathbf{U}(s) = e^{-\sigma s}\mathbf{U}(s) + \Upsilon(s)\mathbf{T}(s, \boldsymbol{\epsilon}(\mathbf{s})),\ 0 < \sigma <<< 1,\ \sigma \longrightarrow 0^+, \tag{12.102}$$

$$\mathbf{T}(s, \mathbf{E}(\mathbf{s})) = \mathcal{L}\left\{\mathbf{T}\left(t, \boldsymbol{\epsilon}(t), \boldsymbol{\epsilon}^{(1)}(t), ..., \int_{t_0=0}^{t} \boldsymbol{\epsilon}(t) dt\right)\right\}, \tag{12.103}$$

$$\Upsilon(s) = G_{ISOPU}(s)^T\left\{G_{ISOPU}(s) G_{ISOPU}(s)^T\right\}^{-1}. \tag{12.104}$$

Proof. *Necessity.* Let the ISO plant described by (12.91), (12.92) be controlled by the natural tracking control $\mathbf{U}$ and exhibit tracking on $\mathfrak{D}^1 \text{x} \mathfrak{Y}_d^1$ determined by (12.17), (12.18). Theorem 364, (Section 12.1), guarantees (12.5) in which ε is replaced by $\boldsymbol{\epsilon}$, i.e., it proves (12.30), (Section 12.1). The plant is trackable on $\mathfrak{D}^1 \text{x} \mathfrak{Y}_d^1$. Theorem 298, i.e., the rank condition (9.51), (Subsection 9.4.2), implies necessity of the conditions 1) and 2). We solve (12.91) for $\mathbf{X}$, which is possible due to $detA \neq 0$ that results from the plant stability (Claim 155, Section 8.2),

$$\mathbf{X}_P(t){=}A_P^{-1}\left(\frac{d\mathbf{X}_P(t)}{dt} - B_P\mathbf{U}(t) - L_P\mathbf{D}(t)\right). \tag{12.105}$$

This and equation (12.30) permit us to set (12.92) into the following form:

$$\begin{aligned}\mathbf{Y}(t) \quad &= \quad C_P A_P^{-1}\left(\frac{d\mathbf{X}_P(t)}{dt} - B_P\mathbf{U}(t) - L_P\mathbf{D}(t)\right) + H_P\mathbf{U}(t) + \\ &+D_P\mathbf{D}(t) - \mathbf{T}\left(t, \boldsymbol{\epsilon}(t), \boldsymbol{\epsilon}^{(1)}(t), ..., \int_{t_0=0}^{t} \boldsymbol{\epsilon}(t)dt\right).\end{aligned}$$

Since the control $\mathbf{U}(t)$ is natural control, it is continuous in $t \in \mathfrak{T}_0$. This, the system linearity and $[\mathbf{D}(.), \mathbf{Y}_d(.)] \in \mathfrak{D}^1 \text{x} \mathfrak{Y}_d^1$ imply [Principle 9, (Section 1.1), implies] continuity of all system variables in $t \in \mathfrak{T}_0$, which holds also at $t = 0$ due to (12.30),

$$\mathbf{D}(t) = \mathbf{D}(t^-),\ \mathbf{X}_P(t) = \mathbf{X}_P(t^-),\ \mathbf{X}_P^{(1)}(t) = \mathbf{X}_P^{(1)}(t^-),\ \mathbf{Y}(t) = \mathbf{Y}(t^-),\ \forall t \in \mathfrak{T}_0. \tag{12.106}$$

Equations (12.93), (12.94) and (12.105) yield

$$\mathbf{Y}(t^-) = C_P A_P^{-1}\left(\frac{d\mathbf{X}_P(t^-)}{dt} - B_P\mathbf{U}(t^-) - D_P\mathbf{D}(t^-)\right) + H_P\mathbf{U}(t^-) + L_P\mathbf{D}(t^-).$$

Altogether,

$$\begin{aligned}\mathbf{Y}(t) - \mathbf{Y}(t^-) = C_P A_P^{-1}\left(\frac{d\mathbf{X}_P(t)}{dt} - B_P\mathbf{U}(t) - D_P\mathbf{D}(t)\right) + H_P\mathbf{U}(t) + \\ +L_P\mathbf{D}(t) - \mathbf{T}\left(t, \boldsymbol{\epsilon}(t), \boldsymbol{\epsilon}^{(1)}(t), ..., \int_{t_0=0}^{t} \boldsymbol{\epsilon}(t)dt\right) - \\ -C_P A_P^{-1}\left(\frac{d\mathbf{X}_P(t^-)}{dt} - B_P\mathbf{U}(t^-) - D_P\mathbf{D}(t^-)\right) - H_P\mathbf{U}(t^-) - L_P\mathbf{D}(t^-) = \mathbf{0}_N,\end{aligned}$$

i.e., due to (12.106),

$$\left[-C_P A_P^{-1}B_P + H_P\right]\left[\mathbf{U}(t) - \mathbf{U}(t^-)\right] = \mathbf{T}\left(t, \boldsymbol{\epsilon}(t), \boldsymbol{\epsilon}^{(1)}(t), ..., \int_{t_0=0}^{t} \boldsymbol{\epsilon}(t)dt\right). \tag{12.107}$$

Let

$$\mathbf{U}(t) = \left[H_P - C_P A_P^{-1}B_P\right]^{\mathbf{T}}\mathbf{W}(t) \tag{12.108}$$

so that (12.107) becomes

$$\left[-C_P A_P^{-1} B_P + H_P\right] \left[H_P - C_P A_P^{-1} B_P\right]^{\mathbf{T}} \left[\mathbf{W}(t) - \mathbf{W}(t^-)\right] =$$
$$= \mathbf{T}\left(t, \boldsymbol{\epsilon}(t), \boldsymbol{\epsilon}^{(1)}(t), ..., \int_{t_0=0}^{t} \boldsymbol{\epsilon}(t)dt\right). \tag{12.109}$$

The conditions 1) and 2) ensure

$$\det\left\{\left[-C_P A_P^{-1} B_P + H_P\right] \left[H_P - C_P A_P^{-1} B_P\right]^T\right\} \neq 0.$$

This permits us to solve (12.109) for $[\mathbf{W}(t) - \mathbf{W}(t^-)]$,

$$\mathbf{W}(t) - \mathbf{W}(t^-) = \left\{\left[-C_P A_P^{-1} B_P + H_P\right] \left[H_P - C_P A_P^{-1} B_P\right]^T\right\}^{-1} \bullet$$
$$\bullet \mathbf{T}\left(t, \boldsymbol{\epsilon}(t), \boldsymbol{\epsilon}^{(1)}(t), ..., \int_{t_0=0}^{t} \boldsymbol{\epsilon}(t)dt\right).$$

This and (12.108) yield

$$\mathbf{U}(t) - \mathbf{U}(t^-) = \underbrace{\left[\underbrace{H_P - C_P A_P^{-1} B_P}_{G_{ISOPUo}}\right]^T \left\{\begin{array}{c} \left[-C_P A_P^{-1} B_P + H_P\right] \bullet \\ \bullet \left[H_P - C_P A_P^{-1} B_P\right]^T \end{array}\right\}^{-1} \bullet}_{\Upsilon}$$
$$\bullet \mathbf{T}\left(t, \boldsymbol{\epsilon}(t), \boldsymbol{\epsilon}^{(1)}(t), ..., \int_{t_0=0}^{t} \boldsymbol{\epsilon}(t)dt\right),$$

which solves (12.107) for $[\mathbf{U}(t) - \mathbf{U}(t^-)]$. We can set this into the following form:

$$\mathbf{U}(t) - \mathbf{U}(t^-) = \Upsilon T\left(t, \boldsymbol{\varepsilon}, \boldsymbol{\varepsilon}^{(1)}, ..., \int_{t_0=0}^{t} \boldsymbol{\varepsilon} dt\right),$$

in view of the condition 2) applied for $s = 0$ and (12.96). This proves (12.100), (12.101).

We can prove this also in the complex domain by starting with the plant transfer function matrix $G_{ISOPU}(s)$ relative to control $\mathbf{U}$ [Equation (9.28), Subsection 9.3.2],

$$G_{ISOPU}(s) = C_P \left(sI_{n_P} - A_P\right)^{-1} B_P + H_P.$$

Laplace transform of the output vector $\mathbf{Y}(t)$ is determined in (7.85)-(7.87) (Section 7.4) by

$$\mathbf{Y}(s) = F_{ISOP}(s) \begin{bmatrix} \mathbf{D}(s) \\ \mathbf{U}(s) \\ \mathbf{X}_{P0} \end{bmatrix} =$$
$$= G_{ISOPD}(s)\mathbf{D}(s) + G_{ISOPU}(s)\mathbf{U}(s) + G_{ISOPXo}(s)\mathbf{X}_{P0}.$$

Equation (12.30) enables us to write

$$\mathbf{Y}(s) = G_{ISOPD}(s)\,\mathbf{D}(s) + G_{ISOPU}(s)\,\mathbf{U}(s) + G_{ISOPXo}(s)\,\mathbf{X}_{P0} - \\ -\mathbf{T}(s, \epsilon(\mathbf{s})).$$

We multiply $\mathbf{Y}(s)$ by $e^{-\sigma s}$, $0 < \sigma <<<< 1$, $\sigma \longrightarrow 0^+$, in the ideal case $\sigma = 0$,

$$e^{-\sigma s}\mathbf{Y}(s) = e^{-\sigma s}\left[G_{ISOPD}(s)\,\mathbf{D}(s) + G_{ISOPU}(s)\,\mathbf{U}(s) + G_{ISOPXo}(s)\,\mathbf{X}_{P0}\right].$$

Hence,

$$\left(1 - e^{-\sigma s}\right)\mathbf{Y}(s) = \left(1 - e^{-\sigma s}\right)\left[G_{ISOPD}(s)\,\mathbf{D}(s) + G_{ISOPXo}(s)\,\mathbf{X}_{P0}\right] + \\ + G_{ISOPU}(s)\left[\mathbf{U}(s) - e^{-\sigma s}\mathbf{U}(s)\right] - \mathbf{T}(s, \epsilon(\mathbf{s})).$$

We replace $(1 - e^{-\sigma s})$ by zero, $(1 - e^{-\sigma s}) = 0$, due to $0 < \sigma <<< 1$, $\sigma \longrightarrow 0^+$, which reduces the preceding equation to

$$G_{ISOPU}(s)\left[\mathbf{U}(s) - e^{-\sigma s}\mathbf{U}(s)\right] = \mathbf{T}(s, \epsilon(\mathbf{s})).$$

The condition 2) transforms this equation into (12.102) as follows. Let

$$\mathbf{U}(s) = G^T_{ISOPU}(s)\,W(s) \Longrightarrow \\ G_{ISOPU}(s)\,G^T_{ISOPU}(s)\left[\mathbf{W}(s) - e^{-\sigma s}\mathbf{W}(s)\right] = \mathbf{T}(s, \epsilon(\mathbf{s})) \Longrightarrow \\ \left[\mathbf{W}(s) - e^{-\sigma s}\mathbf{W}(s)\right] = \left[G_{ISOPU}(s)\,G^T_{ISOPU}(s)\right]^{-1}\mathbf{T}(s, \epsilon(\mathbf{s})) \Longrightarrow \\ G^T_{ISOPU}(s)\left[\mathbf{W}(s) - e^{-\sigma s}\mathbf{W}(s)\right] = \\ = \underbrace{G^T_{ISOPU}(s)\left[G_{ISOPU}(s)\,G^T_{ISOPU}(s)\right]^{-1}}_{\Upsilon(s)}\mathbf{T}(s, \epsilon(\mathbf{s})) \Longrightarrow \\ \left[\mathbf{U}(s) - e^{-\sigma s}\mathbf{U}(s)\right] = \Upsilon(s)\,\mathbf{T}(s, \epsilon(\mathbf{s})) \Longrightarrow \\ \mathbf{U}(s) = e^{-\sigma s}\mathbf{U}(s) + \Upsilon(s)\,\mathbf{T}(s, \epsilon(\mathbf{s})).$$

Sufficiency. Let all the conditions hold. The plant (12.91), (12.92) is natural trackable on $\mathfrak{D}^1 \mathrm{x} \mathfrak{Y}^1_d$ due to Theorem 302 (Subsection 9.4.2). The control defined under the condition 3) is continuous and independent of the plant mathematical description (12.91), (12.92) and of the disturbance **D**. It is natural control.

Equations (12.91), (12.92), (12.100), (12.101) permit

$$\mathbf{Y}(t) = C_P A_P^{-1}\left(\frac{d\mathbf{X}_P(t)}{dt} - B_P\mathbf{U}(t) - D_P\mathbf{D}(t)\right) + H_P\mathbf{U}(t) + L_P\mathbf{D}(t) =$$

$$= C_P A_P^{-1}\frac{d\mathbf{X}_P(t)}{dt}\left(-C_P A_P^{-1}B_P + H_P\right)\mathbf{U}(t) - C_P A_P^{-1}D_P\mathbf{D}(t) + L_P\mathbf{D}(t) =$$

$$= C_P A_P^{-1}\frac{d\mathbf{X}_P(t)}{dt} + \left(-C_P A_P^{-1}B_P + H\right)\mathbf{U}(t^-) +$$

$$+\mathbf{T}\left(t, \boldsymbol{\epsilon}(t), \boldsymbol{\epsilon}^{(1)}(t), ..., \int_{t_0=0}^{t}\boldsymbol{\epsilon}(t)dt\right) - C_P A_P^{-1}D_P\mathbf{D}(t) + L_P\mathbf{D}(t) =$$

$$= C_P A_P^{-1}\frac{d\mathbf{X}_P(t)}{dt} + \mathbf{Y}(t^-) - C_P A_P^{-1}\frac{d\mathbf{X}_P(t^-)}{dt} + C_P A_P^{-1}D_P\mathbf{D}(t^-) -$$

$$-L_P\mathbf{D}(t^-) + \mathbf{T}\left(t, \boldsymbol{\epsilon}(t), \boldsymbol{\epsilon}^{(1)}(t), ..., \int_{t_0=0}^{t}\boldsymbol{\epsilon}(t)dt\right) -$$

$$-C_P A_P^{-1}D_P\mathbf{D}(t) + L_P\mathbf{D}(t).$$

Continuity of all variables on $\mathfrak{T}_0$ including $t = 0$ due to continuity of

$$\mathbf{T}\left(t, \boldsymbol{\epsilon}(t), \boldsymbol{\epsilon}^{(1)}(t), ..., \int_{t_0=0}^{t}\boldsymbol{\epsilon}(t)dt\right)$$

in view of (12.5), (i.e., Principle 9), $t^- = t$-σ and $0 < \sigma <<<, 1, \sigma \longrightarrow 0^+$, i.e., (12.106), simplify the preceding equation to

$$\mathbf{T}\left(t, \boldsymbol{\epsilon}(t), \boldsymbol{\epsilon}^{(1)}(t), ..., \int_{t_0=0}^{t}\boldsymbol{\epsilon}(t)dt\right) = \mathbf{0}_N, \ \forall t \in \mathfrak{T}_0.$$

This proves that the plant exhibits tracking on $\mathfrak{D}^1 \mathrm{x} \mathfrak{Y}_d^1$ determined by (12.30), which implies the validity of the tracking algorithm (12.17), (12.18) due to Theorem 364 (Section 12.1).

In the complex domain we start with

$$\mathbf{Y}(s) = G_{ISOPD}(s)\mathbf{D}(s) + G_{ISOPU}(s)\mathbf{U}(s) + G_{ISOPXo}(s)\mathbf{X}_{P0},$$

and apply (12.102)-(12.104),

$$\mathbf{Y}(s) = G_{ISOPD}(s)\mathbf{D}(s) + G_{ISOPU}(s)e^{-\sigma s}\mathbf{U}(s) +$$
$$+G_{ISOPXo}(s)\mathbf{X}_{P0} + \mathbf{T}(s, \boldsymbol{\epsilon}(\mathbf{s})),$$

together with

$$e^{-\sigma s}\mathbf{Y}(s) = G_{ISOPD}(s)e^{-\sigma s}\mathbf{D}(s) + G_{ISOPU}(s)e^{-\sigma s}\mathbf{U}(s) +$$
$$+G_{ISOPXo}(s)e^{-\sigma s}\mathbf{X}_{P0}, \ 0 < \sigma <<<, 1, \sigma \longrightarrow 0^+.$$

The result is

$$G_{ISOPU}(s)e^{-\sigma s}\mathbf{U}(s) = e^{-\sigma s}\mathbf{Y}(s) - G_{ISOPD}(s)e^{-\sigma s}\mathbf{D}(s) -$$
$$-G_{ISOPXo}(s)e^{-\sigma s}\mathbf{X}_{P0}$$

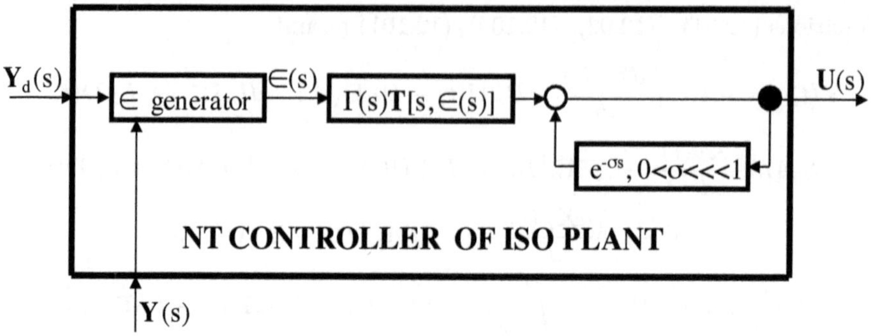

Figure 12.5: The full block diagram of the Natural Tracking (*NT*) Controller of the *ISO* plant (12.91), (12.92) in which $\Gamma(s) = \Upsilon(s)$ (12.102).

so that

$$\mathbf{Y}(s) = G_{ISOPD}(s)\,\mathbf{D}(s) + e^{-\sigma s}\mathbf{Y}(s) - G_{ISOPD}(s)\,e^{-\sigma s}\mathbf{D}(s) - \\ -G_{ISOPXo}(s)\,e^{-\sigma s}\mathbf{X}_{P0} + G_{ISOPXo}(s)\,\mathbf{X}_{P0} + \mathbf{T}(s, \boldsymbol{\epsilon}(\mathbf{s})).$$

Since $0 < \sigma <<<, 1$, $\sigma \longrightarrow 0^+$, then this equation reduces to

$$\mathbf{T}(s, \boldsymbol{\epsilon}(\mathbf{s})) = \mathbf{0}_N.$$

This and Theorem 364 prove that the plant exhibits tracking on $\mathfrak{D}^1 \mathrm{x} \mathfrak{Y}_d^1$ determined by (12.17), (12.18) ■

The full block diagram of the *Natural Tracking* (*NT*) *Controller* of the *ISO* plant (12.91), (12.92) is in Fig. 12.5.

12.3.2 Control synthesis for specific tracking qualities

The control synthesis that uses directly the definitions of various tracking properties via the *k-th* order target set $\Upsilon^k_{ISO}(t; \mathbf{D}; \mathbf{U}; \mathbf{Y}^k_d)$ (Subsection 8.4.6) is too complex and cumbersome. The definitions demand the use of the distance of the system motions from the target set, which cannot be simply used and which has not led so far to simple applicable control algorithms.

It seems that the simpler approach is to determine the desired motion $\mathbf{X}_{Pd}(t)$ induced by the desired output behavior $\mathbf{Y}_d(t)$ and then to apply the tracking definitions in terms of $\mathbf{X}_{Pd}(t)$ rather then via the *k-th* order target set $\Upsilon^k_{ISO}(t; \mathbf{D}; \mathbf{U}; \mathbf{Y}^k_d)$.

Theorem 393 ***NTC for the first order linear elementwise exponential tracking Example 366 (Section 12.1)***

Let (12.19) through (12.25) (Section 12.1) be valid.

In order for the ISO plant (12.91), (12.92) to be controlled by the natural tracking control $\mathbf{U}$ *and to exhibit tracking on* $\mathfrak{D}^1 x \mathfrak{Y}_d^1$ *determined by*

$$\mathbf{T}\left(t,\varepsilon,\varepsilon^{(1)}\right) = T_1\varepsilon^{(1)}(t) + K_0\varepsilon(t) = \mathbf{0}_N, \ \forall t \in \mathfrak{T}_0, \ if \ \mathbf{T}\left(0,\varepsilon_0^k,\mathbf{0}_N\right) = \mathbf{0}_N,$$
$$\varepsilon^1\left(t;\varepsilon_0^1\right) = -\mathbf{f}^1(t), \ \forall t \in \mathfrak{T}_0, \ if \ \mathbf{T}\left(0,\varepsilon_0^1,\mathbf{0}_N\right) \neq \mathbf{0}_N,$$
$$T_1 = diag\{\tau_1 \ \tau_2 \ ...\tau_N\} > O_N, \ K_0 = diag\{k_{01} \ k_{02} \ ...k_{0N}\} > O_N, \quad (12.110)$$

it is necessary and sufficient that

1) $N \leq r$,
2) $rank G_{ISOUP}(s) = rank G_{ISOPUo} = N$, *and*
3) *the control obeys*

$$\mathbf{U}(t) = \mathbf{U}(t^-) + \Upsilon\left[T_1\boldsymbol{\epsilon}(t) + K_0\boldsymbol{\epsilon}(t)\right], \quad (12.111)$$

$$\Upsilon = G_{ISOPUo}^T\left\{G_{ISOPUo}G_{ISOPUo}^T\right\}^{-1}, \quad (12.112)$$

where (12.25) defines $\boldsymbol{\epsilon}(t;\boldsymbol{\epsilon}_0)$, *or equivalently*

$$\mathbf{U}(s) = e^{-\sigma s}\mathbf{U}(s) + \Upsilon(s)\mathbf{T}(s,\boldsymbol{\epsilon}(\mathbf{s})), \ 0 < \sigma <<< 1, \ \sigma \longrightarrow 0^+, \quad (12.113)$$

$$\mathbf{T}(s,\boldsymbol{\epsilon}(\mathbf{s})) = \mathcal{L}\left\{T_1\boldsymbol{\epsilon}(t) + K_0\boldsymbol{\epsilon}(t)\right\}, \quad (12.114)$$

$$\Upsilon(s) = G_{ISOPU}(s)^T\left\{G_{ISOPU}(s)G_{ISOPU}(s)^T\right\}^{-1}. \quad (12.115)$$

Proof. The tracking algorithm (12.17), (12.18) reduces to (12.110). The conditions of Theorem 392 hold so that it implies Theorem 393 ∎

Theorem 394 ***NTC for the higher order linear elementwise exponential tracking Example 367 (Section 12.1)***

Let (12.19) through (12.25) (Section 12.1) be valid.

In order for the ISO plant(12.91), (12.92) to be controlled by the natural tracking control $\mathbf{U}$ *and to exhibit tracking on* $\mathfrak{D}^\eta x \mathfrak{Y}_d^\eta$ *determined by the tracking algorithm (12.116),*

$$\mathbf{T}\left(t,\varepsilon,\varepsilon^{(1)},..,\varepsilon^{(\eta)},\int_{t_0=0}^{t}\varepsilon dt\right) = \sum_{k=0}^{k=\eta\leq\nu-1} E_k\varepsilon^{(k)}(t) = E^{(\eta)}\varepsilon^{\eta}(t) = \mathbf{0}_N,$$
$$\forall t \in \mathfrak{T}_0 \ if \ \mathbf{T}\left(0,\varepsilon_0^k,\mathbf{0}_N\right) = \mathbf{0}_N,$$
$$\varepsilon^{\eta}\left(t;\varepsilon_0^{\eta}\right) = -\mathbf{f}^{\eta}(t), \ \forall t \in \mathfrak{T}_0 \ if \ \mathbf{T}\left(0,\varepsilon_0^{\eta},\mathbf{0}_N\right) \neq \mathbf{0}_N, \quad (12.116)$$

with the matrices $E_k \in \mathfrak{R}^{N x N}$ *such that the real parts* $\mathrm{Re}\, s_i$ *of the roots* s_i *of its characteristic polynomial* $f(s)$,

$$f(s) = \det\left(E^{(\eta)}S_N^{(\eta)}(s)\right), \quad (12.117)$$

are negative,

$$f(s_i) = \det\left(E^{(\eta)}S_N^{(\eta)}(s_i)\right) = 0 \Longrightarrow \mathrm{Re}\, s_i < 0, \ \forall i = 1,2,...,\eta N, \quad (12.118)$$

it is necessary and sufficient that

1) $N \leq r$,
2) $rankG_{ISOUP}(s) = rankG_{ISOPUo} = N$, *and*
3) *the control obeys*

$$\mathbf{U}(t) = \mathbf{U}(t^-) + \Upsilon \left[E^{(\eta)}\boldsymbol{\epsilon}^{\eta}(t)\right], \tag{12.119}$$

$$\Upsilon = G^T_{ISOPUo}\left\{G_{ISOPUo}G^T_{ISOPUo}\right\}^{-1}, \tag{12.120}$$

where (12.25) defines $\boldsymbol{\epsilon}(t;\boldsymbol{\epsilon}_0)$*, or equivalently*

$$\mathbf{U}(s) = e^{-\sigma s}\mathbf{U}(s) + \Upsilon(s)\,\mathbf{T}(s, \boldsymbol{\epsilon}(\mathbf{s})),\ 0 < \sigma <<< 1,\ \sigma \longrightarrow 0^+, \tag{12.121}$$

$$\mathbf{T}(s, \boldsymbol{\epsilon}(\mathbf{s})) = E^{(\eta)}S_N^{(\eta)}(s), \tag{12.122}$$

$$\Upsilon(s) = G_{ISOPU}(s)^T\left\{G_{ISOPU}(s)G_{ISOPU}(s)^T\right\}^{-1}, \tag{12.123}$$

Proof. Theorem 392 yields Theorem 394. Since $\mathfrak{D}^{\eta}\mathrm{x}\mathfrak{Y}^{\eta}_d \subset \mathfrak{D}^1\mathrm{x}\mathfrak{Y}^1_d$, the tracking algorithm (12.17), (12.18) reduces to (12.116) and the conditions of Theorem 392 are satisfied ■

Theorem 395 ***NTC for the sharp elementwise stablewise tracking with the finite vector reachability time*** $\boldsymbol{\tau}_R^{2N}$***, Example 368 (Section 12.1)***

Let (12.19) through (12.25) (Section 12.1) be valid.

In order for the plant (12.91), (12.92) to be controlled by the natural tracking control $\mathbf{U}$ *and to exhibit tracking on* $\mathfrak{D}^1\mathrm{x}\mathfrak{Y}^1_d$ *determined by the tracking algorithm*

$$\mathbf{T}\left(\mathbf{t}^N, \varepsilon, \varepsilon^{(1)}\right) = T_1\varepsilon^{(1)}(\mathbf{t}^N) + K_0 sign\varepsilon(\mathbf{0}_N) = \mathbf{0}_N, \tag{12.124}$$
$$\forall \mathbf{t}^N \in \mathfrak{T}_0^N \ if\ \mathbf{T}\left(\mathbf{0}_N, \varepsilon_0^1, \mathbf{0}_N\right) = \mathbf{0}_N,$$
$$\varepsilon^1\left(\mathbf{t}^{2N}; \varepsilon_0^1\right) = -\mathbf{f}^1\left(\mathbf{t}^{2N}\right),\ \forall \mathbf{t}^{2N} \in \mathfrak{T}_0^{2N},\ if\ \mathbf{T}\left(\mathbf{0}_N, \varepsilon_0^1, \mathbf{0}_N\right) \neq \mathbf{0}_N,$$

for the elementwise stablewise tracking with the finite vector reachability time $\boldsymbol{\tau}_R^{2N}$,

$$\boldsymbol{\tau}_R^N = T_1 K_0^{-1}\left|\varepsilon_0\right|,\ \boldsymbol{\tau}_R^{2N} = \begin{bmatrix} \boldsymbol{\tau}_R^N \\ \boldsymbol{\tau}_R^N \end{bmatrix},$$

it is necessary and sufficient that

1) $N \leq r$,
2) $rankG_{ISOUP}(s) = rankG_{ISOPUo} = N$, *and*
3) *the control obeys*

$$\mathbf{U}(t) = \mathbf{U}(t^-) + \Upsilon\left[T_1\boldsymbol{\epsilon}^{(1)}(t) + K_0 sign\boldsymbol{\epsilon}(0)\right], \tag{12.125}$$

$$\Upsilon = G^T_{ISOPUo}\left\{G_{ISOPUo}G^T_{ISOPUo}\right\}^{-1}, \tag{12.126}$$

where (12.25) defines $\boldsymbol{\epsilon}(t;\boldsymbol{\epsilon}_0)$, *or equivalently*

$$\mathbf{U}(s) = e^{-\sigma s}\mathbf{U}(s) + \Upsilon(s)\,\mathbf{T}(s,\boldsymbol{\epsilon}(\mathbf{s})),\ 0<\sigma<<<1,\ \sigma\longrightarrow 0^+, \tag{12.127}$$

$$\mathbf{T}(s,\boldsymbol{\epsilon}(\mathbf{s})) = \mathfrak{L}\left\{T_1\boldsymbol{\epsilon}^{(1)}(t) + K_0 sign\boldsymbol{\epsilon}(0)\right\}, \tag{12.128}$$

$$\Upsilon(s) = G_{ISOPU}(s)^T\left\{G_{ISOPU}(s)G_{ISOPU}(s)^T\right\}^{-1}. \tag{12.129}$$

Proof. The tracking algorithm (12.124) is a special case of the tracking algorithm (12.17), (12.18). All conditions of Theorem 392 are valid. It implies Theorem 395 ■

Theorem 396 ***The NTC synthesis for the first power smooth elementwise stablewise tracking with the finite vector reachability time*** $\boldsymbol{\tau}_R^{2N}$ ***Example 369 (Section 12.1)***

Let (12.19) through (12.25) (Section 12.1) be valid.

In order for the ISO plant (12.91), (12.92) to be controlled by the natural tracking control $\mathbf{U}$ *and to exhibit tracking on* $\mathfrak{D}^1\mathrm{x}\mathfrak{Y}_d^1$ *determined by*

$$\mathbf{T}\left(\mathbf{t}^N,\varepsilon,\varepsilon^{(1)}\right) = T_1\varepsilon^{(1)}(\mathbf{t}^N) + 2K_0\left|E\left(\mathbf{t}^N\right)\right|^{1/2} sign\varepsilon_0 = \mathbf{0}_N,$$
$$\forall\mathbf{t}^N\in\mathfrak{T}_0^N\ if\ \mathbf{T}\left(\mathbf{0}_N,\varepsilon_0^1,\mathbf{0}_N\right) = \mathbf{0}_N,$$
$$\varepsilon^1\left(\mathbf{t}^{2N};\varepsilon_0^1\right) = -\mathbf{f}^1\left(\mathbf{t}^{2N}\right),\ \forall\mathbf{t}^{2N}\in\mathfrak{T}_0^{2N},\ if\ \mathbf{T}\left(\mathbf{0}_N,\varepsilon_0^1,\mathbf{0}_N\right)\neq\mathbf{0}_N,$$
$$\left|E\left(\mathbf{t}^N\right)\right|^{1/2} = diag\left\{\left|\varepsilon_1(t)\right|^{1/2}\ \ \left|\varepsilon_2(t)\right|^{1/2}\ \ ...\ \ \left|\varepsilon_N(t)\right|^{1/2}\right\}, \tag{12.130}$$

for the first power smooth elementwise stablewise tracking with the finite vector reachability time $\boldsymbol{\tau}_R^{2N}$,

$$\boldsymbol{\tau}_R^N = T_1K_0^{-1}\left|E_0\right|^{1/2}\mathbf{1}_N,\ \boldsymbol{\tau}_R^{2N} = \begin{bmatrix}\boldsymbol{\tau}_R^N\\ \boldsymbol{\tau}_R^N\end{bmatrix},$$

it is necessary and sufficient that

1) $N\leq r$,
2) $rankG_{ISOUP}(s) = rankG_{ISOPUo} = N$, *and*
3) *the control obeys*

$$\mathbf{U}(t) = \mathbf{U}(t^-) + \Upsilon\left[T_1\boldsymbol{\epsilon}(t) + 2K_0\left|\Xi(t)\right|^{1/2} sign\boldsymbol{\epsilon}_0\right], \tag{12.131}$$

$$\Upsilon = G_{ISOPUo}^T\left\{G_{ISOPUo}G_{ISOPUo}^T\right\}^{-1}, \tag{12.132}$$

where (12.25) defines $\boldsymbol{\epsilon}(t;\boldsymbol{\epsilon}_0)$, *or equivalently*

$$\mathbf{U}(s) = e^{-\sigma s}\mathbf{U}(s) + \Upsilon(s)\,\mathbf{T}(s,\boldsymbol{\epsilon}(\mathbf{s})),\ 0<\sigma<<<1,\ \sigma\longrightarrow 0^+, \tag{12.133}$$

$$\mathbf{T}(s,\boldsymbol{\epsilon}(\mathbf{s})) = \mathfrak{L}\left\{T_1\boldsymbol{\epsilon}(t) + 2K_0\left|\Xi(t)\right|^{1/2} sign\boldsymbol{\epsilon}_0\right\}, \tag{12.134}$$

$$\Upsilon(s) = G_{ISOPU}(s)^T\left\{G_{ISOPU}(s)G_{ISOPU}(s)^T\right\}^{-1}. \tag{12.135}$$

Proof. The tracking algorithm (12.130) is a special form of the tracking algorithm (12.17), (12.18). Theorem 392 implies Theorem 396 because all its conditions hold ■

We defined

$$K = diag\{k_1\ k_2\ ...k_N\},\ k_i \in \{2,\ 3,\ ...\},\ \forall i = 1, 2,\ ...\ , N,$$
$$\left|E\left(\mathbf{t}^N\right)\right|^{I-K^{-1}} = diag\left\{\left|\varepsilon_1(t)\right|^{1-k_1^{-1}}\ \left|\varepsilon_2(t)\right|^{1-k_2^{-1}}\ ...\ \left|\varepsilon_N(t)\right|^{1-k_N^{-1}}\right\}$$

in (12.37), (Example 370).

Theorem 397 ***The NTC synthesis for the higher power smooth elementwise stablewise tracking with the finite vector reachability time τ_R^{2N}, Example 370 (Section 12.1)***

Let (12.19) through (12.25) (Section 12.1) be valid.

In order for the ISO plant (12.91), (12.92) to be controlled by the natural tracking control $\mathbf{U}$ *and to exhibit tracking on* $\mathfrak{D}^1 \mathrm{x} \mathfrak{Y}_d^1$ *determined by the tracking algorithm*

$$\mathbf{T}\left(\mathbf{t}^N, \boldsymbol{\varepsilon}, \boldsymbol{\varepsilon}^{(1)}\right) = T_1 \boldsymbol{\varepsilon}^{(1)}(\mathbf{t}^N) + K_0 \left|E\left(\mathbf{t}^N\right)\right|^{I-K^{-1}} sign \boldsymbol{\varepsilon}_0 = \mathbf{0}_N,$$
$$\forall \mathbf{t}^N \in \mathfrak{T}_0^N\ if\ \mathbf{T}\left(\mathbf{0}_N, \boldsymbol{\varepsilon}_0^1, \mathbf{0}_N\right) = \mathbf{0}_N,$$
$$\boldsymbol{\varepsilon}^1\left(\mathbf{t}^{2N}; \boldsymbol{\varepsilon}_0^1\right) = -\mathbf{f}^1\left(\mathbf{t}^{2N}\right),\ \forall \mathbf{t}^{2N} \in \mathfrak{T}_0^{2N},\ if\ \mathbf{T}\left(\mathbf{0}_N, \boldsymbol{\varepsilon}_0^1, \mathbf{0}_N\right) \neq \mathbf{0}_N \quad (12.136)$$

for the higher power smooth elementwise stablewise tracking with the finite vector reachability time $\boldsymbol{\tau}_R^{2N}$,

$$\boldsymbol{\tau}_R^N = T_1 K_0^{-1} \left|E_0\right|^{K^{-1}} \mathbf{1}_N,\ \boldsymbol{\tau}_R^{2N} = \begin{bmatrix} \boldsymbol{\tau}_R^N \\ \boldsymbol{\tau}_R^N \end{bmatrix},$$

it is necessary and sufficient that

1) $N \leq r$,

2) $rankG_{ISOUP}(s) = rankG_{ISOPUo} = N$, *and*

3) the control obeys

$$\mathbf{U}(t) = \mathbf{U}(t^-) + \Upsilon\left[T_1 \boldsymbol{\epsilon}(t) + K_0 \left|\Xi(t)\right|^{I-K^{-1}} sign \boldsymbol{\epsilon}_0\right], \quad (12.137)$$

$$\Upsilon = G_{ISOPUo}^T \left\{G_{ISOPUo} G_{ISOPUo}^T\right\}^{-1}, \quad (12.138)$$

where (12.25) defines $\boldsymbol{\epsilon}(t; \boldsymbol{\epsilon}_0)$, *or equivalently*

$$\mathbf{U}(s) = e^{-\sigma s}\mathbf{U}(s) + \Upsilon(s)\,\mathbf{T}(s, \boldsymbol{\epsilon}(\mathbf{s})),\ 0 < \sigma <<< 1,\ \sigma \longrightarrow 0^+, \quad (12.139)$$
$$\mathbf{T}(s, \boldsymbol{\epsilon}(\mathbf{s})) = \mathfrak{L}\left\{T_1 \boldsymbol{\epsilon}(t) + K_0 sign \boldsymbol{\epsilon}(t)\right\}, \quad (12.140)$$
$$\Upsilon(s) = G_{ISOPU}(s)^T \left\{G_{ISOPU}(s) G_{ISOPU}(s)^T\right\}^{-1}. \quad (12.141)$$

Proof. Theorem 392 implies Theorem 397 because the tracking algorithm (12.17), (12.18) takes the form of the tracking algorithm (12.136) so that all conditions of Theorem 392 hold ■

Theorem 398 ***The NTC synthesis for the sharp absolute error vector value tracking elementwise and stablewise with the finite vector reachability time τ_R^{2N}, Example 371 (Section 12.1)***

Let (12.19) through (12.25) (Section 12.1) be valid.

In order for the ISO plant (12.91), (12.92) to be controlled by the natural tracking control $\mathbf{U}$ *and to exhibit tracking on* $\mathfrak{D}^1\mathrm{x}\mathfrak{Y}_d^1$ *determined by the tracking algorithm*

$$\mathbf{T}\left(\mathbf{t}^N, \boldsymbol{\varepsilon}, \boldsymbol{\varepsilon}^{(1)}\right) = T_1\Sigma\left(\boldsymbol{\varepsilon}, \boldsymbol{\varepsilon}^{(1)}\right)\boldsymbol{\varepsilon}^{(1)} + K_0 sign\left|\boldsymbol{\varepsilon}_0\right| = \mathbf{0}_N,$$
$$\forall \mathbf{t}^N \in \mathfrak{T}_0^N \ if \ \mathbf{T}\left(0, \boldsymbol{\varepsilon}_0^1, \mathbf{0}_N\right) = \mathbf{0}_N,$$
$$\boldsymbol{\varepsilon}^1\left(\mathbf{t}^{2N}; \boldsymbol{\varepsilon}_0^1\right) = -\mathbf{f}^1\left(\mathbf{t}^{2N}\right), \ \forall \mathbf{t}^{2N} \in \mathfrak{T}_0^{2N}, \ if \ \mathbf{T}\left(\mathbf{0}_N, \boldsymbol{\varepsilon}_0^1, \mathbf{0}_N\right) \neq \mathbf{0}_N, \quad (12.142)$$

for the sharp absolute error vector value tracking elementwise stablewise with the finite vector reachability time $\boldsymbol{\tau}_R^{2N}$,

$$\boldsymbol{\tau}_R^N = T_1 K_0^{-1}\left|\boldsymbol{\varepsilon}_0\right|, \ \boldsymbol{\tau}_R^{2N} = \begin{bmatrix} \boldsymbol{\tau}_R^N \\ \boldsymbol{\tau}_R^N \end{bmatrix},$$

it is necessary and sufficient that

1) $N \leq r$,
2) $rankG_{ISOUP}(s) = rankG_{ISOPUo} = N$, *and*
3) *the control obeys*

$$\mathbf{U}(t) = \mathbf{U}(t^-) + \Upsilon\left[T_1\Sigma\left(\boldsymbol{\epsilon}, \boldsymbol{\epsilon}^{(1)}\right)\boldsymbol{\epsilon}^{(1)} + K_0 sign\left|\boldsymbol{\epsilon}_0\right|\right], \quad (12.143)$$
$$\Upsilon = G_{ISOPUo}^T\left\{G_{ISOPUo}G_{ISOPUo}^T\right\}^{-1}, \quad (12.144)$$

where (12.25) defines $\boldsymbol{\epsilon}\left(t; \boldsymbol{\epsilon}_0\right)$, *or equivalently*

$$\mathbf{U}(s) = e^{-\sigma s}\mathbf{U}(s) + \Upsilon\left(s\right)\mathbf{T}\left(s, \boldsymbol{\epsilon}(\mathbf{s})\right), \ 0 < \sigma <<< 1, \ \sigma \longrightarrow 0^+, \quad (12.145)$$
$$\mathbf{T}\left(s, \boldsymbol{\epsilon}(\mathbf{s})\right) = \mathfrak{L}\left\{T_1\Sigma\left[\boldsymbol{\epsilon}(t), \boldsymbol{\epsilon}^{(1)}(t)\right]\boldsymbol{\epsilon}^{(1)}(t) + K_0 sign\left|\boldsymbol{\epsilon}_0\right|\right\}, \quad (12.146)$$
$$\Upsilon\left(s\right) = G_{ISOPU}(s)^T\left\{G_{ISOPU}(s)G_{ISOPU}(s)^T\right\}^{-1}. \quad (12.147)$$

Proof. Theorem 398 results from Theorem 392 because the tracking algorithm (12.142) is the tracking algorithm (12.17), (12.18) in this case, and all conditions of Theorem 392 are fulfilled ■

Theorem 399 ***NTC synthesis for the exponential absolute error vector value tracking elementwise and stablewise with the finite vector reachability time τ_R^{2N}, Example 372 (Section 12.1)***

Let (12.19) through (12.25) (Section 12.1) be valid.

In order for the ISO plant (12.91), (12.92) to be controlled by the natural tracking control **U** *and to exhibit tracking on* $\mathfrak{D}^1 x \mathfrak{Y}_d^1$ *determined by*

$$\mathbf{T}\left(\varepsilon,\varepsilon^{(1)}\right) = T_1 D^+ \left|\varepsilon\right| + K\left(\left|\varepsilon\right| + K_0 sign\left|\varepsilon_0\right|\right) =$$

$$= T_1\Sigma\left(\varepsilon,\varepsilon^{(1)}\right)\varepsilon^{(1)} + K\left(\left|\varepsilon\right| + K_0 sign\left|\varepsilon_0\right|\right) = \mathbf{0}_N,$$

$$\forall \mathbf{t}^N \in \mathfrak{T}_0^N \ if \ \mathbf{T}\left(\mathbf{0}_N, \varepsilon_0^1, \mathbf{0}_N\right) = \mathbf{0}_N,$$

$$\varepsilon^1\left(\mathbf{t}^{2N};\varepsilon_0^1\right) = -\mathbf{f}^1\left(\mathbf{t}^{2N}\right), \ \forall \mathbf{t}^{2N} \in \mathfrak{T}_R^{2N}, \ if \ \mathbf{T}\left(\mathbf{0}_N, \varepsilon_0^1, \mathbf{0}_N\right) \neq \mathbf{0}_N, \quad (12.148)$$

for the exponential absolute error vector value tracking elementwise and stablewise with the finite vector reachability time $\boldsymbol{\tau}_R^{2N}$ *(12.42), (12.43) (Section 12.1), it is necessary and sufficient that*

1) $N \leq r$,
2) $rankG_{ISOUP}(s) = rankG_{ISOPUo} = N$, *and*
3) *the control obeys*

$$\mathbf{U}(t) = \mathbf{U}(t^-) + \Upsilon\left[T_1\Sigma\left(\boldsymbol{\epsilon},\boldsymbol{\epsilon}^{(1)}\right)\boldsymbol{\epsilon}^{(1)} + K\left(\left|\boldsymbol{\epsilon}\right| + K_0 sign\left|\boldsymbol{\epsilon}_0\right|\right) = \mathbf{0}_N\right], \quad (12.149)$$

$$\Upsilon = G_{ISOPUo}^T\left\{G_{ISOPUo}G_{ISOPUo}^T\right\}^{-1}, \quad (12.150)$$

where (12.25) defines $\boldsymbol{\epsilon}\left(t;\boldsymbol{\epsilon}_0\right)$*, or equivalently*

$$\mathbf{U}(s) = e^{-\sigma s}\mathbf{U}(s) + \Upsilon\left(s\right)\mathbf{T}\left(s,\boldsymbol{\epsilon}(\mathbf{s})\right), \ 0 < \sigma <<< 1, \ \sigma \longrightarrow 0^+, \quad (12.151)$$

$$\mathbf{T}\left(s,\boldsymbol{\epsilon}(\mathbf{s})\right) = \mathfrak{L}\left\{T_1\Sigma\left(\boldsymbol{\epsilon},\boldsymbol{\epsilon}^{(1)}\right)\boldsymbol{\epsilon}^{(1)} + K\left(\left|\boldsymbol{\epsilon}\right| + K_0 sign\left|\boldsymbol{\epsilon}_0\right|\right)\right\}, \quad (12.152)$$

$$\Upsilon\left(s\right) = G_{ISOPU}(s)^T\left\{G_{ISOPU}(s)G_{ISOPU}(s)^T\right\}^{-1}. \quad (12.153)$$

Proof. All conditions of Theorem 392 hold in view of the tracking algorithm (12.148) that is the tracking algorithm (12.17), (12.18) in this case. Hence, Theorem 392 implies directly Theorem 399 ■

Comment 400 *The control gain* Υ*, equivalently* $\Upsilon\left(s\right)$*, is the same for all considered tracking algorithms. They are expressed in terms of the plant transfer function matrix* $G_{ISOPU}(s)$*, equivalently* G_{ISOPUo}*, relative to control, respectively. The control gain* Υ*, equivalently* $\Upsilon\left(s\right)$*, is the same as for trackability of the ISO plant (12.91), (12.92).*

The control gain Υ*, hence* $\Upsilon\left(s\right)$*, resolves the complex problem of channeling the control action on the MIMO ISO plant (12.91), (12.92).*

The NTC algorithms do not need any information about the plant state vector, about the plant internal dynamics or about the form and the value of the disturbance vector function **D**(.).

Comment 401 *The higher order linear or nonlinear NTC algorithms need the measurement of the higher output error vector derivatives that are not often accessible for the measurement. For this reason we presented herein the nonlinear NTC algorithms of only the first order. They ensure high tracking qualities including the finite reachability time with the full robustness relative to the fluctuations of the plant internal dynamics and relative to the external disturbances.*

Chapter 13

NTC versus LTC

13.1 General consideration

In order to compare *Natural Tracking Control* (*NTC*) with *Lyapunov Tracking Control* (*LTC*) we accept the tracking operator $\mathbf{T}(.)$ to determine the tracking algorithm (12.5) (Section 12.1),

$$\mathbf{T}\left(t,\boldsymbol{\varepsilon},\boldsymbol{\varepsilon}^{(1)},...,\boldsymbol{\varepsilon}^{(k)},\int_{t_0=0}^{t}\boldsymbol{\varepsilon}dt\right)=\mathbf{T}\left(t,\boldsymbol{\varepsilon}^{k},\int_{t_0=0}^{t}\boldsymbol{\varepsilon}dt\right)=\mathbf{0}_N,\forall t\in\mathfrak{T}_0,$$
$$k\in\{1,2,...\}. \tag{13.1}$$

In view of Theorem 364 (Section 12.1) we accepted the modified tracking goal 359 (Section 12.1):

$$\mathbf{T}\left(\mathbf{t}^N,\boldsymbol{\varepsilon}^k,\int_{t_0^N=\mathbf{0}_N}^{\mathbf{t}^N}E d\mathbf{t}^N\right)=\mathbf{0}_N,\ \forall\mathbf{t}^N\in\mathfrak{T}_0^N\ if\ \mathbf{T}\left(\mathbf{0}_N;\boldsymbol{\varepsilon}_0^k,\mathbf{0}_N\right)=\mathbf{0}_N,$$
$$\boldsymbol{\varepsilon}^k\left(\mathbf{t}^{(k+1)N};\boldsymbol{\varepsilon}_0^k\right)=-\mathbf{f}^k\left(\mathbf{t}^{(k+1)N};\mathbf{f}_0^k\right),\ \left\{\begin{array}{c}\forall t\mathbf{1}_{(k+1)N}\in\mathfrak{T}_0^N,\ if\\ \mathbf{T}\left(\mathbf{0}_N,\boldsymbol{\varepsilon}_0^k,\mathbf{0}_N\right)\neq\mathbf{0}_N\end{array}\right\},$$
$$k\in\{0,1,2,...\}, \tag{13.2}$$

where we refer to (12.19) through (12.25) (Section 12.1) so that:

$$\boldsymbol{\epsilon}^k(\mathbf{t}^{(k+1)N})=\boldsymbol{\varepsilon}^k(\mathbf{t}^{(k+1)N})+$$
$$+\left\{\begin{array}{cc}\mathbf{0}_{(k+1)N}, & \left(\begin{array}{c}\forall\mathbf{t}^{(k+1)N}\in\mathfrak{T}_0^{(k+1)N}\\ if\ \ \mathbf{T}\left(\mathbf{0}_{(k+1)N},\boldsymbol{\varepsilon}_0^k,\mathbf{0}_N\right)=\mathbf{0}_N,\end{array}\right)\\ \mathbf{f}^k(\mathbf{t}^{(k+1)N};\mathbf{f}_0^k), & \left(\begin{array}{c}\forall\mathbf{t}^{(k+1)N}\in\mathfrak{T}_0^{(k+1)N}\\ if\ \ \mathbf{T}\left(\mathbf{0}_{(k+1)N},\boldsymbol{\varepsilon}_0^k,\mathbf{0}_N\right)\neq\mathbf{0}_N,\end{array}\right)\end{array}\right\}. \tag{13.3}$$

This inspires us to introduce both $\mathbf{T}\left(t, \boldsymbol{\epsilon}^k, \int_{t_0=0}^{t} \boldsymbol{\epsilon} dt\right)$ and the time-varying set $\mathcal{T}^k(t)$ in terms of $\boldsymbol{\epsilon}^k$ such that (13.2) holds,

$$\mathbf{T}\left(\mathbf{t}^N, \boldsymbol{\epsilon}\left(\mathbf{t}^N\right), \boldsymbol{\epsilon}^{(1)}\left(\mathbf{t}^N\right), ..., \boldsymbol{\epsilon}^{(k)}\left(\mathbf{t}^N\right), \int_{\mathbf{t}_0^N=\mathbf{0}_N}^{\mathbf{t}^N} \Xi\left(\mathbf{t}^N\right) d\mathbf{t}^N\right) =$$
$$= \mathbf{T}\left(\mathbf{t}^N, \boldsymbol{\epsilon}^k\left(\mathbf{t}^{(k+1)N}\right), \int_{\mathbf{t}_0^N=\mathbf{0}_N}^{\mathbf{t}^N} \Xi\left(\mathbf{t}^N\right) d\mathbf{t}^N\right) = \mathbf{0}_N, \ \forall \mathbf{t}^{(k+1)N} \in \mathfrak{T}_0^{(k+1)N}, \tag{13.4}$$

$$\mathcal{T}^k(t) = \left\{\boldsymbol{\epsilon}^k : \mathbf{T}\left(t, \boldsymbol{\epsilon}^k, \int_{t_0=0}^{t} \boldsymbol{\epsilon} dt\right) = \mathbf{0}_N\right\} \subset \mathfrak{R}^{(k+1)N}, \forall t \in \mathfrak{T}_0, \tag{13.5}$$

and the following form of a tentative Lyapunov function $v(.)$ on $\mathfrak{R}^N$:

$$v(\mathbf{T}) = \frac{1}{2}\|\mathbf{T}\|^2. \tag{13.6}$$

It is globally positive definite and radially unbounded on $\mathfrak{R}^N$ as the function of the vector $\mathbf{T} \in \mathfrak{R}^N$. However, we can treat it as the function of $\boldsymbol{\epsilon}^k$,

$$v(\boldsymbol{\epsilon}^k) = \frac{1}{2}\left\|\mathbf{T}\left(t, \boldsymbol{\epsilon}^k, \int_{t_0=0}^{t} \boldsymbol{\epsilon} dt\right)\right\|^2, \ k \in \{1, 2, ...\}. \tag{13.7}$$

It is globally positive (not necessarily positive definite [232]) with respect to the time-varying set $\mathcal{T}^k(t)$ (13.5) that is the time-varying hypersurface in $\mathfrak{R}^{(k+1)N}$. For the sake of simplicity, let the operator $\mathbf{T}(.)$ be time-invariant, hence independent of the integral of $\boldsymbol{\epsilon}$,

$$\mathbf{T}\left(t, \boldsymbol{\epsilon}^k\right) = \mathbf{T}\left(\boldsymbol{\epsilon}^k\right), \forall t \in \mathfrak{T}_0. \tag{13.8}$$

Now, the set $\mathcal{T}^k(t)$ is also time-invariant,

$$\mathcal{T}^k = \left\{\boldsymbol{\epsilon}^k : \mathbf{T}\left(\boldsymbol{\epsilon}^k\right) = \mathbf{0}_N\right\} \subset \mathfrak{R}^{(k+1)N}, \tag{13.9}$$

and the function $v(.)$ (13.7), (13.8) is globally positive definite on $\mathfrak{R}^{(k+2)N}$ relative to the time-invariant set $\mathcal{T}^k$ (13.9), which is the time-invariant hypersurface in $\mathfrak{R}^{(k+2)N}$,

$$\mathbf{T}\left(\boldsymbol{\epsilon}^k\right) = \mathbf{T}\left(\boldsymbol{\epsilon}, \boldsymbol{\epsilon}^{(1)}, \ ..., \ \boldsymbol{\epsilon}^{(k)}\right) = \mathbf{0}_N. \tag{13.10}$$

Examples 366 through 372 (Section 12.1) illustrate (13.10).

Note 402 *The Euler derivative of* $v(\boldsymbol{\epsilon}^k)$,

$$v(\boldsymbol{\epsilon}^k) = \frac{1}{2}\left\|\mathbf{T}\left(\boldsymbol{\epsilon}^k\right)\right\|^2, \tag{13.11}$$

reads

$$v^{(1)}\left[\mathbf{T}(\boldsymbol{\epsilon}^k)\right]=\frac{dv\left[\mathbf{T}(\boldsymbol{\epsilon}^k)\right]}{dt}=\mathbf{T}^T\left(\boldsymbol{\epsilon}^k\right)\frac{d\mathbf{T}\left(\boldsymbol{\epsilon}^k\right)}{dt}=\mathbf{T}^T\left(\boldsymbol{\epsilon}^k\right)\frac{d\mathbf{T}\left[\mathbf{Y}_d^k(t)-\mathbf{Y}^k(t)\right]}{dt}. \tag{13.12}$$

The gradient $\operatorname{grad} v\left[\mathbf{T}(\boldsymbol{\epsilon}^k)\right]$ *of* $v\left[\mathbf{T}(\boldsymbol{\epsilon}^k)\right]$ *is* $\mathbf{T}\left(\boldsymbol{\epsilon}^k\right)$,

$$\operatorname{grad} v\left[\mathbf{T}(\boldsymbol{\epsilon}^k)\right]=\mathbf{T}\left(\boldsymbol{\epsilon}^k\right).$$

Note 403 *The zero vector value of* $\operatorname{grad} v\left[\mathbf{T}(\boldsymbol{\epsilon}^k)\right]$ *is the zero vector value of* $\mathbf{T}\left(\boldsymbol{\epsilon}^k\right)$,

$$\operatorname{grad} v\left[\mathbf{T}(\boldsymbol{\epsilon}^k)\right]=\mathbf{0}_N \Longleftrightarrow \mathbf{T}\left(\boldsymbol{\epsilon}^k\right)=\mathbf{0}_N.$$

This is what the control should ensure, which is the validity of the tracking algorithm (13.2), equivalently (13.1) due to Theorem 364 (Section 12.1). The annulation of $\operatorname{grad} v\left[\mathbf{T}(\boldsymbol{\epsilon}^k)\right]$ *of Lyapunov function* $v(\mathbf{T})=2^{-1}\|\mathbf{T}\|^2$ *(13.6) is useful and helpful. It overcomes fully the problem of the zero vector value of the* $\operatorname{grad} v(.)$ *of Lyapunov function* $v(.)$ *in general (Comment 327 in Subsection 11.2.1).*

Theorem 404 ***Positive invariance of the set*** $\mathcal{T}^k$ ***(13.9)***

Let the Euler derivative $v^{(1)}(\boldsymbol{\epsilon}^k)$ *(13.12) of* $v(\boldsymbol{\epsilon}^k)$ *(13.11) be negative definite relative to the set* $\mathcal{T}^k$ *(13.9). If* $\boldsymbol{\epsilon}_0^k=\boldsymbol{\epsilon}^k(0)\in\mathcal{T}^k$ *then* $\boldsymbol{\epsilon}^k(t;\boldsymbol{\epsilon}_0^k)$ *rests for ever in the set* $\mathcal{T}^k$,

$$\boldsymbol{\epsilon}_0^k\in\mathcal{T}^k \Longrightarrow \boldsymbol{\epsilon}^k(t;\boldsymbol{\epsilon}_0^k)\in\mathcal{T}^k,\ \forall t\in\mathfrak{T}_0. \tag{13.13}$$

Proof. Let the conditions of the theorem statement be valid. The function $v(.)$ (13.11) is positive definite with respect to the set $\mathcal{T}^k$ (13.9). Its Euler derivative $v^{(1)}(\boldsymbol{\epsilon}^k)$ (13.12) is negative definite relative to the set $\mathcal{T}^k$. From the stability theory it is well known that the set $\mathcal{T}^k$ is then positive invariant relative to $\boldsymbol{\epsilon}^k(t;\boldsymbol{\epsilon}_0^k)$, i.e., (13.13) holds ■

This theorem and Property 352, i.e., (12.7) (Section 12.1), imply the following.

Corollary 405 *If* $\boldsymbol{\epsilon}_0^k=\mathbf{0}_{(k+1)N}$, *then* $\boldsymbol{\epsilon}^k(t;\mathbf{0}_{(k+1)N})$ *rests forever in the set* $\mathcal{T}^k$ *(13.9),*

$$\boldsymbol{\epsilon}^k(t;\mathbf{0}_{(k+1)N})\in\mathcal{T}^k,\ \forall t\in\mathfrak{T}_0. \tag{13.14}$$

Comment 406 *Theorem 404 and its Corollary 405 hold for time-invariant both linear and nonlinear dynamic systems. They are directly applicable to the synthesis of the sliding-mode control. They enable avoiding the control discontinuity on the boundary of the set* $\mathcal{T}^k$ *(13.9) when we use* $\mathbf{Y}_R$ *instead of* $\mathbf{Y}_d$ *and* $\boldsymbol{\epsilon}$ *instead of* $\boldsymbol{\varepsilon}$, *Solution 360, 362 (Section 12.1).*

Note 407 *Theorem 404 and its Corollary 405 hold for time-invariant both linear and nonlinear dynamic systems. They show that we should design Lyapunov tracking control (LTC)* $\mathbf{U}(.)$ *so that Euler derivative* $v^{(1)}(\boldsymbol{\epsilon}^k)$ *(13.12) of* $v(\boldsymbol{\epsilon}^k)$ *(13.11) is negative definite relative to the set* $\mathcal{T}^k$ *(13.9), which we will do in what follows.*

13.2 The *IO* plant

We consider first *LTC* of the *IO* plant (2.15) (Subsection 2.1.2). Let $k = \nu$-1. Now, the tracking algorithm (13.5) becomes

$$\mathbf{T}\left(\boldsymbol{\epsilon}^{\nu-1}\right) = \mathbf{T}\left(\boldsymbol{\epsilon}, \boldsymbol{\epsilon}^{(1)}, \; ..., \; \boldsymbol{\epsilon}^{(\nu-1)}\right) = \mathbf{0}_N. \tag{13.15}$$

It induces $v(\boldsymbol{\epsilon}^k) = v(\boldsymbol{\epsilon}^{\nu-1})$,

$$v(\boldsymbol{\epsilon}^{\nu-1}) = \left\|\mathbf{T}\left(\boldsymbol{\epsilon}^{\nu-1}\right)\right\|^2 \tag{13.16}$$

and

$$\frac{dv(\boldsymbol{\epsilon}^{\nu-1})}{dt} = 2\mathbf{T}^{\mathbf{T}}\left(\boldsymbol{\epsilon}^{\nu-1}\right) \frac{d\mathbf{T}\left[\mathbf{Y}_d^{\nu-1}(t) - \mathbf{Y}^{\nu-1}(t)\right]}{dt}. \tag{13.17}$$

In order to continue effectively we accept a precise form of the tracking operator $\mathbf{T}(.)$, e.g.,

$$\mathbf{T}\left(\boldsymbol{\epsilon}^{\nu-1}\right) = \sum_{k=0}^{k=\nu-1} E_k \boldsymbol{\epsilon}^{(k)}(t) = E^{(\nu-1)} \boldsymbol{\epsilon}^{\nu-1}(t) = \mathbf{0}_N, \; \forall t \in \mathfrak{T}_0,$$

Example 367 (Section 12.1). From (2.15) follows

$$\mathbf{Y}^{(\nu)}(t) = A_{P\nu}^{-1}\left[C_{Pu}^{(\mu_{Pu})}\mathbf{U}^{\mu_{Pu}}(t) + D_{Pd}^{(\mu_{Pd})}\mathbf{D}^{\mu_{Pd}}(t) - A_P^{(\nu-1)}\mathbf{Y}^{\nu-1}(t)\right],$$

so that

$$\frac{d\mathbf{T}\left[\mathbf{Y}_d^{\nu-1}(t) - \mathbf{Y}^{\nu-1}(t)\right]}{dt} = \mathbf{Y}_d^{\nu}(t) -$$

$$-A_{P\nu}^{-1}\left[C_{Pu}^{(\mu_{Pu})}\mathbf{U}^{\mu_{Pu}}(t) + D_{Pd}^{(\mu_{Pd})}\mathbf{D}^{\mu_{Pd}}(t) - A_P^{(\nu-1)}\mathbf{Y}^{\nu-1}(t)\right],$$

and

$$\frac{dv(\boldsymbol{\epsilon}^{\nu-1})}{dt} = 2\mathbf{T}^T\left(\boldsymbol{\epsilon}^{\nu-1}\right)\left\{\begin{array}{c} \mathbf{Y}_d^{\nu}(t) - \\ -A_{P\nu}^{-1}\left[\begin{array}{c} C_{Pu}^{(\mu_{Pu})}\mathbf{U}^{\mu_{Pu}}(t) + \\ +D_{Pd}^{(\mu_{Pd})}\mathbf{D}^{\mu_{Pd}}(t) - A_P^{(\nu-1)}\mathbf{Y}^{\nu-1}(t) \end{array}\right] \end{array}\right\}.$$

Let

$$C_{Pu}^{(\mu_{Pu})}\mathbf{U}^{\mu_{Pu}}(t) = A_{P\nu}\left\{\begin{array}{c} \mathbf{Y}_d^{\nu}(t) + \beta\mathbf{T}\left(\boldsymbol{\epsilon}^{\nu-1}\right) + \\ S\left[\mathbf{T}\left(\boldsymbol{\epsilon}^{\nu-1}\right)\right]\left|A_{P\nu}^{-1}\right|\left|D_{Pd}^{(\mu_{Pd})}\right|\mathbf{D}_M^{\mu_{Pd}} \end{array}\right\} + A_P^{\nu-1}\mathbf{Y}^{\nu-1}(t). \tag{13.18}$$

Inequality in (11.10) (Subsection 11.2) defines $\mathbf{D}_M^{\mu_{Pd}}$. The definition (13.18) of the control vector function $\mathbf{U}(.)$ transforms $v^{(1)}(\boldsymbol{\epsilon}^{\nu-1})$ into

$$\frac{dv(\boldsymbol{\epsilon}^{\nu-1})}{dt} \leq -2\mathbf{T}^T\left(\boldsymbol{\epsilon}^{\nu-1}\right)\left\{\begin{array}{c} S\left[\mathbf{T}\left(\boldsymbol{\epsilon}^{\nu-1}\right)\right]\left|A_{P\nu}^{-1}\right|\left|D_{Pd}^{(\mu_{Pd})}\right|\mathbf{D}_M^{\mu_{Pd}} + \\ +\beta\mathbf{T}\left(\boldsymbol{\epsilon}^{\nu-1}\right) \end{array}\right\} +$$

$$+\left|\mathbf{T}^T\left(\boldsymbol{\epsilon}^{\nu-1}\right)\right|\left|A_{P\nu}^{-1}\right|\left|D_{Pd}^{(\mu_{Pd})}\right|\left|\mathbf{D}^{\mu_{Pd}}(t)\right| \leq -2\beta\mathbf{T}^T\left(\boldsymbol{\epsilon}^{\nu-1}\right)\mathbf{T}\left(\boldsymbol{\epsilon}^{\nu-1}\right) =$$

$$= -2\beta\left\|\mathbf{T}\left(\boldsymbol{\epsilon}^{\nu-1}\right)\right\|^2,$$

i.e.,

$$\frac{dv(\boldsymbol{\epsilon}^{\nu-1})}{dt} \leq -2\beta v(\boldsymbol{\epsilon}^{\nu-1}),\ \forall \boldsymbol{\epsilon}^{\nu-1} \in \mathfrak{R}^{\nu N}.$$

The solution is

$$v\left[\boldsymbol{\epsilon}^{\nu-1}(t)\right]) \leq e^{-2\beta t} v(\boldsymbol{\epsilon}_0^{\nu-1}),\ \forall \left(\boldsymbol{\epsilon}_0^{\nu-1}, t\right) \in \mathfrak{R}^{\nu N} \mathrm{x} \mathfrak{T}_0,$$

equivalently

$$\left\|\mathbf{T}\left[\boldsymbol{\epsilon}^{\nu-1}(t)\right]\right\| \leq e^{-\beta t}\left\|\mathbf{T}\left(\boldsymbol{\epsilon}_0^{\nu-1}\right)\right\|,\ \forall \left(\boldsymbol{\epsilon}_0^{\nu-1}, t\right) \in \mathfrak{R}^{\nu N} \mathrm{x} \mathfrak{T}_0. \tag{13.19}$$

This furnishes

$$\left\|\boldsymbol{\epsilon}^{\nu-1}(t)\right\| \leq e^{-\beta t}\left\|\boldsymbol{\epsilon}_0^{\nu-1}\right\|,\ \forall \left(\boldsymbol{\epsilon}_0^{\nu-1}, t\right) \in \mathfrak{R}^{\nu N} \mathrm{x} \mathfrak{T}_0, \tag{13.20}$$

where $k = \lambda_m^{-1/2} \lambda_M^{1/2}$, λ_m and λ_M are the minimal and the maximal eigenvalue of the NxN square matrix $E^{(\nu-1)^T} E^{(\nu-1)}$.

If $\mathbf{T}\left(0, \boldsymbol{\varepsilon}_0^{\nu-1}, \mathbf{0}_N\right) = \mathbf{0}_N$, then (13.3) yields $\boldsymbol{\epsilon}^{\nu-1}(t) = \boldsymbol{\varepsilon}^{\nu-1}(t)$, $\forall t \in \mathfrak{T}_0$, so that $\boldsymbol{\epsilon}^{\nu-1}(0) = \boldsymbol{\varepsilon}^{\nu-1}(0)$. Hence, $\mathbf{T}\left[\boldsymbol{\epsilon}^{\nu-1}(0)\right] = \mathbf{T}\left[\boldsymbol{\varepsilon}^{\nu-1}(0)\right] = \mathbf{0}_N$. This and (13.19) imply $\mathbf{T}\left[\boldsymbol{\epsilon}^{\nu-1}(t)\right] = \mathbf{T}\left[\boldsymbol{\varepsilon}^{\nu-1}(t)\right] = \mathbf{0}_{\nu N}$, $\forall t \in \mathfrak{T}_0$.

If $\mathbf{T}\left(\mathbf{0}_{(k+1)N}, \boldsymbol{\varepsilon}_0^k, \mathbf{0}_N\right) \neq \mathbf{0}_N$, then (13.3) yields $\boldsymbol{\epsilon}^k(\mathbf{t}^{(k+1)N}) = \boldsymbol{\varepsilon}^k(\mathbf{t}^{(k+1)N}) + \mathbf{f}^k(\mathbf{t}^{(k+1)N}; \mathbf{f}_0^k)$ so that $\boldsymbol{\epsilon}^{\nu-1}(0) = \mathbf{0}_{\nu N}$ due to $\mathbf{f}_0^{\nu-1} = -\boldsymbol{\varepsilon}_0^{\nu-1}$ (12.20) (Section 12.1). This yields $\mathbf{T}\left[\boldsymbol{\epsilon}^{\nu-1}(0)\right] = \mathbf{0}_N$ and, together with (13.19), implies again $\mathbf{T}\left[\boldsymbol{\epsilon}^{\nu-1}(t)\right] = \mathbf{0}_{\nu N}$, $\forall t \in \mathfrak{T}_0$.

Altogether, (13.4), i.e., (12.30) (Section 12.1), holds. Theorem 364 (Section 12.1) proves the validity of (13.2). The control $\mathbf{U}$ (13.18) forces the *IO* plant (2.15) to exhibit tracking determined by (13.2).

Conclusion 408 ***LTC versus NTC for the*** *IO* ***plant (2.15)***

We compare Lyapunov control $\mathbf{U}$ *(13.18) with NTC (12.68) (Section 12.2):*

Both LTC and NTC are robust relative to the disturbance. The former needs information about $\left|A_{P\nu}^{-1}\right| \left|D_{Pd}^{(\mu_{Pd})}\right| \mathbf{D}_M^{\mu_{Pd}}$*, but the latter does not need any information about it or about* $\mathbf{D}^{\mu_{Pd}}(t)$*.*

LTC requires full and exact information about the plant matrix A_P^ν *that describes the plant internal dynamics. NTC does not need any information about* A_P^ν*.*

The measurability of $\mathbf{Y}^{\nu-1}(t)$ *is necessary for the implementation of LTC. The measurability of* $\mathbf{Y}^1(t)$ *only is necessary for the implementation of the NTC.*

The LTC algorithm (13.18) is more complex than the NTC algorithm (12.68). For the implementation of the former the control should be determined as the solution to the vector differential equation (13.18). NTC is directly and completely determined by (12.68).

LTC introduces the global negative feedback in the control system. NTC uses also the global negative feedback, but it exploits the benefit of the local unit positive feedback in the controller.

13.3 The *ISO* plant

For the *ISO* plant (2.33), (2.34) (Subsection 2.2.2) we accept the tracking algorithm (Example 366, Section 12.1)

$$\mathbf{T}\left(t,\varepsilon,\varepsilon^{(1)}\right)=T_1\varepsilon^{(1)}(t)+K_0\varepsilon(t)=\mathbf{0}_N,\ \forall t\in\mathfrak{T}_0,\ if\ \mathbf{T}\left(0,\varepsilon_0^k,\mathbf{0}_N\right)=\mathbf{0}_N,$$
$$\varepsilon^1\left(t;\varepsilon_0^1\right)=-\mathbf{f}^1(t),\ \forall t\in\mathfrak{T}_0,\ if\ \mathbf{T}\left(0,\varepsilon_0^1,\mathbf{0}_N\right)\neq\mathbf{0}_N,$$
$$T_1=diag\left\{\tau_1\,\tau_2\,...\tau_N\right\}>O_N,\ K_0=diag\left\{k_{01}\,k_{02}\,...k_{0N}\right\}>O_N. \tag{13.21}$$

This induces

$$\mathbf{T}\left(\boldsymbol{\epsilon}^1\right)=\mathbf{T}\left(\boldsymbol{\epsilon},\boldsymbol{\epsilon}^{(1)}\right)=\mathbf{0}_N,\ \forall t\in\mathfrak{T}_0,$$

$$v(\boldsymbol{\epsilon}^1)=\left\|\mathbf{T}\left(\boldsymbol{\epsilon}^1\right)\right\|^2,$$

and

$$\frac{dv(\boldsymbol{\epsilon}^1)}{dt}=2\mathbf{T}^T\left(\boldsymbol{\epsilon}^1\right)\frac{d\mathbf{T}\left[\mathbf{Y}_d^1(t)-\mathbf{Y}^1(t)\right]}{dt}.$$

It is easy to verify that

$$\frac{dv(\boldsymbol{\epsilon}^1)}{dt}=2\mathbf{T}^T\left(\boldsymbol{\epsilon}^1\right)\left\{E_0\mathbf{Y}_d^{(1)}(t)+E_1\mathbf{Y}_d^{(2)}(t)-E_0\mathbf{Y}^{(1)}(t)-E_1\mathbf{Y}^{(2)}(t)\right\}=$$

$$=2\mathbf{T}^T\left(\boldsymbol{\epsilon}^1\right)\left\{\begin{array}{c}E_0\mathbf{Y}_d^{(1)}(t)+E_1\mathbf{Y}_d^{(2)}(t)-\\-E_0\left[\begin{array}{c}C_P\left[A_P\mathbf{X}_P(t)+B_P\mathbf{U}(t)+L_P\mathbf{D}(t)\right]\\+H_P\mathbf{U}^{(1)}(t)+D_P\mathbf{D}^{(1)}(t)\end{array}\right]-\\-E_1\left\langle\begin{array}{c}C_P\left[\begin{array}{c}A_P\left[A_P\mathbf{X}_P(t)+B_P\mathbf{U}(t)+L_P\mathbf{D}(t)\right]\\+B_P\mathbf{U}^{(1)}(t)+L_P\mathbf{D}^{(1)}(t)\end{array}\right]\\+H_P\mathbf{U}^{(2)}(t)+D_P\mathbf{D}^{(2)}(t)\end{array}\right\rangle\end{array}\right\}=$$

$$=-2\mathbf{T}^T\left(\boldsymbol{\epsilon}^1\right)\bullet$$
$$\bullet\left\{\begin{array}{c}-E_0\mathbf{Y}_d^{(1)}(t)-E_1\mathbf{Y}_d^{(2)}(t)+\left(E_0C_PA_P+E_1C_PA_P^2\right)\mathbf{X}_P(t)+\\+\left(EC_PB_P+E_1C_PA_PB_P\right)\mathbf{U}(t)+\left(E_0H_P+E_1C_PB_P\right)\mathbf{U}^{(1)}(t)+\\+E_1H_P\mathbf{U}^{(2)}(t)+\left(E_0C_PL_P+E_1C_PA_PL_P\right)\mathbf{D}(t)+\\+\left(E_0D_P+E_1C_PL_P\right)\mathbf{D}^{(1)}(t)+E_1D_P\mathbf{D}^{(2)}(t)\end{array}\right\}.$$

Let the control be the solution to

$$E_1H_P\mathbf{U}^{(2)}(t)+\left(E_0H_P+E_1C_PB_P\right)\mathbf{U}^{(1)}(t)+\left(E_0C_P+E_1C_PA_P\right)B_P\mathbf{U}(t)=$$
$$=E_0\mathbf{Y}_d^{(1)}(t)+E_1\mathbf{Y}_d^{(2)}(t)-\left(E_0C_PA_P+E_1C_PA_P^2\right)\mathbf{X}_P(t)+$$
$$+S\left[\mathbf{T}\left(\boldsymbol{\epsilon}^1\right)\right]\left\{\left|\begin{array}{c}\left(E_0C_PL_P+E_1C_PA_PL_P\right)^T\\\left(E_0D_P+E_1C_PL_P\right)^T\\\left(E_1D_P\right)^T\end{array}\right|^T\mathbf{D}_M^2\right\}+\mathbf{T}\left(\boldsymbol{\epsilon}^1\right), \tag{13.22}$$

where

$$\mathbf{D}_M^2 \geq \sup \left| \begin{array}{c} \mathbf{D}(t) \\ \mathbf{D}^{(1)}(t) \\ \mathbf{D}^{(2)}(t) \end{array} \right|, \ \forall t \in \mathfrak{T}_0.$$

This reduces $v^{(1)}\left(\boldsymbol{\epsilon}^1(t)\right)$ to

$$\frac{dv(\boldsymbol{\epsilon}^1)}{dt} \leq -2\beta v(\boldsymbol{\epsilon}^1), \ \forall \boldsymbol{\epsilon}^1 \in \mathfrak{R}^{2N}.$$

The final form of the solution reads

$$\left\| \mathbf{T}\left[\boldsymbol{\epsilon}^{\nu-1}(t)\right] \right\| \leq e^{-\beta t} \left\| \mathbf{T}\left(\boldsymbol{\epsilon}_0^{\nu-1}\right) \right\|, \ \forall \left(\boldsymbol{\epsilon}_0^{\nu-1}, t\right) \in \mathfrak{R}^{\nu N} \mathrm{x} \mathfrak{T}_0. \tag{13.23}$$

This furnishes

$$\left\| \boldsymbol{\epsilon}^1(t) \right\| \leq k e^{-\beta t} (\left\| \boldsymbol{\epsilon}_0^1 \right\|, \ \forall \left(\boldsymbol{\epsilon}_0^1, t\right) \in \mathfrak{R}^{2N} \mathrm{x} \mathfrak{T}_0, \tag{13.24}$$

where $k = \sqrt{E^{1^T} E^1} = \sqrt{K_0^2 + T_1^2}$ *for* $E^1 = \left[K_0 \ \ T_1\right]^T$. Inequalities (13.23), (13.24) verify (13.19), (13.20). By adjusting the proof of Case 13.2 from (13.23) and (13.24) to this case, we prove that the control $\mathbf{U}$ (13.22) forces the *ISO* plant (2.33), (2.34) to exhibit tracking determined by (13.21).

Conclusion 409 ***LTC versus NTC for the ISO plant*** *12.91), (12.92)*

We compare LTC (13.22) with NTC (12.111) or with (12.119) (Subsection 12.3):

Both LTC and NTC can ensure the demanded tracking control quality defined by the given tracking algorithm of the form (13.1) or (13.21).

Both LTC and NTC are robust relative to the disturbance. The former demands information about all plant matrices and about $\mathbf{D}_M^2$*, but the latter does not need any information about them.*

LTC requires full and exact information about the plant matrices A_P, B_P, C_P *and* H_P *that describe and express the internal dynamics, as well as they determine the state to output relationship. The demand for the full rank of* $G_{ISOPU}(s)$ *to be equal to the dimension* N *of the output vector* $\mathbf{Y}$*, the dimension of which should not exceed the dimension* r *of the control vector* $\boldsymbol{U}$*, is necessary and sufficient information for the synthesis and implementation of NTC.*

LTC algorithm (13.22) is more complex than NTC algorithm (12.111), i.e., (12.119). For the implementation of the former the control should be determined as the solution to the vector differential equation (13.22). NTC is directly and completely determined by (12.111), i.e., (12.119). Lyapunov tracking controller structure and dynamic complexity largely exceed those of the natural tracking controller.

LTC introduces the global negative feedback in the control system. NTC uses also the global negative feedback, but it exploits additionally the advantage of the local unit positive feedback.

Lyapunov tracking controller can work when it is disconnected from the plant. Natural tracking controller can work properly only in the negative feedback closed loop control system and in the feedforward-negative feedback control system, but it cannot operate disconnected.

Part V

CONCLUSION

Chapter 14

On $\boldsymbol{F}$(s)

The book deals with the IO and ISO mathematical models of the time-invariant linear continuous-time systems. The study presented for them can be developed to their *Input-Internal dynamics-Output* (IIO) mathematical models introduced in [148].

In order to be clear and precise in using different notions (e.g., system regime, system desired regime, nominal control), we have presented their definitions and the procedures on how to determine them. These issues are very simple, but constitute the basis of the control systems theory, which is often ignored in the literature on the first control systems course.

Every dynamic physical system *transfers and transmits simultaneously actions and influences of both the input vector and all initial conditions.* The system transfer function matrix $G(s)$ does not and cannot express and/or describe how the linear time-invariant (continuous-time, discrete-time or hybrid) system transforms the actions and influence of nonzero *initial conditions* on the system into its internal and output behavior. This lack of $G(s)$ is the consequence of its definition and validity *only for zero initial conditions.*

Mathematically considered, the obstacle to treating the transmission of the influence of nonzero initial conditions is the double sum in initial conditions of Laplace transform of the *n-th* order *Input-Output* (IO) linear system differential equation. This obstacle has been removed by solving this mathematical problem of how to put the double sum in the equivalent form to $G(s)\mathbf{I}(s)$ that characterizes the product of the system transfer function matrix $G(s)$ and Laplace transform $\mathbf{I}(s)$ of the input vector $\mathbf{I}(t)$. Once this has been solved, we have become able to determine the complex domain description independent of the input vector and of all initial conditions, which completely expresses and describes how the system transfers, transmits and transforms influences of both the input vector and all initial conditions on the system internal and output behavior. Such description is *the system full transfer function matrix* $F(s)$. It has the same characteristics as the system transfer function matrix $G(s)$:

- the independence of the input vector,
- the independence of all initial conditions,

- the invariance relative to the input vector and to all initial conditions,
- the system order, dimension, structure and parameters completely determine $F(s)$.

After presenting the definitions of *the system full transfer function matrix* $F(s)$ and of its submatrices for every type of the system mathematical models, we presented and proved how we can easily determine them by using *the same mathematical knowledge that we apply to determine the system transfer function matrix* $G(s)$. In this context we discovered the existence of row (non)degenerate, column (non)degenerate and (non)degenerate matrix functions.

We should use the full transfer function matrix $F_{IOCS\varepsilon}(s)$ of the control system rather than only its transfer function matrix $G_{IOCSd}(s)$ or $G_{IOCS\varepsilon yd}(s)$ (Theorem 306, Section 10.2, and Theorem 310, Section 10.3). The latter are insufficient for the analysis or synthesis of the control system from the tracking point of view, while the former is both necessary and sufficient.

Moreover, *the system full transfer function matrix* $F(s)$ expresses channeling information and system structure through its submatrices. We explained the physical meaning of $F(s)$.

This book discovers how the use of *the system full transfer function matrix* $F(s)$ permits new results on the pole-zero cancellation, which is permissible if, and only if, it is possible in *the system full transfer function matrix* $F(s)$, or at least in its appropriate submatrix that links the system reaction with its cause.

Chapter 15

On tracking and trackability

The goal of tracking is the achievement of the desired plant behavior under simultaneous actions of the external disturbances and of all initial conditions on the plant, while none of the stability concepts aims at such a goal. The tracking goal is the primary control goal. Tracking is the crucial control issue. Stability is the essential dynamic systems topic in general and the control systems topic in particular.

Various types of the plant controllability and of the disturbance compensation do not clarify whether the plant has a property to enable the existence of a control that can force the plant output *to track* (sufficiently accurately, precisely, with a good quality) any of its desired output behaviors $\mathbf{Y}_d(.)$ from a given functional family $\mathfrak{Y}_d$ under the simultaneous action of an arbitrary disturbance $\mathbf{D}(.)$ from a functional family $\mathfrak{D}^i$ and under the influence of arbitrary initial conditions. The plant property that does enable the existence of such control is the plant ***trackability***. Among various concepts of the plant trackability presented herein the particular importance for tracking control synthesis is the plant **natural trackability**. It discovers the plant ability to enable the existence, synthesis and implementation of the tracking control without knowing the real form and instantaneous vector value of the disturbance vector $\mathbf{D}(t)$, without using information about the plant internal dynamics, hence about the whole plant mathematical model.

The book presents definitions of perfect and imperfect trackability and of natural trackability. The relationships among them are discovered herein.

The book resolves completely the problems of the necessary and sufficient conditions for the trackability and for the natural trackability properties. The conditions have simple algebraic forms in the complex domain.

The book provides the fundamentals of the tracking theory and of the trackability theory. It presents the tracking concept in Lyapunov sense, i.e., Lyapunov tracking concept, and the concept of tracking with finite, either scalar or vector, reachability time. It discovers the great richness of different tracking types that reflect various tracking qualities. They show essential differences between stability and tracking, which are mutually independent properties in general.

Stability concepts and properties concern dynamical systems, including control systems. Stability theory is a part of the qualitative theory of dynamical systems, not only of control systems. However, tracking concepts and properties are tied with control systems. Tracking theory should be the fundamental of the control theory. Future research and the technical need linked with specific engineering and customer requirements related to the given plant will broaden the tracking theory including the trackability theory.

Chapter 16

On tracking control

The linear tracking control ($LITC$) synthesis in the complex domain demands the use of *the system full transfer function matrix* $F(s)$ (rather than of the system transfer function matrix $G(s)$ that is insufficient). This illustrates the theoretical and practical importance and the wide engineering applications of *the system full transfer function matrix* $F(s)$.

Various new Lyapunov tracking control (LTC) algorithms are established herein in order to guarantee the corresponding tracking properties. Scalar Lyapunov functions are the basis for some of them, and the vector Lyapunov function is the background for others. When they are linked with the tracking algorithm, then the problem, which is caused by the annulment of Lyapunov function gradient, does not exist. Beside, the use of the subsidiary error vector ϵ, instead of the real error vector ε, permits us to avoid the problem of control discontinuity and chattering.

By analyzing information that nature uses to generate control we introduced the concept of the **Natural Tracking Control** (NTC).

NTC algorithms are associated with the tracking properties. The necessary and sufficient conditions are proved for NTC synthesis to guarantee the chosen tracking property.

NTC is superior over other control types considered herein.

The natural trackability conditions are a part of the necessary and sufficient conditions for NTC of the plant.

The NTC shows what is theoretically excellent, or the best possible, tracking control under the specific conditions and relative to the accepted tracking algorithm.

The local unit positive feedback without delay opens the question on the real engineering significance of the NTC.

The nanotechnology and the future technological development justify the ignorance of the signal transmission delay through the local unit positive feedback in the controller.

The unit gain, the zero delay and the positive sign of the NT controller local feedback ensure the infinite gain to the NT controller. Such controller should not be ON when disconnected from the control system because it would then blow up immediately. When it is connected with the plant in the closed loop feedback control system, its infinite gain enables the full compensation of the unknown and unpredictable disturbance action and simultaneously of all unpredictable initial conditions. It creates the fully robust and adaptable control.

The harmonized actions of the local unit positive feedback in the controller and of the global negative feedback in the overall control system enable the excellent, or the best possible, tracking quality.

The very high overall gain Γ of the NT controller replaces its infinite overall gain when we replace the unit gain of the local feedback in the controller by its gain g smaller than one but very close to one, $0<<g<1$. For example, $g = 0.9999999999$ [249]-[254] implies the high overall gain Γ to be

$$\Gamma = \frac{1}{1-g} = \frac{1}{1-0.9999999999} = 1,0000000000 = 10^{10}.$$

Future research will create new LTC and NTC algorithms related to specific plants.

Chapter 17

Recommendation

The author recommends the basic notions and the basic discoveries of this book (system regimes, desired regime, nominal control, the system full transfer function matrix $F(s)$, linear tracking control synthesis by using $F(s)$) to become the parts of the first course on linear control time-invariant continuous-time systems. Other issues and results (tracking properties, trackability properties, the conditions for them, Lyapunov tracking control synthesis and natural tracking control synthesis) should enrich and refine the content of the advanced linear control systems courses.

Therefore, the author hopes ***the twenty-first century linear control systems courses***:

- ***will incorporate both***

  \- *the system full (complete) transfer function matrix* $F(s)$ as the basic system dynamic characteristic in the complex domain, as well as its applications to various issues of control systems, e.g., to the system complete response, pole-zero cancellation, the stability theory, and to the optimal control theory, which are the basic issues of the dynamic systems theory in general and of the control theory in particular,

  and

  \- *the tracking and trackability theories* as the fundamentals of the control theory and of the control engineering, which express the primary control goal;

- ***will refine the study in the complex domain of the qualitative system properties by using the system full transfer function matrix*** *$F(s)$ instead of the system transfer function matrix $G(s)$;*

- ***will devote more attention to the basic system phenomena such as system desired regimes and the nominal control***;

and

- ***will pay attention to the differences between***

 - *the transfer function matrix realization and the system realization that reduces to the full transfer function matrix realization,*

and

 - *the irreducible complex rational matrix functions and the degenerate complex rational matrix functions.*

The accompanying book [148] discovers and proves the following:

- The incompetence of the system transfer function matrix $G(s)$ for Lyapunov stability tests. Only *the system full transfer function matrix* $F(s)$, or at least its adequate submatrix that is the submatrix related to the internal initial conditions, is competent for Lyapunov stability tests.

- Starting with the fact that the initial conditions are seldom equal to zero, the concept of system stability under bounded inputs (BI stability) and zero initial conditions is broadened to system stability under bounded inputs (BI stability) and nonzero initial conditions. This led to new BI stability properties and criteria. The latter are expressed in terms of the submatrices of *the system full transfer function matrix* $F(s)$, or directly in terms of *the system full transfer function matrix* $F(s)$. The system transfer function matrix $G(s)$ is inapplicable to such BI stability tests.

The discoveries and the new results presented herein, which continue those of [148], as well as those presented in [148], open new directions for further research. This book and [148] are complementary.

Bibliography

[1] A. B. Açìkmeşe and M. Corles, "Robust output tracking for uncertain/nonlinear systems subject to almost constant disturbances", *Automatica*, vol. 38, pp. 1919-1926, 2002.

[2] J. K. Acka, "Finding the transfer function of time invariant linear system without computing (sI-A)$^{-1}$", *Int. J. Control*, Vol. 62, No.6, pp. 1517-1522, 1995.

[3] T. Ahmed-Ali and F. Lamnabhi-Lagarrigue, "Tracking control of nonlinear systems with disturbance attenuation", *C. R. Acad. Sci.*, Paris, France, t. 325, Série I, pp. 329-338, 1997.

[4] G. Ambrosino, G. Celentano, and F. Garofalo, "Robust model tracking control for a class of nonlinear plants", *IEEE Transactions on Automatic Control*, Vol. AC-30, No. 3, pp. 275-279, 1985.

[5] B. D. O. Anderson, "A note on transmisson zeros of a transfer function matrix", *IEEE Transactions on Automatic Control*, Vol. AC-21, No. 4, pp. 589-591, August 1976.

[6] B. D. O. Anderson and J. B. Moore, *Linear Optimal Control*, Englewood Cliffs, NJ, USA: Prentice Hall, 1971.

[7] N. P. I. Aneke, H. Nijmeijer and A. G. de Jager, "Tracking control of second-order chained form systems by cascaded backstepping", *Int. J. Robust and Nonlinear Control*, Vol. 13, pp. 95-115, 2003.

[8] P. J. Antsaklis and A. N. Michel, *Linear Systems*, New York, NY, USA: McGraw Hill Company, Inc., 1997.

[9] M. Athanassiades, "Bang-bang control for tracking systems", *IRE Transactions on Automatic Control*, Vol. AC-7, No. 3, pp. 77-78, 1962.

[10] S. Barnett, "Some topics in algebraic systems theory: A survey", in *Recent Mathematical Developments in Control*, Ed. D. J. Bell, New York, NY, USA: Academic Press, pp. 323-344, 1973.

[11] Y. Bar-Shalom, "Tracking Methods in a Multitarget Environment", *IEEE Transactions on Automatic Control*, Vol. AC-23, No. 4, pp. 618-626, August 1978.

[12] Y. Bar-Shalom and T. E. Fortmann, *Tracking and Data Association*, Boston, MA, USA: Academic Press, Inc., 1988.

[13] R. Bellman, "Vector Lyapunov functions", *J.S.I.A.M. Control*, Ser. A, Vol. 1, No.1, pp. 32-34, 1962.

[14] J. E. Bertram and P. E. Sarachik, "On optimal computer control", *Proc. of the First International Congress of the Federation of Automatic Control*, London: Butterworths, pp. 419-422, 1961.

[15] G. Besançon, "Global output feeddback tracking control for a class of Lagrangian systems", *Automatica*, Vol. 36, pp. 1915-1921, 2000.

[16] S. P. Bhattacharyya, "Frequency domain conditions for disturbance rejection", *IEEE Transactions on Automatic Control*, Vol. AC-25, No. 6, 1211-1213, December 1980.

[17] S. P. Bhattacharyya, "Transfer function conditions for output feedback disturbance rejection", *IEEE Transactions on Automatic Control*, Vol. AC-27, No. 4, pp. 974-977, August 1982.

[18] S. P. Bhattacharyya, A. C. del Nero Gomes and J. W. Howze, "The structure of robust disturbance rejection", *IEEE Transactions on Automatic Control*, Vol. AC-28, No. 9, 874-881, September 1983.

[19] P. Borne, G. Dauphin-Tanguy, J.-P. Richard, F. Rotella and I. Zambettakis, *Commande et Optimisation des Processus*, Paris, France: Éditions TECHNIP, 1990.

[20] R. W. Brockett and M. D. Mesarović, "The reproducibility of multivariable systems", *J. Mathematics Analysis and Applications*, Vol. 1, pp. 548-563, 1965.

[21] W. L. Brogan, *Modern Control Theory*, New York, NY, USA: Quantum Publishers, Inc., 1974.

[22] G. S. Brown and D. P. Campbell, *Principles of Servomechanisms*, New York, NY, USA: Wiley, 1948.

[23] F. M. Callier and C. A. Desoer, *Linear System Theory*, New York, NY, USA: Springer-Verlag, 1991.

[24] F. M. Callier and C. A. Desoer, *Multivariable Feedback Systems*, New York, NY, USA: Springer-Verlag, 1982

[25] G. E. Carlson, *Signal and Linear Systems Analysis and Matlab*, second edition, New York, NY, USA: Wiley, 1998

[26] Y.-C. Chang, "Robust tracking control for nonlinear MIMO systems via fuzzy approaches", *Automatica*, Vol. 36, pp. 1535-1545, 2000.

[27] B. Chatterjec, "Nonlinear feedback in servo systems", *IRE Transactions on Automatic Control*, Vol. AC-5, No. 4, pp. 329-330, 1960.

[28] B. Chen and J. K. Tugnait, "Tracking of multiple maneuvering targets in clutter using IMM/JPDA filtering and fixed-lag smoothing", *Automatica*, Vol. 37, pp. 239-249, 2001.

[29] C.-T. Chen, *Linear System Theory and Design*, New York, NY, USA: Holt, Rinehart and Winston, Inc., 1984.

[30] Y.-C. Chen and S. Chang, "Output tracking design of affine nonlinear plant via variable structure system", *IEEE Transactions on Automatic Control*, Vol. 37, No. 11, pp. 1823-1828, November 1992.

[31] Y.-C. Chen, P.-L. Lin and S. Chang, "Design of output tracking via variable structure system: for plants with redundant inputs", *IEE Proceedings-D*, Vol. 139, No. 4 pp. 421-428, July 1992,.

[32] X. P. Cheng and R. V. Patel, "Neural network based tracking conrol of a flexible macro-micro manipulator system", *Neural Networks*, Vol. 16, pp. 271-286, 2003.

[33] H. Chestnut and R. W. Mayer, *Servomechanisms and Regulating System Design*, New York: Wiley, 1955.

[34] P. D. Christofides, A. R. Teel and P. Daoutidis, "Robust semi-global output tracking for nonlinear singularly perturbed systems", *Int. J. Control*, Vol. 65, No. 4, pp. 639-666, 1996.

[35] D. Chwa, "Sliding-mode tracking control of nonholonomic wheeled mobile robots in polar coordinates", *IEEE Transactions on Control Systems Technology*, Vol. 12, No. 4, pp. 637-644, July 2004.

[36] F. E. Daum, "Bounds on performance for multiple target tracking", *IEEE Transactions on Automatic Control*, Vol. 35, No. 4, pp. 443-446, 1990.

[37] R. Davies, C. Edwards and S. K. Spurgeon, "Robust tracking with a sliding mode", in *Variable Structure and Lyapunov Control*, Ed. A. S. I. Zinober, London: Springer Verlag, pp. 51-73, 1994.

[38] E. J. Davison, "The robust decentralized control of a servomechanism problem", *IEEE Transactions on Automatic Control*, Vol. AC-21, No. 1, pp. 14-24, 1976.

[39] E. J. Davison, "The robust decentralized control of a servomechanism problem for composite systems with input-output interconnections", *IEEE Transactions on Automatic Control*, Vol. AC-24, No. 4, pp. 325-327, 1979.

[40] E. J. Davison, "The robust decentralized servomechanism problem with extra stabilizing control agents", *IEEE Transactions on Automatic Control*, Vol. AC-22, No. 3, pp. 256-258, 1977.

[41] E. J. Davison and B. R. Copeland, "Gain margin and time lag tolerance constraints applied to the stabilization problem and robust servomechanism problem", *IEEE Transactions on Automatic Control*, Vol. AC-30, No. 3, pp. 229-239, 1985.

[42] E. J. Davison and I. Ferguson, "The design of controllers for the multivariable robust servomechanism problem using parameter optimization methods", *IEEE Transactions on Automatic Control*, Vol. AC-26, No. 1, pp. 93-110, 1981.

[43] E. J. Davison and A. Goldenberg, "Robust control of a general servomechanism problem: The servo-compensator", *Automatica*, Vol. 11, pp. 461-471, 1975.

[44] E. J. Davison and P. Patel, "Application of the robust servomechanism controller to systems with periodic tracking/disturbance signals", *Int. J. Control*, Vol. 47, No. 1, pp. 111-127, 1988.

[45] E. J. Davison and B. M. Scherzinger, "Perfect control of the robust servomechanism problem", *IEEE Transactions on Automatic Control*, Vol. AC-32, No. 8, pp. 689-702, August 1987.

[46] J. J. D'Azzo and C. H. Houpis, *Linear Control System Analysis & Design*, New York, NY, USA: McGraw-Hill Book Company, 1988.

[47] L. Debnath, *Integral Transformations and Their Applications*, Boca Raton, FL, USA: CRC Press, 1995.

[48] A. Denker and K. Ohnishi, "Robust tracking control of mechatronic arms", *IEEE/ASME Transactions on Mechatronics*, Vol. 1, No. 2, pp. 181-188, June 1996.

[49] C. A. Desoer, *Notes for a Second Course on Linear Systems*, New York, NY, USA: Van Nostrand Reinhold Company, 1970.

[50] C. A. Desoer and C.-A. Lin, "Tracking and disturbance rejection of MIMO nonlinear systems with PI controller", *IEEE Transactions on Automatic Control*, Vol. AC-30, No. 9, pp. 861-867, 1985.

[51] C. A. Desoer and J. D. Schulman, "Zeros and poles of matrix transfer functions and their dynamical interpretation", *IEEE Transactions on Circuits and Systems*, Vol. CAS-21, No. 1, pp. 3-8, January 1974.

[52] C. A. Desoer and M. Vidyasagar, *Feedback Systems: Input-Output Properties*, New York: Academic Press, 1975.

[53] C. A. Desoer and Y. T. Wang, "The robust non-linear servomechanism problem", *International Journal of Control*, Vol. 29, No. 5, pp. 803-828, 1979.

[54] S. Di Gennaro, "Output attitude tracking for flexible spacecraft", *Automatica*, Vol. 38, pp. 1719-1726, 2002.

[55] C. Edwards and S. K. Spurgeon, "Robust output tracking using a sliding mode controller/observer scheme", *International Journal of Control*, Vol. 64, No. 5, pp. 967-983, 1996.

[56] O. I. Elgerd, *Control Systems Theory*, New York, NY, USA: McGraw-Hill Book Company, 1967.

[57] A. Emami-Naeini and G. F. Franklin, "Deadbeat control and tracking of discrete-time systems", *IEEE Transactions on Automatic Control*, Vol. AC-27, No. 1, pp. 176-180, 1982.

[58] E. Fabian and W. M. Wonham, "Decoupling and disturbance rejection", *IEEE Transactions on Automatic Control*, Vol. AC-20, No. 4, 399-401, 1975.

[59] D. R. Fannin, W. H. Tranter and R. E. Ziemer, *Signals & Systems Continuous and Discrete*, fourth edition, Englewood Cliffs, NJ: Prentice Hall, 1998.

[60] I. Flügge-Lotz and C. F. Taylor, "Synthesis of a nonlinear control system", *IRE Transactions on Automatic Control*, Vol. 1, No. 1, pp. 3-9, May 1956.

[61] T. E. Fortmann, Y. Bar-Shalom, M. Scheffe and S.Gelfand, "Detection thresholds for tracking in clutter-A connection between estimation and signal processing", *IEEE Transactions on Automatic Control*, Vol. AC-30, No. 3, pp. 221-228, 1985.

[62] T. E. Fortmann and K. L. Hitz, *An Introduction to Linear Control Systems*, New York, NY, USA: Marcel Dekker, Inc., 1977.

[63] G. F. Franklin, "Design of ripple-free multivariable robust servomechanisms", *IEEE Transactions on Automatic Control*, Vol. AC-31, No. 7, pp. 661-664, 1986.

[64] M. I. Freedman and R. Glassey, "Tracking via feedback for systems with irrational transfer function", *SIAM J. Control*, Vol. 9, No. 3, pp. 317-338, August 1971.

[65] L.-C. Fu and T.-L. Liao, "Globally stable robust tracking of nonlinear systems using variable structure control and with an application to a robotic manipulator", *IEEE Transactions on Automatic Control*, Vol. 35, No. 12, pp. 1345-1350, December 1990.

[66] E. G. Gilbert, "Controllability and observability in multivariable control systems", *SIAM J. Control*, Vol. 1, pp. 128-151, 1963.

[67] S. Gopalswamy and J. K. Hedrick, "Tracking nonlinear non-minimum phase systems using sliding control", *International Journal of Control*, Vol. 57, No. 5, pp. 1141-1158, 1993.

[68] O. M. Grasselli, S. Longhi, A. Tornambè, "Robust output regulation and tracking for linear periodic systems under structured uncertainties", *Automatica*, Vol. 32, No. 7, pp. 1015-1019, 1996.

[69] J. W. Grizzle, M. D. Di Benedetto and F. Lamnabhi-Lagarrigue, "Necessary conditions for asymptotic tracking in nonlinear systems", *IEEE Transactions on Automatic Control*, Vol. AC-39, No. 9, pp.1782-1794, September 1994.

[70] Ly. T. Grouyitch, *Automatique: dynamique linéaire*, Notes de cours, Belfort, France : Ecole Nationale d'Ingénieurs de Belfort, 1997.

[71] Ly. T. Grouyitch, *Automatique: dynamique linéaire*, Lecture Notes, Belfort, France: University of Technology Belfort-Montbeliard, 1999, 2000.

[72] Ly. T. Grouyitch, *Conduite des systèmes*, Notes de cours SY 98, Belfort, France: Université de Technologie de Belfort-Montbeliard, 2000.

[73] Lj. T. Grujić, "Adaptive tracking control for a class of plants with uncertain parameters and non-linearities", *International Journal of Adaptive Control and Signal Processing*, Vol. 2, pp. 49-71, 1988.

[74] Lj. T. Grujić, "Algebraic conditions for absolute tracking control of continuous-time Lurie systems", *Proceedings of the Conference on Linear Algebra in Signals, Systems, and Control*, (Boston, MA, August 12-14, 1986), published as *Linear Algebra in Signals, Systems, and Control*, Eds. B. N. Datta, C. R. Johnson, M. A. Kaashoek, R. J. Plemmons, and E. D. Sontag, Philadelphia, PA, USA: SIAM, pp. 535-555, 1988.

[75] Lj. T. Grujić, "Algebraic conditions for absolute tracking control of Lurie systems", *Int. J. Control*, Vol. 48, No. 2, pp. 729-754, 1988.

[76] Lj. T. Grujić, "Algorithms for CAD of Continuous-Time Non-Stationary Non-Linear Tracking Systems via the Output-Space", *ACTA Press*, Anaheim, pp. 58-61, 1985.

[77] Lj. T. Grujić, "Algorithms for CAD of Discrete-Time Non-Stationary Non-Linear Tracking Systems via the State-Space", *ACTA Press*, Anaheim, pp. 135-138, 1985.

[78] Lj. T. Grujić, *Automatique-Dynamique Linéaire*, Lecture Notes, Belfort, France: Ecole Nationale d'Ingénieurs de Belfort, 1994-1998.

[79] Lj. T. Grujić, "Complete exact solution to the Lyapunov stability problem: Time-varying nonlinear systems with differentiable motions", *Nonlinear Analysis, Theory, Methods & Applications*, Vol. 22, No. 8, pp. 971-981, 1994.

[80] Lj. T. Grujić, "Consistent Lyapunov methodology for time-invariant nonlinear systems", *Avtomatika i Telemekhanika* (in Russian), No. 12, pp. 35-73, December 1997.

[81] Lj. T. Grujić (Ly. T. Gruyitch), *Continuous Time Control Systems*, Lecture notes for the course "DNEL4CN2: Control Systems", Durban, RSA: Department of Electrical Engineering, University of Natal, South Africa, 1993.

[82] Lj. T. Grujić, "Exact determination of a Lyapunov function and the asymptotic stability domain", *Int. J. Systems Sc.*, Vol. 23, No. 11, pp. 1871-1888, 1992.

[83] Lj. T. Grujić, "Exact solutions for asymptotic stability: Non-linear systems", *Int. J. Non-Linear Mechanics*, Vol. 30, No. 1, pp. 45-56, 1995.

[84] Lj. T. Grujić, "Exponential quality of time-varying dynamical systems: stability and tracking", Ch. 5, in *Advances in Nonlinear Dynamics*, Vol. 5, Eds. S. Sivasundaram and A. A. Martynyuk, Amsterdam: Gordon and Breach Science Publishers Ltd., pp. 51-61, 1997.

[85] Lj. T. Grujić, "Natural trackability and control: Multiple time scale systems", *Proceedings of the IFAC Conference: Control of Industrial Systems*, Eds. Lj. T. Grujić, P. Borne, A. El Moudni and M. Ferney, Vol. 2, London: Pergamon, Elsevier, pp. 669-674, 1997.

[86] Lj. T. Grujić, "Natural trackability and control: Perturbed robots", *Proceedings of the IFAC Conference: Control of Industrial Systems*, Eds. Lj. T. Grujić, P. Borne, A. El Moudni and M. Ferney, Vol. 3, London: Pergamon, Elsevier, pp. 1641-1646, 1997.

[87] Lj. T. Grujić, "Natural trackability and tracking control of robots", *IMACS-IEEE-SMC Multiconference CESA'96: Symposium on Control, Optimization and Supervision*, Vol. 1, Lille, France, pp. 38-43, 1996.

[88] Lj. T. Grujić, "New approach to asymptotic stability: Time-varying nonlinear systems", *International J. of Mathematics and Mathematical Sciences*, Vol. 20, No. 2, pp. 347-366, 1997.

[89] Lj. T. Grujić, "New Lyapunov methodology and exact construction of a Lyapunov function: Exponential stability", *Problems of the Nonlinear Analysis in Engineering Systems*, Kazan, Russia, Vol. 1, pp. 9-16, 1995.

[90] Lj. T. Grujić, "Non-linear singularly perturbed tracking systems", *Proc. AMSE Conference on Modelling and Simulation*, Paris, pp. 116-123, 1982.

[91] Lj. T. Grujić, "Novel Lyapunov stability methodology for nonlinear systems: Complete solutions", *Nonlinear Analysis, Theory, Methods & Applications: Proc. 2nd Second World Congress of Nonlinear Analysts*, London: Elsevier Science Ltd., Vol. 30, No. 8, pp. 5315-5325, 1997.

[92] Lj. T. Grujić, "On general solutions of non-linear tracking for stationary systems", *AI 83 IASTED Symposium*, Lille, pp. 49-53, 1983.

[93] Lj. T. Grujić, "On large-scale systems stability", in *Computing and computers for control systems*, Eds. P. Borne et al., J. C. Baltzer AG, Scientific Publishing Co., IMACS, pp. 201-206, 1989.

[94] Lj. T. Grujić, "On non-linear tracking domain estimates: Continuous-time systems", *AI 83 IASTED Symposium*, Lille, pp. 65-66, 1983.

[95] Lj. T. Grujić, "On non-linear tracking domain estimates: Discrete-time systems", *AI 83 IASTED Symposium*, Lille, pp. 59-63, 1983.

[96] Lj. T. Grujić, "On non-linear tracking phenomena and problems", *AI 83 IASTED Symposium*, Lille, pp. 45-48, 1983.

[97] Lj. T. Grujić, "On solutions to Lyapunov stability problems", *Facta, Universitatis, Series: Mechanics, Automatic Control and Robotics*, University of Nish, Serbia, Yugoslavia, Vol. 1, No. 2, pp. 121-138, 1992.

[98] Lj. T. Grujić, "On the non-linear tracking systems theory: I-Phenomena, concepts and problems via ouptut space", *Automatika*, Zagreb, Vol. 27, No. 1-2, pp. 3-8, 1986.

[99] Lj. T. Grujić, "On the non-linear tracking systems theory: II-Phenomena, concepts and problems via state space", *Automatika*, Zagreb, Vol. 27, No. 1-2, pp. 9-16, 1986.

[100] Lj. T. Grujić, "On the non-linear tracking systems theory: III-Liapunov-like approach via the output-space: Continuous-time", *Automatika*, Zagreb, Vol. 27, No. 3-4, pp. 99-104, 1986.

[101] Lj. T. Grujić, "On the non-linear tracking systems theory: IV-Liapunov-like approach via the state-space: Continuous-time", *Automatika*, Zagreb, Vol. 27, No. 3-4, pp. 105-116, 1986.

[102] Lj. T. Grujić, "On the non-linear tracking systems theory: V-Liapunov-like approach via the output-space: Discrete-time", *Automatika*, Zagreb, Vol. 27, No. 5-6, pp. 197-202, 1986.

[103] Lj. T. Grujić, "On the non-linear tracking systems theory: VI-Liapunov-like approach via the state-space: Discrete-time", *Automatika*, Zagreb, Vol. 27, No. 5-6, pp. 203-211, 1986.

[104] Lj. T. Grujić, "On the theory and synthesis of non-linear non-stationary tracking singularly perturbed systems", *Control Theory and Advanced Technology*, MITA Press, Tokyo, Japan, Vol. 4, No. 4, pp. 395-409, 1988.

[105] Lj. T. Grujić, "On the theory of nonlinear systems tracking with guaranteed performance index bounds: Application to robot control", *Proceedings of the 1989 IEEE International Conference on Robotics and Automation*, Scottsdale, AZ: IEEE-Computer Society Press, Vol. 3, pp. 1486-1490, May 14-19, 1989.

[106] Lj. T. Grujić, "On the tracking problem for nonlinear systems", in *Applied Control*, Ed. S. G. Tzafests, New York, NY, USA: Marcel Dekker, pp. 325-343, 1993.

[107] Lj. T. Grujić, "On the tracking theory with a prespecified quality" (in Serbo-Croatian), *Zastava*, Vol. VIII, No. 28-29, pp. 18-22, Oct. 1990.

[108] Lj. T. Grujić, "On tracking domain estimates of large-scale systems", *AI 83 IASTED Symposium*, Lille, pp. 55-57, 1983.

[109] Lj. T. Grujić, "On tracking control of singularly perturbed systems", *Proc. IMACS-IFAC Symposium: "Modelling and Simulation for Control of Lumped and Distributed Parameter Systems"*, Villeneuve d'Ascq, pp. 565-568, June 1-3, 1986.

[110] Lj. T. Grujić, "On tracking domains of continuous-time non-linear control systems", *Proc. 1982 American Control Conference*, AACC-IEEE, New York, 1982, pp. 670-674. Also: *R.A.I.R.O. Automatique / Systems Analysis and Control*, Vol. 16, No. 4, pp. 311-327, 1982.

[111] Lj. T. Grujić, "Phenomena, concepts and problems of automatic tracking: Continuous-time stationary non-linear systems with variable inputs" (in Serbo-Croatian), *Proc. of the First International Seminar "AUTOMATON and ROBOT"*, USAUM Srbije i "OMO", Belgrade, pp. 307-330, 1985.

[112] Lj. T. Grujić, "Phenomena, concepts and problems of automatic tracking: Discrete-time stationary non-linear systems with variable inputs" (in Serbo-Croatian), *Proc. of the First International Seminar "AUTOMATON and ROBOT"*, USAUM Srbije i "OMO", Belgrade, pp. 401-422, 1985.

[113] Lj. T. Grujić, "Solutions to Lyapunov stability problems: Nonlinear systems with differentiable motions", *Computational and Applied Mathematics II: Differential Equations*, Eds. W.F Ames and P.J. van der Houwen, Amsterdam, Elsevier, pp. 39-47, 1992.

[114] Lj. T. Grujić, "Solutions to Lyapunov stability problems of sets: Nonlinear systems with differentiable motions", *International Journal of Mathematics and Mathematical Sciences*, Orlando, FL, Vol. 17, No. 1, pp. 103-112, 1994.

[115] Lj. T. Grujić, "Solutions to Lyapunov stability problems: Nonlinear systems with continuous motions", *International Journal of Mathematics and Mathematical Sciences*, Orlando, FL, Vol. 17, No. 3, pp. 587-596, 1994.

[116] Lj. T. Grujić, "Solutions to Lyapunov stability problems: Nonlinear systems with globally differentiable motions", *The Lyapunov functions method and applications,* Eds. P. Borne and V. Matrosov, J.C. Baltzer AG, Scientific Publishing Co, IMACS, pp. 19-27, 1990.

[117] Lj. T. Grujić, "Solutions to Lyapunov stability problems via O-uniquely bounded sets", *Control-Theory and Advanced Technology*, Tokyo, Japan, pp. 1069-1091, 1995.

[118] Lj. T. Grujić, "Stability versus tracking in automatic control systems" (in Serbo-Croatian), *Proc. JUREMA 29*, Part 1, Zagreb, pp. 1-4, 1984.

[119] Lj. T. Grujić, "State-space domains of singularly perturbed systems", *Proc. 12th World Congress on Scientific Computation*, IMACS, Paris, France, pp. 83-91, July 18-22, 1988.

[120] Lj. T. Grujić, "Synthesis of automatic tracking systems and the output-space: Continuous-time stationary non-linear systems with variable inputs" (in Serbo-Croatian), *Proc. of the First International Seminar "AUTOMATON and ROBOT"*, USAUM Srbije i "OMO", Belgrade, pp. 331-370, 1985.

[121] Lj. T. Grujić, "Synthesis of automatic tracking systems and the output-space: Discrete-time stationary non-linear systems with variable inputs" (in Serbo-Croatian), *Proc. of the First International Seminar "AUTOMATON and ROBOT"*, USAUM Srbije i "OMO", Belgrade, pp. 423-448, 1985.

[122] Lj. T. Grujić, "Synthesis of automatic tracking systems and the state-space: Continuous-time stationary non-linear systems with variable inputs" (in Serbo-Croatian), *Proc. of the First International Seminar "AUTOMATON and ROBOT"*, USAUM Srbije i "OMO", Belgrade, pp. 371-400, 1985.

[123] Lj. T. Grujić, "Synthesis of automatic tracking systems and the state-space: Discrete-time stationary non-linear systems with variable inputs" (in Serbo-Croatian), *Proc. of the First International Seminar "AUTOMATON and ROBOT"*, USAUM Srbije i "OMO", Belgrade, pp. 449-476, 1985.

[124] Lj. T. Grujić, "Time-varying continuous nonlinear systems: Uniform asymptotic stability", *International Journal of Systems Science*, Vol. 26, No. 5, pp. 1103-1127, 1995; "Corrigendum", ibid, Vol. 27, No. 7, p. 689, 1996.

[125] Lj. T. Grujić, "Tracking analysis for non-stationary non-linear systems", *Proc. 1986 IEEE International Conference on Robotics and Automaton*, San Francisco, Vol. 2, pp. 713-721, 1986.

[126] Lj. T. Grujić, "Tracking control obeying prespecified performance index", *Proc. 12th World Congress on Scientific Computation*, IMACS, Paris, France, pp. 332-336, July 18-22, 1988.

[127] Lj. T. Grujić, "Tracking versus stability: Theory", *Proc. 12th World Congress on Scientific Computation*, IMACS, Paris, pp. 319-327, July 18-22, 1988.

[128] Lj. T. Grujić, "Tracking with prespecified index limits: Control synthesis for non-linear objects", *Proc. II International Seminar and Symposium: "AUTOMATON and ROBOT"*, SAUM and IEE, Belgrade, pp. S-20-S-52, 1987.

[129] Lj. T. Grujić and Z. Janković, "Synthesis of tracking control for a plane motion", *Proceedings of the Second Conference on Systems, Automatic Control and Measurement* (in Serbo-Croatian), SAUM, Belgrade, Serbia, pp. 477-492, 1986.

[130] Lj. T. Grujić and Z. Janković, "Synthesis of tracking control for a process", *Proceedings of the Second Conference on Systems, Automatic Control and Measurement* (in Serbo-Croatian), SAUM, Belgrade, Serbia, pp. 305-324, 1986.

[131] Lj. T. Grujić, A. A. Martynyuk and M. Ribens-Pavella, *Large-Scale Systems under Structural and Singular Perturbations* (in Russian, Kiev: Naukova Dumka, 1984), Berlin, Germany: Springer Verlag, 1987.

[132] Lj. T. Grujić and W. P. Mounfield, Jr., "Natural tracking control of linear systems", *Proceedings of the 13th IMACS World Congress on Computation and Applied Mathematics*, Eds. R. Vichnevetsky and J. J. H. Miller, Trinity College, Dublin, Ireland, Vol. 3, pp. 1269-1270, July 22-26, 1991.

[133] Lj. T. Grujić and W. P. Mounfield, "Natural tracking control of linear systems", in *Mathematics of the Analysis and Design of Process Control*, Eds. P. Borne, S.G. Tzafestas and N.E. Radhy, London: Elsevier Science Publishers B. V., IMACS, pp. 53-64, 1992.

[134] Lj. T. Grujić and W. P. Mounfield, *Natural Tracking Controller*, US Patent No 5,379,210, Jan. 3, 1995.

[135] Lj. T. Grujić and W. P. Mounfield, "Natural tracking PID process control for exponential tracking", *American Institute of Chemical Engineers Journal*, Vol. 38, No. 4, pp. 555-562, 1992.

[136] Lj. T. Grujić and W. P. Mounfield, "PD-control for stablewise tracking with finite reachability time: Linear continuous time MIMO systems with state-space description", *International Journal of Robust and Nonlinear Control*, England, Vol. 3, pp. 341-360, 1993.

[137] Lj. T. Grujić and W. P. Mounfield, "PD natural tracking control of an unstable chemical reaction", *Proc. 1993 IEEE International Conference on Systems, Man and Cybernetics*, Le Touquet, France, Vol. 2, pp. 730-735, 1993.

[138] Lj. T. Grujić and W. P. Mounfield, "PID natural tracking control of a robot: Theory", *Proc. 1993 IEEE International Conference on Systems, Man and Cybernetics*, Le Touquet, France, Vol. 4, pp. 323-327, 1993.

[139] Lj. T. Grujić and W. P. Mounfield, "Stablewise tracking with finite reachability time: Linear time-invariant continuous-time MIMO systems", *Proc. of the 31st IEEE Conference on Decision and Control*, Tucson, AZ, USA, pp. 834-839, 1992.

[140] Lj. T. Grujić and W. P. Mounfield, Jr., "Tracking control of time-invariant linear systems described by IO differential equations", *Proceedings of the 30th IMACS Conference on Decision and Control*, Brighton, England, Vol. 3, pp. 2441-2446, December 11-13, 1991.

[141] Lj. T. Grujić and W. P. Mounfield, "Ship roll stabilization by natural tracking control: Stablewise tracking with finite reachability time", *Proc. 3rd IFAC Workshop on Control Applications in Marine Systems*, Trondheim, Norway, pp. 202-207, May 10-12, 1995.

[142] Lj. T. Grujić and Z. Novaković, "Feedback principle via Liapunov function concept rejects robot control synthesis drawback", *Proceedings of the Sixth IASTED International Symposium on Modelling, Identification and Control*, Grindelwald, pp. 227-230, February 17-20, 1987.

[143] Lj. T. Grujić and Z. Novaković, "Output robot control: Stablewise tracking with a requested reachability time", *Proc. IEEE International Workshop on Intelligent Robots and Systems*, Tokyo, Japan, pp. 85-90, Oct. 31-Nov. 2, 1988.

[144] Lj. T. Grujić and Z. Novaković, "Theory of robust adaptive exponential tracking control. Application to robots without using inverse mechanics", *Proceedings of Fifth Yale Workshop on Applications of Adaptive Systems Theory*, Center for Systems Science, Yale University, New Haen, CT, USA, pp. 237-243, May 20-22, 1987.

[145] Lj. T. Grujić and B. Porter, "Continuous-time tracking systems incorporating Lur'e plants with single non-linearities", *Int. J. Systems Science*, Vol. 11, No. 2, pp. 177-189, 1980.

[146] Lj. T. Grujić and B. Porter, "Discrete-time tracking systems incorporating Lur'e plants with mutliple non-linearities", *Int. J. Systems Science*, Vol. 11, No. 12, pp. 1505-1520, 1980.

[147] Lj. T. Grujić and Z. Ribar, "Synthesis of time-varying Lurie structurally variable tracking systems", *Proceedings of the 13th IMACS World Congress on Computation and Applied Mathematics*, pp. 1267-1268, July 22-26, 1991.

[148] Ly. T. Gruyitch, *Advances in the Linear Dynamic Systems Theory. Time-Invariant Continuous-Time Systems,* in print, Llumina Press, Tamarac, FL, USA, http://www.llumina.com, 2013.

[149] Ly. T. Gruyitch, "Aircraft natural control synthesis: Vector Lyapunov function approach", *Actual Problems of Airplane and Aerospace Systems: Processes, Models, Experiments*, Vol. 2, No. 6, Kazan, Russia, and Daytona Beach, FL, USA, pp. 1-9, 1998.

[150] Ly. T. Gruyitch, *Conduite des systèmes*, Lecture Notes: Notes de cours SY 98, Belfort: University of Technology Belfort-Montbeliard, 2000, 2001.

[151] Ly. T. Gruyitch, "Consistent Lyapunov methodology for exponential stability: PCUP approach", in *Advances in Stability Theory at the End of the 20th Century*, Ed. A. A. Martynyuk, London: Taylor and Francis, pp. 107-120, 2003.

[152] Ly. T. Gruyitch, "Consistent Lyapunov methodology: Non-differentiable non-linear systems", *Nonlinear Dynamics and Systems Theory*, Vol. 1, No. 1, pp. 1-22, 2001.

[153] Ly. T. Gruyitch, "Consistent Lyapunov methodology, time-varying nonlinear systems and sets", *Nonlinear Analysis, Theory and Applications*, Vol. 39, pp. 413-446, 2000.

[154] Ly. T. Gruyitch, *Contrôle commande des processus industriels*, Lecture Notes: Notes de cours SY 51, Belfort: University of Technology Belfort-Montbeliard, 2002, 2003.

[155] Ly. T. Gruyitch, *Einstein's Relativity Theory. Correct, Paradoxical, and Wrong*, ISBN 1-4122-0211-6, Trafford, Victoria, Canada, http://www.trafford.com/06-2239, 2006.

[156] Ly. T. Gruyitch, "Exponential quality of time-varying dynamical systems: Stability and tracking" in *Advances in Nonlinear Dynamics-Stability and Control: Theory, Methods and Applications*, Eds. S. Sivasundaram and A. A. Martynyuk, Amsterdam, Holland: Gordon and Breach Science Publishers, Chapter 5, pp. 51-61, 1997.

[157] Ly. T. Gruyitch, "Exponential stabilizing natural tracking control of robots: Theory", *Proceedings of the Third ASCE Specialty Conference on Robotics for Challenging Environments*, Albuquerque, NM, USA, (Eds. Laura A. Demsetz, Raymond H. Bryne and John P. Wetzel), Reston, VA, USA: American Society of Civil Engineers (ASCE), pp. 286-292, April 26-30, 1998.

[158] Ly. T. Gruyitch, "Gaussian generalisations of the relativity theory fundaments with applications", *Proceedings of the VII International Conference: Physical Interpretations of Relativity Theory*, Ed. M. C. Duffy, British Society for the Philosophy of Science, London, pp. 125 -136, September 15-18, 2000.

[159] Ly. T. Gruyitch, "Global natural θ-tracking control of Lagrangian systems", *Proceedings of the American Control Conference*, San Diego, CA, USA, pp. 2996-3000, June 1999.

[160] Ly. T. Gruyitch, "Natural control of robots for fine tracking", *Proceedings of the 38th Conference on Decision and Control*, Phoenix, AZ, USA, pp. 5102-5107, December 1999.

[161] Ly. T. Gruyitch, "Natural tracking control synthesis for Lagrangian systems",, *V International Seminar on Stability and Oscillations of Nonlinear Control Systems*, Russian Academy of Sciences, Moscow, pp. 115-120, June 3-5, 1998.

[162] Ly. T. Gruyitch, "New Development of vector Lyapunov functions and airplane control synthesis", Chapter 7 in *Advances in Dynamics and Control*, Ed. S. Sivasundaram, Boca Raton, FL, USA: Chapman & Hall/CRC, pp. 89-102, 2004.

[163] Ly. T. Gruyitch, "On tracking theory with embedded stability: Control duality resolution", *Proceedings of the 40th IEEE Conference on Decision and Control*, Orlando, FL, USA, pp. 4003-4008, December 2001.

[164] Ly. T. Gruyitch, "Physical continuity and uniqueness principle. Exponential natural tracking control", *Neural, Parallel & Scientific Computations*, Vol. 6, pp. 143-170, 1998.

[165] Ly. T. Gruyitch, "A physical principle and consistent Lyapunov methodology: Time-invariant nonlinear systems", *Proc. International Conference on Advances in Systems, Signals, Control and Computers*, Vol. 1, Durban, South Africa, pp. 42-50, 1998.

[166] Ly. T. Gruyitch, "Robot global tracking with finite vector reachability time", *Proceedings of the European Control Conference*, Karlsruhe, Germany, Paper # 132, pp. 1-6, August 31-September 3, 1999.

[167] Ly. T. Gruyitch, "Robust prespecified quality tracking control synthesis for 2D systems", *Proc. International Conference on Advances in Systems, Signals, Control and Computers*, Vol. 3, Durban, South Africa, pp. 171-175, 1998.

[168] Ly. T. Gruyitch, *Systèmes d'asservissement industriels*, Lecture Notes: Notes de cours SY 40, Belfort: Universite de Technologie de Belfort-Montbeliard, 2001.

[169] Ly. T. Gruyitch, "Time and uniform relativity theory fundaments", *Problems of Nonlinear Analysis in Engineering Systems*, Vol. 7, No. 2(14), Kazan, Russia, pp. 1-29, 2001.

[170] Ly. T. Gruyitch, *Time and Time Fields. Modeling, Relativity, and Systems Control*, ISBN 1-4251-0726-5, Trafford, Victoria, Canada, http://www.trafford.com/06-2484, 2006.

[171] Ly. T. Gruyitch, *Time. Fields, Relativity, and Systems*, ISBN 1-59526-671-2, LCCN 2006909437, Llumina, Coral Springs, Florida, USA, http://www.llumina.com/store/timefieldsrelativity.htm, 2006.

[172] Ly. T. Gruyitch, "Time, relativity and physical principle: Generalizations and applications", *Proc. V International Conference: Physical Interpretations of Relativity Theory*, Ed. M. C. Duffy, pp. 134-170, London, September 11-14 , 1998 (also in: *Nelinijni Koluvannya*, Vol. 2, No. 4, pp. 465-489, Kiev, Ukraine, 1999).

[173] Ly. T. Gruyitch, "*Time*, Systems and Control", invited, submitted and accepted paper has had 50 pages, its abstract was published in *Abstracts of the papers of the VIII International seminar "Stability and oscillations of nonlinear control systems*, Ed. V. N. Thai, Moscow: IPU RAN, ISBN 5-201-14972-3, June 2-4, 2004.

[174] Ly. T. Gruyitch, "Time, systems, and control: Qualitative properties and methods", Chapter 2 in *Stability and Control of Dynamical Systems with Applications*, Eds. D. Liu and P. J. Antsaklis, Boston: Birkhâuser, pp. 23-46, 2003.

[175] Ly. T. Gruyitch, "Vector Lyapunov function synthesis of aircraft control", *Proceedings of INPAA-98: Second International Conference on Nonlinear Problems in Aviation & Aerospace*, Ed. Seenith Sivasundaram, ISBN: 0 9526643 1 3, Cambridge UK: European Conference Publications, Vol. 1, pp. 253-260, 1999.

[176] Ly. T. Gruyitch and W. P. Mounfield, Jr., "Absolute output natural tracking control: MIMO Lurie systems", *Proceedings of the 14th Triennial World Congress*, Beijing, P. R. China, Pergamon-Elsevier Science, Vol. C, pp. 389-394, July 5-9, 1999.

[177] Ly. T. Gruyitch and W. P. Mounfield, Jr., "Constrained natural tracking control algorithms for bilinear DC shunt wound motors", *Proceedings of the 40th IEEE Conference on Decision and Control*, Orlando, FL USA, pp. 4433-4438, December 2001.

[178] Ly. T. Gruyitch and W. P. Mounfield, Jr., "Elementwise stablewise tracking with finite reachability time: linear time-invariant continuous-time MIMO systems", *International Journal of Systems Science*, Vol. 33, No. 4, pp. 277-299, 2002.

[179] Ly. T. Gruyitch and W. P. Mounfield, Jr., "Robust elementwise exponential tracking control: IO linear systems", *Proceedings of the 36th IEEE Conference on Decision and Control*, San Diego, CA, USA, pp. 3836-3841, December 1997.

[180] Ly. T. Gruyitch and W. Pratt Mounfield, Jr., "Stablewise absolute output natural tracking control with finite reachability time: MIMO Lurie systems", *CD ROM Proceedings of the 17th IMACS World Congress, Invited session IS-2: Tracking Theory and Control of Nonlinear Systems*, Paris, France, pp. 1-17, July 11-15, 2005; *Mathematics and computers in simulation*, Vol. 76, pp. 330-344, 2008.

[181] Ly. T. Gruyitch, J.-P. Richard, P. Borne and J.-C. Gentina, *Stability Domains*, Boca Raton, FL, USA: Chapman & Hall/CRC, 2004.

[182] S. C. Gupta and R. J. Solem, "Accurate error analysis of a satellite tracking control loop system", *International Journal of Control*, Vol. 2, No. 6, pp. 539-549, 1965.

[183] I. J. Ha and E. G. Gilbert, "Robust Tracking in Nonlinear Systems", *IEEE Transactions on Automatic Control*, Vol. AC-32, No. 9, pp. 763-771, 1987.

[184] W. Hahn, *Stability of Motion*, New York: Springer-Verlag, 1967.

[185] S. Hara and T. Sugie, "Independent parametrization of two-degree-of-freedom compensators in general robust tracking systems", *IEEE Transactions on Automatic Control*, Vol. 33, No. 1, pp. 59-67, 1988.

[186] M. L. J. Hautus, "Controllability and observability conditions of linear autonomous systems", *Proceedings of the Nederland Academy of Science: Mathematics*, Ser. A, Vol. 72, pp. 443-448, 1969.

[187] M. A. Henson and D. E. Seborg, "An internal model control strategy for nonlinear systems", *AIChE Journal*, Vol. 37, No. 7, pp. 1065-1081, 1991.

[188] R. M. Hirschorn, "Output tracking in multivariable nonlinear systems", *IEEE Transactions on Automatic Control*, Vol, AC-26, No. 6, pp. 593-595, 1981.

[189] R. M. Hirschorn and J. H. Davis, "Global output tracking for nonlinear systems", *SIAM Journal of control and optimization*, Vol. 26, No. 6, pp. 1321-1330, 1988.

[190] P.-Y. Huang and B.-S. Chen, "Robust tracking of linear MIMO time-varying systems", *Automatica*, Vol. 30, No. 5, pp. 817-830, 1994.

[191] R. O. Hughes, "Optimal control of sun tracking solar concentrators", *Journal of Dynamic Systems, Measurement, and Control*, Vol. 101, No. 2, pp. 157-161, 1979.

[192] C.-L. Hwang, Y.-M. Chen and C. Jan, "Trajectory tracking of large-displacement piezoelectric actuators using a nonlinear observer-based variable structure control", *IEEE Transactions on Control Systems Technology*, Vol. 13, No. 1, pp. 56-66, January 2005.

[193] M. Indri and A. Tornambè, "Robust asymptotic and disturbance rejection of a class of MIMO linear systems", *IFAC Conference: System Structure and Control*, Nantes, pp. 416-421, 1995.

[194] A. Isidori and C. I. Byrnes, "Output regulation of nonlinear systems", *IEEE Transactions on Automatic Control*, Vol. 35, No. 2, pp. 131-140, 1990.

[195] S. Jayasuriya, "Robust tracking for a class of uncertain linear systems", *Int. J. Control*, Vol. 45, No. 3, pp. 875-892, 1987.

[196] S. Jayasuriya and C.-D. Kee, "Circle-type criterion for synthesis of robust tracking controllers", *Int. J. Control*, Vol. 48, No. 3, pp. 865-886, 1988.

[197] T. Kaczorek, "Dead-beat servo problem for 2-dimensional linear systems", *International Journal of Control*, Vol. 37, No. 6, pp. 1349-1353, 1983.

[198] T. Kailath, *Linear Systems*, Englewood Cliffs, NJ, USA: Prentice-Hall, Inc., 1980.

[199] R. E. Kalman, "Algebraic structure of linear dynamical systems, I. The module of Σ", *Proceedings of the National Academy of Science: Mathematics*, USA NAS, Vol. 54, pp. 1503-1508, 1965.

[200] R. E. Kalman, "Canonical structure of linear dynamical systems", *Proceedings of the National Academy of Science: Mathematics*, USA NAS, Vol. 48, pp. 596-600, 1962.

[201] R. E. Kalman, "Mathematical description of linear dynamical systems", *J.S.I.A.M. Control*, Ser. A, Vol. 1, No. 2, pp. 152-192, 1963.

[202] R. E. Kalman, "On the general theory of control systems", *Proceedings of the First International Congress on Automatic Control*, pp. 481-491, London: Butterworth, 1960.

[203] R. E. Kalman, Y. C. Ho and K. S. Narendra, "Controllability of linear dynamical systems", *Contributions to Differential Equations*, Vol. 1, No. 2, pp. 189-213, 1963.

[204] D. E. Koditschek, "Application of a new Lyapunov function to global adaptive attitude tracking", *Proceedings of the 27th IEEE Conference on Decision and Control*, pp. 63-68, 1988.

[205] A. Kojima and S. Ishijim, "H_∞ preview tracking in output feedback setting", *International Journal of Robust and Nonlinear Control*, Vol. 14, pp. 627-641, 2004.

[206] A. Kökösy, *Poursuite Pratique de Systemes de commande Automatique des Robots Industriels*, Ph. D. dissertation, Belfort, France: University of Belfort-Montbeliard, 1999.

[207] A. Kőkősy, "Practical tracking with settling time: Bounded control for robot motion", *Proceedings of the 14th IFAC Triennial World Congress*, Beijing, P. R. China, C-2a-11-4, pp. 377–382, 1999.

[208] A. Kőkősy, "Practical tracking with settling time: Criteria and algorithms", *Proceedings of the VI International SAUM Conference on Systems, Automatic Control and Measurement*, Nish, Serbia, Yugoslavia, pp. 296–301, September 28–30, 1998.

[209] A. Kőkősy, "Practical tracking with vector settling and vector reachability time", *Proceedings of the IFAC Workshop on Motion Control*, Grenoble, France, pp, 297–302, Sept. 21–23, 1998.

[210] A. Kőkősy, "Robot control: Practical tracking with reachability time", *Proceedings of the International Conference on System, Signals, Control, Computers*, Vol. 3, Durban, RSA, pp. 186-190, 1998.

[211] B. Kouvaritakis, "Gain margins and root locus asymptotic brhaviour in multivariable design. Part I. The properties of the Markov parameters and the use of high feedback gain", *Int. J. Control*, Vol. 27, No. 5, pp. 705-724, 1978.

[212] N. N. Krasovskii, *Some Problems of the Theory of Stability of Motion*, in Russian, Moscow: FIZMATGIZ, 1959.

[213] N. N. Krasovskii, *Stability of Motion*, Stanford, CA, USA: Stanford University Press, 1963.

[214] B. C. Kuo, *Automatic Control Systems*, Englewood Cliffs, NJ, USA: Prentice-Hall, Inc., 1967

[215] B. C. Kuo, *Automatic Control Systems*, Englewood Cliffs, NJ, USA: Prentice-Hall, Inc., 1987.

[216] H. Kwakernaak and R. Sivan, *Linear Optimal Control Systems*, New York: Wiley-Interscience, 1972.

[217] W. H. Kwon and D. G. Byun, "Receding horizon tracking control as a predictive control and its stability properties", *International Journal of Control*, Vol. 50, No. 5, pp. 1807-1824, 1989.

[218] J. P. LaSalle, *The Stability of Dynamical Systems*, Philadelphia, PA, USA: Society for Industrial and Applied Mathematics, 1976.

[219] H. Lauer, R. Lesnick and L. E. Matson, *Servomechanism Fundamentals*, New York, NY, USA: McGraw-Hill Book Company, Inc., 1947.

[220] D. V. Lazitch, *Analysis and Synthesis of Practical Tracking Automatic Control* (in Serb), D. Sci. Dissertation, Faculty of Mechanical Engineering, University of Belgrade, Belgrade, Serbia, 1995.

[221] D. V. Lazitch, "Uniform exponential practical automatic control tracking" (in Serb), *Proceedings of the V Conference on Systems, Automatic Control and Measurement (SAUM)*, Novi Sad, pp. 68-70, October 2-3, 1995.

[222] D. V. Lazitch, "Uniform practical automatic control tracking" (in Serb), *Proceedings of the V Conference on Systems, Automatic Control and Measurement (SAUM)*, Novi Sad, pp. 53-57, October 2-3, 1995.

[223] D. V. Lazitch, "Uniform practical automatic control tracking with the vector reachability time" (in Serb), *Proceedings of the V Conference on Systems, Automatic Control and Measurement (SAUM)*, Novi Sad, pp. 63-67, October 2-3, 1995.

[224] D. V. Lazitch, "Uniform practical automatic control tracking with the vector settling time", (in Serb), *Proceedings of the V Conference on Systems, Automatic Control and Measurement (SAUM)*, Novi Sad, pp. 58-62, October 2-3, 1995.

[225] M. D. Leviner and D. M. Dawson, "Position and force tracking control of rigid-link electrically driven robots actuated by switched reluctance motors", *International Journal of Systems Science*, Vol. 26, No. 8, pp. 1479-1500, 1995.

[226] C.-M. Lin and T.-D. Meng, "Simultaneous deadbeat tracking control of two plants", *Int. J. Control*, Vol. 67, No. 6, pp. 921-931, 1997.

[227] J.-S. Liu and K. Yuan, "On tracking control for affine nonlinear systems by sliding mode", *Systems and Control Letters*, Vol. 13, pp. 439-443, 1989.

[228] P.-T. Liu and P. L. Bongiovanni, "On a passive vehicle tracking problem and max-minimization", *IEEE Transactions on Automatic Control*, Vol. AC-28, No. 2, pp. 269-304, 1983.

[229] A. G. Loukianov, B. Castillo-Toledo, J. E. Hernández and E. Núñez-Perez, "On the problem of tracking for a class of linear systems with delays and sliding modes", *International Journal of Control*, Vol. 76, No. 9/10, pp. 942-958, 2003.

[230] R. Lozano, "Independent tracking and regulation adaptive control with forgetting factor", *Automatica*, Vol. 18, No. 4, pp. 455-459, 1982.

[231] J. C. Lozier, "A steady state approach to the theory of saturable servo systems", *IRE Transactions on Automatic Control*, Vol. 1, No. 1, pp. 19-39, May 1956.

[232] A. M. Lyapunov, *The General Problem of Stability of Motion* (in Russian), Kharkov Mathematical Society, Kharkov, 1892; in Academician A. M. Lyapunov: "Collected Papers", U.S.S.R. Academy of Science, Moscow, II, pp. 5-263, 1956. French translation: "Problème général de la stabilité du mouvement", *Ann. Fac. Toulouse*, Vol. 9, pp. 203-474, 1907; also in *Annals of Mathematics Study*, No. 17, Princeton University Press, 1949. English translation: *Intern. J. Control*, Vol. 55, pp. 531-773, 1992; also the book, Taylor and Francis, London, 1992.

[233] L. A. MacColl, *Fundamental Theory of Servomechanisms*, New York: D. Van Nostrand Company, Inc., 1945.

[234] J. M. Maciejowski, *Multivariable Feedback Systems,* Wokingham, England: Addison-Wesley Publishing Company, *1989.*

[235] L. Marconi and A. Isidori, "Mixed internal model-based and feedforward control for robust tracking in nonlinear systems", *Automatica*, Vol. 36, pp. 993-1000, 2000.

[236] V. M. Matrosov, "To the theory of stability of motion" (in Russian), *Prikl. Math. Mekh.* Vol. 26, No. 5, pp. 885-895, 1962.

[237] V. M. Matrosov, *Vector Lyapunov Function Method: Analysis of Dynamical Properties of Nonlinear Systems* (in Russian), Moscow, Russia: FIZMATLIT, 2001.

[238] V. M. Matrosov, "Vector Lyapunov functions in the analysis of nonlinear interconnected systems", *Proc. Symp. Mathematica*, Bologna, Italy, Vol. 6, pp. 209-242, 1971.

[239] G. P. Matthews and R. A. DeCarlo, "Decentrelized tracking for a class of interconnected nonlinear systems using variable structure control", *Automatica*, Vol. 24, No. 2, pp.187-193, 1988.

[240] J. L. Melsa and D. G. Schultz, *Linear Control Systems*, New York, NY, USA: McGraw-Hill Book Company, 1969.

[241] A. N. Michel and C. J. Herget, *Algebra and Analysis for Engineers and Scientists*, Boston, MA, USA: Birkhäuser, 2007.

[242] A. N. Michel, L. Hou and D. Liu, *Stability of Dynamical Systems. Continuous, Discontinuous and Discrete Systems*, Boston-Basel-Berlin: Birkhäuser, 2008.

[243] A. N. Michel and R. K. Miller, *Qualitative Analysis of Large-Scale Dynamical Systems*, New York, NY, USA: Academic Press, 1977

[244] D. E. Miller and E. J. Davison, "The self-tuning robust servomechanism problem", *IEEE Transactions on Automatic Control*, Vol. AC-34, No. 5, pp. 511-523, May 1989.

[245] R. K. Miller and A. N. Michel, *Ordinary Differential Equations*, New York, NY, USA: Academic Press, 1982.

[246] B. R. Milojković and Lj. T. Grujić, *Automatic Control* (in Serbo-Croatian), Belgrade, Serbia: Faculty of Mechanical Engineering, University of Belgrade, 1977.

[247] N. Minamide, "Design of a deadbeat adaptive tracking system", *International Journal of Control*, Vol. 39, No. 1, pp. 63 – 81, 1984.

[248] T. Mori and M. Kuwahara, "Estimate for the root-location of linear systems via the Lyapunov matrix equation", *Journal of the Franklin Institute*, Vol. 314, No. 2, pp. 123-127, 1982.

[249] W. P. Mounfield, Jr. and Lj. T. Grujić, "High-gain natural tracking control of linear systems", *Proceedings of the 13th IMACS World Congress on Computation and Applied Mathematics*, Eds. R. Vichnevetsky and J. J. H. Miller, Trinity College, Dublin, Ireland, Vol. 3, pp. 1271-1272, July 22-26, 1991.

[250] W. P. Mounfield, Jr. and Lj. T. Grujić, "High-gain natural tracking control of time-invariant systems described by IO differential equations", *Proceedings of the 30th Conference on Decision and Control*, Brighton, England, pp. 2447-2452, 1991.

[251] W. P. Mounfield and Lj. T. Grujić, "High-gain PI control of an aircraft lateral control system", *Proc. 1993 IEEE International Conference on Systems, Man and Cybernetics*, Le Touquet, Vol. 2, pp. 736-741, 1993.

[252] W. P. Mounfield, Jr. and Lj. T. Grujić, "High-gain PI natural tracking control for exponential tracking of linear MIMO systems with state-space description", *International Journal of Control*, Vol. 25, No. 11, pp. 1793-1817, 1994.

[253] W. P. Mounfield, Jr. and Lj. T. Grujić, "High-gain PI natural tracking control for exponential tracking of linear single-output systems with state-space description", *RAIRO-Automatique, Productique, Informatique Industrielle (APII)*, Vol. 26, pp. 125-146, 1992.

[254] W. P. Mounfield and Lj. T. Grujić, "Natural tracking control for exponential tracking: Lateral high-gain PI control of an aircraft system with state-space description", *Neural, Parallel & Scientific Computations*, Vol. 1, No. 3, pp. 357-370, 1993.

[255] W. P. Mounfield and Lj. T. Grujić, "PID-Natural tracking control of a robot: Application", *Proc. 1993 IEEE International Conference on Systems, Man and Cybernetics*, Le Touquet, Vol. 4, pp. 328-333, 1993.

[256] W. P. Mounfield and Lj. T. Grujić, "Robust natural tracking control for multi-zone space heating systems", *Proc. 14th IMACS World Congress*, Vol. 2, pp. 841-843, 1994.

[257] W. P. Mounfield, Jr. and Ly. T. Gruyitch, "Control of aircrafts with redundant control surfaces: Stablewise tracking control with finite reachability time", *Proceedings of the Second International Conference on Nonlinear Problems in Aviation and Aerospace*, European Conference Publishers, Cambridge, Vol. 2, 1999, pp. 547-554, 1999.

[258] W. P. Mounfield, Jr. and Ly. T. Gruyitch, "Elementwise stablewise finite reachability time natural tracking control of robots", *Proceedings of the 14th Triennial World Congress*, Beijing, P. R. China, Pergamon-Elsevier Science, Vol. B, pp. 31-36, July 5-9, 1999.

[259] N. G. Nath, A. K. Bhattacharyya, and A. K. Choudhury, "Dead-beat control of third-order servo-systems", *International Journal of Control*, Vol. 2, No. 4, pp. 385-394, 1965.

[260] D. B. Nauparac, *Analysis of the Tracking Theory on the Real Electrohydraulic Servosystem* (in Serb), M. Sci. Thesis, Faculty of Mechanical Engineering, University of Belgrade, Belgrade Serbia, 1993.

[261] N. N. Nedić and D. H. Pršić, "Pneumatic and hydraulic time varying desired motion control using natural tracking control", *SAI-Avtomatika i Informatika '2000*, Sofia, Vol. 3, pp. 37-40, 24 -26 ocktomvri, 2000.

[262] N. N. Nedić and D. H. Pršić, "Pneumatic position control using natural tracking law", *IFAC Workshop on Trends in Hydraulic and Pneumatic Components and Systems*, Chicago, IL, USA, pp. 1-13, Nov. 8-9, 1994.

[263] N. N. Nedić and D. H. Pršić, "Time variable speed control of pump controlled hydraulic motor using natural tracking control", *IFAC-IFIP-IMACS International Conference on Control of Industrial Systems*, Belfort, France, pp. 197-202, May 20-22, 1997.

[264] Sir Isaac Newton, *Mathematical Principles of Natural Philosophy-Book I. The Motion of Bodies*, Encyclopaedia Britannica, Inc., Chicago, IL, USA, William Benton, 1952 (the first publication: 1687).

[265] M.-L. Ni and Y. Chen, "Decentralized stabilization and output tracking of large-scale uncertain systems", *Automatica*, Vol. 32, No. 7, pp. 1077-1080, 1996.

[266] S. Nicosia and P. Tomei, "Tracking control with disturbance attenuation for robot manipulators", *Int. J. of Adaptive Control and Signal Processing*, Vol. 10, pp. 443-449, 1996.

[267] Z. Novaković and Lj. T. Grujić, "Robot control: Stablewise tracking with reachability time under load variations", *Proceedings of the Sixth IASTED International Symposium on Modelling, Identification and Control*, Grindelwald, Switzerland, pp. 223-226, February 17-20, 1987.

[268] K. Ogata, *Modern Control Engineering*, Englewood Cliffs, NJ, USA: Prentice Hall, 1970.

[269] K. Ogata, *State Space Analysis of Control Systems*, Englewood Cliffs, NJ, USA: Prentice Hall, 1967.

[270] D. H. Owens, *Feedback and Multivariable Systems*, Stevenage, Herts: Peter Peregrinus Ltd., 1978.

[271] K. Özçaldiran and V. Eldem, "A complete classification of minimal realizations of nonsingular and strictly proper transfers under similarity transformations", *Automatica*, Vol. 33, No. 5, pp. 835-850, 1997.

[272] M. Pachter, P. R. Chandler and M. Mears, "Reconfigurable tracking control with saturation", *Journal of Guidance, Control and Dynamics*, Vol. 18, No. 5, pp. 1016-1022, 1995.

[273] V. Parra-Vega, S. Arimoto, Y.-H. Liu, G. Hirzinger and P. Akella, "Dynamic sliding PID control for tracking of robot manipulators: Theory and experiments", *IEEE Transactions on Robotics and Automation*, Vol. 19, No. 6, pp. 967-976, December 2003.

[274] K. M. Passino, "Disturbance rejection in nonlinear systems: Examples", *IEEE Proceedings*, Vol. 136, No. 6, pp. 317-323, 1989.

[275] J. B. Plant, Y. T. Chan and D. A. Redmond ,"A discrete tracking control law for nonlinear plants", *IFAC Control Science and Technology, 8th Triennial World Congress*, Kyoto, Japan, Vol. 2, pp. 55-60, 1981.

[276] B. Porter, "Fast-sampling tracking systems incorporating Lur'e plants with multiple nonlinearities", *Int. J. Control*, Vol. 34, No. 2, pp. 333-344, 1981.

[277] B. Porter, "High-gain tracking systems incorporating Lur'e plants with multiple nonlinearities", *Int. J. Control*, Vol. 34, No. 2, pp. 345-358, 1981.

[278] B. Porter, "High-gain tracking systems incorporating Lur'e plants with multiple switching nonlinearities", *VIII IFAC World Congress*, Kyoto, Japan, Session 3, pp. I-72-I-77, 1981.

[279] B. Porter, T. R. Crossley and A. Bradshaw, "Synthesis of disturbance-rejection controllers for linear multivariable continuous-time systems", *Israel Journal of Technology*, Vol. 13, pp. 25-30, 1975.

[280] B. Porter and Lj. T. Grujić, "Continuous-time tracking systems incorporating Lur'e plants with multiple non-linearities", *Int. J. Systems Science*, Vol. 11, No. 7, pp. 827-840, 1980.

[281] B. Porter and Lj. T. Grujić, "Discrete-time tracking systems incorporating Lur'e plants with single non-linearities", *Third IMA Conference on Control Theory*, Academic Press, London, pp. 115-133, 1981.

[282] H. M. Power and R. J. Simpson, *Introduction to Dynamics and Control*, London: McGraw-Hill Book Company (UK) Limited, 1978.

[283] F. D. Priscoli, "Sufficient conditions for robust tracking in nonlinear systems", *Int. J. Control*, Vol. 67, No. 5, pp. 825-836, 1997.

[284] J. G. Proakis and D. G. Manolakis, *Digital Signal Processing Principals: Algorithms and Applications*, third edition, Englewood Cliffs, NJ, USA: Prentice Hall, 1996.

[285] D. Prshitch and N, Neditch, "Pneumatic cylinder control by using natural control" (in Serb), *HIPNEF'93*, Belgrade, Serbia, pp. 127-132, 1993.

[286] D. B. Reid, "An algorithm for tracking multiple targets", *IEEE Transactions on Automatic Control*, Vol. AC-24, No. 6, pp. 843-854, December 1979.

[287] W. T. Reid, "Some elementary properties of proper values and proper vectors of matrix functions", *SIAM J. Appl. Math.*, Vol. 18, No. 2, pp. 259-266, 1970.

[288] Z. B. Ribar, D. V. Lazic and M. R. Jovanovic, "Application of practical exponential tracking in fluid transportation industry", *Proceedings of the 14th International Conference on Material Handling and Warehousing*, Belgrade, Serbia, pp. 5.65-5.70, December 11-12, 1996.

[289] Z. B. Ribar, M. R. Yovanovitch and R. Z. Yovanovitch, "Application of practical linear tracking in process industry" (in Serb), *Proceedings of the XLI Conference ETRAN*, Zlatibor, Serbia, Notebook 1, pp. 444-447, June 3-6, 1997.

[290] M. Rios-Bolívar, A. S. I. Zinober and H. Sira-Ramírez, "Dynamical adaptive sliding mode output tracking control of a class of nonlinear systems", *Int. J. Robust and Nonlinear Control*, Vol. 7, pp. 387-405, 1997.

[291] H. H. Rosenbrock, *State-Space and Multivariable Theory*, London: Thomas Nelson and Sons Ltd., 1970.

[292] R. Saeks and J. Murray, "Feedback system design: The tracking and disturbance rejection problems", *IEEE Transactions on Automatic Control*, Vol. AC-26, No. 3, pp. 203-217, 1981.

[293] W. E. Schmitendorf, "Methods for obtaining robust tracking control laws", *Automatica*, Vol. 23, No. 5, pp. 675-677, 1987.

[294] H. Seraji, "Transfer function matrix", *Electronic Letters*, Vol. 23, No.4, pp. 256-257, 1987.

[295] I. A. Shkolnikov and Y. B. Shtessel, "Tracking in a class of nonminimum-phase systems with nonlinear internal dynamics via sliding mode control using method of system center", *Automatica*, Vol. 38, pp. 837-842, 2002.

[296] Y. B. Shtessel, "Nonlinear nonminimum phase output tracking via dynamic sliding manifolds", *J. Franklin Inst.*, 335B, No. 5, pp. 841-850, 1998.

[297] Y. B. Shtessel, "Nonlinear output tracking in conventional and dynamic sliding manifolds", *IEEE Transactions on Automatic Control*, Vol. 42, No. 9, pp. 1282-1286, September 1997.

[298] D. D. Šiljak, *Large-scale dynamic systems: Stability and structure*, New York: North Holland, 1978.

[299] C. Silvestre, A. Pascoal and I. Kaminer, "On the design of gain-scheduled trajectory tracking controllers", *Int. J. Robust and Nonlinear Control*, Vol. 12, pp. 797-839, 2002.

[300] R. E. Skelton, *Dynamic Systems Control. Linear Systems Analysis and Synthesis*, New York: John Wiley & Sons, 1988.

[301] J. J. Slotine and S. S. Sastry, "Tracking control of non-linear systems using sliding surfaces, with application to robot manipulators", *Int, J. Control*, Vol. 38, No. 2, pp. 465-492, 1983.

[302] E. D. Sontag, "Further facts about input to state stabilization", *IEEE Transactions on Automatic Control*, Vol. 35, No. 3, pp. 473-476, 1990.

[303] E. D. Sontag, *Mathematical Control Theory: Deterministic Finite Dimensional Systems*, New York: Springer, 1990.

[304] E. D. Sontag, "Remarks on stabilization and input-to-state stability", *Proc. IEEE Conference on Decision and Control*, Tampa, FL, USA, *IEEE Publications*, pp. 1376-1378, Dec. 1989.

[305] M. T. Söylemez and İ. Üstoğlu, "Polynomial control systems", *IEEE Control Systems Magazine*, Vol. 27, No. 4, pp.124-137, August 2007.

[306] S. K. Spurgeon and X. Y. Lu, "Output tracking using dynamic sliding mode techniques", *International Journal of Robust and Nonlinear Control*, Vol. 7, pp. 407-427, 1997.

[307] M. Y. Stoychitch, *Practical Tracking of Digital Control Systems* (in Serb), D. Sci. Dissertation, Faculty of Mechanical Engineering, University of Banya Luka, Banya Luka, Republic of Serb, 2004.

[308] T. Sugie and M. Vidyasagar, "Further results on the robust tracking problem in two-degree-of-freedom control systems", *Systems & Control Letters*, Vol. 13, pp. 101-108, 1989.

[309] A. I. Talkin, "Adaptive servo tracking", *IRE Transactions on Automatic Control*, Vol. 6, No. 2, pp.167-172, May 1961.

[310] S. Tarbouriech, C. Pittet and C. Burgat, "Output tracking problem for systems with input saturations via nonlinear integrating actions", *Int. J. Robust and Nonlinear Control*, Vol. 10, pp. 489-512, 2000.

[311] B. O. S. Teixeira, M. A. Santillo, R. S. Erwin and D. S. Bernstein, "Spacecraft tracking using sampled-data Kalman filters", *IEEE Control Systems Magazine,* Vol. 28, No. 4, pp. 78-94, August 2008.,

[312] H. T. Toivonen and J. Pensar, "A worst-case approach to optimal tracking control with robust performance", *Int. J. Control*, Vol. 65, No. 1, pp. 17-32, 1996.

[313] A. Uraz and F. L. N-Nagy, "Matrix formulation for partial-fraction expansion of transfer functions", *Journal of the Franklin Institute,* Vol. 297, No. 2, pp. 81-87, 1974.

[314] A. F. Vaz and E, J. Davison, "The structured robust decentralized servomechanism problem for interconnected systems", *Automatica*, Vol. 25, No. 2, pp. 267-272, 1989.

[315] S.-S. Wang and B-S Chen, "Simultaneous deadbeat tracking controller synthesis", *International Journal of Control*, Vol. 44, No. 6, pp. 1579-1586, 1986.

[316] P. E. Wellstead and P. Zanker, "Servo self-tuners", *International Journal of Control*, Vol. 30, No. 1, pp. 27-36, 1979.

[317] J. C. West, *Textbook of Servomechanisms*, London: English Universities Press, 1953.

[318] D. M. Wiberg, *State Space and Linear Systems*, New York: McGraw-Hill Book Company, 1971.

[319] W. A. Wolovich, *Linear Multivariable Systems*, New York: Springer-Verlag, 1974.

[320] W. M. Wonham, *Linear Multivariable Control. A Geometric Approach*, Berlin: Springer-Verlag, 1974.

[321] J.-M. Yang and J.-H. Kim, "Sliding mode control for trajectory tracking nonholonomic wheeled mobile robots", *IEEE Transactions on Robotics and Automation*, Vol. 15, No. 3, pp. 578-587, June 1999.

[322] R. Yokoyama, "Transfer function matrix of linear, multi-input and multi-output systems", *Int. J. Control*, Vol. 18, No. 2, pp. 369-375, 1973.

[323] T. Yoshikawa, T. Sugie and H. Hanafusa, "Synthesis of robust tracking systems with specified transfer matrices", *Int. J. Control*, Vol. 43, No. 4, pp. 1201-1214, 1986.

[324] K. Youcef-Toumi and O. Ito, "A time delay controller for systems with unknown dynamics", *Journal of Dynamic Systems, Measurement, and Control*, Vol. 112, pp. 133-142, 1990.

[325] M. R. Yovanovitch, *Practical Tracking Automatic Control of the Axial Piston Hydraulic Motors* (in Serb), M. Sci. Thesis, Faculty of Mechanical Engineering, University of Belgrade, Belgrade, Serbia, 1998.

[326] W.-S. Yu and Y.-H. Chen, "Decoupled variable structure control design for trajectory tracking on mechatronic arms", *IEEE Transactions on Control Systems Technology*, Vol. 13, No. 5, pp. 798-806, September 2005.

[327] S. Y. Zhang and C. T. Chen, "Design of compensators for robust tracking and disturbance rejection", *IEEE Transactions on Automatic Control*, Vol. AC-30, No. 7, pp. 684-687, July 1985.

[328] J. Zhao and I. Kanellakopoulos, "Flexible backstepping design for tracking and disturbance attenuation", *International Journal of Robust and Nonlinear Control*, Vol. 8, pp. 331-348, 1998.

Part VI

APPENDIXES

Appendix A

Notation

The meaning of the notation is explained in the text at its first use.

A.1 Abbreviations

BI *Bounded-Input*

BIBO *Bounded-Input-Bounded-Output*

CS *control system*

FRT *finite (scalar) reachability time*

FVRT *finite vector reachability time*

IISO *Input-Initial State-Output*

IO *Input-Output*

IO system *the Input-Output system defined by (2.9)*

IS *Input-State*

ISO *Input-State-Output*

ISO system *the Input-State-Output system defined by (2.29), (2.30)*

LITC *Linear Tracking Control*

LTC *Lyapunov Tracking Control*

LY *Lyapunov's*

LY stability concept *Lyapunov stability concept*

MIMO *Multiple-Input-Multiple-Output*

NTC *Natural Tracking Control*

PCUP *Physical Continuity and Uniqueness Principle*

SISO *Single-Input-Single-Output*

TCUP *Time Continuity and Uniqueness Principle*

VLF *Vector Lyapunov Function*

A.2 Indexes

A.2.1 Subscripts

cnd *colomn nondegenerate*

CR *controller*

CS *control system*

d *desired*

e *equilibrium*

i *the i-th*

irr *irreducible*

j *the j-th*

nd *nondegenerate*

P *plant*

R *Rosenbrock*

rd *reduced*

rnd *row nondegenerate*

zero *the zero value*

0 the subscript 0 (zero) associated with a variable (.) denotes its initial value $(.)_0$; however, if $(.)\subset \mathfrak{T}$, then the subscript 0 (zero) associated with (.) denotes the *time* set $\mathfrak{T}_0$

A.2.2 SUPERSCRIPT

k *k-dimensional*, $k \in \{1, 2, ...m-1, ..., n, ...\}$

A.3 Letters

Lower case block or italic letters are used for scalars. Lower case bold block letters denote vectors. Upper case block letters denote matrices, or points. Upper case fraktur letters designate sets or spaces.

The notation $;t_{(.)0}$ will be omitted as an argument of a variable if, and only if, a choice of the initial moment $t_{(.)0}$ does not have any influence on the value of the variable.

A.3.1 Calligraphic letters

$\mathcal{CR}$ *controller,*

$\mathcal{CS}$ *control system,*

$\mathcal{L}^{\mp}\{\mathbf{I}(t)\}$ *left* $(-)$, *right* $(+)$, *respectively, Laplace transform of* $\mathbf{I}(t)$,

$$\mathcal{L}^{\mp}\{\mathbf{I}(t)\} = \mathbf{I}^{\mp}(s) = \int_{0^{\mp}}^{\infty} \mathbf{I}(t)e^{-st}dt = lim\left[\int_{\mp\zeta}^{\infty} \mathbf{I}(t)e^{-st}dt : \zeta \longrightarrow 0^{+}\right],$$

$\mathcal{L}^{-1}\{\mathbf{I}(s)\}$ *inverse Laplace transform of* $\mathbf{I}(s)$,

$$\mathcal{L}^{-1}\{\mathbf{I}(s)\} = \mathbf{I}(t) = \frac{1}{2\pi j}\int_{c-j\infty}^{c+j\infty} \mathbf{I}(s)e^{st}ds,$$

$\mathcal{P}$ *plant*

A.3.2 Fraktur letters

Capital fraktur letters are used for spaces or for sets.

$\mathfrak{B} \subseteq \mathfrak{R}^n$ *a nonempty subset of* $\mathfrak{R}^n$.

$\mathfrak{B}_\xi(\mathbf{z})$ *an open hyperball with the radius* ξ *centered at the point* $\mathbf{z}$ in the corresponding space,

$$\mathfrak{B}_\xi(\mathbf{z}) = \{\mathbf{w} : \ \|\mathbf{w} - \mathbf{z}\| < \xi\}$$

$\mathfrak{B}_\xi$ *an open hyperball with the radius* ξ *centered at the origin* of the corresponding space,

$$\mathfrak{B}_\xi = \mathfrak{B}_\xi(\mathbf{0})$$

$\mathfrak{C}$ *the set of all complex numbers*, or *the family of all continuous functions on* $\mathfrak{T}_0$

$\mathfrak{C}^{ki}(\mathfrak{S})$ *the family of all functions defined, continuous and k-times continuously differentiable on the set* $\mathfrak{S}$, $\mathfrak{S} \subseteq \mathfrak{R}^i$, $\mathfrak{C}^{ki}\left(\mathfrak{R}^i\right) = \mathfrak{C}^k\left(\mathfrak{R}^i\right) = \mathfrak{C}^{ki}$

$\mathfrak{C}^k = \mathfrak{C}^k(\mathfrak{T}_0)$ *the family of all functions defined, continuous and k-times continuously differentiable on* $\mathfrak{T}_0$, $\mathfrak{C}^0 = \mathfrak{C}$

$\mathfrak{C}^0(\mathfrak{S})$ *the family of all functions defined and continuous on the set* $\mathfrak{S}$, $\mathfrak{C}^0\left(\mathfrak{R}^i\right) = \mathfrak{C}^{0,i}$

$\mathfrak{C}^{k-}\left(\mathfrak{R}^i\right)$ *the family of all functions defined everywhere and k-times continuously differentiable on* $\mathfrak{R}^i$-$\{\mathbf{0}_i\}$, *which have defined and continuous derivatives at the origin* $\mathbf{0}_i$ of $\mathfrak{R}^i$ up to the order $(k-1)$, which are defined and continuous at *at the origin* $\mathbf{0}_i$ and have defined the left and the right *k-th* order derivative *at the origin* $\mathbf{0}_i$

$\mathfrak{C}^-(\mathfrak{T}_0)$ *the family of all piecewise continuous functions defined at every* $t \in \mathfrak{T}_0$

$\mathfrak{D}^k$ a given, or to be determined, *family of all bounded k-times continuously differentiable on* $\mathfrak{T}_0$ *permitted disturbance vector total functions* $\mathbf{D}(.)$, or *deviation functions* $\mathbf{d}(.)$, of which Laplace transforms are strictly proper real rational complex functions, , $\mathfrak{D}^k \subset \mathfrak{C}^k \cap \mathfrak{L}$

$\mathfrak{D}^k_-$ a subfamily of $\mathfrak{D}^k$, $\mathfrak{D}^k_- \subset \mathfrak{D}^k$, such that the real part of every pole of Laplace transform $\mathbf{D}(s)$ of every $\mathbf{D}(.) \in \mathfrak{D}^k_-$ is negative, $\mathfrak{D}_- = \mathfrak{D}^0_-$

$\mathfrak{D}^0 = \mathfrak{D}$ the *family of all bounded continuous permitted disturbance vector total functions* $\mathbf{D}(.)$, of which Laplace transforms are strictly proper real rational complex functions, $\mathfrak{D} \subset \mathfrak{C} \cap \mathfrak{L}$

$\mathfrak{I}^k$ a given, or to be determined, family of all *bounded and k-times continuously differentiable permitted input vector functions* $\mathbf{I}(.) \in \mathfrak{C}^k \cap \mathfrak{L}$ $\mathfrak{I}^k \subset \mathfrak{C}^k \cap \mathfrak{L}$

$\mathfrak{I}^0 = \mathfrak{I}$ the family of all bounded continuous permitted input vector functions $\mathbf{I}(.)$

$\mathfrak{I}^k_-$ a subfamily of $\mathfrak{I}^k$, $\mathfrak{I}^k_- \subset \mathfrak{I}^k$, such that the real part of every pole of Laplace transform $\mathbf{I}(s)$ of every $\mathbf{I}(.) \in \mathfrak{I}^k_-$ is negative, $\mathfrak{I}_- = \mathfrak{I}^0_-$

$\mathfrak{L}$ a family of all vector functions $\mathbf{I}(.)$, of which Laplace transforms are strictly proper real rational functions,

$$\mathfrak{L} = \left\{ \mathbf{I}(.) : \left(\begin{array}{c} \exists \gamma(\mathbf{I}) \in \mathfrak{R}^+ \Longrightarrow \|\mathbf{I}(t)\| < \gamma(\mathbf{I}),\ \forall t \in \mathfrak{T}_0, \\ \mathcal{L}^{\mp}\{\mathbf{I}(t)\} = \mathbf{I}^{\mp}(s) = \left[I_1^{\mp}(s)\ \ I_2^{\mp}(s)\ \ \ldots\ \ I_M^{\mp}(s) \right]^T, \\ I_k^{\mp}(s) = \dfrac{\sum\limits_{j=0}^{j=\zeta_k} a_{kj} s^j}{\sum\limits_{j=0}^{j=\psi_k} b_{kj} s^j}, 0 \leq \zeta_k < \psi_k,\ \forall k = 1, 2, ..., M \end{array} \right) \right\}$$

$\mathfrak{R}$ *the set of all real numbers*

$\mathfrak{R}^+$ *the set of all positive real numbers*

$\mathfrak{R}_+$ *the set of all nonnegative real numbers*

$\mathfrak{R}^{\nu N}$ *the extended output space of the IO system*, which is simultaneously the space of its internal dynamics, i.e., its *internal dynamics space*

$\mathfrak{R}^n$ an *n-dimensional real vector space, the state space of the ISO system*

$\mathfrak{R}^{N\nu} \backslash \mathfrak{B}_\varepsilon$ *the set of all vectors* $\mathbf{Y}^{\nu-1}$ *in* $\mathfrak{R}^{N\nu}$ *out of* $\mathfrak{B}_\varepsilon$,

$$\mathfrak{R}^{N\nu} \backslash \mathfrak{B}_\varepsilon = \left\{ \mathbf{Y}^{\nu-1} :\ \mathbf{Y}^{\nu-1} \in \mathfrak{R}^{N\nu},\ \mathbf{Y}^{\nu-1} \notin \mathfrak{B}_\varepsilon \right\}$$

$\mathfrak{T}$ *the accepted reference time set*, the arbitrary element of which is an arbitrary moment t and the *time* unit of which is second s, $1_t = s$, $t\langle s \rangle$,

$$\mathfrak{T} = \{t : t[T]\langle s \rangle,\ numt \in \mathfrak{R},\ dt > 0\},\ \ inf\ \mathfrak{T} = -\infty,\ \ sup\ \mathfrak{T} = \infty$$

$\mathfrak{T}_{0\mp}$ *the subset of* $\mathfrak{T}$, which has the minimal element $min\mathfrak{T}_{0\mp}$ that is the initial instant $t_{0\mp}$, $num\ t_{0\mp} = 0^{\mp}$,

$$\mathfrak{T}_{0\mp} = \{t : t \in \mathfrak{T},\ t \geq t_{0\mp},\ num\ t_{0\mp} \in \mathfrak{R}\}, \mathfrak{T}_{0\mp} \subset \mathfrak{T},$$
$$min\mathfrak{T}_{0\mp} = t_{0\mp} \in \mathfrak{T},\ \ sup\ \mathfrak{T}_{0\mp} = \infty$$

$\mathfrak{T}_0^i$ *the i-th order set product of* $\mathfrak{T}_0$,

$$\mathfrak{T}_0^i = \underbrace{\mathfrak{T}_0 \mathrm{x} \mathfrak{T}_0 \mathrm{x} \ldots \mathrm{x} \mathfrak{T}_0}_{i-times}$$

$\mathfrak{T}_A$ *the attainability time set related to the output space*,

$$\mathfrak{T}_A = \{t :\ 0 \leq t \leq \tau_A\} \subset \mathfrak{T}_0 \Longleftrightarrow 0 \leq \tau_A < \infty$$

$\mathfrak{T}_A^N$ *the vector attainability time set related to the output space*,

$$\mathfrak{T}_A^N = \left\{ \mathbf{t}^N :\ \mathbf{0}_N \leq \mathbf{t}^N \leq \boldsymbol{\tau}_A \right\} \subset \mathfrak{T}_0^N \Longleftrightarrow \mathbf{0}_N \leq \boldsymbol{\tau}_A < \infty \mathbf{1}_N$$

$\mathfrak{T}_{A\infty}$ *the postattainability time set related to the output space,*

$$\mathfrak{T}_{A\infty} = \{t: \ t \geq \tau_A\} \subset \mathfrak{T}_0 \Longleftrightarrow 0 \leq \tau_A < \infty,$$
$$\mathfrak{T}_{A\infty} = \{\infty\} \Longleftrightarrow \tau_A = \infty$$

$\mathfrak{T}^N_{A\infty}$ *the vector postattainability time set related to the output space,*

$$\mathfrak{T}^N_{A\infty} = \left\{\mathbf{t}^N: \ \mathbf{t}^N \geq \boldsymbol{\tau}_A\right\} \subset \mathfrak{T}^N_0 \Longleftrightarrow \mathbf{0}_N \leq \boldsymbol{\tau}_A < \infty\mathbf{1}_N,$$
$$\mathfrak{T}^N_{A\infty} = \{\infty\}^N = \{\infty\mathbf{1}_N\} \Longleftrightarrow \boldsymbol{\tau}_A = \infty\mathbf{1}_N$$

$\mathfrak{T}^{(k+1)N}_A$ *the general attainability time set related to the output space,*

$$\mathfrak{T}^{(k+1)N}_A = \left\{\mathbf{t}^{(k+1)N}: \ \mathbf{0}_{(k+1)N} \leq \mathbf{t}^{(k+1)N} \leq \boldsymbol{\tau}^{(k+1)N}_A\right\}$$

$\mathfrak{T}^{(k+1)N}_{A\infty}$ *the general postattainability time set related to the output space,*

$$\mathfrak{T}^{(k+1)N}_{A\infty} = \left\{\mathbf{t}^{(k+1)N}: \ \boldsymbol{\tau}^{(k+1)N}_A \leq \mathbf{t}^{(k+1)N} < \infty\mathbf{1}_{(k+1)N}\right\}$$

$\mathfrak{T}_R$ *the reachability time set related to the output space,*

$$\mathfrak{T}_R = \{t: 0 \leq t \leq \tau_R\}$$

$\mathfrak{T}_{R\infty}$ *the postreachability time set related to the output space,*

$$\mathfrak{T}_{R\infty} = \{t: \tau_R \leq t < \infty\}$$

$\mathfrak{T}^{(k+1)N}_R$ *the vector prereachability time set related to the output space,*

$$\mathfrak{T}^{(k+1)N}_R = \left\{\mathbf{t}^{(k+1)N}: \ \mathbf{0}_{(k+1)N} \leq \mathbf{t}^{(k+1)N} \leq \boldsymbol{\tau}^{(k+1)N}_R\right\}$$

$\mathfrak{T}^N_R$ *the vector prereachability time set for $k = 0$ related to the output space,*

$$\mathfrak{T}^N_R = \left\{\mathbf{t}^N: \ \mathbf{0}_N \leq \mathbf{t}^N \leq \boldsymbol{\tau}^N_R\right\} = \left[\mathbf{0}_N, \boldsymbol{\tau}^N_R\right] =$$
$$= [0, \tau_{R1}] \times [0, \tau_{R21}] \times ... \times [0, \tau_{RN}]$$

$\mathfrak{T}^{(k+1)N}_{R\infty}$ *the vector postreachability time set related to the output space,*

$$\mathfrak{T}^{(k+1)N}_{R\infty} = \left\{\mathbf{t}^{(k+1)N}: \ \boldsymbol{\tau}^{(k+1)N}_{R[k]} \leq \mathbf{t}^{(k+1)N} < \infty\mathbf{1}_{(k+1)N}\right\}$$

$\mathfrak{T}^N_{R\infty}$ *the vector postreachability time set for $k = 0$ related to the output space,*

$$\mathfrak{T}^N_{R\infty} = \left\{t\mathbf{1}_N : t \in \mathfrak{T}_0, \ t\mathbf{1}_N \geq \boldsymbol{\tau}^N_R\right\} = [\boldsymbol{\tau}^N_R, \infty\mathbf{1}_N[=$$
$$= [\tau_{R1}, \infty[\times[\tau_{R21}, \infty[\times ... \times[\tau_{RN}, \infty[$$

$\mathfrak{T}_R^n$ *the vector prereachability time set for* $k = 0$ *related to the state space,*

$$\mathfrak{T}_R^n = \{\mathbf{t}^n : \mathbf{0}_n \leq \mathbf{t}^n \leq \boldsymbol{\tau}_R^n\}$$

$\mathfrak{T}_{R\infty}^n$ *the vector postreachability time set for* $k = 0$ *related to the state space,*

$$\mathfrak{T}_{R\infty}^n = [\boldsymbol{\tau}_R^n, \infty\mathbf{1}_n[$$

$\mathfrak{U}^k$ a given, or to be determined, *family of all bounded and k-times continuously differentiable realizable control vector functions* $\mathbf{U}(.)$, $\mathfrak{U}^k \subseteq \mathfrak{C}^k$

$\mathfrak{U}^0 = \mathfrak{U}$ the *family of all bounded continuous realizable control vector functions* $\mathbf{U}(.)$, $\mathfrak{U}^0 \subseteq \mathfrak{C}^0$

$\mathfrak{V}_\mathbf{c} \subseteq \mathfrak{R}^N$ the largest open connected neighborhood of $\boldsymbol{\varepsilon} = \mathbf{0}_N$ such that a vector function $\mathbf{v}(.)$ and the set $\mathfrak{V}_\mathbf{c}$ obey

$$\mathbf{v}(\boldsymbol{\varepsilon}) < \mathbf{c},\ \forall \boldsymbol{\varepsilon} \in \mathfrak{V}_\mathbf{c},\ \mathbf{c} \in \mathfrak{R}^{+^N} \tag{A.1}$$

$Cl\mathfrak{V}_\mathbf{c}$ the closure of the set $\mathfrak{V}_\mathbf{c}$; $\partial\mathfrak{V}_\mathbf{c}$ is its boundary if the boundary exists

$\mathfrak{Y}_d^k$ a given, or to be determined, *family of all k-times continuously differentiable realizable desired total output vector functions* $\mathbf{Y}_d(.)$, $\mathfrak{Y}_d^k \subset \mathfrak{C}^{kN} \cap \mathfrak{L}$, of which Laplace transforms are strictly proper real rational complex functions

$\mathfrak{Y}_{d-}^k$ a subfamily of $\mathfrak{Y}_d^k$, $\mathfrak{Y}_{d-}^k \subset \mathfrak{Y}_d^k$, such that the real part of every pole of Laplace transform $\mathbf{Y}_d(s)$ of every $\mathbf{Y}_d(.) \in \mathfrak{Y}_{d-}^k$ is negative, $\mathfrak{Y}_{d-} = \mathfrak{Y}_{d-}^0$

$\mathfrak{Y}_d = \mathfrak{Y}_d^0$ the *family of all continuous realizable desired total output vector functions* $\mathbf{Y}_d(.)$, $\mathfrak{Y}_d = \mathfrak{Y}_d^0 \subseteq \mathfrak{C}^{0d}$, of which Laplace transforms are strictly proper real rational complex functions

$\mathfrak{Y}_{dy}^0 = \mathfrak{Y}_{dy}$ *the known family of all k-times continuously differentiable realizable desired deviation output vector functions* $\mathbf{y}_d(.) = \mathbf{Y}_d(.)\text{-}\mathbf{Y}_d(.) = \mathbf{0}_N$, i.e., it is singleton

$$\mathfrak{Y}_{dy} = \{\mathbf{0}_N\},\ \forall \mathbf{Y}_d(.) \in \mathfrak{Y}_d^k$$

A.3.3 Greek letters

α *a nonnegative integer*

β *a nonnegative integer*

γ *max* $\{\beta, \mu\}$

δ_{ij} *the Kronecker delta,* $\delta_{ij} = 1$ for $i = j$, and $\delta_{ij} = 0$ for $i \neq j$

$\boldsymbol{\delta}^j \in \mathfrak{R}^{+^j}$ *elementwise positive vector* $\boldsymbol{\delta}^j$ of the dimension j

$\Delta_{i(j)}$ *a positive real number, or* $\Delta_{i(j)} = \infty$, *associated with the j-th derivative of* Y_i *and of* Y_{di}, and it is taken for the entry of the positive kN

vector $\mathbf{\Delta}^{kN}$, when tracking is considered in the extended output space $\mathfrak{R}^{kN}$,

$$\mathbf{\Delta}_{(j)}^{N}=\begin{bmatrix}\Delta_{1,(j)}\\ \Delta_{2,(j)}\\ ...\\ \Delta_{N,(j)}\end{bmatrix},\ \forall j=0,1,2,...,k,\ \Delta_{i,(0)}\equiv\Delta_i,\ \mathbf{\Delta}_{(0)}^{N}\equiv\mathbf{\Delta}^{N},$$

$$\mathbf{\Delta}^{kN}=\begin{bmatrix}\mathbf{\Delta}_{(0)}^{N}\\ \mathbf{\Delta}_{(1)}^{N}\\ ...\\ \mathbf{\Delta}_{(k-1)}^{N}\end{bmatrix}=\begin{bmatrix}\mathbf{\Delta}^{N}\\ \mathbf{\Delta}_{(1)}^{N}\\ ...\\ \mathbf{\Delta}_{(k-1)}^{N}\end{bmatrix},\ k\in\{1,2,..,\nu\}$$

$\mathbf{\Delta}_{IO[k]}^{\nu N}$ *a vector associated with* $\mathbf{Y}^{\nu-1}$, i.e., related to the IO system, which is defined by

$$\mathbf{\Delta}_{IO(k)}^{\nu N}=[\Delta_1\ ...\ \Delta_N\ \ \Delta_{1(1)}\ ...\ \Delta_{N(1)}\ ....\ \Delta_{1(\nu-1)}\ ...\ \Delta_{N(\nu-1)}]^T,$$

$$\Delta_{i(j)}\left\{\begin{array}{c}\in\mathfrak{R}^{+},\ \forall i=1,2,...,N,\ \forall j=0,1,...,k,\\ =\infty,\ \forall i=1,2,...,N,\ \forall j=k+1,k+2,...,\nu-1\end{array}\right\}$$

$\boldsymbol{\varepsilon}$ *the output error vector* $\boldsymbol{\varepsilon}\in R^N$ *(2.24) (Subsection 2.1.3),*

$$\boldsymbol{\varepsilon}=\mathbf{Y_d}-\mathbf{Y}=-\mathbf{y},\ \boldsymbol{\varepsilon}=[\varepsilon_1\ \varepsilon_2\ ...\ \varepsilon_N\ ]^T$$

$\boldsymbol{\varepsilon}^j\in\mathfrak{R}^{+^j}$ *a vector* of the dimension j

$\boldsymbol{\varepsilon}_{[k]}^{\nu N}\in\mathfrak{R}^{+^{\nu N}}$ *a vector* of the dimension νN defined by

$$\boldsymbol{\varepsilon}_{[k]}^{\nu N}=\begin{bmatrix}\varepsilon_1\ \varepsilon_2\ ...\ \varepsilon_N\ \varepsilon_{1(1)}\ \varepsilon_{2(1)}\ ...\ \varepsilon_{N(1)}\ ...\ \varepsilon_{1(k)}\ \varepsilon_{2(k)}\ ...\\ ...\varepsilon_{N(k)}\ \infty\ \ \infty\ ...\ \infty\end{bmatrix}^T,$$

$$k\in\{0,1,2,...,\nu-1\},$$

$$\varepsilon_{i(j)}\left\{\begin{array}{c}\in\mathfrak{R}^{+},\ i=1,2,...,N,\ j=0,1,...,k,\\ =\infty,\ i=1,2,...,N,\ j=k+1,\ k+2,...,\nu\end{array}\right\}$$

$\boldsymbol{\epsilon}$ *the subsidiary output error vector* $\boldsymbol{\epsilon}$ relative to the reference output vector $\mathbf{Y}_R$ (12.25) (Section 12.1),

$$\boldsymbol{\epsilon}=\mathbf{Y}_R-\mathbf{Y},\ \boldsymbol{\epsilon}=[\epsilon_1\ \ \epsilon_2\ \ ...\ \epsilon_N\ ]^T$$

Ξ the diagonal matrix associated with $\boldsymbol{\epsilon}$,

$$\Xi=diag\{\epsilon_1\ \ \epsilon_2\ \ ...\ \epsilon_N\}$$

θ *a nonnegative integer*

$\lambda_m(H)$ *the minimal eigenvalue of the square matrix H*

$\lambda_M(H)$ *the maximal eigenvalue of the square matrix H*

μ *a nonnegative integer*

ν *a nonnegative integer*

$\sigma\left(\varepsilon_i^{(k)}, \varepsilon_i^{(k+1)}\right)$ is defined by

$$\sigma\left(\varepsilon_i^{(k)}, \varepsilon_i^{(k+1)}\right) = \left\{ \begin{array}{c} -1,\ \varepsilon_i^{(k)} < 0, \forall \varepsilon_i^{(k+1)} \in \mathfrak{R};\ \varepsilon_i^{(k)} = 0\ and\ \varepsilon_i^{(k+1)} < 0, \\ 0,\ \varepsilon_i^{(k)} = 0\ and\ \varepsilon_i^{(k+1)} = 0, \\ 1,\ \varepsilon_i^{(k)} > 0, \forall \varepsilon_i^{(k+1)} \in \mathfrak{R};\ \varepsilon_i^{(k)} = 0\ and\ \varepsilon_i^{(k+1)} > 0 \end{array} \right\}$$

$\Sigma\left(\boldsymbol{\varepsilon}^{(k)}, \boldsymbol{\varepsilon}^{(k+1)}\right)$ is defined by

$$\Sigma\left(\boldsymbol{\varepsilon}^{(k)}, \boldsymbol{\varepsilon}^{(k+1)}\right) = diag\left\{\sigma\left(\varepsilon_1^{(k)}, \varepsilon_1^{(k+1)}\right)\ \sigma\left(\varepsilon_2^{(k)}, \varepsilon_2^{(k+1)}\right)\ ...\ \sigma\left(\varepsilon_N^{(k)}, \varepsilon_N^{(k+1)}\right)\right\}$$

$\Sigma\left(\boldsymbol{\varepsilon}^{v}\right)$ is defined by

$$\Sigma\left(\boldsymbol{\varepsilon}^{v}\right) = blockdiag\left\{\Sigma\left(\boldsymbol{\varepsilon}, \boldsymbol{\varepsilon}^{(1)}\right)\ \Sigma\left(\boldsymbol{\varepsilon}^{(1)}, \boldsymbol{\varepsilon}^{(2)}\right)\ ...\ \Sigma\left(\boldsymbol{\varepsilon}^{(\nu-1)}, \boldsymbol{\varepsilon}^{(\nu)}\right)\right\}$$

τ a subsidiary notation for *time* t

$\tau_A \in \mathfrak{T}_0 \cup \{\infty\}$ *the attainability time*

$\boldsymbol{\tau}_A^N = \left[\tau_{A1}\ \ \tau_{A2}\ ...\ \tau_{AN}\right]^T$ *the vector attainability time*

$\tau_R \in \mathfrak{T}_0$ *the reachability time*

$\tau_{Ri} \in \mathfrak{T}_0$ *the reachability time associated with the i-th output variable*

$\boldsymbol{\tau}_R^p \in \mathfrak{T}_0^p$ *the vector reachability time*, the *i-th* entry of which is τ_{Ri},

$$\boldsymbol{\tau}_R^p = \boldsymbol{\tau}_{R(0)}^p = \begin{bmatrix} \tau_{R1} \\ \tau_{R2} \\ ... \\ \tau_{Rp} \end{bmatrix} = \begin{bmatrix} \tau_{R1,(0)} \\ \tau_{R2,(0)} \\ ... \\ \tau_{Rp,(0)} \end{bmatrix} \in \mathfrak{T}_0^m$$

$\tau_{Ri,(j)} \in \mathfrak{T}_0$ *the reachability time associated with the j-th derivative of the i-th output variable*

$\boldsymbol{\tau}_{R(j)}^N \in \mathfrak{T}_0^N$ *the vector reachability time associated with the j-th derivative* $\mathbf{Y}^{(j)}(t)$ *of the output vector* $\mathbf{Y}(t)$,

$$\boldsymbol{\tau}_{R(j)}^N = \begin{bmatrix} \tau_{R1,(j)} \\ \tau_{R2,(j)} \\ ... \\ \tau_{RN,(j)} \end{bmatrix} \in \mathfrak{T}_0^N,\ \boldsymbol{\tau}_{R(0)}^N = \boldsymbol{\tau}_R^N$$

$\boldsymbol{\tau}_{R[k]}^{(k+1)N} \in \mathfrak{T}_0^{(k+1)N}$ *the vector reachability time associated with the output vector* $\mathbf{Y}(t)$ *and its first k derivatives* $\mathbf{Y}(t)$, $\mathbf{Y}^{(1)}(t)$, ... , $\mathbf{Y}^{(k)}(t)$ *in the extended output space* $\mathfrak{R}^{(k+1)N}$,

$$\boldsymbol{\tau}_{R[k]}^{(k+1)N} = \begin{bmatrix} \boldsymbol{\tau}_{R(0)}^N \\ \boldsymbol{\tau}_{R(1)}^N \\ \boldsymbol{\tau}_{R(2)}^N \\ ... \\ \boldsymbol{\tau}_{R(k)}^N \end{bmatrix} = \begin{bmatrix} \boldsymbol{\tau}_{R}^N \\ \boldsymbol{\tau}_{R(1)}^N \\ \boldsymbol{\tau}_{R(2)}^N \\ ... \\ \boldsymbol{\tau}_{R(k)}^N \end{bmatrix} \in \left(Int\mathfrak{T}_0\right)^{(k+1)N},$$

$$k \in \{0, 1, 2, .., \nu - 1\},\ \boldsymbol{\tau}_{R[0]}^N = \boldsymbol{\tau}_R^N$$

$\boldsymbol{\tau}_{R[k]}^{\nu N}$ the reachability times are defined only for the first derivatives of the output variables, while their higher derivatives are free, $Y_i^{(j)}(t) \in \mathfrak{R}$, $\forall j = k+1, k+2, ..., \nu$-1, so that formally we let for the $\tau_{Ri(j)} = \infty$, $\forall j = k+1$, $k+2, ..., \nu$-1,

$$\boldsymbol{\tau}_{R[k]}^{\nu N} = \begin{bmatrix} \boldsymbol{\tau}_{R[k]}^{(k+1)N} \\ \boldsymbol{\tau}_{\mathfrak{R}k}^{\nu} \end{bmatrix} \Longrightarrow \boldsymbol{\tau}_{R[\nu-1]}^{\nu N} = \begin{bmatrix} \boldsymbol{\tau}_{R(0)}^{N} \\ \boldsymbol{\tau}_{R(1)}^{N} \\ \boldsymbol{\tau}_{R(2)}^{N} \\ ... \\ \boldsymbol{\tau}_{R(\nu-1)}^{N} \end{bmatrix} = \begin{bmatrix} \boldsymbol{\tau}_{R}^{N} \\ \boldsymbol{\tau}_{R(1)}^{N} \\ \boldsymbol{\tau}_{R(2)}^{N} \\ ... \\ \boldsymbol{\tau}_{R(\nu-1)}^{N} \end{bmatrix},$$

$$\boldsymbol{\tau}_{\mathfrak{R}k}^{\nu} \in \mathfrak{R}^{\nu-k-1}, \ k < \nu - 1$$

$\boldsymbol{\tau}_{R[\nu-1]}^{\nu N}$ *the finite vector reachability time related to the IO plant output behavior treated via the internal dynamics space,*

$$\boldsymbol{\tau}_{R[\nu-1]}^{\nu N} = \left[\tau_{R1}...\tau_{RN} \ \ \tau_{R1(1)}... \ \tau_{RN(1)} \ ..... \ \tau_{R1(\nu-1)}... \ \tau_{RN(\nu-1)}\right]^T,$$

$\boldsymbol{\tau}_{R[0]}^{N} = \boldsymbol{\tau}_{R(0)}^{N} = \boldsymbol{\tau}_{R}^{N}$

$\boldsymbol{\tau}_{R}^{n} \in \mathfrak{T}_0^n$ *the finite vector reachability time related to the ISO plant output behavior treated via the state space,*

$$\boldsymbol{\tau}_{R}^{n} = \left[\tau_{R1} \ \ \tau_{R1} \ ... \ \tau_{Rn}\right]^T$$

Υ^k *the k-th order target set*, e.g., for the *ISO* system see 8.71 (Subsection 8.4.4)

Υ *the zero order order target set* called simply *the target set*

ϕ *the empty set*

ρ *a natural number*

$\boldsymbol{\chi}(.; t_0, \mathbf{x}_0) : \mathfrak{T}_0 \text{x} \mathfrak{T} \text{x} \mathfrak{R}^n \rightarrow \mathfrak{R}^n$ *a motion of a dynamic system*, which passes through $\mathbf{x}_0$ at t_0

$\boldsymbol{\chi}(t; t_0, \mathbf{x}_0) \in R^n$ *the instantaneous vector value of the motion* $\boldsymbol{\chi}(.; t_0, \mathbf{x}_0)$ *at a moment t*, $\boldsymbol{\chi}(t_0; t_0, \mathbf{x}_0) \equiv \mathbf{x}_0$

A.3.4 Roman letters

$A \in \mathfrak{R}^{n\text{x}n}$ *the matrix describing the internal dynamics of the an ISO system*

$A^{(\nu)} \in \mathfrak{R}^{N\text{x}(\nu+1)N}$ *the extended matrix describing the IO system internal dynamics*, $A^{(\nu)} = \left[A_0 \vdots \ A_1 \vdots ... \vdots \ A_\nu\right]$

$A_{CR}^{(\nu)} \in \mathfrak{R}^{r\text{x}r}$ *the matrix describing the an IO controller internal dynamics*

$A_P^{(\nu)} \in \mathfrak{R}^{N\text{x}(\nu_P+1)N}$ *the extended matrix describing the IO plant internal dynamics*, $A_P^{(\nu)} = \left[A_{P0} \vdots \ A_{P1} \vdots ... \vdots \ A_{P\nu_P}\right]$

$B \in R^{n_C \text{x} 2N}$ *the ISO controller matrix describing the transmission of the influence of* $\mathbf{I}(t)$ *on the controller state,* $B = \left[B_C \vdots L_C \right]$

$B \in R^{n_{CS} \text{x}(d+N)}$ *the ISO control system matrix describing the transmission of the influence of* $\mathbf{I}(t)$ *on the control system state,* $B = \left[L_{CS} \vdots P_{CS} \right]$

$B \in R^{n_P \text{x}(r+d)}$ *the ISO plant matrix describing the transmission of the influence of* $\mathbf{I}(t)$ *on the plant state,* $B = \left[B_P \vdots L_P \right]$

$B^{(\mu)} \in \mathfrak{R}^{N\text{x}(\mu+1)M}$ *the IO system extended matrix describing the transmission of the influence of* $\mathbf{I}^{\mu}(t)$ *on the system output,* $B^{(\mu)} = \left[B_0 \vdots B_1 \vdots ... \vdots B_\mu \right]$

$B_C \in R^{n_C \text{x} N}$ *the ISO controller matrix describing the transmission of the influence of* $\mathbf{Y}_d(t)$ *on the controller state*

$B_{CR}^{(\nu)} \in \mathfrak{R}^{r\text{x}2N(\nu+1)}$ *the controller extended matrix describing the transmission of the influence of* $\mathbf{I}^{\nu}(t)$ *on the controller output,* $\mathbf{U}(t)$, $B_{CR}^{(\nu)} = [B_{CR0} \vdots B_{CR1} \vdots ... \vdots B_{CR\nu}]$, $B_{CRk} = \left[P_{CRk} \vdots -Q_{CRk} \right] \in \mathfrak{R}^{r\text{x}2N}$

$B_P \in R^{n_P \text{x} r}$ *the ISO plant matrix describing the transmission of the influence of* $\mathbf{U}(t)$ *on the plant state*

$B_P^{(\nu)} \in \mathfrak{R}^{N\text{x}(\nu+1)(d+r)}$ *the IO plant extended matrix describing the transmission of the influence of* $\mathbf{I}^{\nu}(t)$ *on the plant output,* $B_P^{(\nu)} = [B_{P0} \vdots B_{P1} \vdots ... \vdots B_{P\nu}]$, $B_{Pk} = \left[C_{Puk} \vdots D_{P_d k} \right] \in \mathfrak{R}^{N\text{x}(d+r)}$

$C \in \mathfrak{R}^{N\text{x}n}$ *the matrix relating the ISO system output to its state*

$C_C \in R^{r\text{x}n_C}$ *the matrix relating the ISO controller output to its state*

$C_{CS} \in R^{N\text{x}n_{CS}}$ *the matrix relating the ISO control system output to its state*

$C_P \in R^{N\text{x}n_P}$ *the matrix relating the IO plant output to its state*

$C_{Pu}^{(\nu_{Pu})} \in \mathfrak{R}^{N\text{x}(\nu_{Pu}+1)r}$ *the IO plant extended matrix describing the transmission of the influence of* $\mathbf{U}^{\nu_{Pu}}(t)$ *on the plant output,* $C_{Pu}^{(\nu_{Pu})} = [C_{P_u 0} \vdots C_{P_u 1} \vdots ... \vdots C_{P_u \nu_{Pu}}]$

$Cl\mathfrak{B}(t)$ *the closure of the set* $\mathfrak{B}(t)$

d *a natural number*

$\mathbf{D} \in \mathfrak{R}^d$ *the total disturbance vector*

$\mathbf{D}_N \in \mathfrak{R}^d$ *the nominal disturbance vector*

$D \in R^{N\text{x}M}$ *the ISO system matrix describing the transmission of the influence of* $\mathbf{I}(t)$ *on the system output*

$D \in R^{r\text{x}2N}$ *the ISO controller matrix describing the transmission of the influence of* $\mathbf{I}(t)$ *on the controller output,* $D = \left[H_C \vdots D_C \right]$

$D \in R^{N\times(d+N)}$ *the ISO control system matrix describing the transmission of the influence of* $\mathbf{I}(t)$ *on the control system output,* $D = \left[D_{CS} \vdots Q_{CS}\right]$

$D \in R^{N\times(d+r)}$ *the ISO plant matrix describing the transmission of the influence of* $\mathbf{I}(t)$ *on the plant output,* $D = \left[H_P \vdots D_P\right]$

$D_C \in R^{rxN}$ *the ISO controller matrix describing the transmission of the influence of* $\mathbf{Y}(t)$ *on the controller output*

$D_{CS} \in R^{Nxd}$ *the ISO control system matrix describing the transmission of the influence of* $\mathbf{D}(t)$ *on the control system output*

$D_P \in R^{Nxd}$ *the ISO plant matrix describing the transmission of the influence of* $\mathbf{D}(t)$ *on the plant output*

$D_{Pd}^{(\nu_{Pd})} \in \mathfrak{R}^{N\times(\nu_{Pd}+1)d}$ *the IO plant matrix describing the transmission of the influence of* $\mathbf{D}^{\nu_{Pd}}(t)$ *on the plant output,* $D_{Pd}^{(\nu_{Pd})} = [D_{P_d0} \vdots\ D_{P_d1} \vdots ... \vdots D_{P_d\nu_{Pd}}]$

$e^{-tK_0T_1^{-1}} \in \mathfrak{R}_+^{N\times N}$ *the diagonal exponential matrix,*

$$e^{-tK_0T_1^{-1}} = \exp\left(-tK_0T_1^{-1}\right) = diag\left\{e^{-tk_{01}^{-1}\tau_{11}} \quad e^{-tk_{021}^{-1}\tau_{12}} \quad . . . \ e^{-tk_{0N}^{-1}\tau_{1N}} \quad \right\}$$

$E \in \mathfrak{R}^{N\times N}$ *the diagonal matrix composed of the entries* ε_k *of the error vector* $\boldsymbol{\varepsilon}$,

$$E = diag\left\{\varepsilon_1 \ \ \varepsilon_2 \ . \ . \ . \ \ \varepsilon_N\right\}$$

$E_k \in \mathfrak{R}^{N\times N}$ *a matrix,*

$E^{(\nu)} \in \mathfrak{R}^{N\times N(\nu+1)}$ *the extended matrix describing the output dynamics of the IIO system,*

$$E^{(\nu)} = \left[E_0 \vdots E_1 \vdots ... \ E_\nu\right]$$

$\mathbf{f}(.) : \mathfrak{T}_0 \longrightarrow \mathfrak{R}^N$ *a subsidiary vector function,*

$$\mathbf{f}(.) : \mathfrak{T}_0 \longrightarrow \mathfrak{R}^N, \ \mathbf{f}(t) \in \mathfrak{C}^l(\mathfrak{T}_0), \ l \in \{1, 2, ...\},$$
$$\mathbf{f}^{(i)}(0) = -\boldsymbol{\varepsilon}^{(i)}(0), \ \mathbf{f}^{(i)}(t) = \mathbf{0}_N, \ \forall t \in \mathfrak{T}_{A\infty}, \ \forall i = 0, 1, .., l$$

$F(s)$ *the full (complete) transfer function matrix of a time-invariant continuous-time linear dynamical system*

$F_{IO}(s)$ *the full IO transfer function matrix of the IO system*

$F_{IOISO}(s)$ *the full transfer function matrix obtained from the IO mathematical model of the given ISO system*

$F_{ISO}(s) \in \mathfrak{C}^{N\times(M+n)}$ *the full ISO transfer function matrix of the ISO system*

$F_{ISOIS}(s) \in \mathfrak{C}^{n\times(M+n)}$ *the full (complete) IS transfer function matrix of the ISO system*

$G = G^T \in R^{p\times p}$ *the symmetric matrix of the quadratic form* $v(\mathbf{w}) = \mathbf{w}^T G \mathbf{w}$,

$G(s)$ *the transfer function matrix of a time-invariant continuous-time linear dynamical system,*

$G_{IO}(s)$ *the IO transfer function matrix of the IO system*

$G_{IO_0}(s) \in \mathfrak{C}^{N \times (\mu M + \nu N)}$ *the ICO transfer function matrix relative to all initial conditions of the IO system*

$G_{IOi_0}(s) \in \mathfrak{C}^{N \times \mu M}$ *the IICO transfer function matrix relative to* $\mathbf{I}_{0(\mp)}^{\mu-1}$ *of the IO system*

$G_{IOy_0}(s) \in \mathfrak{C}^{N \times \nu N}$ *the OICO transfer function matrix relative to* $\mathbf{Y}_{0(\mp)}^{\nu-1}$ *of the IO system*

$G_{IOISO}(s) \in \mathfrak{C}^{N \times N}$ *the transfer function obtained from the IO mathematical model of the given ISO system*

$G_{ISO}(s) \in \mathfrak{C}^{N \times M}$ *the ISO transfer function matrix of the ISO system*

$G_{ISOIS}(s) \in \mathfrak{C}^{n \times M}$ *the IS transfer function matrix of the ISO system*

$G_{ISOSS}(s) \in \mathfrak{C}^{n \times n}$ *the ISS transfer function matrix of the ISO system*

$G_{ISOx_0}(s) \in \mathfrak{C}^{N \times n}$ *the ISCO transfer function matrix relative to* $\mathbf{X}_{0\mp}$ *of the ISO system*

$H \in R^{N \times r}$ *a matrix*

$H = H^T \in R^{p \times p}$ *the symmetric matrix of the quadratic form* $v(\mathbf{w}) = \mathbf{w}^T H \mathbf{w}$

$H_C \in R^{r \times N}$ *an ISO controller matrix describing the transmission of the influence of* $\mathbf{Y}_d(t)$ *on the controller output*

$H_P \in R^{N \times r}$ *an ISO system matrix describing the transmission of the influence of* $\mathbf{U}(t)$ *on the system output*

i *an arbitrary natural number,* or *the imaginary unit* $\sqrt{-1}$

I_i *the i-th order identity matrix*

$\mathbf{I} \in \mathfrak{R}^M$ *the input deviation vector,* $\mathbf{I} = [I_1 \quad I_2 \ ... \ I_M]^T$

$\mathbf{I}^{\mu}(t) \in \mathfrak{R}^{(\mu+1)M}$ *the extended input vector at a moment t,*

$$\mathbf{I}^{\mu}(t) = \left[\mathbf{I}^T(t) \vdots \mathbf{I}^{(1)^T}(t) \vdots ... \vdots \mathbf{I}^{(\mu)^T}(t) \right]^T$$

$\mathbf{I}_{0\mp}^{\mu-1} \in \mathfrak{R}^{\mu M}$ *the initial extended input vector at the initial moment* $t_0 = 0$,

$$\mathbf{I}_{0\mp}^{\mu-1} = \mathbf{I}^{\mu-1}(0^{\mp}) = \left[\mathbf{I}_{0(\mp)}^T \vdots \mathbf{I}_{0(\mp)}^{(1)^T} \vdots ... \vdots \mathbf{I}_{0(\mp)}^{(\mu-1)^T} \right]^T \in \mathfrak{R}^{\mu M}$$

I, I *the identity matrix of the n-th order,* I$= I = diag\{1 \ 1 \ ... \ 1\} \in \mathfrak{R}^{n \times n}$, or *the total input variable*

I_N *the identity matrix of the N-th order,* $I_N = diag\{1 \ 1 \ ... \ 1\} \in \mathfrak{R}^{N \times N}$

$\mathbf{I} \in \mathfrak{R}^M$ *the total input vector,* $\mathbf{I} = [I_1 \quad I_2 \ ... \ I_M]^T$

$\mathbf{I}_N \in \mathfrak{R}^M$ *the nominal input vector,* $\mathbf{I}_N = [I_{N1} \quad I_{N2} \ ... \ I_{NM}]^T$

$Int\, \mathfrak{T}_0$ *the interior of* $\mathfrak{T}_0$,

$$Int\, \mathfrak{T}_0 = \{t : t \in \mathfrak{T}_0, \ t > t_0 = 0\}, \ Int\, \mathfrak{T}_0 \subset \mathfrak{T}_0,$$
$$min\ (Int\, \mathfrak{T}_0) = t_0 = 0 \in \mathfrak{T}_0, \ \ sup\ (Int\, \mathfrak{T}_0) = \infty$$

$(Int\mathfrak{T}_0)^i$ *the i-th order set product of the interior of* $\mathfrak{T}_0$,

$$(Int\mathfrak{T}_0)^i = \underbrace{Int\mathfrak{T}_0 \text{x} Int\mathfrak{T}_0 \text{x} ... \text{x} Int\mathfrak{T}_0}_{i-times}$$

$Int\ \mathfrak{S}$ the interior of the set $\mathfrak{S}$

$\operatorname{Im} s$ the imaginary part of $s = \sigma + j\omega$, $\operatorname{Im} s = j\omega$

j *an arbitrary natural number*, or $j = \sqrt{-1}$ is the imaginary unit

$J \in \mathfrak{R}^{n\text{x}M}$ *a matrix*

k *an arbitrary natural number*

K_0 a constant diagonal matrix,

$$K_0 = diag\{k_{01}\ \ k_{02}\ \ ...\ \ k_{0N}\}$$

$K \in \mathfrak{R}^{N\text{x}M}$ *a matrix*

$L \in R^{N\text{x}d}$ *a matrix*

$L_c \in R^{n_C\text{x}N}$ *the ISO controller matrix describing the transmission of the influence of* $\mathbf{Y}(t)$ *on the controller state*

$L_{CS} \in R^{n_{CS}\text{x}d}$ *the ISO control system matrix describing the transmission of the influence of* $\mathbf{D}(t)$ *on the control system state*

$L_P \in R^{n_P\text{x}d}$ *the ISO plant matrix describing the transmission of the influence of* $\mathbf{D}(t)$ *on the plant state*

$M(.)$ *a complex valued matrix function* of any type

$M(s)$ *a complex valued matrix* of any type

m *a nonnegative integer*, $m \in \{1, \nu\}$, $m = \nu \geq 1$ for the IO plant, and $m = 1$ for the ISO plant

n *a natural number*

N *a natural number*, if N is the dimension of the output vector and if n is the dimension of the state vector then $N \leq n$

O *the origin of* $\mathfrak{R}^n$, or *the zero matrix of the appropriate order*

p *a natural number*

$P \in R^{n\text{x}N}$ *a matrix*

$P_{CS} \in R^{n_{CS}\text{x}N}$ *an ISO control system matrix describing the transmission of the influence of* $\mathbf{Y}_d(t)$ *on the control system state*

$P_k \in \mathfrak{R}^{\rho\text{x}M}$ *a matrix*

$P_{CRk} \in \mathfrak{R}^{r\text{x}N}$ *the IO controller matrix*

$P_{CR}^{(\nu_{Cyd})} \in \mathfrak{R}^{r\text{x}(\nu_{Cyd}+1)N}$ *the IO controller extended matrix describing the transmission of the influence of* $\mathbf{Y}_d^{(\nu_{Cyd})}(t)$ *on the controller output*, $P_{CR}^{(\nu_{Cyd})} = \left[P_{CR0} \vdots\ P_{CR1} \vdots ... \vdots P_{CR\nu_{Cyd}}\right]$

$Q_{CRk} \in \mathfrak{R}^{r\text{x}N}$ *the IO controller matrix*

$Q_{CR}^{(\nu_{Cy})} \in \mathfrak{R}^{r\text{x}(\nu_{Cy}+1)N}$ *the IO controller extended matrix describing the transmission of the influence of* $\mathbf{Y}^{(\nu_{Cyd})}(t)$ *on the controller output*, $Q_{CR}^{(\nu_{Cy})} = \left[Q_{CR0} \vdots\ Q_{CR1} \vdots ... \vdots Q_{CR\nu_{Cy}}\right]$

q *a natural number*

$Q \in R^{N\mathrm{x}N}$ *a matrix*

$Q_{CS} \in R^{N\mathrm{x}N}$ *the ISO control system matrix describing the transmission of the influence of* $\mathbf{Y}_d(t)$ *on the control system output*

$Re\ s$ *the real part of* $s = \sigma + j\omega$, $Re\ s = \sigma$

s *the basic time unit*: *second*, or *a complex variable* $s = \sigma + j\omega$, *or a complex number*,

$sign(.) : \mathfrak{R} \rightarrow \{-1, 0, 1\}$ *the scalar signum function*,

$$sign(x) = |x|^{-1} x \ if \ x \neq 0, \ and \ sign(0) = 0$$

$sign\,(.) : \mathfrak{R} \rightarrow \{-1, 0, 1\}^N$ *the vector signum function*,

$$sign\boldsymbol{\varepsilon} = [sign\varepsilon_1 \ \ sign\varepsilon_2 \ ... \ sign\varepsilon_N]^T$$

$S\,(\boldsymbol{\varepsilon}) : \mathfrak{R}^N \rightarrow \{-1, 0, 1\}^{N\mathrm{x}N}$ *the matrix signum function*,

$$S\,(\boldsymbol{\varepsilon}) = diag\,\{sign\varepsilon_1 \ \ sign\varepsilon_2 \ ... \ sign\varepsilon_N\}$$

$S\,(E) : \mathfrak{R}^{N\mathrm{x}N} \rightarrow \{-1, 0, 1\}^{N\mathrm{x}N}$ *the matrix signum function*,

$$S\,(E) = diag\,\{sign\varepsilon_1 \ \ sign\varepsilon_2 \ ... \ sign\varepsilon_N\} = S\,(\boldsymbol{\varepsilon})$$

$S_i^{(k)}(.) : \mathfrak{C} \longrightarrow \mathfrak{C}^{\ i(k+1)\mathrm{x}i}$ *the matrix function of* s *defined by (7.23) in Section 7.2.1*:

$$S_i^{(k)}(s) = \left[s^0 I_i \vdots s^1 I_i \vdots s^2 I_i \vdots ... \vdots s^k I_i \right]^T \in \mathfrak{C}^{\ i(k+1)\mathrm{x}i},$$
$$(k, i) \in \{(\mu, M)\,, \ (\nu, N)\}$$

S_{yIO} *the family of all realizable desired output responses* $\mathbf{Y}_d(.)$ *of the IO system*

S_{yISO} *the family of all realizable desired output responses* $\mathbf{Y}_d(.)$ *of the ISO system*

t *time* (*temporal variable*), or *an arbitrary time value* (*an arbitrary moment, an arbitrary instant*); and formally mathematically t denotes for short also the numerical *time* value *numt* if it does not create a confusion,

$$t[\mathrm{T}]\,\langle s \rangle\,, \ numt \in \mathfrak{R}, \ dt > 0 \ , \ or \ equivalently\text{: } t \in \mathfrak{T}.$$

It has been the common attitude to use the letter t for *time,* for its arbitrary temporal value and for its numerical value *numt*, e.g., $t = 0$ is used in the sense $numt = 0$. We do the same throughout the book if there is not any confusion because we can replace t everywhere by $t1_t^{-1}$, $(t1_t^{-1}) \in \mathfrak{R}$, that we denote again by t, $numt = num\,(t1_t^{-1})$,

t_0 a conventionally accepted *initial value of time* (*initial instant, initial moment*), $t_0 \in \mathfrak{T}$, $numt_0 = 0$, i.e., simply $t_0 = 0$ in the sense $numt_0 = 0$,

$t_{\inf}$ *the first instant*, which has not happened, $t_{\inf} = -\infty$,

$t_{\sup}$ *the last instant*, which will not occur, $t_{\sup} = \infty$,

$t_{ZeroTotal}$ *the total zero value of time*, which has not existed and will not happen,

t_{zero} *a conventionally accepted relative zero value of time*,

$\mathbf{t}^{(k+1)N}$ *the time vector*,

$$\mathbf{t}^{(k+1)N} = t\mathbf{1}_{(k+1)N} = [t\ t...t]^T \in \mathfrak{T}_0^{(k+1)N},$$

so that

$$\mathbf{Y}^k(\mathbf{t}^{(k+1)N}) = \mathbf{Y}_d^k(\mathbf{t}^{(k+1)N}),\ \forall \mathbf{t}^{(k+1)N} \in [\boldsymbol{\tau}_R^k,\ \infty\mathbf{1}_{(k+1)N}[,$$

means

$$Y_i^{(k)}(t) = Y_{di}^{(k)}(t),\ \forall t \in [\tau_{Ri(k)},\ \infty[,\ \forall i = 1, 2, ..., N,$$

T *the temporal dimension, the time dimension*, which is the physical dimension of *time*

$T \in \mathfrak{R}^+$ *the period of a periodic behavior*

T^N *the time diagonal matrix*,

$$T^N = diag\{t\ \ t\ \ ...\ \ t\} \in \mathfrak{T}_0^{N\times N},$$

$\mathbf{T}\left(t, \boldsymbol{\varepsilon}, \boldsymbol{\varepsilon}^{(1)}, ..., \int_{t_0=0}^{t} \boldsymbol{\varepsilon} dt\right) \in \mathfrak{R}^N$ the vector tracking operator $\mathbf{T}(.)$ defines a tracking algorithm

$T_1^N \in \mathfrak{R}^{+N\times N}$ *a constant time matrix*,

$$T_1 = diag\{\tau_1\ \ \tau_2\ \ .\ .\ .\ \ \tau_N\}$$

$T_k \in \mathfrak{R}^{N\times M}$ *a matrix*

$T^{(\mu)} \in \mathfrak{R}^{N\times M(\mu+1)}$ *the extended matrix describing the action of the extended input vector* $\mathbf{i}^\mu$ *on the output dynamics of the IIO system*, $T^{(\mu)} = \left[T_0 \vdots T_1 \vdots ...\ T_\mu\right]$

$u \in \mathfrak{R}^+$ *the control variable*,

$\mathbf{U}_F$ *the total full control vector* partitioned into $\mathbf{U}_S$ and $\mathbf{U}_T$ if, and only if, the plant is not stable,

$$\mathbf{U_F} = \mathbf{U}_S + \mathbf{U}_T,$$

if the plant is stable, then we set

$$\mathbf{U_F} = \mathbf{U}$$

$\mathbf{U}_S$ *the total stabilizing control vector*

$\mathbf{U}_T$ *the total tracking control vector*

$\mathbf{U} \in \mathfrak{R}^r$ *the total tracking control vector* $\mathbf{U}_T$*, which is the controller output vector,*

$$\mathbf{U} = \mathbf{U}_T$$

$\mathbf{U}_N \in \mathfrak{R}^r$ *the total nominal control vector, which is the controller nominal output vector*

$v(.) : \mathfrak{R}^p \to \mathfrak{R}$ *a quadratic form,* $v(\mathbf{w}) = \mathbf{w}^T \mathbf{W} \mathbf{w}$

$\mathbf{V}_{IOo}$ *vector that contains all (input and output) initial conditions of the IO system,*

$$\mathbf{V}_{IOo} = \begin{bmatrix} I_0 \\ I_0^{(1)} \\ \dots \\ I_0^{(\mu-1)} \\ Y_0 \\ Y_0^{(1)} \\ \dots \\ Y_0^{(\nu-1)} \end{bmatrix} \in \mathfrak{R}^{\mu M + \nu N}$$

$\mathbf{V}_{IO}(s)$ *Laplace transform of all actions on the IO system*; composed of Laplace transform $\mathbf{I}(s)$ of the input vector $\mathbf{I}(t)$ and of all (input and output) initial conditions,

$$\mathbf{V}_{IO}(s) = \begin{bmatrix} \mathbf{I}(s) \\ \mathbf{V}_{IOo} \end{bmatrix} \in \mathfrak{R}^{(\mu+1)M + \nu N}$$

$\mathbf{w} \in \mathfrak{R}^p$ *a subsidiary real valued vector,*

$$\mathbf{w} = [w_1 \ \ w_2 \ \ ... \ \ wp]^T \in \left\{ \left[\mathbf{r}^{\alpha-1^T} \ \ \mathbf{y}^{\nu-1^T} \right]^T, \ \mathbf{x}, \ \mathbf{y}^{\nu-1} \right\},$$
$$p \in \{\rho, \ n, \ N\}$$

$W = W^T \in R^{pxp}$ *the symmetric matrix of the quadratic form* $v(\mathbf{w})$, $v(\mathbf{w}) = \mathbf{w}^T W \mathbf{w}$, $W \in \{G = G^T, \ H = H^T\}$

$X \in \mathfrak{R}$ *a real valued scalar state variable*

$\mathbf{X} \in \mathfrak{R}^n$ *the total state vector of the ISO system,*

$$\mathbf{X} = [X_1 \ \ X_2 \ \ ... \ \ X_n]^T$$

$\mathbf{X}_P(.; \mathbf{X}_{P0}; \mathbf{D}_N; \mathbf{U}_N; \mathbf{Y}^k)$ *the plant motion,*

$$\mathbf{X}_P(0; \mathbf{X}_{P0}; \mathbf{D}_N; \mathbf{U}_N; \mathbf{Y}^k) \equiv \mathbf{X}_{P0},$$

its instantaneous value $\mathbf{X}_P(t; \mathbf{X}_{P0}; \mathbf{D}_N; \mathbf{U}_N; \mathbf{Y}^k)$ is *the plant state (at the moment t),*

$$\mathbf{X}_P(t; \mathbf{X}_{P0}; \mathbf{D}_N; \mathbf{U}_N; \mathbf{Y}^k) \equiv \mathbf{X}_P(t; \mathbf{X}_{P0})$$

$\mathbf{X}_{Pd}(.; \mathbf{X}_{Pd0}; \mathbf{D}_N; \mathbf{U}_N; \mathbf{Y}_d^k)$ *the plant desired motion*, its instantaneous value $\mathbf{X}_{Pd}(t; \mathbf{X}_{Pd0}; \mathbf{D}_N; \mathbf{U}_N; \mathbf{Y}_d^k)$ is *the plant desired state (at the moment t),*

$$\mathbf{X}_{Pd}(t; \mathbf{X}_{Pd0}; \mathbf{D}_N; \mathbf{U}_N; \mathbf{Y}_d^k) \equiv \mathbf{X}_{Pd}(t; \mathbf{X}_{Pd0})$$

$\mathbf{X}_C \in \mathfrak{R}^{n_C}$ *the state vector of the ISO controller*

$\mathbf{X}_{CS} \in \mathfrak{R}^{n_{CS}}$ *the state vector of the ISO control system*
$\mathbf{X}_{P} \in \mathfrak{R}^{n_{P}}$ *the state vector of the ISO plant*
$\mathbf{X}_{N} \in \mathfrak{R}^{n}$ *the total nominal state vector of the ISO system,*

$$\mathbf{X}_N = [X_{N1} \;\; X_{N2} \;\; ... \;\; X_{Nn}]^T$$

$Y \in \mathfrak{R}$ *a real valued scalar output variable*
$Y_L \in \mathfrak{R}$ *Lyapunov output variable*
$\mathbf{Y} \in \mathfrak{R}^N$ *a real total valued vector output, the total output vector of both the plant and of its control system,* $\mathbf{Y} = [Y_1 \;\; Y_2 \;\; ... \;\; Y_N]^T$
$\mathbf{Y}_{CR} \in R^r$ *the ISO controller output vector*
$\mathbf{Y}_d \in \mathfrak{R}^N$ *a desired (a nominal) total valued vector output, the desired total output vector of both the plant and of its control system,* $\mathbf{Y}_d = [Y_{d1} \;\; Y_{d2} \;\; ... \;\; Y_{dN}]^T$
$\mathbf{Y}_{CS} \in R^N$ *the ISO control system output vector*
$\mathbf{Y}_{P} \in R^N$ *the ISO plant output vector*
$\mathbf{Y}_{R}(.)$ a *reference output vector function (12.22),*

$$\mathbf{Y}_R(t) = \mathbf{Y}_d(t) - \mathbf{f}(t), \;\; \forall t \in \mathfrak{T}_0$$

$\mathbf{Y}_{0\mp}^{\nu-1} \in \mathfrak{R}^{\nu N}$ *the initial extended output vector at the initial moment* $t_0 = 0$,

$$\mathbf{Y}_{0\mp}^{\nu-1} = \mathbf{Y}^{\nu-1}(0^{\mp}) = \left[\mathbf{Y}_{0(\mp)}^{T} \vdots \mathbf{Y}_{0(\mp)}^{(1)^T} \vdots ... \vdots \mathbf{Y}_{0(\mp)}^{(\nu-1)^T}\right]^T,$$

$$\mathbf{Y}_{0\mp}^{0} = \mathbf{Y}^0(0^{\mp}) = \mathbf{Y}_{0\mp} = \mathbf{Y}(0^{\mp})$$

$Z_k^{(\varsigma-1)}(.) : \mathfrak{C} \rightarrow \mathfrak{C}^{(\varsigma+1)k\mathrm{x}\varsigma k}$ *the matrix function of s defined by (3.8) in Subsection 3.3.2:*

$$Z_k^{(\varsigma-1)}(s) = \begin{bmatrix} O_k & O_k & O_k & ... & O_k \\ s^0 I_k & O_k & O_k & ... & O_k \\ ... & ... & ... & ... & ... \\ s^{\varsigma-1} I_k & s^{\varsigma-2} I_k & s^{\varsigma-3} I_k & ... & s^0 I_k \end{bmatrix}, \; \varsigma \geq 1,$$

$$Z_k^{(\varsigma-1)}(s) \in \mathfrak{C}^{(\varsigma+1)k\mathrm{x}\varsigma k}, \;\; (\varsigma, k) \in \{(\mu, M), \;\; (\nu, N)\}$$

See Note 63 (Subsection 3.3.2) on $Z_k^{(\zeta-1)}(.)$ for $\zeta \leq 0$.

A.4 Names and symbols

A.4.1 Names

Input-Output (IO) systems are described by (2.1) *(Section 2.1),*

Input-State-Output (ISO) systems are described by *the state space equation* (2.29) and by *the output equation* (2.30) *(Section 2.2),*

Stable (stability) matrix is a square matrix is *stable (stability) matrix* if, and only if, the real parts of all its eigenvalues are negative.

A.4.2 Symbols and vectors

(.) *an arbitrary variable*, or *an index*,

$|(.)|: \mathfrak{R} \rightarrow \mathfrak{R}_+$ *the absolute value (module) of a (complex valued) scalar variable* (.)

$|(.)|: \mathfrak{R}^N \rightarrow \mathfrak{R}_+^N$ *the absolute value (module) of a (complex valued) vector variable* (.),

$$|\varepsilon| = [|\varepsilon_1| \quad |\varepsilon_2| \ \ldots \ |\varepsilon_N|]^T,$$

$$\left[|\varepsilon_0|^{K^{-1}} - tT_1^{-1}K_0\mathbf{1}_N\right]^K = \begin{bmatrix} \left[|\varepsilon_{10}|^{k_1^{-1}} - t\tau_{11}^{-1}k_{01}\right]^{k_1} \\ \left[|\varepsilon_{20}|^{k_2^{-1}} - t\tau_{12}^{-1}k_{02}\right]^{k_2} \\ \cdots\cdots\cdots \\ \left[|\varepsilon_{N0}|^{k_N^{-1}} - t\tau_{1N}^{-1}k_{0N}\right]^{k_N} \end{bmatrix} \in \mathfrak{R}^N$$

$|E|^{I-K^{-1}} \in \mathfrak{R}_+^{N\times N}$,

$$|E|^{I-K^{-1}} = diag\left\{|\varepsilon_1|^{1-k_1^{-1}} \quad |\varepsilon_2|^{1-k_2^{-1}} \ \ldots \ |\varepsilon_N|^{1-k_N^{-1}}\right\}$$

$||.||: \mathfrak{R}^n \rightarrow \mathfrak{R}_+$ *an accepted norm on* $\mathfrak{R}^n$, which is the *Euclidean norm on* $\mathfrak{R}^n$ if, and only if, not stated otherwise:

$$||\mathbf{x}|| = ||\mathbf{x}||_2 = \sqrt{\mathbf{x}^T\mathbf{x}} = \sqrt{\sum_{i=1}^{i=n} x_i^2}$$

$||.||_1: \mathfrak{R}^n \rightarrow \mathfrak{R}_+$ the *taxicab norm* or *Manhattan norm:*

$$||\mathbf{x}||_1 = \sum_{i=1}^{i=n} |x_i|$$

$\langle 1_{..}\rangle$ shows *the units 1... of a physical variable*

$[\,\alpha, \beta\,] \subset \mathfrak{R}$ *a compact interval*, $[\alpha,\beta] = \{x: \ x \in \mathfrak{R}, \ \alpha \leq x \leq \beta\}$

$[\,\alpha, \beta\,[\, \subseteq \mathfrak{R}$ *a left closed, right open interval*, $[\alpha,\beta[= \{x: x \in \mathfrak{R}, \ \alpha \leq x < \beta\}$

$]\,\alpha, \beta\,] \subseteq \mathfrak{R}$ *a left open, right closed interval*, $[\alpha,\beta[= \{x: x \in \mathfrak{R}, \ \alpha < x \leq \beta\}$

$]\,\alpha, \beta\,[\, \subseteq \mathfrak{R}$ *an open interval*, $]\alpha,\beta[= \{x: \ x \in \mathfrak{R}, \ \alpha < x < \beta\}$

$(\,\alpha, \beta\,) \subseteq \mathfrak{R}$ *a general interval*, $(\,\alpha, \beta\,) \in \{[\alpha,\beta], \ [\alpha,\beta[, \ ]\alpha,\beta], \ ]\alpha,\beta[\}$

$\lambda_i(A)$ *the eigenvalue* $\lambda_i(A)$ *of the matrix* A

[A..] shows *the physical dimension* A... *of a physical variable*

$\left[A_1 \vdots A_2 \vdots \ldots \vdots A_\nu\right]$ *a structured matrix* composed of the submatrices A_1, A_2, ..., A_ν

$\mathbf{0}_k = [0 \ 0 \ldots 0]^T \in \mathfrak{R}^k$, *the elementwise zero vector*, $\mathbf{0}_n = \mathbf{0}$

$\mathbf{1}_k = [1 \ 1 \ldots 1]^T \in \mathfrak{R}^k$, *the elementwise unity vector*, $\mathbf{1}_n = \mathbf{1}$

$\forall$ *for every*
$\exists$ *there exist(s)*
$\exists!$ *there exists exactly one*
$\in$ *belong(s) to, are (is) members (a member) of*, respectively
$\subset$ *a proper subset of* (it cannot be equal to)
$\subseteq$ *a subset of* (it can be equal to)
$\sqrt{-1}$ *the imaginary unit* denoted by i, $i = \sqrt{-1}$
$adjA$ *the adjoint matrix of the nonsingular square matrix* A,

$$detA \neq 0 \Longrightarrow AadjA = (detA)\,I$$

$detA$ *the determinant of the matrix* A, $detA = |A|$
A^{-1} *the inverse matrix of the nonsingular square matrix* A,

$$detA \neq 0 \Longrightarrow A^{-1} = \frac{adjA}{detA}$$

$d(\mathbf{v}, \mathfrak{S})$ *the scalar distance of a vector* $\mathbf{v}$ *from a set* $\mathfrak{S}$,

$$d(\mathbf{v}, \mathfrak{S}) = \inf[\|\mathbf{v} - \mathbf{w}\| : \mathbf{w} \in \mathfrak{S}]$$

$dim\ \mathbf{z}$ *the mathematical dimension of a vector* $\mathbf{z}$, $\mathbf{z} \in \mathfrak{R}^n \Longrightarrow dim\ \mathbf{z} = n$

$ddim$ *the dynamical dimension of a system* composed of the system order and the system dimension

$DenF(s)$ *the denominator matrix polynomial of the real rational matrix* $F(s) = [DenF(s)]^{-1}\,NumF(s)$, or $NumF(s)\,[DenF(s)]^{-1}$

$Dist\,[\mathbf{v}, \mathfrak{S}]$ *the vector* (i.e., *the elementwise*) *distance of a vector* $\mathbf{v}$ *from a set* $\mathfrak{S}$,

$$Dist(\mathbf{v}, \mathfrak{S}) = \inf[|\mathbf{v} - \mathbf{w}| : \mathbf{w} \in \mathfrak{S}],$$

where the infimum (inf) holds elementwise

$deg\left[adj\left(\sum_{k=0}^{k=\nu} A_k s^k\right)\right]$ *the greatest power of* s *over all elements of*

$$adj\left(\sum_{k=0}^{k=\nu} A_k s^k\right)$$

$deg\left(\sum_{k=0}^{k=\mu} B_k s^k\right)$ *the greatest power of* s *over all elements of* $\sum_{k=0}^{k=\mu} B_k s^k$

$deg\left[det\left(\sum_{k=0}^{k=\nu} A_k s^k\right)\right]$ *the greatest power of* s *in*

$$det\left(\sum_{k=0}^{k=\nu} A_k s^k\right)$$

$D_r^k \mathbf{Y}(t)$ *the k-th order right-hand side derivative of* $\mathbf{Y}(t)$ at $t \in \mathfrak{T}_0$

E *the diagonal matrix associated with the vector* $\boldsymbol{\varepsilon}$,

$$\boldsymbol{\varepsilon} = [\varepsilon_1\ \ \varepsilon_2\ \ .\ .\ .\ \varepsilon_N]^T = \boldsymbol{\varepsilon}^{(0)} \Longrightarrow E = diag\,\{\varepsilon_1\ \ \varepsilon_2\ \ .\ .\ .\ \varepsilon_N\} = E^{(0)},$$

E^k *the bloc diagonal matrix associated with the vector* $\boldsymbol{\varepsilon}^k$,

$$\boldsymbol{\varepsilon}^{(i)} = \left[\varepsilon_1^{(i)} \ \varepsilon_2^{(i)} \ \ldots \ \varepsilon_N^{(i)}\right]^T \Longrightarrow \boldsymbol{\varepsilon}^k = \left[\boldsymbol{\varepsilon}^{(0)T} \ \boldsymbol{\varepsilon}^{(1)T} \ \ldots \ \boldsymbol{\varepsilon}^{(k)T}\right]^T \Longrightarrow$$

$$E^{(i)} = diag\left\{\varepsilon_1^{(i)} \ \varepsilon_2^{(i)} \ \ldots \ \varepsilon_N^{(i)}\right\} \Longrightarrow E^k = blocdiag\left\{E^{(0)} \ E^{(1)} \ \ldots \ E^{(k)}\right\},$$

$|\boldsymbol{\varepsilon}|, |E|$ *the vector and matrix absolute values hold elementwise,*

$$|\boldsymbol{\varepsilon}| = [|\varepsilon_1| \ |\varepsilon_2| \ \ldots \ |\varepsilon_N|]^T, \ |E| = diag\{|\varepsilon_1| \ |\varepsilon_2| \ \ldots \ |\varepsilon_N|\},$$

$gradv\left(\mathbf{y}^{\nu-1}\right)$ is *the gradient of* $v\left(\mathbf{y}^{\nu-1}\right)$,

$$gradv\left(\mathbf{y}^{\nu-1}\right) = \left[\frac{\partial v\left(\mathbf{y}^{\nu-1}\right)}{\partial y_1} .. \frac{\partial v\left(\mathbf{y}^{\nu-1}\right)}{\partial y_N} ..... \frac{\partial v\left(\mathbf{y}^{\nu-1}\right)}{\partial y_1^{(\nu-1)}} .. \frac{\partial v\left(\mathbf{y}^{\nu-1}\right)}{\partial y_N^{(\nu-1)}}\right]^T$$

$Im\lambda_i(A)$ *the imaginary part* of the eigenvalue $\lambda_i(A)$ of the matrix A

inf *infimum*

max *maximum*

min *minimum*

$\min(\delta, \Delta)$ denotes *the smaller between* δ *and* Δ,

$$\min(\delta, \Delta) = \left\{ \begin{array}{c} \delta, \ \delta \leq \Delta, \\ \Delta, \ \Delta \leq \delta \end{array} \right\}$$

mddim *the minimal dynamical dimension of a system*

$NumF(s)$ *the numerator matrix polynomial of the real rational matrix* $F(s) = [DenF(s)]^{-1} NumF(s)$, or $NumF(s) [DenF(s)]^{-1}$

numx *the numerical value of* x; if $x = 50V$, then $numx = 50$

phdim $x(.)$ *the physical dimension of a variable* $x(.)$,

$$x(.) = t \Longrightarrow phdim \ x(.) = phdim \ t = \mathrm{T}, \ but \ dim \ t = 1$$

$Re\lambda_i(A)$ *the real part* of the eigenvalue $\lambda_i(A)$ of the matrix A

sup *supremum*,

$|X|$ *the absolute value of* $X \in \Re$, $|X| = X$ for $X \geq 0$ and $|X|$ =-X for $X \leq 0$, so that

$$\left|\mathbf{Y}^{(j)}(\mathbf{t}^{(k+1)N}) - \mathbf{Y}_d^{(j)}(\mathbf{t}^{(k+1)N})\right| = \left| \begin{array}{c} Y_1^{(j)}(t) - Y_{d1}^{(j)}(t) \\ Y_2^{(j)}(t) - Y_{d2}^{(j)}(t) \\ \cdots \\ Y_N^{(j)}(t) - Y_{dN}^{(j)}(t) \end{array} \right|,$$

and

$$\left|\mathbf{Y}^k(\mathbf{t}^{(k+1)N}) - \mathbf{Y}_d^k(\mathbf{t}^{(k+1)N})\right| = \left| \begin{array}{c} \mathbf{Y}(\mathbf{t}^{(k+1)N}) - \mathbf{Y}_d(\mathbf{t}^{(k+1)N}) \\ \mathbf{Y}^{(1)}(\mathbf{t}^{(k+1)N}) - \mathbf{Y}_d^{(1)}(\mathbf{t}^{(k+1)N}) \\ \cdots \\ \mathbf{Y}^{(k)}(\mathbf{t}^{(k+1)N}) - \mathbf{Y}_d^{(k)}(\mathbf{t}^{(k+1)N}) \end{array} \right|,$$

together with

$$\left|\mathbf{Y}_0^k - \mathbf{Y}_{d0}^k\right| < \boldsymbol{\Delta}^k$$

which signifies

$$\left|Y_{i0}^{(j)} - Y_{di0}^{(j)}\right| < \Delta_{i(j)},\ \forall i = 1, 2, ..., N,\ \forall j = 0, 1, 2, ..., k,$$

and

$$\mathbf{Y}^{\nu-1}(\mathbf{t}^{(k+1)N}) \in \boldsymbol{\Upsilon}^k(t; \mathbf{Y}_d^k),\ \forall \mathbf{t}^{(k+1)N} \in [\boldsymbol{\tau}_R^k,\ \infty \mathbf{1}_{(k+1)N}[,$$
$$\Longleftrightarrow$$
$$Y_i^{(j)}(t) = Y_{di}^{(j)}(t),\ \forall t \in [\tau_{Ri(j)},\ \infty[,\ \forall i = 1, 2, ..., N,\ \forall j = 0, 1, 2, ..., k,$$
$$k \in \{0, 1, 2, .., \nu - 1\},$$

$\mathbf{w} \neq \boldsymbol{\varepsilon}$ the elementwise vector inequality,

$$\mathbf{w} = [w_1\ \ w_2\ \ .\ .\ .\ \ w_N]^T,$$
$$\mathbf{w} \neq \boldsymbol{\varepsilon} \Longleftrightarrow w_i \neq v_i,\ \forall i = 1, 2, ..., N.$$

A.5 Units

$1_{(.)}$ *the unit of a physical variable* (.)

1_t *the time unit of the reference time axis* T, $1_t = s$

Appendix B

From *IO* system to *ISO* system

What follows is well known in the control theory, but is not available in many books on the first course of control systems. In order to transform the *IO* system (2.1) (Section 2.1),

$$\sum_{k=0}^{k=\nu} A_k \mathbf{Y}^{(k)}(t) = \sum_{k=0}^{k=\mu} B_k \mathbf{I}^{(k)}(t),\ det A_\nu \neq 0,\ t \in \mathfrak{T},\ \nu \geq 1,\ 0 \leq \mu \leq \nu, \tag{B.1}$$

into the equivalent *ISO* system (2.29), (2.30) (Section 2.2),

$$\frac{d\mathbf{X}(t)}{dt} = A\mathbf{X}(t) + B\mathbf{I}(t),\ t \in \mathfrak{T}, \tag{B.2}$$

$$\mathbf{Y}(t) = C\mathbf{X}(t) + D\mathbf{I}(t),\ t \in \mathfrak{T}, \tag{B.3}$$

let mathematically, without any physical justification or meaning, subsidiary vector variables $\mathbf{X}_1$, $\mathbf{X}_2$, ... $\mathbf{X}_\nu$ be

$$\mathbf{X}_1 = \mathbf{Y} - B_\nu \mathbf{I}, \tag{B.4}$$

$$\mathbf{X}_2 = \dot{\mathbf{X}}_1 + A_{\nu-1}\ \mathbf{Y} - B_{\nu-1}\mathbf{I}, \tag{B.5}$$

$$\mathbf{X}_3 = \dot{\mathbf{X}}_2 + A_{\nu-2}\ \mathbf{Y} - B_{\nu-2}\mathbf{I}, \tag{B.6}$$

$$\mathbf{X}_4 = \dot{\mathbf{X}}_3 + A_{\nu-3}\ \mathbf{Y} - B_{\nu-3}\mathbf{I}, \tag{B.7}$$

$$\ldots. \tag{B.8}$$

$$\mathbf{X}_{\nu-2} = \dot{\mathbf{X}}_{\nu-3} + A_3\ \mathbf{Y} - B_3\mathbf{I} \tag{B.9}$$

$$\mathbf{X}_{\nu-1} = \dot{\mathbf{X}}_{\nu-2} + A_2\ \mathbf{Y} - B_2\mathbf{I} \tag{B.10}$$

$$\mathbf{X}_\nu = \dot{\mathbf{X}}_{\nu-1} + A_1\ \mathbf{Y} - B_1\mathbf{I}. \tag{B.11}$$

Hence,

$$\begin{aligned}
\mathbf{X}_1 &= \mathbf{Y} - B_\nu \mathbf{I}, \\
\overset{\bullet}{\mathbf{X}}_1 &= \mathbf{X}_2 - A_{\nu-1}\ \mathbf{Y} + B_{\nu-1}\mathbf{I}, \\
\overset{\bullet}{\mathbf{X}}_2 &= \mathbf{X}_3 - A_{\nu-2}\ \mathbf{Y} + B_{\nu-2}\mathbf{I}, \\
\overset{\bullet}{\mathbf{X}}_3 &= \mathbf{X}_4 - A_{\nu-3}\ \mathbf{Y} + B_{\nu-3}\mathbf{I} \\
&\ldots. \\
\overset{\bullet}{\mathbf{X}}_{\nu-3} &= \mathbf{X}_{\nu-2} - A_3\ \mathbf{Y} + B_3\mathbf{I}, \\
\overset{\bullet}{\mathbf{X}}_{\nu-2} &= \mathbf{X}_{\nu-1} - A_2\ \mathbf{Y} + B_2\mathbf{I}, \\
\overset{\bullet}{\mathbf{X}}_{\nu-1} &= \mathbf{X}_\nu - A_1\ \mathbf{Y} + B_1\mathbf{I}.
\end{aligned} \tag{B.12}$$

The solution of the first equation in (B.12) in $\mathbf{Y}$ reads

$$\mathbf{Y} = \mathbf{X}_1 + B_\nu \mathbf{I}. \tag{B.13}$$

This permits us to replace $\mathbf{Y}$ by $\mathbf{X}_1 + B_\nu \mathbf{I}$ in all other equations (B.12):

$$\begin{aligned}
\overset{\bullet}{\mathbf{X}}_1 &= \mathbf{X}_2 - A_{\nu-1}\ (\mathbf{X}_1 + B_\nu \mathbf{I}) + B_{\nu-1}\mathbf{I}, \\
\overset{\bullet}{\mathbf{X}}_2 &= \mathbf{X}_3 - A_{\nu-2}\ (\mathbf{X}_1 + B_\nu \mathbf{I}) + B_{\nu-2}\mathbf{I}, \\
\overset{\bullet}{\mathbf{X}}_3 &= \mathbf{X}_4 - A_{\nu-3}\ (\mathbf{X}_1 + B_\nu \mathbf{I}) + B_{\nu-3}\mathbf{I} \\
&\ldots. \\
\overset{\bullet}{\mathbf{X}}_{\nu-3} &= \mathbf{X}_{\nu-2} - A_3\ (\mathbf{X}_1 + B_\nu \mathbf{I}) + B_3\mathbf{I}, \\
\overset{\bullet}{\mathbf{X}}_{\nu-2} &= \mathbf{X}_{\nu-1} - A_2\ (\mathbf{X}_1 + B_\nu \mathbf{I}) + B_2\mathbf{I}, \\
\overset{\bullet}{\mathbf{X}}_{\nu-1} &= \mathbf{X}_\nu - A_1\ (\mathbf{X}_1 + B_\nu \mathbf{I}) + B_1\mathbf{I}.
\end{aligned}$$

These equations can be rewritten as

$$\begin{aligned}
\overset{\bullet}{\mathbf{X}}_1 &= -A_{\nu-1}\mathbf{X}_1 + \mathbf{X}_2 + (B_{\nu-1} - A_{\nu-1}B_\nu)\,\mathbf{I}, \\
\overset{\bullet}{\mathbf{X}}_2 &= -A_{\nu-2}\ \mathbf{X}_1 + \mathbf{X}_3 + (B_{\nu-2} - A_{\nu-2}B_\nu)\,\mathbf{I}, \\
\overset{\bullet}{\mathbf{X}}_3 &= -A_{\nu-3}\mathbf{X}_1 + \mathbf{X}_4 + (B_{\nu-3} - A_{\nu-3}\ B_\nu)\,\mathbf{I}, \\
&\ldots.. \\
\overset{\bullet}{\mathbf{X}}_{\nu-3} &= -A_3\ \mathbf{X}_1 + \mathbf{X}_{\nu-2} + (B_3 - A_3B_\nu)\,\mathbf{I}, \\
\overset{\bullet}{\mathbf{X}}_{\nu-2} &= -A_2\ \mathbf{X}_1 + \mathbf{X}_{\nu-1} + (B_2 - A_2B_\nu)\,\mathbf{I}, \\
\overset{\bullet}{\mathbf{X}}_{\nu-1} &= -A_1\ \mathbf{X}_1 + \mathbf{X}_\nu + (B_1 - A_1B_\nu)\,\mathbf{I}.
\end{aligned} \tag{B.14}$$

These equations permit the elimination of the derivatives $\overset{\bullet}{\mathbf{X}}_1$, $\overset{\bullet}{\mathbf{X}}_2$, ... $\overset{\bullet}{\mathbf{X}}_{\nu-1}$ from (B.5) through (B.11):

$$\mathbf{X}_1 = \mathbf{Y} - B_\nu \mathbf{I},$$

$$\mathbf{X}_2 = \mathbf{Y}^{(1)} - B_\nu \mathbf{I}^{(1)} + A_{\nu-1}\ \mathbf{Y} - B_{\nu-1}\mathbf{I},$$

$$\mathbf{X}_3 = \mathbf{Y}^{(2)} - B_\nu \mathbf{I}^{(2)} + A_{\nu-1}\ \mathbf{Y}^{(1)} - B_{\nu-1}\mathbf{I}^{(1)} + A_{\nu-2}\ \mathbf{Y} - B_{\nu-2}\mathbf{I},$$

$$\mathbf{X}_4 = \begin{pmatrix} \mathbf{Y}^{(3)} - B_\nu \mathbf{I}^{(3)} + A_{\nu-1}\ \mathbf{Y}^{(2)} - B_{\nu-1}\mathbf{I}^{(2)} + A_{\nu-2}\ \mathbf{Y}^{(1)} - \\ -B_{\nu-2}\mathbf{I}^{(1)} + A_{\nu-3}\ \mathbf{Y} - B_{\nu-3}\mathbf{I} \end{pmatrix},$$

$$\ldots\ldots$$

$$\mathbf{X}_{\nu-2} = \begin{pmatrix} \mathbf{Y}^{(\nu-3)} - B_\nu \mathbf{I}^{(\nu-3)} + A_{\nu-1}\ \mathbf{Y}^{(\nu-4)} - B_{\nu-1}\mathbf{I}^{(\nu-4)} + . \\ \ldots + A_{\nu-2}\ \mathbf{Y}^{(1)} - B_{\nu-2}\mathbf{I}^{(1)} + A_3\ \mathbf{Y} - B_3\mathbf{I} \end{pmatrix}$$

$$\mathbf{X}_{\nu-1} = \begin{pmatrix} \mathbf{Y}^{(\nu-2)} - B_\nu \mathbf{I}^{(\nu-2)} + A_{\nu-1}\ \mathbf{Y}^{(\nu-3)} - B_{\nu-1}\mathbf{I}^{(\nu-3)} + \\ +A_{\nu-2}\ \mathbf{Y}^{(\nu-4)} - B_{\nu-2}\mathbf{I}^{(\nu-4)} + \ldots \\ +A_3\ \mathbf{Y}^{(1)} - B_3\mathbf{I}^{(1)} + A_2\ \mathbf{Y} - B_2\mathbf{I} \end{pmatrix}$$

$$\mathbf{X}_\nu = \begin{pmatrix} \mathbf{Y}^{(\nu-1)} - B_\nu \mathbf{I}^{(\nu-1)} + A_{\nu-1}\ \mathbf{Y}^{(\nu-2)} - B_{\nu-1}\mathbf{I}^{(\nu-2)} + \\ +A_{\nu-2}\ \mathbf{Y}^{(\nu-3)} - B_{\nu-2}\mathbf{I}^{(\nu-3)} + \ldots + A_3\ \mathbf{Y}^{(2)} - B_3\mathbf{I}^{(2)} + \\ +A_2\ \mathbf{Y}^{(1)} - B_2\mathbf{I}^{(1)} + A_1\ \mathbf{Y} - B_1\mathbf{I} \end{pmatrix}.$$

The first derivative of the last equation reads

$$\begin{gathered} \overset{\bullet}{\mathbf{X}}_\nu = \mathbf{Y}^{(\nu)} + A_{\nu-1}\ \mathbf{Y}^{(\nu-1)} + A_{\nu-2}\ \mathbf{Y}^{(2)} + \ldots + A_3\ \mathbf{Y}^{(3)} + \\ +A_2\ \mathbf{Y}^{(2)} + A_1\ \mathbf{Y}^{(1)} - B_\nu \mathbf{I}^{(\nu)} - B_{\nu-1}\mathbf{I}^{(\nu-1)} - B_{\nu-2}\mathbf{I}^{(\nu-2)} .. \\ .. - B_3\mathbf{I}^{(3)} - B_2\mathbf{I}^{(2)} - B_1\mathbf{I}^{(1)} = \\ = \sum_{k=1}^{k=\nu} A_k \mathbf{Y}^{(k)}(t) - \sum_{k=1}^{k=\mu} B_k \mathbf{I}^{(k)}(t). \end{gathered}$$

This and (B.1) yield

$$\begin{gathered} \sum_{k=1}^{k=\nu} A_k \mathbf{Y}^{(k)}(t) - \sum_{k=1}^{k=\mu} B_k \mathbf{I}^{(k)}(t) = \\ = B_0 \mathbf{I}(t) - A_0 \mathbf{Y}(t) = B_0 \mathbf{I}(t) - A_0 (\mathbf{X}_1 \ + B_\nu \mathbf{I}) \ = \\ = -A_0 \mathbf{X}_1 + (B_0 - A_0 B_\nu)\, \mathbf{I}. \end{gathered}$$

Therefore,

$$\overset{\bullet}{\mathbf{X}}_\nu = -A_0 \mathbf{X}_1 + (B_0 - A_0 B_\nu)\, \mathbf{I}.$$

This and (B.14) lead to

$$\underbrace{\begin{bmatrix} \overset{\bullet}{\mathbf{X}}_1 \\ \overset{\bullet}{\mathbf{X}}_2 \\ \overset{\bullet}{\mathbf{X}}_3 \\ \cdots \\ \overset{\bullet}{\mathbf{X}}_{\nu-3} \\ \overset{\bullet}{\mathbf{X}}_{\nu-2} \\ \overset{\bullet}{\mathbf{X}}_{\nu-1} \\ \overset{\bullet}{\mathbf{X}}_{\nu} \end{bmatrix}}_{\overset{\bullet}{\mathbf{x}}} = \begin{bmatrix} -A_{\nu-1}\mathbf{X}_1 + \mathbf{X}_2 + (B_{\nu-1} - A_{\nu-1}B_\nu)\mathbf{I} \\ -A_{\nu-2}\ \mathbf{X}_1 + \mathbf{X_3} + (B_{\nu-2} - A_{\nu-2}B_\nu)\mathbf{I} \\ -A_{\nu-3}\ \mathbf{X}_1 + \mathbf{X_4} + (B_{\nu-3} - A_{\nu-3}B_\nu)\mathbf{I} \\ \cdots\cdot \\ -A_3\ \mathbf{X}_1 + \mathbf{X_{\nu-2}} + (B_3 - A_3B_\nu)\mathbf{I} \\ -A_2\ \mathbf{X}_1 + \mathbf{X}_{\nu-1} + (B_2 - A_2B_\nu)\mathbf{I} \\ -A_1\ \mathbf{X}_1 + \mathbf{X}_\nu + (B_1 - A_1B_\nu)\mathbf{I}. \\ -A_0\mathbf{X}_1 + (B_0 - A_0B_\nu)\mathbf{I} \end{bmatrix} =$$

$$= \underbrace{\begin{bmatrix} -A_{\nu-1} & I_N & O_N & O_N & \dots & O_N & O_N & O_N \\ -A_{\nu-2} & O_N & I_N & O_N & \dots & O_N & O_N & O_N \\ -A_{\nu-3} & O_N & O_N & I_N & \dots & O_N & O_N & O_N \\ \dots & \dots & \dots & \dots & \dots & \dots & \dots & \dots \\ -A_3 & O_N & O_N & O_N & \dots & I_N & O_N & O_N \\ -A_2 & O_N & O_N & O_N & \dots & O_N & I_N & O_N \\ -A_1 & O_N & O_N & O_N & \dots & O_N & O_N & I_N \\ -A_0 & O_N & O_N & O_N & \dots & O_N & O_N & O_N \end{bmatrix}}_{A} \underbrace{\begin{bmatrix} \mathbf{X}_1 \\ \mathbf{X}_2 \\ \mathbf{X}_3 \\ \\ \mathbf{X_{\nu-3}} \\ \mathbf{X_{\nu-2}} \\ \mathbf{X}_{\nu-1} \\ \mathbf{X}_\nu \end{bmatrix}}_{\mathbf{X}} +$$

$$+ \underbrace{\begin{bmatrix} B_{\nu-1} - A_{\nu-1}B_\nu \\ B_{\nu-2} - A_{\nu-2}B_\nu \\ B_{\nu-3} - A_{\nu-3}B_\nu \\ \cdots\cdot \\ B_3 - A_3B_\nu \\ B_2 - A_2B_\nu \\ B_1 - A_1B_\nu \\ B_0 - A_0B_\nu \end{bmatrix}}_{B} \mathbf{I}. \tag{B.15}$$

The equivalent form of $\mathbf{Y} = \mathbf{X}_1 + B_\nu\mathbf{I}$ (B.13) is

$$\mathbf{Y} = \mathbf{X}_1 + B_\nu\mathbf{I}. = \underbrace{\begin{bmatrix} I_N & O_N & O_N & \dots & O_N & O_N & O_N \end{bmatrix}}_{C}\mathbf{X} + \underbrace{[B_\nu]}_{D}\mathbf{I}.$$

This and (B.15) imply the final form of the equivalent ISO system:

$$\frac{d\mathbf{X}}{dt} = A\mathbf{X} + \mathbf{BI},\ \mathbf{Y} = C\mathbf{X} + D\mathbf{I}. \tag{B.16}$$

The ISO realization of the IO system (B.1), i.e., of (2.9), is the quadruple (A, B, C, D).

Appendix C

From ISO system to IO system

The ISO system (2.29), (2.30) (Section 2.2),

$$\frac{d\mathbf{X}(t)}{dt} = A\mathbf{X}(t) + B\mathbf{I}(t),\ t \in \mathfrak{T}, \tag{C.1}$$

$$\mathbf{Y}(t) = C\mathbf{X}(t) + D\mathbf{I}(t),\ t \in \mathfrak{T}, \tag{C.2}$$

can be transformed into the IO system as follows. Laplace transform of (C.1), (C.2) under all zero initial conditions reads

$$\begin{gathered}
(sI - A)\,\mathbf{X}(s) = B\mathbf{I}(s) \Longrightarrow \\
\mathbf{X}(s) = (sI - A)^{-1}\,B\mathbf{I}(s) \\
\mathbf{Y}(s) = \left[C\,(sI - A)^{-1}\,B + D\right]\mathbf{I}(s) = \\
\mathbf{Y}(s) = \left[\frac{C adj\,(sI - A)\,B}{\det\,(sI - A)} + D\right]\mathbf{I}(s) \\
\Longrightarrow \\
[\det\,(sI - A)]\,\mathbf{Y}(s) = [C adj\,(sI - A)\,B + D\det\,(sI - A)]\,\mathbf{I}(s).
\end{gathered}$$

For the sake of simplicity let

$$f(s) = \det\,(sI - A) = \sum_{i=0}^{i=n} c_i s^i,\ c_n = 1,$$

$$C adj\,(sI - A)\,B + D\det\,(sI - A) = \sum_{i=0}^{i=n} K_i s^i,\ K_i \in \mathfrak{R}^{N\times M}.$$

Altogether,

$$\left(\sum_{i=0}^{i=n} c_i s^i\right)\mathbf{Y}(s) = \left(\sum_{i=0}^{i=n} K_i s^i\right)\mathbf{I}(s) \Longrightarrow$$
$$\sum_{i=0}^{i=n} c_i\left[s^i\mathbf{Y}(s)\right] = \sum_{i=0}^{i=n} K_i\left[s^i\mathbf{I}(s)\right].$$

Inverse Laplace transform of the last equation for all zero initial conditions reads

$$\sum_{i=0}^{i=n} c_i\mathbf{Y}^{(i)}(t) = \sum_{i=0}^{i=n} K_i\mathbf{I}^{(i)}(t).$$

This is the *IO* mathematical model of the *ISO* system (C.1), (C.2). It suggests

$$C^{(n)} = [c_0 I \;\; c_1 I \; ... \; c_n I] \in \mathfrak{R}^{N\mathrm{x}(n+1)N},$$
$$K^{(n)} = [K_0 \;\; K_1 \; ... \; K_n] \in \mathfrak{R}^{N\mathrm{x}(n+1)M},$$

so that

$$C^{(\nu)}\mathbf{Y}^n(t) = K^{(n)}\mathbf{I}^n(t).$$

This is the compact form of the *IO* mathematical model of the *ISO* system (C.1), (C.2). The *IO* realization of the *ISO* system (C.1), (C.2) is the quadruple $(n,\, n,\, C^{(\nu)},\, K^{(\nu)})$.

Appendix D

Proof of Theorem 64

Proof. What follows is from [148].

Necessity. Let a vector function $\mathbf{I}^*(.)$ be nominal for the *IO* plant (2.1), i.e., for (2.9), relative to its desired response $\mathbf{Y}_d(.)$. Definition 45 (Subsection 3.3.1) holds. It and (2.1), i.e., (2.9), imply

$$\sum_{k=0}^{k=\nu} A_k \mathbf{Y}_d^{(k)}(t) = \sum_{k=0}^{k=\mu} B_k \mathbf{I}^{*(k)}(t), \ \forall t \in \mathfrak{T}_0,$$

$$A^{(\nu)} \mathbf{Y}_d^{\nu}(t) = B^{(\mu)} \mathbf{I}^{*\mu}(t), \ \forall t \in \mathfrak{T}_0.$$

These equations are (3.9) and (3.10) in another forms, respectively. Their Laplace transforms solved in $\mathbf{I}^*(s)$ are given in (3.11), (3.12), respectively. Since they are solvable in $\mathbf{I}^{*(k)}(.)$, it follows that the conditions 1) and 2) hold.

Sufficiency. Let the conditions 1) and 2) be valid. The input vector function $\mathbf{I}^*(.)$ to the *IO* plant (2.1), i.e., (2.9),

$$A^{(\nu)} \mathbf{Y}^{\nu}(t) = B^{(\mu)} \mathbf{I}^{*\mu}(t), \ \forall t \in \mathfrak{T}_0,$$

satisfies (3.9), hence (3.10):

$$A^{(\nu)} \mathbf{Y}_d^{\nu}(t) = B^{(\mu)} \mathbf{I}^{*\mu}(t), \ \forall t \in \mathfrak{T}_0. \tag{D.1}$$

These equations and

$$\boldsymbol{\varepsilon} = \mathbf{Y}_d - \mathbf{Y} \tag{D.2}$$

yield

$$A^{(\nu)} \boldsymbol{\varepsilon}^{\nu}(t) = \mathbf{0}_N, \ \forall t \in \mathfrak{T}_0. \tag{D.3}$$

Definition 45 requires $\boldsymbol{\varepsilon}^{\nu}(0) = \mathbf{0}_{N+1}$, which implies the trivial solution $\boldsymbol{\varepsilon}(t) = \mathbf{0}_{N+1}$, $\forall t \in \mathfrak{T}_0$, of (D.3). This and $\boldsymbol{\varepsilon} = \mathbf{Y}_d - \mathbf{Y}$ prove $\mathbf{Y}(t) = \mathbf{Y}_d(t)$, $\forall t \in \mathfrak{T}_0$.

Let the input vector function $\mathbf{I}^*(.)$ to the *IO* plant (2.1), i.e., to (2.9), obey (3.11), equivalently (3.12). Laplace transforms of (2.1) and of (2.9) read for the input vector function $\mathbf{I}(.) = \mathbf{I}^*(.)$:

$$\mathbf{I}^*(s) = \left(\sum_{k=0}^{k=\mu} B_k s^k\right)^T \left[\left(\sum_{k=0}^{k=\mu} B_k s^k\right)\left(\sum_{k=0}^{k=\mu} B_k s^k\right)^T\right]^{-1} \bullet$$
$$\bullet \left\langle \sum_{k=0}^{k=\mu} B_k \left[\sum_{i=1}^{i=k} s_d^{k-i} \mathbf{I}^{*(i-1)}(0)\right] + \sum_{k=0}^{k=\nu} A_k \left[s^k \mathbf{Y}(s) - \sum_{i=1}^{i=k} s^{k-i} \mathbf{Y}^{(i-1)}(0)\right]\right\rangle,$$

i.e.,

$$\mathbf{I}^*(s) = \left(B^{(\mu)} S_M^{(\mu)}(s)\right)^T \left[\left(B^{(\mu)} S_M^{(\mu)}(s)\right)\left(B^{(\mu)} S_M^{(\mu)}(s)\right)^T\right]^{-1} \bullet$$
$$\bullet \left\langle B^{(\mu)} Z_M^{(\mu-1)}(s) \mathbf{I}^{*\mu-1}(0) + A^{(\nu)} \left[S_N^{(\nu)}(s)\mathbf{Y}(s) - Z_N^{(\nu-1)}(s)\mathbf{Y}^{\nu-1}(0)\right]\right\rangle.$$

These equations multiplied on the left by $\left(\sum_{k=0}^{k=\mu} B_k s^k\right)$, i.e., by $B^{(\mu)} S_M^{(\mu)}(s)$, respectively, and (3.11), (3.12) imply, respectively,

$$\sum_{k=0}^{k=\nu} A_k \left[s^k \mathbf{Y}_d(s) - \sum_{i=1}^{i=k} s^{k-i} \mathbf{Y}_d^{(i-1)}(0)\right] = \sum_{k=0}^{k=\nu} A_k \left[s^k \mathbf{Y}(s) - \sum_{i=1}^{i=k} s^{k-i} \mathbf{Y}^{(i-1)}(0)\right]$$

and

$$A^{(\nu)} \left[S_N^{(\nu)}(s)\mathbf{Y}_d(s) - Z_N^{(\nu-1)}(s)\mathbf{Y}_d^{\nu-1}(0)\right] =$$
$$= A^{(\nu)} \left[S_N^{(\nu)}(s)\mathbf{Y}(s) - Z_N^{(\nu-1)}(s)\mathbf{Y}^{\nu-1}(0)\right].$$

Definition 45 requires $\mathbf{Y}^{\nu-1}(0) = \mathbf{Y}_d^{\nu-1}(0)$ that reduces the preceding equations to

$$\sum_{k=0}^{k=\nu} A_k s^k \left(\mathbf{Y}_d(s) - \mathbf{Y}(s)\right) = \sum_{k=0}^{k=\nu} A_k s^k \boldsymbol{\varepsilon}(s) = \mathbf{0}_N,$$
$$A^{(\nu)} S_N^{(\nu)}(s)\left(\mathbf{Y}_d(s) - \mathbf{Y}(s)\right) = A^{(\nu)} S_N^{(\nu)}(s)\boldsymbol{\varepsilon}(s) = \mathbf{0}_N.$$

These equations imply $\boldsymbol{\varepsilon}(s) = \mathbf{0}_N$ due to Condition 18 (Section 2.1); hence, $\boldsymbol{\varepsilon}(s) = \mathbf{0}_N$, $\forall s \in \mathfrak{C}$, which is equivalent to $\boldsymbol{\varepsilon}(t) = \mathbf{0}_N$, $\forall t \in \mathfrak{T}_0$, i.e., $\mathbf{Y}(t) = \mathbf{Y}_d(t)$, $\forall t \in \mathfrak{T}_0$ ■

Appendix E

Proof of Theorem 67

Proof. Laplace transform of (2.15) (Subsection 2.1.2) for $\mathbf{U}(.) = \mathbf{U}^*(.)$ and for $\mathbf{Y}(.) = \mathbf{Y}_d(.)$ reads

$$\left[C_{Pu}^{(\mu_{Pu})} S_r^{(\mu_{Pu})}(s)\right] \mathbf{U}^*(s) - C_{Pu}^{(\mu_{Pu})} Z_r^{(\mu_{Pu}-1)}(s) \mathbf{U}_0^{*(\mu_{Pu}-1)} =$$
$$= \left[A_P^{(\nu)} S_N^{(\nu)}(s)\right] \mathbf{Y}_d(s) - A_P^{(\nu)} Z_N^{(\nu-1)}(s) \mathbf{Y}_0^{(\nu-1)} -$$
$$- \left[D_{Pd}^{(\mu_{Pd})} S_d^{(\mu_{Pd})}(s)\right] \mathbf{D}(s) + D_{Pd}^{(\mu_{Pd})} Z_d^{(\mu_{Pd}-1)}(s) \mathbf{D}_0^{(\mu_{Pd}-1)},$$

or equivalently,

$$\left[C_{Pu}^{(\mu_{Pu})} S_r^{(\mu_{Pu})}(s)\right] \mathbf{U}^*(s) = C_{Pu}^{(\mu_{Pu})} Z_r^{(\mu_{Pu}-1)}(s) \mathbf{U}_0^{*(\mu_{Pu}-1)} +$$
$$+ \left[A_P^{(\nu)} S_N^{(\nu)}(s)\right] \mathbf{Y}_d(s) - A_P^{(\nu)} Z_N^{(\nu-1)}(s) \mathbf{Y}_0^{(\nu-1)} -$$
$$- \left[D_{Pd}^{(\mu_{Pd})} S_d^{(\mu_{Pd})}(s)\right] \mathbf{D}(s) + D_{Pd}^{(\mu_{Pd})} Z_d^{(\mu_{Pd}-1)}(s) \mathbf{D}_0^{(\mu_{Pd}-1)}. \tag{E.1}$$

The matrix $C_{Pu}^{(\mu_{Pu})} S_r^{(\mu_{Pu})}(s)$ is rectangular matrix $\left[C_{Pu}^{(\mu_{Pu})} S_r^{(\mu_{Pu})}(s)\right] \in \mathfrak{C}^{Nxr}$. For the solvability of (E.1) in $\mathbf{U}^*(s)$ it is necessary and sufficient that it has the maximal rank, which is the maximal rank of $C_{Pu}^{(\mu_{Pu})}$. Its maximal rank cannot exceed either N or r,

$$rank \left[C_{Pu}^{(\mu_{Pu})} S_r^{(\mu_{Pu})}(s)\right] \leq \max rank \left[C_{Pu}^{(\mu_{Pu})} S_r^{(\mu_{Pu})}(s)\right] =$$
$$= \max rank C_{Pu}^{(\nu_{Pu})} = \min \{N, r\}. \tag{E.2}$$

Analogously,

$$rank \left[C_{Pu}^{(\mu_{Pu})} S_r^{(\mu_{Pu})}(s)\right]^T \leq \max rank \left[C_{Pu}^{(\mu_{Pu})} S_r^{(\mu_{Pu})}(s)\right]^T =$$
$$= \max rank C_{Pu}^{(\nu_{Pu})^T} = \min \{N, r\}. \tag{E.3}$$

Since $\left[C_{Pu}^{(\mu_{Pu})}S_r^{(\mu_{Pu})}(s)\right]\left[C_{Pu}^{(\mu_{Pu})}S_r^{(\mu_{Pu})}(s)\right]^T \in \mathfrak{C}^{N\times N}$, then

$$rank\left\langle\left[C_{Pu}^{(\mu_{Pu})}S_r^{(\mu_{Pu})}(s)\right]\left[C_{Pu}^{(\mu_{Pu})}S_r^{(\mu_{Pu})}(s)\right]^T\right\rangle = rank\left(C_{Pu}^{(\mu_{Pu})}C_{Pu}^{(\nu_{Pu})^T}\right) \leq N,$$

$$\begin{aligned}\max rank\left\langle\left[C_{Pu}^{(\mu_{Pu})}S_r^{(\mu_{Pu})}(s)\right]\left[C_{Pu}^{(\mu_{Pu})}S_r^{(\mu_{Pu})}(s)\right]^T\right\rangle &= \\ = \max rank\left(C_{Pu}^{(\mu_{Pu})}C_{Pu}^{(\nu_{Pu})^T}\right) = N. \end{aligned} \tag{E.4}$$

Analogously, from $\left[C_{Pu}^{(\mu_{Pu})}S_r^{(\mu_{Pu})}(s)\right]^T\left[C_{Pu}^{(\mu_{Pu})}S_r^{(\mu_{Pu})}(s)\right] \in \mathfrak{C}^{r\times r}$ follows

$$rank\left[C_{Pu}^{(\mu_{Pu})}S_r^{(\mu_{Pu})}(s)\right]^T\left[C_{Pu}^{(\mu_{Pu})}S_r^{(\mu_{Pu})}(s)\right] \leq r,$$

$$\begin{aligned}\max rank\left[C_{Pu}^{(\mu_{Pu})}S_r^{(\mu_{Pu})}(s)\right]^T\left[C_{Pu}^{(\mu_{Pu})}S_r^{(\mu_{Pu})}(s)\right] &= \\ = \max rank\left(C_{Pu}^{(\mu_{Pu})^T}C_{Pu}^{(\nu_{Pu})}\right) = r. \end{aligned} \tag{E.5}$$

The following two cases should be distinguished.

Case 410 $N > r$

If we multiply (E.1) on the left by $\left[C_{Pu}^{(\mu_{Pu})}S_r^{(\mu_{Pu})}(s)\right]^T$, *then it becomes*

$$\left[C_{Pu}^{(\mu_{Pu})}S_r^{(\mu_{Pu})}(s)\right]^T\left[C_{Pu}^{(\mu_{Pu})}S_r^{(\mu_{Pu})}(s)\right]\mathbf{U}^*(s) =$$

$$= \left[C_{Pu}^{(\mu_{Pu})}S_r^{(\mu_{Pu})}(s)\right]^T \left\{\begin{array}{c} \left[A_P^{(\nu)}S_N^{(\nu)}(s)\right]\mathbf{Y}_d(s)- \\ -\left[D_{Pd}^{(\mu_{Pd})}S_d^{(\mu_{Pd})}(s)\right]\mathbf{D}(s)- \\ -A_P^{(\nu)}Z_N^{(\nu-1)}(s)\mathbf{Y}_0^{(\nu-1)}+ \\ +D_{Pd}^{(\mu_{Pd})}Z_d^{(\mu_{Pd}-1)}(s)\mathbf{D}_0^{(\mu_{Pd}-1)}+ \\ +C_{Pu}^{(\mu_{Pu})}Z_r^{(\mu_{Pu}-1)}(s)\mathbf{U}_0^{*(\mu_{Pu}-1)} \end{array}\right\}.$$

To be solvable in $\mathbf{U}^*(s)$ *it is necessary and sufficient that*

$$rank\left[C_{Pu}^{(\mu_{Pu})}S_r^{(\mu_{Pu})}(s)\right]^T\left[C_{Pu}^{(\mu_{Pu})}S_r^{(\mu_{Pu})}(s)\right] = rank C_{Pu}^{(\mu_{Pu})^T}C_{Pu}^{(\nu_{Pu})} = r.$$

The solution reads

$$\mathbf{U}^*(s) = \left\langle\left[C_{Pu}^{(\mu_{Pu})}S_r^{(\mu_{Pu})}(s)\right]^T\left[C_{Pu}^{(\mu_{Pu})}S_r^{(\mu_{Pu})}(s)\right]\right\rangle^{-1}\bullet$$

$$\bullet\left[C_{Pu}^{(\mu_{Pu})}S_r^{(\mu_{Pu})}(s)\right]^T \left\{\begin{array}{c} \left[A_P^{(\nu)}S_N^{(\nu)}(s)\right]\mathbf{Y}_d(s)- \\ -\left[D_{Pd}^{(\mu_{Pd})}S_d^{(\mu_{Pd})}(s)\right]\mathbf{D}(s)- \\ -A_P^{(\nu)}Z_N^{(\nu-1)}(s)\mathbf{Y}_0^{(\nu-1)}+ \\ +D_{Pd}^{(\mu_{Pd})}Z_d^{(\mu_{Pd}-1)}(s)\mathbf{D}_0^{(\mu_{Pd}-1)}+ \\ +C_{Pu}^{(\mu_{Pu})}Z_r^{(\mu_{Pu}-1)}(s)\mathbf{U}_0^{*(\mu_{Pu}-1)} \end{array}\right\}. \tag{E.6}$$

When we replace this in (E.1), which should yield the identity, we get

$$\left[C_{Pu}^{(\mu_{Pu})}S_r^{(\mu_{Pu})}(s)\right]\left\langle\left[C_{Pu}^{(\mu_{Pu})}S_r^{(\mu_{Pu})}(s)\right]^T\left[C_{Pu}^{(\mu_{Pu})}S_r^{(\mu_{Pu})}(s)\right]\right\rangle^{-1}\bullet$$

$$\bullet\left[C_{Pu}^{(\mu_{Pu})}S_r^{(\mu_{Pu})}(s)\right]^T\left\{\begin{array}{c}\left[A_P^{(\nu)}S_N^{(\nu)}(s)\right]\mathbf{Y}_d(s)-\\ -\left[D_{Pd}^{(\mu_{Pd})}S_d^{(\mu_{Pd})}(s)\right]\mathbf{D}(s)-\\ -A_P^{(\nu)}Z_N^{(\nu-1)}(s)\mathbf{Y}_0^{(\nu-1)}+\\ +D_{Pd}^{(\mu_{Pd})}Z_d^{(\mu_{Pd}-1)}(s)\mathbf{D}_0^{(\mu_{Pd}-1)}+\\ +C_{Pu}^{(\mu_{Pu})}Z_r^{(\mu_{Pu}-1)}(s)\mathbf{U}_0^{*(\mu_{Pu}-1)}\end{array}\right\}=$$

$$=C_{Pu}^{(\mu_{Pu})}Z_r^{(\mu_{Pu}-1)}(s)\mathbf{U}_0^{*(\mu_{Pu}-1)}+$$

$$+\left[A_P^{(\nu)}S_N^{(\nu)}(s)\right]\mathbf{Y}_d(s)-A_P^{(\nu)}Z_N^{(\nu-1)}(s)\mathbf{Y}_0^{(\nu-1)}-$$

$$-\left[D_{Pd}^{(\mu_{Pd})}S_d^{(\mu_{Pd})}(s)\right]\mathbf{D}(s)+D_{Pd}^{(\mu_{Pd})}Z_d^{(\mu_{Pd}-1)}(s)\mathbf{D}_0^{(\mu_{Pd}-1)}, \qquad (E.7)$$

for which to be possible it is necessary and sufficient that

$$\left[C_{Pu}^{(\mu_{Pu})}S_r^{(\mu_{Pu})}(s)\right]\left\langle\left[C_{Pu}^{(\mu_{Pu})}S_r^{(\mu_{Pu})}(s)\right]^T\left[C_{Pu}^{(\mu_{Pu})}S_r^{(\mu_{Pu})}(s)\right]\right\rangle^{-1}\bullet$$

$$\bullet\left[C_{Pu}^{(\mu_{Pu})}S_r^{(\mu_{Pu})}(s)\right]^T=I_N. \qquad (E.8)$$

However, this is not possible because

$$rank\left\{\begin{array}{c}\left[C_{Pu}^{(\mu_{Pu})}S_r^{(\mu_{Pu})}(s)\right]\bullet\\ \bullet\left\langle\left[C_{Pu}^{(\mu_{Pu})}S_r^{(\mu_{Pu})}(s)\right]^T\left[C_{Pu}^{(\mu_{Pu})}S_r^{(\mu_{Pu})}(s)\right]\right\rangle^{-1}\bullet\\ \bullet\left[C_{Pu}^{(\mu_{Pu})}S_r^{(\mu_{Pu})}(s)\right]^T\end{array}\right\}\leq r<N=rankI_N.$$

If we multiply (E.8) on the right by $\left[C_{Pu}^{(\mu_{Pu})}S_r^{(\mu_{Pu})}(s)\right]$, *the result reads*

$$\left[C_{Pu}^{(\mu_{Pu})}S_r^{(\mu_{Pu})}(s)\right]=\left[C_{Pu}^{(\mu_{Pu})}S_r^{(\mu_{Pu})}(s)\right].$$

From this trivial equation we cannot prove (E.8). The same conclusion follows if we multiply (E.8) on the left by $\left[C_{Pu}^{(\mu_{Pu})}S_r^{(\mu_{Pu})}(s)\right]^T$. *Furthermore, if we multiply (E.1) on the left by* $\left[C_{Pu}^{(\mu_{Pu})}S_r^{(\mu_{Pu})}(s)\right]^T$, *then it becomes*

$$\left[C_{Pu}^{(\mu_{Pu})}S_r^{(\mu_{Pu})}(s)\right]^T \left\{\begin{array}{c} \left[A_P^{(\nu)}S_N^{(\nu)}(s)\right]\mathbf{Y}_d(s)- \\ -\left[D_{Pd}^{(\mu_{Pd})}S_d^{(\mu_{Pd})}(s)\right]\mathbf{D}(s)- \\ -A_P^{(\nu)}Z_N^{(\nu-1)}(s)\mathbf{Y}_0^{(\nu-1)}+ \\ +D_{Pd}^{(\mu_{Pd})}Z_d^{(\mu_{Pd}-1)}(s)\mathbf{D}_0^{(\mu_{Pd}-1)}+ \\ +C_{Pu}^{(\mu_{Pu})}Z_r^{(\mu_{Pu}-1)}(s)\mathbf{U}_0^{*(\mu_{Pu}-1)} \end{array}\right\} =$$

$$= \left[C_{Pu}^{(\mu_{Pu})}S_r^{(\mu_{Pu})}(s)\right]^T \left\{\begin{array}{c} C_{Pu}^{(\mu_{Pu})}Z_r^{(\mu_{Pu}-1)}(s)\mathbf{U}_0^{*(\mu_{Pu}-1)}+ \\ +\left[A_P^{(\nu)}S_N^{(\nu)}(s)\right]\mathbf{Y}_d(s)- \\ -A_P^{(\nu)}Z_N^{(\nu-1)}(s)\mathbf{Y}_0^{(\nu-1)}- \\ -\left[D_{Pd}^{(\mu_{Pd})}S_d^{(\mu_{Pd})}(s)\right]\mathbf{D}(s)+ \\ +D_{Pd}^{(\mu_{Pd})}Z_d^{(\mu_{Pd}-1)}(s)\mathbf{D}_0^{(\mu_{Pd}-1)} \end{array}\right\}. \tag{E.9}$$

This does not prove the following needed identity

$$\left\{\begin{array}{c} \left[A_P^{(\nu)}S_N^{(\nu)}(s)\right]\mathbf{Y}_d(s)-\left[D_{Pd}^{(\mu_{Pd})}S_d^{(\mu_{Pd})}(s)\right]\mathbf{D}(s)- \\ -A_P^{(\nu)}Z_N^{(\nu-1)}(s)\mathbf{Y}_0^{(\nu-1)}+D_{Pd}^{(\mu_{Pd})}Z_d^{(\mu_{Pd}-1)}(s)\mathbf{D}_0^{(\mu_{Pd}-1)}+ \\ +C_{Pu}^{(\mu_{Pu})}Z_r^{(\mu_{Pu}-1)}(s)\mathbf{U}_0^{*(\mu_{Pu}-1)} \end{array}\right\} \equiv$$

$$\equiv \left\{\begin{array}{c} C_{Pu}^{(\mu_{Pu})}Z_r^{(\mu_{Pu}-1)}(s)\mathbf{U}_0^{*(\mu_{Pu}-1)}+\left[A_P^{(\nu)}S_N^{(\nu)}(s)\right]\mathbf{Y}_d(s)- \\ -A_P^{(\nu)}Z_N^{(\nu-1)}(s)\mathbf{Y}_0^{(\nu-1)}-\left[D_{Pd}^{(\mu_{Pd})}S_d^{(\mu_{Pd})}(s)\right]\mathbf{D}(s)+ \\ +D_{Pd}^{(\mu_{Pd})}Z_d^{(\mu_{Pd}-1)}(s)\mathbf{D}_0^{(\mu_{Pd}-1)} \end{array}\right\}. \tag{E.10}$$

because

$$rank\left[C_{Pu}^{(\mu_{Pu})}S_r^{(\mu_{Pu})}(s)\right]\left[C_{Pu}^{(\mu_{Pu})}S_r^{(\mu_{Pu})}(s)\right]^T \leq r < N$$

due to $N > r$ *and (E.4), so that the* NxN *matrix*

$$\left[C_{Pu}^{(\mu_{Pu})}S_r^{(\mu_{Pu})}(s)\right]\left[C_{Pu}^{(\mu_{Pu})}S_r^{(\mu_{Pu})}(s)\right]^T$$

is singular for every $s \in \mathfrak{C}$, *and (E.9) cannot yield (E.10).*

The dimension N *of the output vector* $\mathbf{Y}$ *is bigger than that (i.e.,* r*) of the control vector* $\mathbf{U}$. *There are* N *mutually independent scalar equations and* r *unknown control variables. The equation (3.14) does not have a solution, because* $\mathbf{U}^*(s)$ *determined by (E.6) does not satisfy identically (3.14) (Subsection 3.3.2), which signifies that the control vector function* $\mathbf{U}^*(.)$ *(which is to be nominal for the object (2.15) (Section 2.1) relative to* $\mathbf{Y}_d(.)$ *for any* $\mathbf{D}(.) \in \mathfrak{D}^{\nu_{Pd}}$*) does not exist. It is not possible to find* r *independent functions that should determine a greater number* N *of other independent functions.*

Case 411 $N \leq r$

For the solvability of (E.1) in $\mathbf{U}^*(s)$ *it is necessary and sufficient that* $\left[C_{Pu}^{(\mu_{Pu})} S_r^{(\mu_{Pu})}(s)\right]\left[C_{Pu}^{(\mu_{Pu})} S_r^{(\mu_{Pu})}(s)\right]^T$ *has the maximal rank, which is* N,

$$rank\left[C_{Pu}^{(\mu_{Pu})} S_r^{(\mu_{Pu})}(s)\right]\left[C_{Pu}^{(\mu_{Pu})} S_r^{(\mu_{Pu})}(s)\right]^T = rank C_{Pu}^{(\nu_{Pu})} = N. \tag{E.11}$$

This guarantees nonsingularity of the $N \mathrm{x} N$ *matrix*

$$\left\langle\left[C_{Pu}^{(\mu_{Pu})} S_r^{(\mu_{Pu})}(s)\right]\left[C_{Pu}^{(\mu_{Pu})} S_r^{(\mu_{Pu})}(s)\right]^T\right\rangle,$$

i.e.,

$$\det\left\langle\left[C_{Pu}^{(\mu_{Pu})} S_r^{(\mu_{Pu})}(s)\right]\left[C_{Pu}^{(\mu_{Pu})} S_r^{(\mu_{Pu})}(s)\right]^T\right\rangle \neq 0.$$

Consequently,

$$\exists\left\langle\left[C_{Pu}^{(\mu_{Pu})} S_r^{(\mu_{Pu})}(s)\right]\left[C_{Pu}^{(\mu_{Pu})} S_r^{(\mu_{Pu})}(s)\right]^T\right\rangle^{-1}.$$

The function determined by

$$\mathbf{U}^*(s) = \left[C_{Pu}^{(\mu_{Pu})} S_r^{(\mu_{Pu})}(s)\right]^T \left\langle\left[C_{Pu}^{(\mu_{Pu})} S_r^{(\mu_{Pu})}(s)\right]\left[C_{Pu}^{(\mu_{Pu})} S_r^{(\mu_{Pu})}(s)\right]^T\right\rangle^{-1} \bullet$$
$$\bullet\left\{\begin{array}{c} \left[A_P^{(\nu)} S_N^{(\nu)}(s)\right]\mathbf{Y}_d(s) - A_P^{(\nu)} Z_N^{(\nu-1)}(s)\mathbf{Y}_0^{(\nu-1)} - \\ -\left[D_{Pd}^{(\mu_{Pd})} S_d^{(\mu_{Pd})}(s)\right]\mathbf{D}(s) + D_{Pd}^{(\mu_{Pd})} Z_d^{(\mu_{Pd}-1)}(s)\mathbf{D}_0^{(\mu_{Pd}-1)} + \\ +C_{Pu}^{(\mu_{Pu})} Z_r^{(\mu_{Pu}-1)}(s)\mathbf{U}_0^{*(\mu_{Pu}-1)} \end{array}\right\} \tag{E.12}$$

satisfies (E.1),

$$\left[C_{Pu}^{(\mu_{Pu})} S_r^{(\mu_{Pu})}(s)\right]\mathbf{U}^*(s) \equiv$$

$$\equiv \left[C_{Pu}^{(\mu_{Pu})} S_r^{(\mu_{Pu})}(s)\right]\bullet$$
$$\bullet\left[C_{Pu}^{(\mu_{Pu})} S_r^{(\mu_{Pu})}(s)\right]^T \left\langle\left[C_{Pu}^{(\mu_{Pu})} S_r^{(\mu_{Pu})}(s)\right]\left[C_{Pu}^{(\mu_{Pu})} S_r^{(\mu_{Pu})}(s)\right]^T\right\rangle^{-1} \bullet$$
$$\bullet\left\{\begin{array}{c} \left[A_P^{(\nu)} S_N^{(\nu)}(s)\right]\mathbf{Y}_d(s) - A_P^{(\nu)} Z_N^{(\nu-1)}(s)\mathbf{Y}_0^{(\nu-1)} - \\ -\left[D_{Pd}^{(\mu_{Pd})} S_d^{(\mu_{Pd})}(s)\right]\mathbf{D}(s) + D_{Pd}^{(\mu_{Pd})} Z_d^{(\mu_{Pd}-1)}(s)\mathbf{D}_0^{(\mu_{Pd}-1)} + \\ +C_{Pu}^{(\mu_{Pu})} Z_r^{(\mu_{Pu}-1)}(s)\mathbf{U}_0^{*(\mu_{Pu}-1)} \end{array}\right\} \equiv$$

$$\equiv \left[\begin{array}{c} \left[A_P^{(\nu)} S_N^{(\nu)}(s)\right]\mathbf{Y}_d(s) - A_P^{(\nu)} Z_N^{(\nu-1)}(s)\mathbf{Y}_0^{(\nu-1)} - \\ -\left[D_{Pd}^{(\mu_{Pd})} S_d^{(\mu_{Pd})}(s)\right]\mathbf{D}(s) + D_{Pd}^{(\mu_{Pd})} Z_d^{(\mu_{Pd}-1)}(s)\mathbf{D}_0^{(\mu_{Pd}-1)} + \\ C_{Pu}^{(\mu_{Pu})} Z_r^{(\mu_{Pu}-1)}(s)\mathbf{U}_0^{*(\mu_{Pu}-1)} \end{array}\right] \equiv$$

$$\equiv C_{Pu}^{(\mu_{Pu})} Z_r^{(\mu_{Pu}-1)}(s) \mathbf{U}_0^{*(\mu_{Pu}-1)} +$$
$$+ \left[A_P^{(\nu)} S_N^{(\nu)}(s) \right] \mathbf{Y}_d(s) - A_P^{(\nu)} Z_N^{(\nu-1)}(s) \mathbf{Y}_0^{(\nu-1)} -$$
$$- \left[D_{Pd}^{(\mu_{Pd})} S_d^{(\mu_{Pd})}(s) \right] \mathbf{D}(s) + D_{Pd}^{(\mu_{Pd})} Z_d^{(\mu_{Pd}-1)}(s) \mathbf{D}_0^{(\mu_{Pd}-1)}.$$

Hence, such $\mathbf{U}^*(s)$ *is the solution of (E.1), i.e., its time-domain original* $\mathbf{U}^*(t)$ *is the well-defined unique solution of (3.13). We conclude that for the existence of the well-defined (unique) nominal control vector function* $\mathbf{U}_N(.), \mathbf{U}_N(.) = \mathbf{U}^*(.)$, *for the system (2.15) relative to* $\mathbf{Y}_d(.)$ *for every* $\mathbf{D}(.) \in \mathfrak{D}^{\nu_{Pd}}$, *it is necessary and sufficient that E.11 holds. The object desired output response* $\mathbf{Y}_d(.)$ *is realizable on* $\mathfrak{D}^{\nu_{Pd}}$.

This completes the proof ■

The application of W. A. Wolovich Theorem 67 (Subsection 3.3.2) enables simpler proof and verifies the above proof.

Appendix F

Proof of Theorem 72

Proof. [148]. *Necessity.* Let $[\mathbf{I}^*(.), \mathbf{X}^*(.)]$ be a nominal functional (input and state) vector pair for the *ISO* plant (2.29), (2.30) relative to its desired response $\mathbf{Y}_d(.)$. The plant is in its desired regime relative to $\mathbf{Y}_d(.)$. Definition 71 shows that $[\mathbf{I}(.), \mathbf{X}(.)] = [\mathbf{I}^*(.), \mathbf{X}^*(.)]$ implies $\mathbf{Y}(.) = \mathbf{Y}_d(.)$. This and the *ISO* model (2.29), (2.30) yield the following equations:

$$\frac{d\mathbf{X}^*(t)}{dt} = A\mathbf{X}^*(t) + B\mathbf{I}^*(t),\ \forall t \in \mathfrak{T}_0,$$
$$\mathbf{Y}(t) = \mathbf{Y}_d(t) = C\mathbf{X}^*(t) + D\mathbf{I}^*(t),\ \forall t \in \mathfrak{T}_0, \tag{F.1}$$

which can be easily set in the form of the equations (3.17), (3.18). Application of Laplace transform together with its properties sets the equations (3.17), (3.18) into (3.19).

Sufficiency. We accept that all the conditions of the theorem are valid. We chose $[\mathbf{I}(.), \mathbf{X}(.)] = [\mathbf{I}^*(.), \mathbf{X}^*(.)]$. The equation (3.17) written in the normal state form,

$$\frac{d\mathbf{X}^*(t)}{dt} = A\mathbf{X}^*(t) + B\mathbf{I}^*(t),\ \forall t \in \mathfrak{T}_0,$$

shows that the pair $[\mathbf{I}^*(.), \mathbf{X}^*(.)]$ satisfies (2.29). Furthermore, the equation (2.30) takes the following form:

$$\mathbf{Y}(t) = C\mathbf{X}^*(t) + D\mathbf{I}^*(t),\ \forall t \in \mathfrak{T}_0.$$

It, subtracted from (3.18), yields

$$\mathbf{Y}(t) - \mathbf{Y}_d(t) = \mathbf{0},\ \forall t \in \mathfrak{T}_0,$$

i.e.,

$$\mathbf{Y}(t) = \mathbf{Y}_d(t),\ \forall t \in \mathfrak{T}_0.$$

Laplace transform of the equations (3.17), (3.18) yields (3.19). This completes the proof ■

Appendix G

Proof of Theorem 91

Proof. We refer to Theorem 90 and its Equations (3.33),

$$\begin{bmatrix} -B_P & (sI_n - A_P) \\ H_P & C_P \end{bmatrix} \begin{bmatrix} \mathbf{U}^*(s) \\ \mathbf{X}_P^*(s) \end{bmatrix} = \begin{bmatrix} L_P & I_n & O_{n,N} \\ -D_P & O_n & I_N \end{bmatrix} \begin{bmatrix} \mathbf{D}(s) \\ \mathbf{X}_{P0}^* \\ \mathbf{Y}_d(s) \end{bmatrix},$$
$$\mathbf{X}_{P0}^* \in \mathfrak{R}^n. \tag{G.1}$$

The system matrix

$$\begin{bmatrix} -B_P & (sI_n - A_P) \\ H_P & C_P \end{bmatrix} \in \mathfrak{C}^{(n+N)\mathrm{x}(n+r)}.$$

Case 1: N > r. There is not a solution $[\mathbf{U}^*(s), \mathbf{X}_P^*(s)]$ to (G.1). For details see Appendix E.

Case 2: N = r. For the existence of the unique solution to (G.1) it is necessary and sufficient that

$$\exists s \in \mathfrak{C} \Longrightarrow \det \begin{bmatrix} -B_P & (sI_n - A_P) \\ H_P & C_P \end{bmatrix} \neq 0.$$

The solution is determined by

$$\begin{bmatrix} \mathbf{U}^*(s) \\ \mathbf{X}_P^*(s) \end{bmatrix} = \begin{bmatrix} -B_P & (sI_n - A_P) \\ H_P & C_P \end{bmatrix}^{-1} \begin{bmatrix} L_P & I_n & O_{n,N} \\ -D_P & O_n & I_N \end{bmatrix} \begin{bmatrix} \mathbf{D}(s) \\ \mathbf{X}_{P0}^* \\ \mathbf{Y}_d(s) \end{bmatrix},$$
$$\mathbf{X}_{P0}^* \in \mathfrak{R}^n. \tag{G.2}$$

Case 3: N < r. For the existence of the solution to (G.1) it is necessary and sufficient that (Wolovich Theorem 67 in Subsection 3.3.2)

$$\exists s \in \mathfrak{C} \Longrightarrow rank \begin{bmatrix} -B_P & (sI_n - A_P) \\ H_P & C_P \end{bmatrix} = N. \tag{G.3}$$

It is easy to verify

$$\begin{bmatrix} O_{n,N} & (sI_n - A_P) \\ I_N & C_P \end{bmatrix} \begin{bmatrix} G_u(s) & O_{N,n} \\ -(sI_n - A_P)^{-1} B_P & I_n \end{bmatrix} = \\ = \begin{bmatrix} -B_P & (sI_n - A_P) \\ H_P & C_P \end{bmatrix}.$$

From this follows that for every $s \in \mathfrak{C}$, which is not an eigenvalue of A_P,

$$rank \begin{bmatrix} -B_P & (sI_n - A_P) \\ H_P & C_P \end{bmatrix} = rank G_u(s).$$

This proves the equivalence between (3.35) and (3.36).

The solution is determined by

$$\begin{bmatrix} \mathbf{U}^*(s) \\ \mathbf{X}_P^*(s) \end{bmatrix} = \begin{bmatrix} -B_P & (sI_n - A_P) \\ H_P & C_P \end{bmatrix}^T \bullet \\ \bullet \left\langle \begin{bmatrix} -B_P & (sI_n - A_P) \\ H_P & C_P \end{bmatrix} \begin{bmatrix} -B_P & (sI_n - A_P) \\ H_P & C_P \end{bmatrix}^T \right\rangle^{-1} \bullet \\ \bullet \begin{bmatrix} L_P & I_n & O_{n,N} \\ -D_P & O_n & I_N \end{bmatrix} \begin{bmatrix} \mathbf{D}(s) \\ \mathbf{X}_{P0}^* \\ \mathbf{Y}_d(s) \end{bmatrix}, \\ \mathbf{X}_{P0}^* \in \mathfrak{R}^n. \tag{G.4}$$

We can set (G.1) into the following form:

$$\mathbf{X}_P^*(s) = (sI_n - A_P)^{-1} \left[B_P \mathbf{U}^*(s) + L_P \mathbf{D}(s) + \mathbf{X}_{P0}^* \right] \tag{G.5}$$

$$\mathbf{Y}_d(s) = \underbrace{\left[C_P (sI_n - A_P)^{-1} B_P + H_P \right]}_{G_u(s)} \mathbf{U}^*(s) + \\ + \underbrace{\left[C_P (sI_n - A_P)^{-1} L_P + D_P \right]}_{G_d(s)} \mathbf{D}(s) + \underbrace{C_P (sI_n - A_P)^{-1}}_{G_{xo}(s)} \mathbf{X}_{P0}^*. \tag{G.6}$$

The second equation is solvable in $\mathbf{U}^*(s)$ if, and only if, (Wolovich Theorem 67)

$$rank G_u(s) = rank \left[C_P (sI_n - A_P)^{-1} B_P + H_P \right] = N \leq r. \tag{G.7}$$

This guarantees

$$\exists s \in \mathfrak{C} \Longrightarrow \det G_u(s) G_u^T(s) \neq 0,$$

which enables us to determine the well defined $\mathbf{U}^*(s)$ by

$$\mathbf{U}^*(s) = G_u^T(s) \left[G_u(s) G_u^T(s) \right]^{-1} \left\{ \mathbf{Y}_d(s) - G_d(s)\mathbf{D}(s) - G_{xo}(s)\mathbf{X}_{P0}^* \right\}. \tag{G.8}$$

This, (G.5) and (G.7) prove (3.38) ■

Appendix H

Proof of Lemma 102 (Basic Lemma)

Proof. [148]. Let the conditions of Lemma 102 hold. From (6.3) and (6.4) we determine the original $\mathbf{z}(t)$. Let the $ij\text{-}th$ element of $M(s)$ be $m_{ij}(s)$, the $i\text{-}th$ element of $\mathbf{z}(t)$ be $z_i(t)$, the $i\text{-}th$ element of $\mathbf{Z}(s)$ be $\varsigma_i(s)$, and the $j\text{-}th$ element of $\mathbf{W}(s)$ be $w_j(s)$, so that

$$z_i(t) = \mathcal{L}^{-1}\{\varsigma_i(s)\} = \mathcal{L}^{-1}\left\{\sum_j m_{ij}(s)w_j(s)\right\}. \tag{H.1}$$

Since $M(.)$ is a real rational proper matrix function of s, $\mathbf{Z}(.)$ and $\mathbf{W}(.)$ are real rational proper vector functions of s; then, the same holds for their entries, which can be presented in the factorized forms,

$$m_{ij}(s) = \frac{\prod_{k=1}^{\mu_{ij}}\left(s - s_{mk}^{oij}\right)}{\prod_{k=1}^{\nu_{ij}}\left(s - s_{mk}^{*ij}\right)},\ \mu_{ij} \leq \nu_{ij},\ w_j(s) = \frac{\prod_{k=1}^{\upsilon_j}\left(s - s_{wk}^{oj}\right)}{\prod_{k=1}^{\omega_j}\left(s - s_{wk}^{*j}\right)},\ \upsilon_j \leq \omega_j.$$

These equations set (H.1) into the following form

$$z_i(t) = \mathcal{L}^{-1}\{\varsigma_i(s)\} = \mathcal{L}^{-1}\left\{\sum_j \frac{\prod_{k=1}^{\mu_{ij}}\left(s - s_{mk}^{oij}\right)\prod_{k=1}^{\upsilon_j}\left(s - s_{wk}^{oj}\right)}{\prod_{k=1}^{\nu_{ij}}\left(s - s_{mk}^{*ij}\right)\prod_{k=1}^{\omega_j}\left(s - s_{wk}^{*j}\right)}\right\}.$$

We can conclude as follows. All residua of Heaviside expansion of $\varsigma_i(s) = \mathcal{L}\{z_i(t)\}$ are equal to zero in a pole that is equal to a zero of $\varsigma_i(s)$. They can be cancelled. If $s_{mk}^{oij} = s_{mk}^{*ij}$, $\forall j = 1, 2, .., q$, then they should be cancelled. This proves 1). If $s_{wk}^{oj} = s_{wk}^{*j}$, $\forall j = 1, 2, .., q$, then they should also be cancelled. This proves 2). The equal poles and zeros of $m_{ij}(s)w_j(s)$, $\forall j = 1, 2, .., q$, do not influence $z_i(t)$. They should be cancelled, too. This proves 3). The equal poles and zeros of any entry of $M(s)\mathbf{W}(s)$ do not influence $\mathbf{z}(t)$. They should be cancelled. The result is the row nondegenerate form $[M(s)\mathbf{W}(s)]_{rnd}$ of $M(s)\mathbf{W}(s)$,

$$\mathbf{z}(t) = \mathcal{L}^{-1}\{\mathbf{Z}(s)\} = \mathcal{L}^{-1}\{[M(s)\mathbf{W}(s)]_{rnd}\},$$

i.e., (6.4). They determine the original $\mathbf{z}(t)$. The claim under 4) is correct. Let every zero of every element of every row of $M(s)$ be different from every pole of the corresponding entry of $\mathbf{W}(s)$, and let every pole of every element of every row of $M(s)$ be different from every zero of the corresponding entry of $\mathbf{W}(s)$. Then, the zero-pole cancellation is possible only among zeros and poles of the elements of the rows of $M(s)$ and, independently of them, among zeros and poles of the members of the entries of $\mathbf{W}(s)$. The cross cancellations of the zeros/poles of the elements of the rows of $M(s)$ with poles/zeros of the members of the entries of $\mathbf{W}(s)$ is not possible. After carrying out all possible cancellations in the elements of the rows of $M(s)$ and in the components of the entries of $\mathbf{W}(s)$, we get the row nondegenerate form $[M(s)\mathbf{W}(s)]_{rnd}$ of $M(s)\mathbf{W}(s)$ as the product of the row nondegenenerate forms $M(s)_{rnd}$ and $\mathbf{W}(s)_{rnd}$ of $M(s)$ and $\mathbf{W}(s)$,

$$[M(s)\mathbf{W}(s)]_{rnd} = M(s)_{rnd}\mathbf{W}(s)_{rnd},$$

which is (6.5). This and (6.4) imply (6.6) ■

Appendix I

Proof of Theorem 116

Proof. [148]. a) Left, right, Laplace transform $\mathcal{L}^{\mp}\{.\}$ of the left-hand side of (2.1), i.e., (2.9) (Section 2.1), i.e., of (7.20) (Subsection 7.2.1), yields the following:

$$\mathcal{L}^{\mp}\left\{\sum_{k=0}^{k=\nu} A_k \mathbf{Y}^{(k)}(t)\right\} = \mathcal{L}^{\mp}\left\{A^{(\nu)}\mathbf{Y}^{\nu}(t)\right\} = \overbrace{\left(A_0 s^0 + A_1 s^1 + ... + A_\nu s^\nu\right)}^{A^{(\nu)}S_N^{(\nu)}(s)} \mathbf{Y}^{\mp}(s) -$$

$$\begin{array}{l}
- A_0 O_N - \\
- A_1 \left(s^{1-1}\mathbf{Y}^{(1-1)}(0^{\mp})\right) - \\
- A_2 \left(s^{2-1}\mathbf{Y}^{(1-1)}(0^{\mp}) + s^{2-2}\mathbf{Y}^{(2-1)}(0^{\mp})\right) - \\
\cdots\cdots\cdots\cdots\cdots\cdots\cdots\cdots\cdots\cdots \\
- A_k \left(\begin{array}{c} s^{k-1}\mathbf{Y}^{(1-1)}(0^{\mp}) + s^{k-2}\mathbf{Y}^{(2-1)}(0^{\mp}) + . \\ . + s^{k-i}\mathbf{Y}^{(i-1)}(0^{\mp}) + .. + s^{k-k}\mathbf{Y}^{(k-1)}(0^{\mp}) \end{array}\right) - \\
\cdots\cdots\cdots\cdots\cdots\cdots\cdots\cdots\cdots\cdots \\
- A_\nu \left(s^{\nu-1}\mathbf{Y}^{(1-1)}(0^{\mp}) + s^{\nu-2}\mathbf{Y}^{(2-1)}(0^{\mp}) + ... + s^{\nu-\nu}\mathbf{Y}^{(\nu-1)}(0^{\mp})\right)
\end{array}$$

$$= A^{(\nu)} S_N^{(\nu)}(s)\mathbf{Y}^{\mp}(s) - \overbrace{\left[A_0 \,\vdots\, A_1 \vdots\, ... \vdots\, A_k \,\vdots\, ... \vdots\, A_\nu\right]}^{A^{(\nu)},\ (7.23)} \bullet$$

$$\bullet \left[\begin{array}{c} O_N \\ s^{1-1}\mathbf{Y}^{(1-1)}(0^{\mp}) \\ \cdots\cdots\cdots\cdots\cdots\cdots \\ s^{k-1}\mathbf{Y}^{(1-1)}(0^{\mp}) + s^{k-2}\mathbf{Y}^{(2-1)}(0^{\mp}) + ... + s^{k-k}\mathbf{Y}^{(k-1)}(0^{\mp}) \\ \cdots\cdots\cdots\cdots\cdots\cdots \\ s^{\nu-1}\mathbf{Y}^{(1-1)}(0^{\mp}) + s^{\nu-2}\mathbf{Y}^{(2-1)}(0^{\mp}) + ... + s^{\nu-\nu}\mathbf{Y}^{(\nu-1)}(0^{\mp}) \end{array}\right]$$

$$= \left[A_0 \vdots A_1 \vdots \ldots \vdots A_k \vdots \ldots \vdots A_\nu \right] \bullet$$

$$\bullet \underbrace{\begin{bmatrix} O_N & O_N & \cdots & O_N & \cdots & O_N \\ s^{1-1}I_N & O_N & \cdots & O_N & \cdots & O_N \\ \cdots & \cdots & \cdots & O_N & \cdots & O_N \\ s^{k-1}I_N & s^{k-2}I_N & \cdots & s^{k-k}I_N & \cdots & O_N \\ \cdots & \cdots & \cdots & \cdots & \cdots & \cdots \\ s^{\nu-1}I_N & s^{\nu-2}I_N & \cdots & s^{\nu-2}I_N & \cdots & s^{\nu-\nu}I_N \end{bmatrix}}_{Z_N^{(\nu-1)}(s),\ (7.24)} \bullet$$

$$\bullet \underbrace{\begin{bmatrix} \mathbf{Y}^{(1-1)}(0^{\mp}) \\ \mathbf{Y}^{(2-1)}(0^{\mp}) \\ \cdots \\ \mathbf{Y}^{(k-1)}(0^{\mp}) \\ \cdots \\ \mathbf{Y}^{(\nu-1)}(0^{\mp}) \end{bmatrix}}_{\mathbf{Y}_{0^{\mp}}^{\nu-1}(0^{\mp}),\ (7.26)} = A^{(\nu)}S_N^{(\nu)}(s)\mathbf{Y}^{\mp}(s) - A^{(\nu)}Z_N^{(\nu-1)}(s)\mathbf{Y}_{0^{\mp}}^{\nu-1}. \tag{I.1}$$

By repeating the above procedure applied to

$$\mathcal{L}^{\mp}\left\{\sum_{k=0}^{k=\nu} B_k \mathbf{I}^{(k)}(t)\right\} = \mathcal{L}^{\mp}\left\{B^{(\mu)}\mathbf{I}^{\mu}(t)\right\},$$

and in view of (7.24) (Subsection 7.2.1),

$$\mathcal{L}^{\mp}\left\{\sum_{k=0}^{k=\nu} B_k \mathbf{I}^{(k)}(t)\right\} =$$

$$= \left\{ \begin{array}{c} B^{(\mu)}S_M^{(\mu)}(s)\mathbf{I}^{\mp}(s) - B^{(\mu)}Z_M^{(\mu-1)}(s)\mathbf{I}^{\nu-1}(0^{\mp}),\ \mu \geq 1, \\ B^{(\mu)}S_M^{(\mu)}(s)\mathbf{I}^{\mp}(s) = B_0\mathbf{I}^{\mp}(s),\ \mu = 0. \end{array} \right\} \tag{I.2}$$

These results imply the following compact form of left, right Laplace transform $\mathcal{L}^{\mp}\{.\}$ of (2.1), i.e., of (2.9):

$$A^{(\nu)}S_N^{(\nu)}(s)\mathbf{Y}^{\mp}(s) - A^{(\nu)}Z_N^{(\nu-1)}(s)\mathbf{Y}_{0^{\mp}}^{\nu-1}(0^{\mp}) =$$

$$= \left\{ \begin{array}{c} B^{(\mu)}S_M^{(\mu)}(s)\mathbf{I}^{\mp}(s) - B^{(\mu)}Z_M^{(\mu-1)}(s)\mathbf{I}^{\nu-1}(0^{\mp}),\ \mu \geq 1, \\ B^{(\mu)}S_M^{(\mu)}(s)\mathbf{I}^{\mp}(s) = B_0\mathbf{I}^{\mp}(s),\ \mu = 0. \end{array} \right\}. \tag{I.3}$$

This determines $\mathbf{Y}^{\mp}(s)$ linearly in terms of the vector function

$$\left[\mathbf{I}^{(\mp)T}(s) \vdots \left(\mathbf{I}_{0^{\mp}}^{\nu-1}\right)^T \vdots \left(\mathbf{Y}_{0^{\mp}}^{\nu-1}\right)^T \right]^T$$

as follows:

$$\mathbf{Y}^{\mp}(s) =$$

$$= \underbrace{\left[\begin{array}{c} \left(A^{(\nu)} S_N^{(\nu)}(s)\right)^{-1} \cdot \\ \cdot \left\{ \begin{array}{c} \left[B^{(\mu)} S_M^{(\mu)}(s) \vdots \left(-B^{(\mu)} Z_M^{(\mu-1)}(s)\right) \vdots A^{(\nu)} Z_N^{(\nu-1)}(s) \right], \ \mu \geq 1, \\ \left[B_0 \vdots A^{(\nu)} Z_N^{(\nu-1)}(s) \right], \ \mu = 0. \end{array} \right\} \end{array} \right]}_{F_{IO}(s)} \cdot$$

$$\cdot \underbrace{\left\{ \begin{array}{c} \left[\left(\mathbf{I}^{\mp}(s)\right)^T \vdots \left(i_{0\mp}^{\nu-1}\right)^T \vdots \left(\mathbf{Y}_{0\mp}^{\nu-1}\right)^T \right]^T, \ \mu \geq 1, \\ \left[\left(\mathbf{I}^{\mp}(s)\right)^T \vdots \left(\mathbf{Y}_{0\mp}^{\nu-1}\right)^T \right]^T, \ \mu = 0 \end{array} \right\}}_{\mathbf{V}_{IO}^{\mp}(s)} = F_{IO}(s) \mathbf{V}_{IO}^{\mp}(s). \quad \text{(I.4)}$$

The definition of $F_{IO}(s)$ (Definition 104, Subsection 7.1.1) and this equation prove the statement under a) of the theorem. The statement under b) results directly from a), and the definition of $G_{IO}(s)$ (7.4). The formulae under c) through e) result directly from (7.5) through (7.7) linked with (I.1) through (I.4) ■

Appendix J

Proof of Theorem 142

Proof. [148]. (i) The inverse of left Laplace transform of $\mathbf{Y}^{-}(s)$ leads to the inverse of left Laplace transform of (7.3) (Subsection 7.1.1):

$$\mathbf{Y}(t;\mathbf{Y}_0^{\nu-1};\mathbf{I}) = \mathcal{L}^{-1}\left\{\mathbf{Y}^{-}(s)\right\} = \frac{1}{2\pi j}\int_{c-j\infty}^{c+j\infty}\mathbf{Y}^{-}(s)e^{st}ds =$$

$$= \frac{1}{2\pi j}\int_{c-j\infty}^{c+j\infty}\left\{F_{IO}(s)\left(\left(\mathbf{I}^{-}(s)\right)^T \vdots \left(\mathbf{I}_{0^-}^{\mu-1}\right)^T \vdots \left(\mathbf{Y}_{0^-}^{\nu-1}\right)^T\right)^T\right\}e^{st}ds,$$

where $j = \sqrt{-1}$. Let

$$\Xi(t) = \left[\Xi_I(t) \vdots \Xi_{io}(t) \vdots \Xi_{yo}(t)\right] = \mathcal{L}^{-1}\left\{F_{IO}(s)\right\}, \tag{J.1}$$

$$\left[\Xi_I(t) \vdots \Xi_{io}(t) \vdots \Xi_{yo}(t)\right] = \mathcal{L}^{-1}\left\{\left[G_{IO}(s) \vdots G_{IOio}(s) \vdots G_{IOyo}(s)\right]\right\},$$

$$\mathcal{L}\left\{\left[\Xi_I(t) \vdots \Xi_{io}(t) \vdots \Xi_{yo}(t)\right]\right\} = \left[G_{IO}(s) \vdots G_{IOio}(s) \vdots G_{IOyo}(s)\right], \tag{J.2}$$

which transforms the preceding result as follows:

$$\mathbf{Y}(t;\mathbf{Y}_0^{\nu-1};\mathbf{I}) = \frac{1}{2\pi j}\int_{c-j\infty}^{c+j\infty}\left\{\begin{array}{c}\left[G_{IO}(s) \vdots G_{IOio}(s) \vdots G_{IOyo}(s)\right] \\ \bullet\left[\left(\mathbf{I}^{-}(s)\right)^T \vdots \left(\mathbf{I}_{0^-}^{\mu-1}\right)^T \vdots \left(\mathbf{Y}_{0^-}^{\nu-1}\right)^T\right]^T\end{array}\right\}e^{st}ds =$$

$$= \frac{1}{2\pi j}\int_{c-j\infty}^{c+j\infty}\left\{\left[\begin{array}{c}G_{IO}(s)\mathbf{I}^{-}(s)+ \\ +G_{IOio}(s)\mathbf{I}_{0^-}^{\mu-1}+ \\ +G_{IOyo}(s)\mathbf{Y}_{0^-}^{\nu-1}\end{array}\right]\right\}e^{st}ds =$$

$$= \frac{1}{2\pi j}\int_{c-j\infty}^{c+j\infty} G_{IO}(s)\mathbf{I}^{-}(s)e^{st}ds+$$

$$+\frac{1}{2\pi j}\int_{c-j\infty}^{c+j\infty}\left[G_{IOio}(s)\mathbf{I}_{0^-}^{\mu-1}+G_{IOyo}(s)\mathbf{Y}_{0^-}^{\nu-1}\right]e^{st}ds \Longrightarrow$$

$$\mathbf{Y}(t;\mathbf{Y}_0^{\nu-1};\mathbf{I}) = \int_0^t \Xi_I(\tau)\mathbf{I}(t-\tau)d\tau + \Xi_{io}(t)\mathbf{I}_{0^-}^{\mu-1} + \Xi_{yo}(t)\mathbf{Y}_{0^-}^{\nu-1}.$$

This and (7.107) (Subsection 7.6.1) prove

$$\Xi(t) = \left[\Xi_I(t) \vdots \Xi_{io}(t) \vdots \Xi_{yo}(t)\right] = \Phi_{IO}(t) = \left[\Gamma_{IO}(t) \vdots \Gamma_{IOi_0}(t) \vdots \Gamma_{IOy_0}(t)\right],$$

$$\Phi_{IO}(t) = \mathcal{L}^{-1}\left\{F_{IO}(s)\right\} = \left\{\left[\Gamma_{IO}(t) \vdots \Gamma_{IOi_0}(t) \vdots \Gamma_{IOy_0}(t)\right]\right\}.$$

This proves (7.108) (Subsection 7.6.1).

(ii) Laplace transform of the preceding equations proves (7.109) (Subsection 7.6.1), which completes the proof ■

Appendix K

Proof of Theorem 145

Proof. What follows is from [148]. (i) The inverse of left Laplace transform of (7.16) (Subsection 7.1.2) gives the following:

$$\mathbf{x}(t) = \mathcal{L}^{-1}\left\{\mathbf{X}^{-}(s)\right\} = \mathcal{L}^{-1}\left\{F_{ISOIS}(s)\left[\mathbf{I}^{-^T}(s) \vdots \mathbf{x}_{0^-}^T\right]^T\right\} =$$

$$= \frac{1}{2\pi j}\int_{c-j\infty}^{c+j\infty}\left\{F_{ISOIS}(s)\left[\mathbf{I}^{-^T}(s) \vdots \mathbf{x}_{0^-}^T\right]^T\right\} e^{st}ds =$$

$$= \frac{1}{2\pi j}\int_{c-j\infty}^{c+j\infty}\left\{\int_{0^-}^{\infty}\mathcal{L}^{-1}\left\{F_{ISOIS}(s)\right\}e^{-s\tau}d\tau\right\}\left[\mathbf{I}^{-^T}(s) \vdots \mathbf{x}_{0^-}^T\right] e^{st}ds =$$

$$= \int_{0^-}^{\infty}\mathcal{L}^{-1}\left\{F_{ISOIS}(s)\right\}\left[\frac{1}{2\pi j}\int_{c-j\infty}^{c+j\infty}\left[\mathbf{I}^{-^T}(s) \vdots \mathbf{x}_{0^-}^T\right] e^{s(t-\tau)}ds\right]d\tau =$$

$$= \int_{0^-}^{\infty}\mathcal{L}^{-1}\left\{F_{ISOIS}(s)\right\}\left[\mathbf{i}^T(t-\tau) \vdots \delta(t-\tau)\mathbf{x}_{0^-}^T\right]d\tau.$$

This and (i) of Definition 144 (Subsection 7.6.2) imply

$$\Phi_{ISOIS}(t) = \mathcal{L}^{-1}\left\{F_{ISOIS}(s)\right\},$$

that is, the equation (7.114).

(ii) Left Laplace transform of (7.114) is the equation (7.115) ■

Appendix L

Proof of Theorem 149

Proof. What follows is from [148]. (i) The inverse of left Laplace transform of (7.47) (Subsection 7.2.2) gives the following:

$$\mathbf{y}(t) = \mathcal{L}^{-1}\left\{\mathbf{Y}^{-}(s)\right\} = \mathcal{L}^{-1}\left\{F_{ISO}(s)\left[\mathbf{I}^{-^{T}}(s) \vdots \mathbf{x}_{0^{-}}^{T}\right]^{T}\right\} =$$

$$= \frac{1}{2\pi j}\int_{c-j\infty}^{c+j\infty} \mathbf{Y}^{-}(s)e^{st}ds = \frac{1}{2\pi j}\int_{c-j\infty}^{c+j\infty}\left\{F_{ISO}(s)\left[\mathbf{I}^{-^{T}}(s) \vdots \mathbf{x}_{0^{-}}^{T}\right]^{T}\right\}e^{st}ds =$$

$$= \frac{1}{2\pi j}\int_{c-j\infty}^{c+j\infty}\left\{\int_{0^{-}}^{\infty}\mathcal{L}^{-1}\left\{F_{ISO}(s)\right\}e^{-s\tau}d\tau\right\}\left[\mathbf{I}^{-^{T}}(s) \vdots \mathbf{x}_{0^{-}}^{T}\right]e^{st}ds =$$

$$= \int_{0^{-}}^{\infty}\mathcal{L}^{-1}\left\{F_{ISO}(s)\right\}\left[\frac{1}{2\pi j}\int_{c-j\infty}^{c+j\infty}\left[\mathbf{I}^{-^{T}}(s) \vdots \mathbf{x}_{0^{-}}^{T}\right]e^{s(t-\tau)}ds\right]d\tau =$$

$$= \int_{0^{-}}^{\infty}\mathcal{L}^{-1}\left\{F_{ISO}(s)\right\}\left[\mathbf{i}^{T}(t-\tau) \vdots \delta(t-\tau)\mathbf{x}_{0^{-}}^{T}\right]d\tau.$$

This and (i) of Definition 146 (Subsection 7.6.2) imply

$$\Phi_{ISO}(t) = \mathcal{L}^{-1}\left\{F_{ISO}(s)\right\},$$

that is, the equation (7.118).

(ii) Left Laplace transform of (7.118) is the equation (7.119) ■

Author Index

Subject Index

Milton Keynes UK
Ingram Content Group UK Ltd.
UKHW021831071024
449327UK00021B/1480

9 780367 379995